Electrical Power Systems: Design, Networks and Applications

Electrical Power Systems: Design, Networks and Applications

Edited by
Linda Morand

Willford Press
www.willfordpress.com

Published by Willford Press,
118-35 Queens Blvd., Suite 400,
Forest Hills, NY 11375, USA

ISBN: 978-1-68285-572-0

Cataloging-in-Publication Data

Electrical power systems : design, networks and applications / edited by Linda Morand.
p. cm.
Includes bibliographical references and index.
ISBN 978-1-68285-572-0
1. Electric power systems. 2. Electric power systems--Design and construction.
3. Electric network analysis. I. Morand, Linda.

TK1001 .E44 2019
621.31--dc21

For information on all Willford Press publications
visit our website at www.willfordpress.com

Willford Press

Contents

Preface

An electrical power system is an assembly of electrical components aimed at the supply, transfer and storage of electric power. The field of electrical power systems has undergone a major transition in recent years as the focus from fossil fuels has shifted to renewable energy sources. Further, with the advent of new technologies like power electronics, energy storage, better models for the efficient production and transmission of energy are now available. The applications of electrical power systems are in the areas of photonics, renewable energy generation, automation, robotics, telecommunication, electric and hybrid vehicle technologies among many others. This book provides significant information on the design, networks and applications of power systems. The various studies that are constantly contributing towards advancing technologies and evolution of this field are examined in detail. This book is an essential guide for engineers, academicians and students who wish to pursue this discipline further.

This book is a comprehensive compilation of works of different researchers from varied parts of the world. It includes valuable experiences of the researchers with the sole objective of providing the readers (learners) with a proper knowledge of the concerned field. This book will be beneficial in evoking inspiration and enhancing the knowledge of the interested readers.

In the end, I would like to extend my heartiest thanks to the authors who worked with great determination on their chapters. I also appreciate the publisher's support in the course of the book. I would also like to deeply acknowledge my family who stood by me as a source of inspiration during the project.

Editor

Hybrid Approach for Placement of Type-III Multiple DGs in Distribution Network

Kansal S[1]*, Kumar V[2] and Tyagi B[3]

[1]*Department of Electrical Engineering, Baba Hira Singh Bhattal Institute of Engineering and Technology, Lehragaga-148031, Punjab, India*
[2]*Department of Electrical Engineering, Indian Institute of Technology, Roorkee, India*
[3]*EED, Indian Institute technology, Roorkee, India*

Abstract

This paper proposes the hybridization of analytical method and heuristic search for the optimal placement of type-III DGs in power distribution network for reduction of power loss. The type-III DG is capable injecting both real and reactive powers. In this approach the locations are determined by the application of PSO while the sizes of DGs are evaluated by using the analytical method which is based on the exact loss formula. The reduction of power distribution losses has been achieved by compensation of active and reactive powers. The improvement in bus voltage profile and the optimal power factor of the DGs have also been considered. The proposed technique has been tested on a 33-bus test system and the results are compared.

Keywords: Distributed generation; Particle Swarm Optimization (PSO); Optimal size; Optimal location; Power loss.

Introduction

The newly introduced distributed or decentralized generation units connected to local distribution systems are not dispatchable by central operator, but they can have a significant impact on the power flow, stability, voltage profile, reliability, short circuit level and quality of power supply for customers and electricity suppliers. Optimization techniques should be employed for deregulation of power industry, allowing for the best allocation of the DG.

There are many approaches for deciding the optimum sizing and sitting of distributed generation units in distribution systems. Some of the factors that must be taken into account in the planning process of expanding distribution system with DG are: the number and capacity of DG units, best locations and technology, the network connection, capacity of existing system, protection schemes, among others. Different methodologies and tools have been developed to identify optimal places to install DG capacity and its size. These methodologies are based on analytical tools, optimization programs or heuristic techniques. Most of them find the optimal allocation and size of single DG in order to reduce losses and improve voltage profiles with various techniques [1-3] considered. Others include the placement of multiple DGs with artificial intelligence-based optimization methods and a few go with analytical approach.

In [4], GA based technique along with Optimal Power Flow (OPF) calculations were used to determine the optimum size and location of DG units installed to the system in order.

To minimize the cost of active and reactive power generation. In [5], a GA based method was also proposed to find the optimal placement of DG in the compensated network for restoration the system caused by CLPU condition and to conserve load diversity for reduction in losses, improvement in voltage regulation. In [6], authors proposed a Tabu Search (TS) based method to find the optimal solution of their problem. In [7], the objective was to minimize a multi-objective performance index function using GA. The indices were reflecting the effect of DG insertion on the real and reactive power losses of the system, the voltage profile, and the distribution line loading with different load models. In [8], an analytical method to determine the optimum location–size pair of a DG unit was proposed in order to minimize only the line losses of the power system. In [9], DG units were placed at the most sensitive buses to voltage collapse. The units had the same capacity and were placed one by one.

In [10], a Particle Swarm Optimization (PSO) algorithm was introduced to determine the optimal size and location of DG and Capacitor unit to compensate the active and reactive powers of the distribution system. The evaluation of optimal power factor and improvement in voltage profile has also been considered in this work.

Most of the researches placed DG units with unity power factor. An analytical approach based on exact loss formula was presented to find the optimal size and location of DG to minimize the real power loss [11], although the results violate the voltage constraint. Recently, another fast analytical approach to find the optimal size of DG at optimal power factor to minimize the power losses however only type III has been exploited [12].

The present work develops the comprehensive formula by extending the analytical expression presented in [11] to find the optimal size of multiple DGs supplying real and reactive power and a search to identify best locations and optimal power factor to achieve the objective by compensating the active and reactive powers. Besides, voltage profile enhancement is also examined and the results of the proposed hybrid approach are verified with existing technique.

Mathematical Background

The total power loss has been formulated as is given by (1).This formula is popularly referred as "Exact Loss" formula [13].

***Corresponding author:** Kansal S, Department of Electrical Engineering, Baba Hira Singh Bhattal Institute of Engineering and Technology, Lehragaga-148031, Punjab, India, E-mail: kansal.bhsb@gmail.com

Where,

$$P_L = \sum_{i=1}^{N}\sum_{j=1}^{N}[\alpha_{ij}(P_iP_j + Q_iQ_j) + \beta_{ij}(Q_iP_j - P_iQ_j)] \quad (1)$$

$$\alpha_{ij} = \frac{r_{ij}}{V_iV_j}\cos(\delta_i - \delta_j)$$

$$\beta_{ij} = \frac{r_{ij}}{V_iV_j}\sin(\delta_i - \delta_j)$$

and $Z_{ij} = r_{ij} + jx_{ij}$ is the ij[th] element of [Zbus] matrix

N-Total number of buses.

$$Gbest_m = (gbest_{m,1}, gbest_{m,2}, gbest_{m,3}, ...gbest_{m,n})$$

$$P_i = P_{Gi} - P_{Di} \quad \dots (2)$$

$$Q_i = Q_{Gi} - Q_{Di} \quad \dots (3)$$

P_{Gi} & Q_{Gi} are generated active and reactive powers at i[th] bus respectively;

P_i & Q_i are active and reactive power injections at i[th] bus respectively;

P_{Di} & Q_{Di} are the active and reactive loads at i[th] bus respectively.

Sizing of multiple DGs

The real power loss formula (1) is used to determine the sizes of multiple DGs at respective buses to minimize the power loss.

Considering, $a_{k_i} = (\sin)\tan(\cos^{-1}(PF_{DG_{k_i}}))P_{Di}$ The reactive power output of DG, where n is the number of DGs and is the bus number of ith DG is given by

$$Q_{DG_{k_i}} = a_{k_i}P_{DG_{k_i}} \quad \dots (4)$$

In which (+) sign is for injecting reactive power and (-) sign is consuming reactive power by DG. $PF_{DG_{k_i}}$ is the power factor of DG at kith bus of ith DG, which equal to power factor of system load.

The active and reactive power injected at bus kith, where DG is located, are given by (5) and (6), respectively,

$$P_{k_i} = P_{DG_{k_i}} - P_{D_{k_i}} \quad \dots (5)$$

$$Q_{k_i} = Q_{DG_{k_i}} - Q_{D_{k_i}} = a_{k_i}P_{DG_{k_i}} - Q_{D_{k_i}} \quad \dots (6)$$

$$P_L = \sum_{i=1}^{N}\sum_{j=1}^{N}[\alpha_{ij}((P_{DG_{k_i}} - P_{D_{k_i}})P_j + (a_{k_i}P_{DG_{k_i}} - Q_{D_{k_i}})Q_j) + \beta_{ij}((a_{k_i}P_{DG_{k_i}} - Q_{D_{k_i}})P_j - (P_{DG_{k_i}} - P_{D_{k_i}})Q_j)$$

Differentiate P_L w.r.t. $P_{D_{k_1}}$

$$A_{k_1k_2}P_{DG_{k_1}} + A_{k_1k_2}P_{DG_{k_2}} + ... + A_{k_ik_z}P_{DG_{ki}} + ... + A_{k_nk_z}P_{DG_{k_n}} = B_{k_1}$$

Differentiate P_L w.r.t. $P_{D_{k_1}}$

$$A_{k_1k_2}P_{DG_{k_1}} + A_{k_2k_2}P_{DG_{k_2}} + ... + A_{k_ik_z}P_{DG_{ki}} + ... + A_{k_nk_z}P_{DG_{k_n}} = B_{k_z}$$

Similarly Differentiate P_L w.r.t. P_{DGKI}

$$A_{k_ik_i}P_{DG_{k_1}} + A_{k_2k_i}P_{DG_{k_2}} + ... + A_{k_ik_i}P_{DG_{k_i}} + ... + A_{k_nk_i}P_{DG_{k_n}} = B_{k_i}$$

Where

$$A_{k_ik_j} = \begin{cases} \alpha_{k_ik_j}(1+a_{k_i}^2) & \text{If } k_i = k_j \\ \alpha_{k_ik_j}(1+a_{k_i}a_{k_j}) + \beta_{k_ik_j}(a_{k_i} - a_{k_j}) & k_i \neq k_j \end{cases} \quad ...(7)$$

There will be n equations with n variables, which can be solved as

$$[P_{DG}]_{nx1} = [A]_{nxn}^{-1} \times [B]_{nx1} \quad \dots (8)$$

Where,

$$[P_{DG}]_{n\times 1} = [P_{DG_{k_1}} ... P_{DG_{k_i}} ... P_{DG_{k_n}}]^T \quad \dots (9)$$

$$B_{k_i} = \begin{pmatrix} \alpha_{k_ik_j}(P_{D_{k_i}} + a_{k_i}Q_{D_{k_i}}) + \sum_{j=1, j\neq i(\neq k_j)}^{n} \left(\alpha_{k_ik_j}(P_{D_{k_j}} + a_{k_i}Q_{D_{k_j}}) - \beta_{k_ik_j}(Q_{D_{k_j}} + a_{k_i}P_{D_{k_j}})\right) \\ + \sum_{j=1, j\neq k_1k_2 ... k_n}^{n} \left(\alpha_{k_ik_j}(P_{D_{k_j}} + a_{k_i}Q_{D_{k_j}}) - \beta_{k_ik_j}(Q_{D_{k_j}} + a_{k_i}P_{D_{k_j}})\right) \end{pmatrix} \quad 10)$$

Equation (9) and (4) provides the sizes of multiple DGs at respective bus for the losses to be minimum.

Optimal power factor of DGs

With obtained optimal sizes of P_{DGI} and Q_{DGI} the power factor of the DGs is considered as

$$OPF = \frac{P_{DG_i}}{\sqrt{P_{DG_i}^2 + Q_{DG_i}^2}} \quad (11)$$

Optimal locations of DGs

For single DG placement, it is possible to calculate DG size and to evaluate the loss at every bus by analytical approach. But when it comes to determine combination of N buses in the same network for n DGs, the number of combinations will be NCn, so a search technique or a heuristic method needs to be implemented to find the optimal locations. The optimal locations for the placement of multiple DGs are determined by using PSO technique taking the location and optimal power factor of each DG as the variable.

Problem Formulation

Objective function

The main objective is to minimize the total power loss as given in (1) while meeting the following constraints.

$$MinP_L = \sum_{i=1}^{N}\sum_{j=1}^{N}\left[\alpha_{ij}(P_iP_j + Q_iQ_j) + \beta_{ij}(Q_iP_j + P_iQ_j)\right] \quad (12)$$

The network power flow equation must be satisfied,

The sizing and locations are considered at point load only,

The voltage at every bus in the network should be within the acceptable range (Utility's standard ANSI Std. C84.1-1989) i.e., within permissible limit (±5%) [14],

Vmin ≤ Vi ≤ Vmax $\forall_i \in$ {buses of the network}

Current in a feeder or conductor, must be well within the maximum thermal capacity of the conductor

$I_i \leq I_i^{Rated}$ I_i^{Rated} {Branches of the network}

Here, I_i^{Rated} is current permissible for branch i within safe limit of temperature.

Distributed Generation is defined as the generation of electricity by facilities that are sufficiently smaller than the central generating plants so to allow interconnection at nearly any point in a power system. There is no defined limit on the amount of generation through DG. For example, in [15] and [16] the maximum DG installed capacity limits have been considered as 30% and 50% respectively. Hence, here the total installed capacity of DG in the network has been limited to less than 30% of substation rated capacity plus line losses to maintain the concept of DG against centralized generation similar to [15].

$$S_{DG} \leq 0.30S_T \quad (12)$$

Where S_T is the rating of the transformer

Computational procedure

The proposed approach has been used for determining the optimal placement of multiple DGs, and is given step by step in the following subsections. The backward sweep and forward sweep method of load flow [17] is used for radial network solution.

Particle swarm optimization technique.

Particle Swarm Optimization (PSO) is a population-based optimization technique which provides a population-based search procedure in which individuals called particles change their position (state) with time. In a PSO system, particles fly around in a n-dimensional search space. During flight, each particle adjusts its position according to its own experience (This value is called pbest), and according to the experience of a neighboring particle (This value is called gbest), made use of the best position encountered by itself and its neighbor [18].

Mathematically, the position of particle in an n-dimensional vector is represented as:

$$X_m = \left(x_{m,1}, x_{m,2}, x_{m,3}, \ldots\ldots\ldots\ldots\ldots\ldots x_{m,n}\right) \quad (13)$$

The velocity of this particle is also an n-dimensional vector,

$$V_m = \left(v_{m,1}, v_{m,2}, v_{m,3}, \ldots\ldots\ldots\ldots\ldots\ldots v_{m,n}\right) \quad (14)$$

Alternatively, the best position related to the lowest value (for objective minimization) of the objective function for each particle is

$$Pbest_m = \left(Pbest_{m,1}, Pbest_{m,2}, Pbest_{m,3}, \ldots\ldots\ldots\ldots\ldots\ldots Pbest_{m,n}\right)$$ and the global best position among all the particles or best pbest is denoted as:

$$Gbest_m = \left(gbest_{m,1}, gbest_{m,2}, gbest_{m,3}, \ldots\ldots\ldots\ldots\ldots\ldots gbest_{m,n}\right)$$

During the iteration procedure, the velocity and position of the particles are updated. The population size of swarms and iterations are fixed i.e. the PSO parameters, population size of swarms and iterations are taken 50 and 60 respectively. The population of mth particles Xm (consisting of location and power factor of DG) as well as their velocity V_m in the search space is initialized as given in (13-14). The appropriate values for weights ω_{min} and ω_{max} are 0.4 and 0.9 [19] are set respectively.

The present work extends the analytical expressions presented in [11] to find the optimal sizes of multiple DGs and optimal locations with PSO technique. The number of DGs (n=1, 2, 3...) are considered to minimize the power loss. The computational procedure to find the optimal sizes at locations of multiple DGs is described below.

Step 1: Input line and bus data, and constraints.

Step 2: Enter the number of DG units.

Step 3: Run the load flow for the base case and calculate the losses using (1).

Step 4: Find the size of DGs for each bus using (9) and (4).

Step 5: Initialize random values into particles which correspond to bus numbers or locations of DGs of the given network. Set the iteration counter k=0.

Step 6: Take the first particle as the locations of DGs, there are number of possible locations as the number of particles.

Step 7:Find the fitness value for the every selected location for DGs using (1).

Step 8: Update the weight, velocity and position of each particle.

Step 9: if the iteration number reaches the maximum limit, go to step 10. Otherwise, set the iteration index k=k+1, and go back to step 5.

Step 10: Print out the optimal solution to the target problem. The best position includes the optimal locations and sizes of DGs and representing the corresponding minimum total real power loss.

Numerical results

Test system

The proposed methodology is tested on test system contains 33 buses and 32 branches as shown in Figure 1. It is a radial system with a total load of 3.72 MW and 2.3 MVAR [20] with Beaver conductor. The base voltage for the test system is 12.66 kV. An analytical software tool has been developed in MATLAB environment for the proposed approach to run load flow, calculate losses. The optimal sizes of DGs at optimal locations are determined to achieve the objective. The maximum number of DG units installed considered to be three and the total capacity of the DG units is also assumed to equal to the total load plus line losses. The sizes of DGs are set less than 30% of substation rated capacity plus line losses.

DG placement at optimal power factor

Table 1 shows the placement of multiple DGs supplying both real and reactive powers. The maximum sizes of DGs installed are 30% of the rated capacity of substation plus line losses. The results of the base case and three cases with DG numbers ranging from one to three are determined. The result includes the optimal sizes, optimal locations and optimal power factor of DGs with respect to the total loss. The power

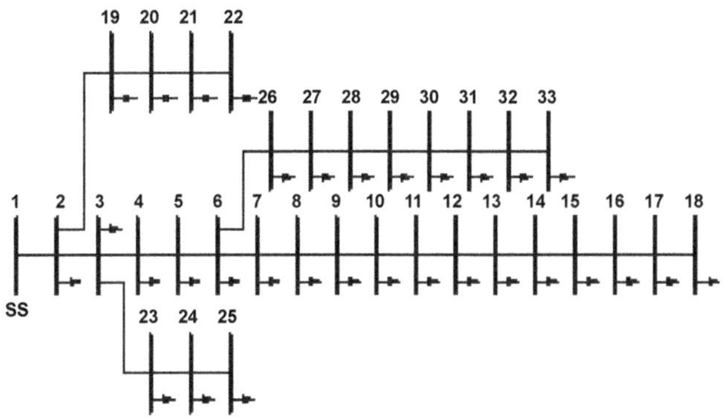

Figure 1: Single line diagram of 33 bus distribution test system.

Cases	Installed DG Schedule				Ploss (kW)	Loss Reduction (%)
No DG					211	0
I DG	Location	30			75.6	64.18
	Size (kVA)	1749				
	OPF	0.86				
2 DGs	Locations	14	30		42.53	79.84
	Size (kVA)	705	1076			
	OPF	0.95	0.77			
3 DGs	Locations	14	30	32	40.88	80.62
	Size (kVA)	706	663	430		
	OPF	0.95	0.67	0..89		

Table 1: Type-III DG Placement for 33-Bus System.

factor of the DG must be opposite to the power factor of bus load. The 33-bus system has a lagging power factor load; hence the power factor of DG must be leading. Consequently, the net total of both active and reactive power of that bus where the DG is placed will decrease.

The loss reduction and schedule of installed DGs are presented in the Table 1. For single DG, the loss reduction by the hybrid approach is 64.18% and by two DGs and three DGs the loss reductions are 79.84% and 80.62% respectively. It is observed that as the number of DGs is increased, the reduction in line losses is more effective.

Comparative study

Although the proposed hybrid approach proved its robustness in solving the test case, additional case of DG without technical size constraints was adopted on 33-bus radial distributed feeder system for comparison purpose. Therefore, the results of the proposed hybrid approach was compared with the solutions obtained based on the Improved analytical method [12]. Table 2 summarizes the optimal solutions achieved by these methods.

Observing Table 2, the optimal placement of the single DG, bus location, optimal power factor and reduction in line losses were nearly identical, with reduction in size of DG. For two DG placement the reduction in line loss was 28.55 kW by the hybrid approach as compared to 44.39 kW by Improved Analytical approach [12].

As seen in the Improved Analytical approach the DGs are placed one by one with fixed OPF, whereas in the proposed hybrid approach the numbers of DGs are placed simultaneously with their OPFs.

For placement of three DGs, the reduction in losses was 11.76 kW as compared to 22.29 kW of Fast Analytical approach. However, if the optimal DG size in the analytical methods were rounded off to the closest practical one, the accuracy of the results would be affected. The proposed hybrid approach avoids this limitation and accuracy of the results is guaranteed.

Voltage profile

Figures 2-4 indicate the minimum and maximum voltages before

Cases	Technique	Installed DG Schedule				Ploss (kW)
No DG						211
I DG Unit	Improved Analytical [12]	Location	6			67.9
		Size (kVA)	3107			
		OPF	0.82			
	Hybrid Approach	Location	6			67.95
		Size (kVA)	3028			
		OPF	0.82			
2 DG units	Improved Analytical [12]	Locations	6	30		44.39
		Size (kVA)	2195	1098		
		OPF	0.82	0.82		
	Hybrid Approach	Locations	13	30		28.55
		Size (kVA)	828	1114		
		OPF	0.91	0.73		
3 DG units	Improved Analytical [12]	Locations	6	30	14	22.29
		Size (kVA)	1098	1098	768	
		OPF	0.82	0.82	0.82	
	Hybrid Approach	Locations	13	24	30	11.76
		Size (kVA)	782	1069	1016	
		OPF	0.91	0.9	0.71	

Table 2: Comparison of DG placement results for the 33-bus system.

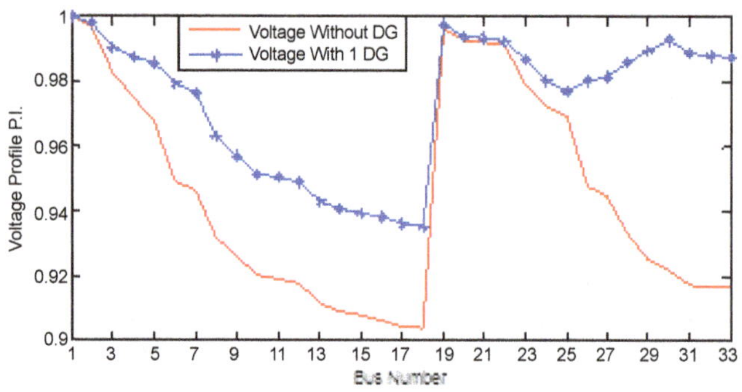

Figure 2: Voltage profile before and after 1DG.

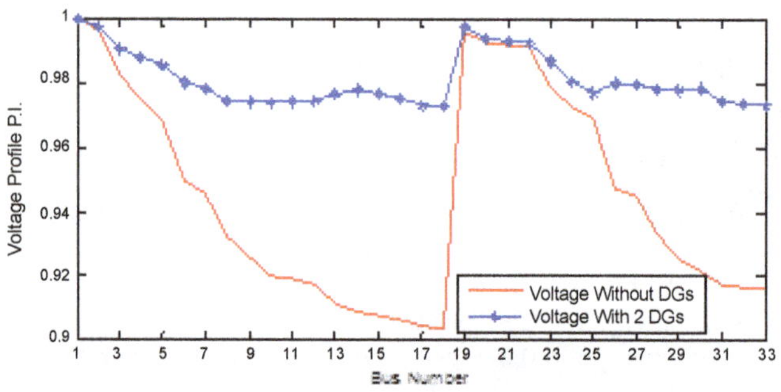

Figure 3: Voltage profile before and after 2DGs.

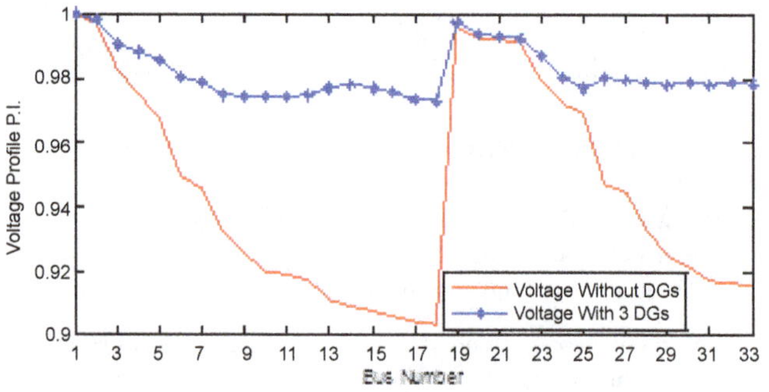

Figure 4: Voltage profile before and after 3DGs.

and after the placement of 1 DG, 2 DGs and 3 DGs for 33-bus test system.

It is seen that in all the cases the voltage profile improves, when the number of DG units installed in the system is increased, while satisfy all the current and voltage constraints.

Conclusion

In the proposed hybrid approach, the sizes of DGs are evaluated by analytical approach and the locations are determined by the application of PSO approach. This paper has presented the allocation of multiple DGs capable of injecting both real and reactive powers for active and reactive power compensation to minimize the line losses in the primary distribution networks. The number of DG units with appropriate sizes at locations can reduce the losses to a considerable amount. The optimal power factor which results minimum power loss has also been determined. The proposed approach of optimal placement of multiple DGs not only reduces the line losses but also minimize the sizes of DGs with satisfaction of the permissible voltage limits. In the age of integrated grid, the placement and analysis of multiple DGs give guidance for optimal operation of power system.

References

1. Abu-Mouti FS, El-Hawary ME (2011) Heuristic Curve-Fitted Technique for Distributed Generation Optimization in Radial Distribution Feeder Systems. IET Generation, Transmission & Distribution 5: 172-180.
2. Ochoa LF, Padilha-Feltrin A, Harrison GP (2006) Evaluating distributed generation impacts with a multiobjective index. IEEE Transactions on Power Delivery 21: 1452–1458.
3. Imrana AM, Kowsalyaa M, Kothari DP (2014) A novel integration technique for optimal network reconfiguration and distributed generation placement in power distribution networks. Electric Power & Energy System 63: 461-472.
4. Mardaneh M, Gharehpetian GB (2004) Siting and Sizing of DG Units Using GA and OPF Based Technique. IEEE Region 10 Conference 3: 331-334.
5. Kumar V, Rohit Kumar, Gupta I, Gupta HO (2010) DG integrated approach for service restoration under cold load pickup. IEEE Transactions Power Delivery 25: 398-406.
6. Katsigiannis YA, Georgilakis PS (2008) Optimal Sizing of Small Isolated Hybrid Power Systems Using Tabu Search. J Optoelectronics and Advanced Materials 10: 1241-1245.
7. Singh D, Singh D, Verma KS (2009) Multiobjective optimization for DG planning with load models. IEEE Transactions on Power Systems 24: 427–436.
8. Gozel T, Hocaoglu MH (2009) An analytical method for the sizing and siting of distributed generators in radial systems. Int J Elec Pow Sys Res 79: 912-918.
9. Hedayati H, Nabaviniaki SA, Akbarimajd A (2009) A method for placement of DG units in distribution networks. IEEE Transactions on Power Delivery 23: 1620-1628.
10. Kansal S, Kumar V, Tyagi B (2012) Composite Active and Reactive Power Compensation of Distribution Networks. Proceedings of IEEE 7th International conference ICIIS-2012, IIT Madras, 1-6.
11. Acharya N, Mahat P, Mithulananthan N (2006) An analytical approach for DG allocation in primary distribution network. Electric Power & Energy System 28: 669-678.
12. Hung DQ, Mithulananthan N (2011) Multiple distributed generators placement in Primary Distribution Networks for loss reduction. IEEE Trans Industrial Electronics 60: 1700-1708.
13. Elgerd IO (1971) Electric energy system theory: an introduction, McGraw- Hill.
14. American National Standards Institute (2006) American National Standard for Electric Power Systems and Equipment—Voltage Ratings (60 Hertz). National Electrical Manufacturers Association.
15. El-Khattam W, Hegazy YG, Salama MMA (2005) An integrated distributed generation optimization model for distribution system planning. IEEE Trans Power Syst 20: 1158-1165.
16. Méndez VH, Rivier J, Gómez T (2006) Assesment of energy distribution losses for increasing penetration of distributed generation. IEEE Trans Power Syst 21: 533-540.
17. Haque MH (1996) Efficient load flow method for distribution systems with radial or mesh configuration. IEE Proc.-Gener Transm Distrib 143: 33-38.
18. Kennedy J, Eberhart R (1995) Particle Swarm Optimizer. IEEE International Conference on Neural Networks , Perth (Australia), IEEE Service Centre Piscataway, NJ, 4: 1942-1948.
19. Eberhart RC, Shi Y (2000) Comparing inertial weights and constriction factor in particle swarm optimization. Proceedings of the 2000 International congress on Evaluating computation, San Diego,Calfornia NJ 1: 84-88.
20. Kashem MA, Ganapathy V, Jasmon GB, Buhari MI (2000) A novel method for loss minimization in distribution networks. Int Conference on Electric Utility Deregulation and Restructuring and Power Technology, London 251-256.

Modeling and Control of a Wind System Based Doubly Fed Induction Generator: Optimization of the Power Produced

Marouan Elazzaoui*

Department of Electrical Engineering, Electronic Power and Control Laboratory, Mohammedia School of Engineering Université Mohammed V-Agdal, Morocco

Abstract

This paper deals with the modeling and control of a wind energy system based on a doubly-fed induction generator. Initially, an MPPT control strategy of the doubly-fed induction generator is presented. Thereafter, the control vector-oriented stator flux is performed. Finally, the simulation results of the wind system using a doubly-fed of 3 MW are presented in the Matlab / Simulink environment.

Keywords: Wind turbine; Doubly-fed induction generator; Converter; Vector control

Introduction

Today, wind energy has become a viable solution for energy production, in addition to other renewable energy sources. While the majority of wind turbines are fixed speed, the number of variable speed wind turbines is increasing [1]. The doubly-fed induction generator with vector control is a machine that has excellent performance and is commonly used in the wind turbine industry [2]. There are many reasons for using a doubly-fed induction generator for a variable speed wind turbine; such as reducing efforts on the mechanical parts, reducing noise and the possibility of control of active and reactive power [3]. The wind system using DFIG and a "back-to-back" converter that connects the rotor of the generator and the network has many advantages (Figure 1).

One advantage of this structure is that the power converters used are dimensioned to pass a fraction of the total power of the system [4,5]. Thereby reducing losses in power electronic components. The performance and power production does not only depend on the DFIG, but also the way in which the two parts of "back-to-back" converter is controlled. The power converter machine side is called «Rotor Side Converter» (RSC) and the grid side power converter is called «Grid Side Converter» (GSC). The machine side power converter controls the active power and reactive power produced by the machine. As for the grid-side converter, it controls the DC bus voltage and line-side power factor. In this paper, we present a technique for controlling the two power converters. We will analyze their dynamic performance by simulations in Matlab / Simulink environment. We start with a model of the wind turbine,then a technique for continued operation at maximum power point tracking (MPPT) will be presented. Subsequently, we present a model of DFIG in the landmark Park, and the general principle of the control of two power converters. We conclude by presenting the simulation results and their interpretation.

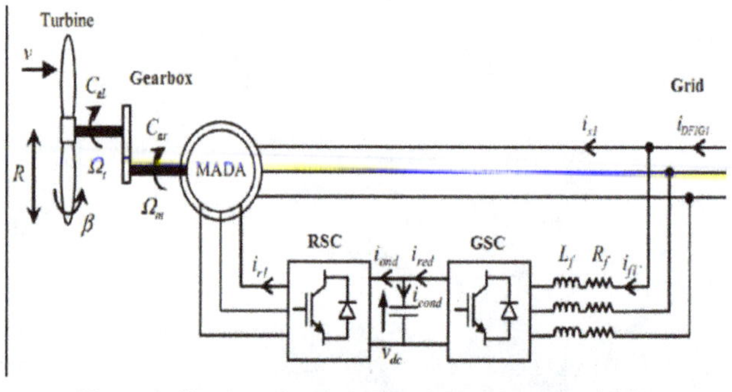

Figure 1: Structure of a wind energy system based on DFIG.

Modeling of Wind Turbine

By applying the theory of momentum and Bernoulli's theorem, we can determine the incident power (the theoretical power) due to wind [6,7]:

$$P_{incident} = \frac{1}{2}.\rho.S.v^2 \quad (1)$$

S : the surface swept by the blades of the turbine m²

ρ : the density of air ($\rho = 1,225 kg/m^2$ at atmospheric pressure).

v : wind speed ($T_{sa} = \frac{1}{2\Omega_t} C_p(\lambda,\beta)\rho S v^3$ m/s)

In a wind energy system, due to various losses, the power extracted from provided on the rotor of the turbine is less than the incident power. The power extracted is expressed by [8]:

$$P_{extract} = \frac{1}{2}.\rho.S.C_p(\lambda,\beta).v^3 \quad (2)$$

C_p (λ,β) is called the power coefficient, which expresses the aerodynamic efficiency of the turbine. It depends on the ratio, which is the ratio between the speed at the end of the blades and the wind speed, and the orientation angle β of the blades. The ratio λ can be expressed by the following equation (8,9):

$$\lambda = \frac{\Omega_t R}{v} \quad (3)$$

Ω: The turbine speed of rotation (rad/s).

R: The length of a blade.

The maximum of power coefficient C_p was determined by Albert

***Corresponding author:** Marouan Elazzaoui, Department of Electrical Engineering, Electronic Power and Control Laboratory, Mohammedia School of Engineering Université Mohammed V-Agdal, Morocco, E-mail: marouan.elazzaoui@gmail.com

Betz as follows:

$$C_{p\_max}(\lambda,\beta)=\frac{16}{27}=0,5926 \quad (4)$$

The power coefficient is intrinsic to the formation of wind turbine and depends on the profiles of the blades. The power coefficient can be modeled with a single equation that depends on the speed ratio λ and the orientation angle β of the blades as follows:

$$C_{p\_max}(\lambda,\beta)=(0.5-0.0167.(\beta-2))\sin\left[\frac{\pi(\lambda+0.1)}{18.5-0.3(\beta-2)}\right]-0.00184.(\lambda-3)(\beta-2) \quad (5)$$

The Figure 2 shows curves of the power coefficient as a function of λ for various β values. This gives a maximum power coefficient of 0.5 for a speed ratio λ which is 9.13 maintaining β at 2°. By setting β and λ respectively to their optimal values, the wind system will provide optimum electrical power.

The aerodynamic torque on the slow axis is expressed by:

$$T_{sa}=\frac{1}{2\Omega_t}C_p(\lambda,\beta)\rho S v^3 \quad (6)$$

Mechanical speed is related to the speed of rotation of the turbine by the coefficient of the gearbox. The torque on the slow axis is connected to the torque on the fast axis (generator side) by the coefficient of the gearbox (Figure 3).

The total inertia J is made up of the inertia of the turbine plotted on the generator rotor and inertia of the generator:

$$J=\frac{J_t}{G^2}+J_g \quad (7)$$

$$\begin{cases} P_s=-v_{sq}\dfrac{M}{L_s}i_{rq} \\ Q_s=v_{sq}^{\,2}-\dfrac{M.v_{sq}}{L_s}i_{rd} \end{cases}$$

J_t : inertia of the turbine.

J_g : inertia of the generator.

To determine the evolution of the mechanical speed from total torque T_{mec} applied to the rotor of DFIG, we apply the fundamental equation of dynamics:

$$J=\frac{d\Omega_m}{dt}=T_{mec}=T_{fa}-T_{em}-f\Omega_m \quad (8)$$

Ω: mechanical speed of DFIG.

T_{fa} : aerodynamic torque on the fast axis of the turbine.

T_{em}: Electromagnetic torque.

$f\Omega_m$: coefficient of friction.

The previous equations used to establish the block diagram of the turbine model.

Maximum Power Extraction

In order to capture the maximum power of the incident wind, permanently must adjust the rotational speed of the turbine to the wind. An erroneous speed measurement therefore inevitably leads to degradation of the received power that is why most wind turbines are controlled without control of the speed. The controller in this case should impose a reference torque to allow DFIG turning at an adjustable speed to ensure optimal operating point of power extraction. In this context, the ratio of the speed of wind λ must be maintained at its optimum value $\lambda=\lambda_{opt}$ on a certain wind speed range. Thus, the power coefficient would be maintained at its maximum value (Figure 4).

Modelling of DFIG

The electrical equations of DFIG in the dq reference can be written equation 9,11:

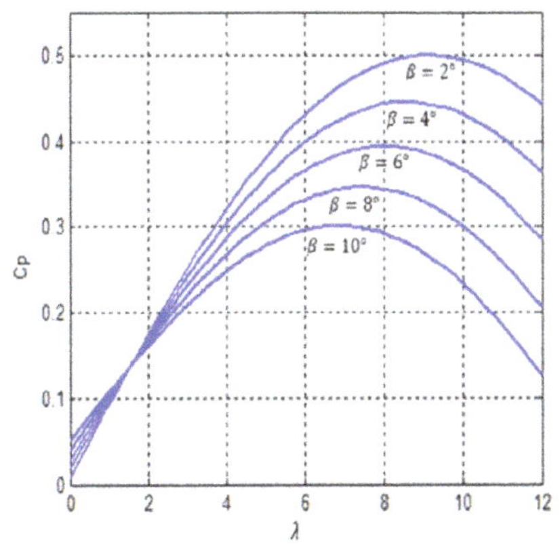

Figure 2: Power coefficient as a function of λ for various β values.

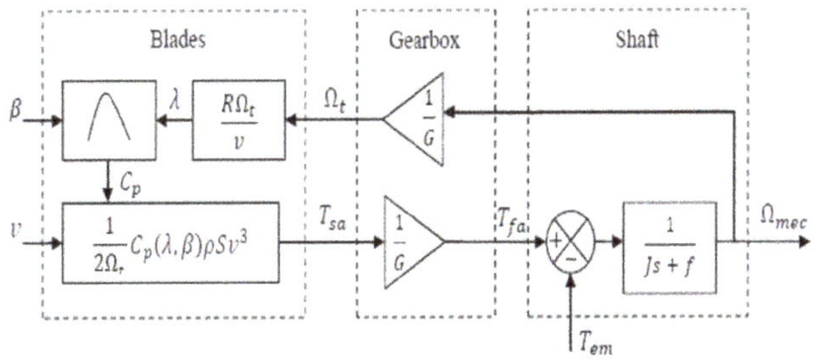

Figure 3: Block diagram of the turbine model.

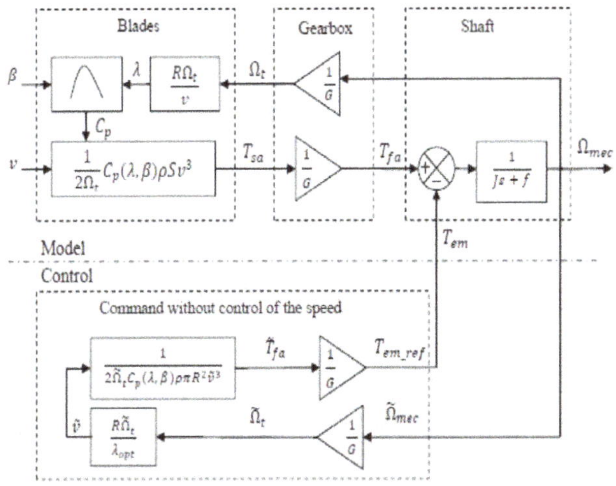

Figure 4: Block diagram without control of the speed.

$$\begin{cases} v_{sd} = R_s i_{sd} + \dfrac{d\varphi_{sd}}{dt} - \omega_s \varphi_{sq} \\ v_{sq} = R_s i_{sq} + \dfrac{d\varphi_{sq}}{dt} - \omega_s \varphi_{sd} \\ v_{rd} = R_r i_{rd} + \dfrac{d\varphi_{rd}}{dt} - \omega_r \varphi_{rq} \\ v_{rq} = R_r i_{rq} + \dfrac{d\varphi_{rq}}{dt} - \omega_r \varphi_{rd} \end{cases} \tag{9}$$

The pulse of the stator currents being constant, the rotor pulse is derived by:

$$\omega_r = \omega_s - p\Omega_m \tag{10}$$

The equations for the flux in dq reference are given by:

$$\begin{cases} \varphi_{sd} = L_s i_{sd} + M i_{rd} \\ \varphi_{sq} = L_s i_{sq} + M i_{rq} \\ \varphi_{rd} = L_r i_{sd} + M i_{sd} \\ \varphi_{rq} = L_r i_{rq} + M i_{sq} \end{cases}$$

With:

$L_s = l_s - M_s, L_r = l_r - M_r$: Cyclic inductances of stator and rotor phase. l_s et l_r inductors own stator and rotor of the machine. M_s et M_r mutual inductances between two stator phases and between two rotor phases of the machine.

M: maximum mutual inductance between stator and rotor stage.

p: number of pairs of poles of the DFIG.

The expression of the electromagnetic torque of the DFIG based on the flow and stator currents is written as follows:

$$T_{em} = p(\varphi_{sd} i_{sd} - \varphi_{sq} i_{sd}) \tag{12}$$

The active and reactive power of the stator and rotor are written as follows (9,11):

$$\begin{cases} P_s = v_{sd} i_{sd} + v_{sq} i_{sq} \\ Q_s = v_{sq} i_{sd} + v_{sd} i_{sq} \\ P_r = v_{rd} i_{rd} + v_{sq} i_{rq} \\ Q_s = v_{rq} i_{rd} + v_{rd} i_{rq} \end{cases} \tag{13}$$

Vector Control of DFIG

DFIG control strategies are based on two different approaches (12):

- Flow control in closed loop, where the frequency and voltage are considered variables (unstable grid).
- Flow control in open loop when the voltage and frequency are constant (stable grid).

In our study, the frequency and voltage are assumed to be constant. We can see from equation (12), the strong coupling between flows and currents. Indeed, the electromagnetic torque is the cross product between flows and stator currents, making the control of DFIG particularly difficult. To simplify ordering, we approximate the model to that of the DC machine which has the advantage of having a natural coupling between flows and currents. For this, we apply vector control, also known order by direction of flow. We choose dq reference linked to the rotating field (Figure 5).

Stator flux φ_s is oriented along the axis. Thus, we can write:

$$\begin{cases} \varphi_{sd} = \varphi_s \\ \varphi_{sq} = 0 \end{cases} \tag{14}$$

$$\begin{cases} \varphi_{sd} = L_s i_{sd} + M i_{rd} = \varphi_s \\ \varphi_{sq} = L_s i_{sq} + M i_{rq} = 0 \end{cases} \tag{15}$$

In the field of production of wind energy, these are average machines and high power which are mainly used. Thus, we neglect the stator resistance. Taking the constant stator flux we can write:

$$\begin{cases} v_{sd} = 0 \\ v_{sq} = V_s = \omega_s \varphi_s \end{cases} \tag{16}$$

To determine the angles necessary for transformation Park of the stator variables θ_s and the rotor variables θ_r, we used a phase locked loop (PLL) as shown in Figure 6. This PLL allows to accurately estimate the frequency and amplitude of the grid (13). The architecture of the controller is shown in Figure 7. It is based on the three-phase model of the electromechanical conversion chain of the wind energy system. From Figure 7, three commands are needed:

- The maximum extraction control wind power by controlling said MPPT (detailed in Section III).
- Control of RSC by controlling the electromagnetic torque and stator reactive power of DFIG.
- Control of GSC by controlling the voltage of the DC bus and the active and reactive power exchanged with the grid.

Control of the rotor side converter (RSC)

The principle of the Control of the rotor side converter is shown in Figure 8. The Controls of electromagnetic torque and stator reactive power will be obtained by controlling the rotor dq axes currents of DFIG. From the equations (15) and (16) we obtain the expression of the stator current:

$$\begin{cases} i_{sd} = \dfrac{v_{sq}}{\omega_s L_s} - \dfrac{M}{L_s} i_{rd} \\ i_{sq} = -\dfrac{M}{L_S} i_{rq} \end{cases} \tag{17}$$

These expressions are then substituted in the equation (11) of the rotor flux which then become:

$$\begin{cases} \varphi_{rd} = \sigma L_r i_{rd} + \dfrac{M v_{sq}}{\omega_s L_s} \\ \varphi_{rq} = \sigma L_r i_{rq} \end{cases} \tag{18}$$

With:

$\sigma = 1 - \dfrac{M^2}{L_r L_s}$: the dispersion coefficient of the DFIG Substituting the expressions of the direct and quadrature components of the rotor flux in the equation (9) we get:

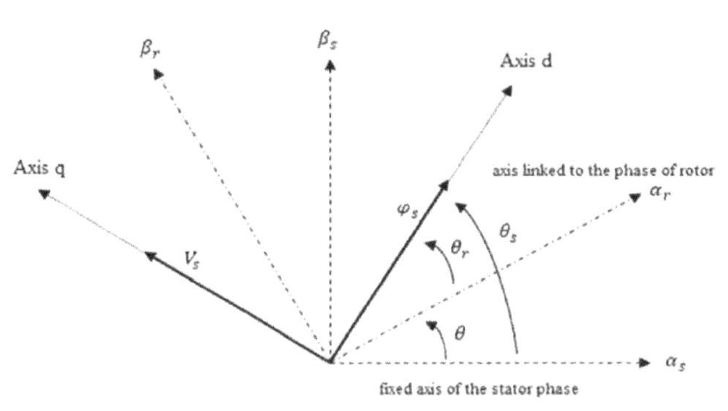

Figure 5: Orientation of the axis d to stator flux.

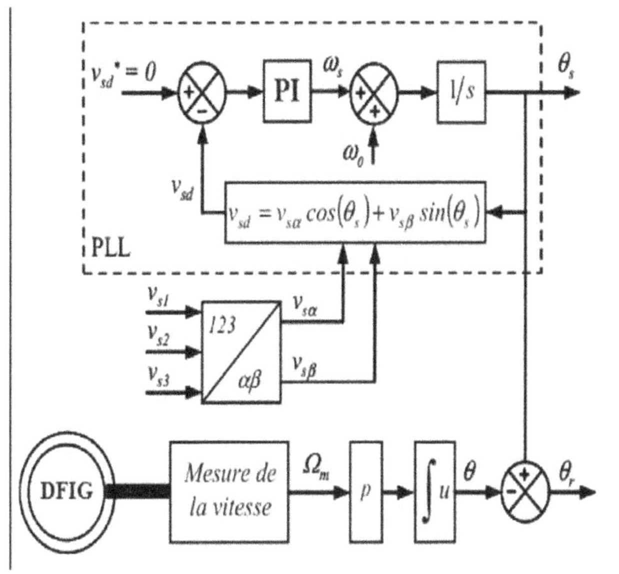

Figure 6: Establishment angles processing using a PLL.

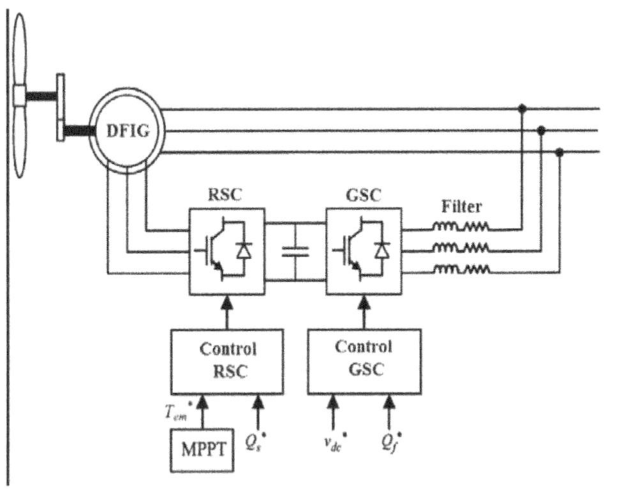

Figure 7: Control architecture of the wind system.

$$\begin{cases} v_{rd} = R_r i_{rd} + \sigma L_r \dfrac{di_{rd}}{dt} + e_{rd} \\ v_{rq} = R_r i_{rq} + \sigma L_r \dfrac{di_{rd}}{dt} + e_{rq} + e_{\phi} \end{cases}$$

With:

$$e_{rd} = -\sigma L_r \omega_r i_{rq}; e_{rq} = \sigma L_r \omega_r i_{rd}; e_{\phi} = \frac{\omega_r M V_{sq}}{\omega_s L_s}$$

The expression of the electromagnetic torque becomes:

$$T_{em} = -\frac{pMv_{sq}}{\omega_s L_s} i_{rq} \tag{20}$$

Stator active and reactive powers are expressed by:

$$\begin{cases} P_s = -v_{sq} \dfrac{M}{L_s} i_{rq} \\ Q_s = v_{sq}^{\;2} - \dfrac{M.v_{sq}}{L_s} i_{rd} \end{cases} \tag{21}$$

These last expressions show that the choice of coordinate system (dq) makes the electromagnetic torque produced by the DFIG, and therefore the stator power, proportional to the current of the rotor axis (q). The stator reactive power, in turn, is proportional to the current of the rotor axis (d) due to a constant imposed by the grid. Thus, these stator powers can be controlled independently of one another. This shows us that we can set up a control rotor currents due to the influence of the couplings, every current can be controlled independently each with its own controller. The reference values for these regulators will be the rotor axis current (q) and the rotor axis current (d). The block diagram of the control loops of the axis rotor currents (dq) is shown in Figure 9, the regulators used are PI correctors.

The rotor current of the reference axis (q) is derived from the MPPT control via the electromagnetic torque reference (Figure 4). The rotor current of the reference axis (d) is, in turn, derived from the control of the stator reactive power.

From the equations (20) and (21) we obtain:

$$i_{rd}^{*} = -\frac{L_s}{M.v_{sq}} Q_s^{*} + \frac{v_{sq}}{\omega_s M} \tag{22}$$

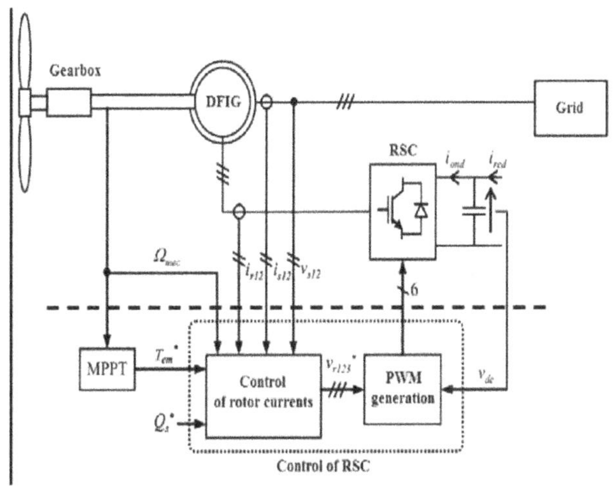

Figure 8: Principle of the Control of the rotor side converter (RSC).

$$i_{rd}^{*} = -\frac{\omega_s L_s}{pMv_{sq}} T_{em}^{*} \tag{23}$$

Figure 10 shows the block diagram of the control of RSC. Figure 11 show the principle of the Control of the rotor side converter performs the following two functions:

- Control currents flowing in the RL filter.
- Control voltage of the DC bus.

Control currents flowing in the RL filter

According to Kirchhoff's laws, the equations of the filter in the three-phase reference voltages are given by:

$$\begin{cases} V_{f1} = -R_f i_{f1} - L_f \dfrac{di_{f1}}{dt} + V_{s1} \\ v_{f2} = -R_f i_{f2} - L_f \dfrac{di_{f2}}{dt} + V_{s2} \\ V_{fa} = -R_f i_{fa} - L_f \dfrac{di_{fa}}{dt} + V_{sa} \end{cases} \tag{24}$$

Applying the Park transformation, we obtain: P_f^{*}

$$\begin{cases} v_{fd} = -R_f i_{fd} - L_f \dfrac{di_{fd}}{dt} + e_{fd} \\ v_{fq} = -R_f i_{fq} - L_f \dfrac{di_{fq}}{dt} - \omega L_f i_{fq} + e_{fq} \end{cases} \tag{25}$$

With:

$$e_{fd} = \omega_s L_f i_{fd}; e_{fq} = -\omega_s L_f i_{fd} + v_{sq}$$

The pattern of binding of GSC to grid in the landmark Park along the stator rotating field shows us that we can put in place a control of the currents flowing in the RL filter is given to the influence of the couplings, every axis can be controlled independently with each its own PI controller. The reference values for these controllers will be current in RL filter axes (dq) (Figure 12). The reference currents i_{fd}^{*} and i_{fq}^{*} are respectively the voltage from the control block of the DC bus and control of reactive power at the GSC connection point to the grid (Figure 11).

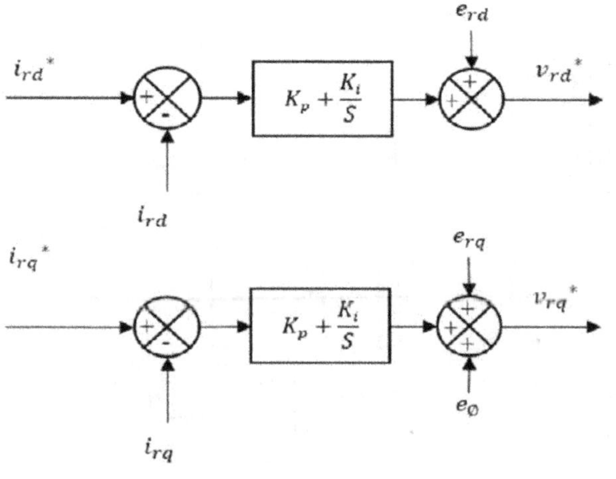

Figure 9: Control of rotor currents.

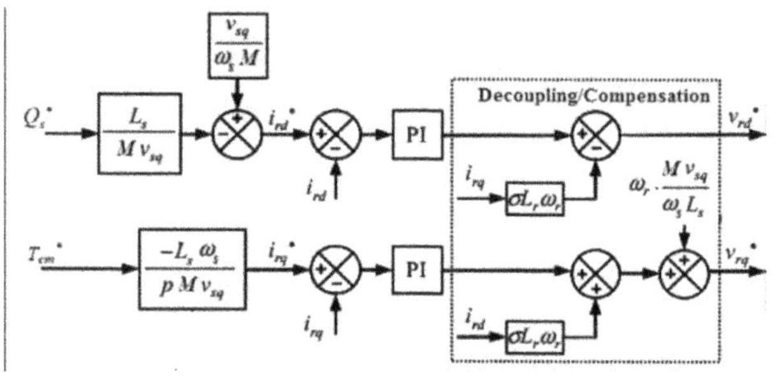

Figure 10: Control of RSC.

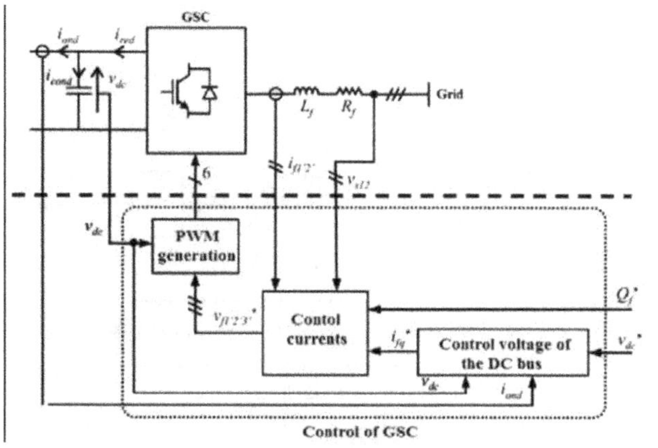

Figure 11: Principle of the Control of the grid side converter (GSC).

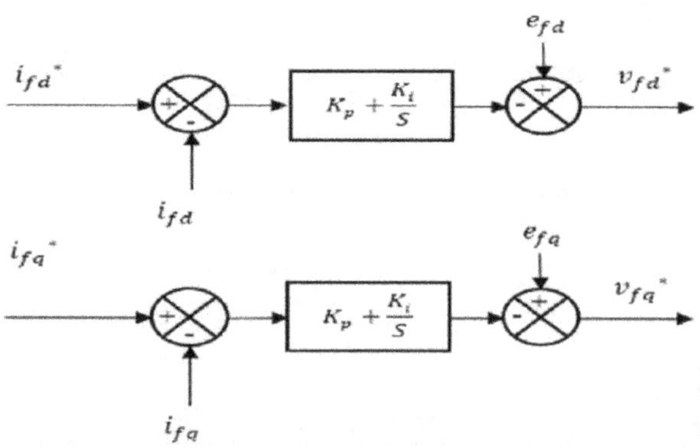

Figure 12: Control currents flowing in the RL filter.

Neglecting losses in the resistance of RL filter and taking the orientation of the coordinate system (dq) connected to the rotary stator field v_{sd}=0 the equations for the powers generated by the GSC are given by:

$$\begin{cases} P_f = v_{sq} i_{fq} \\ Q_f = v_{sq} i_{fd} \end{cases} \tag{26}$$

From these equations, it is possible to impose the active and reactive power reference noted here P_f^{*} and Q_f^{*} imposing the following reference currents:

$$\begin{cases} i_{fd}^{*} = \dfrac{Q_f^{*}}{v_{sq}} \\ i_{fq}^{*} = \dfrac{P_f^{*}}{v_{sq}} \end{cases} \quad (27)$$

Control of the DC bus voltage

We can express the powers involved on the DC bus by:

$$\begin{cases} \mathrm{P}_{red} = v_{dc} i_{red} \\ P_c = v_{dc} i_{cond} \\ P_{ond} = v_{dc} i_{ond} \end{cases} \quad (28)$$

These powers are linked by the relation:

$$\mathrm{P}_{red} = P_c + P_{ond} \quad (29)$$

Neglecting all the Joule losses (losses in the capacitor, the converter and the RL filter), we can write:

$$P_f = \mathrm{P}_{red} = P_c + P_{ond} \quad (30)$$

By adjusting the power P_f, then it is possible to control the power P_c in the capacitor and therefore to regulate the DC bus voltage. To do this, the P_{ond} and powers must be known to determine P_f^*. The reference power for the capacitor is connected to the reference current flowing through the capacitor:

$$P_c^{*} = v_{dc} i_{cond}^{*} \quad (31)$$

Regulating the DC bus voltage is then effected by an external loop (with respect to the inner loop control of current), for maintaining a constant voltage on the DC bus, with a PI controller that generates the reference current i_c^* in the capacitor (Figure 13). Figure 14 shows the block diagram of the control of GSC. This block diagram includes the terms of decoupling and compensation to be able to independently control the (dq) axes currents circulating in the RL filter and the active and reactive power exchanged between the GSC and the grid.

Simulation Results

The simulations of the whole system were performed with Matlab/Simulink, the DC bus reference voltage, denoted V_{dc}^* is set at 1200 V. The reactive power references Q_s^* and Q_f^* are set to 0 VAr, ensuring unity power factor. We present in this section the results of the proposed control. Figure 15 illustrates the profile of the average wind speed used for simulation, while Figure 16 shows the shaft rotational speed derived by the turbine. The figures (Figures 17 and 18) show that the electromagnetic torque and reactive power provided by DFIG follow their references, this is due to control of direct and quadrature components of the rotor current. Figure 19 illustrates the power stator

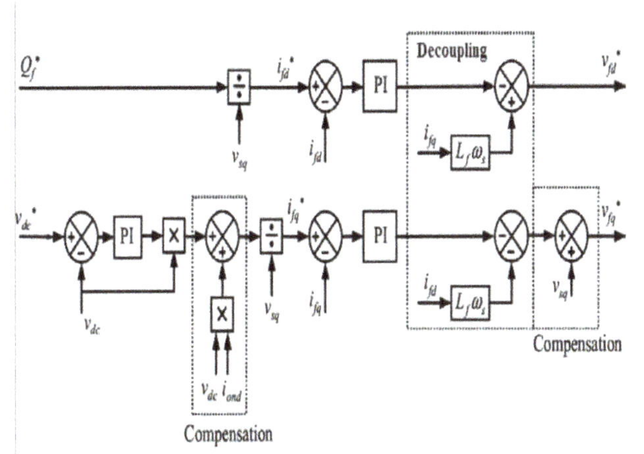

Figure 14: Control of GSC.

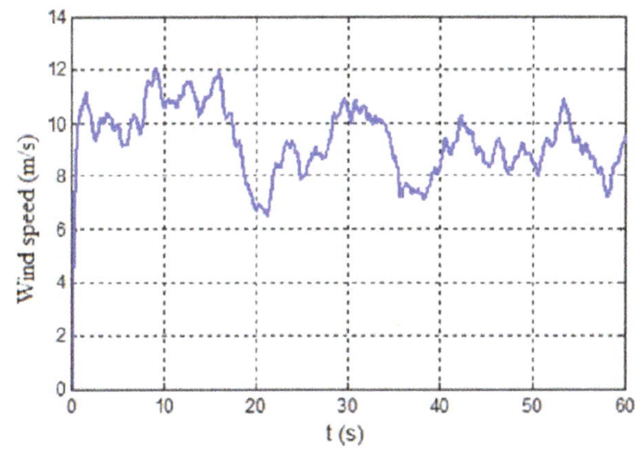

Figure 15: Wind speed.

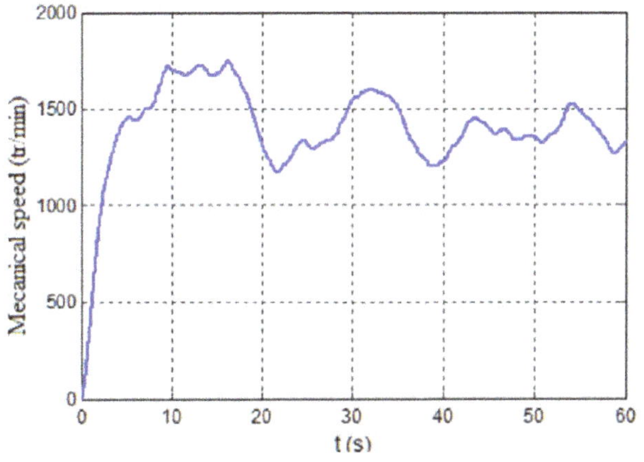

Figure 16: Mechanical speed.

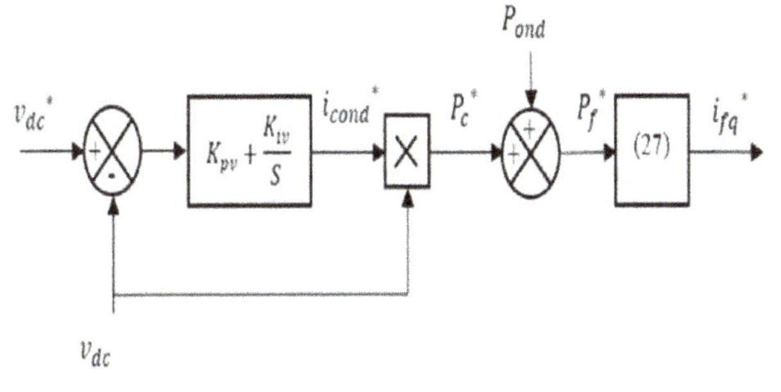

Figure 13: Control loop of the DC bus voltage.

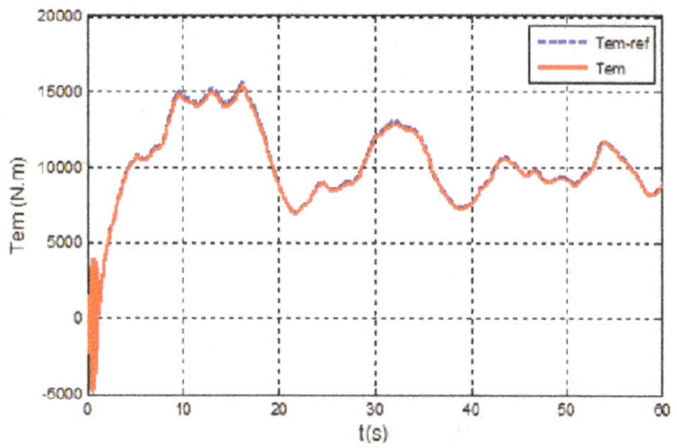

Figure 17: Electromagnetic torque (reference and simulated).

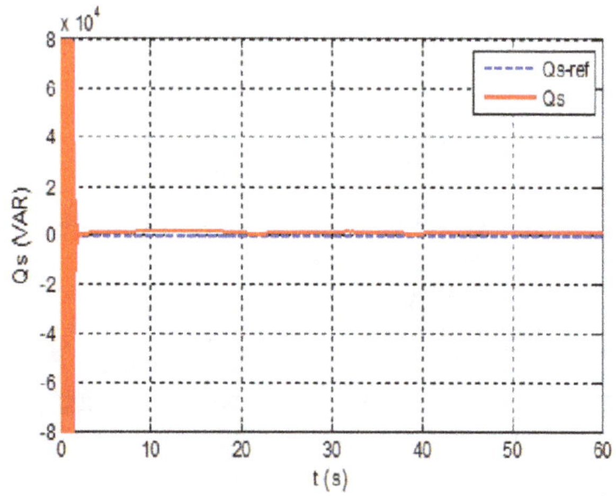

Figure 18: Stator reactive power (reference and simulated).

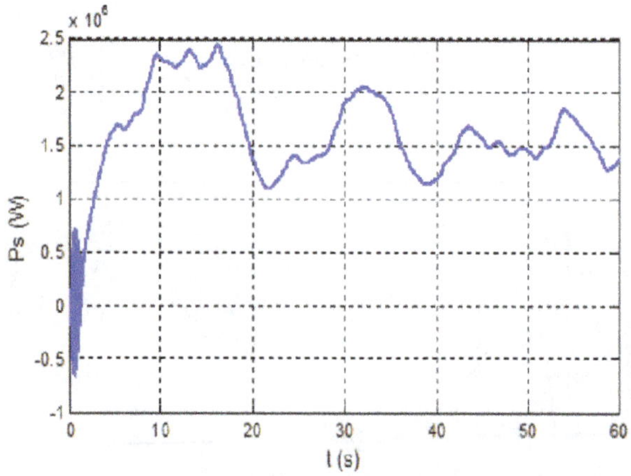

Figure 19: Stator active power.

extracted, and Figure 20 shows that the DC bus voltage is perfectly regulated. We see in Figure 21 that the current delivered by the wind system is in phase with respect to the supply voltage, this confirms that the wind system injects the active power into the grid.

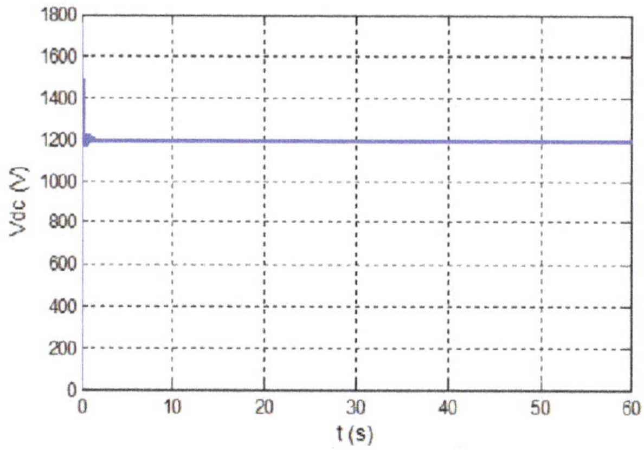

Figure 20: DC bus voltage.

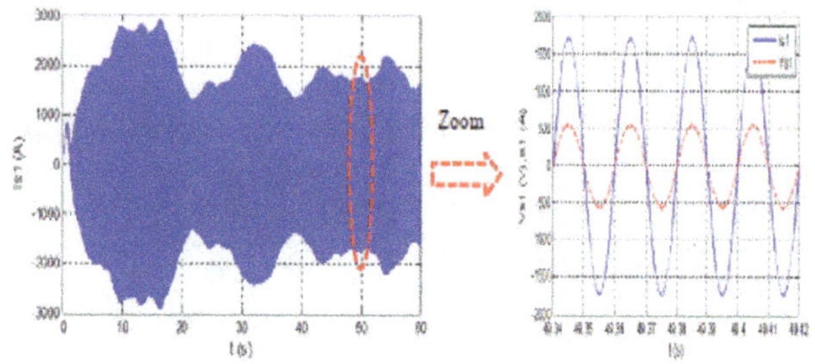

Figure 21: Stator current evolution.

Conclusion

This paper is devoted to modeling, simulation and analysis of a wind turbine operating at variable speed. Stable operation of the wind energy system was obtained with the application of the control direction of the flow. The overall operation of the wind turbine and its control system were illustrated in transient and permanent regimes. DFIG operates in two quadrants. Operation hyposynchronous for positive slip and hypersynchronous operation for a negative slippage. The generator supplied power to the grid with an active power regardless of the mode of operation. The control strategy based on the PI control with correction was tested. Simulation results show that the proposed wind energy system is feasible and has many benefits.

References

1. Ackermann T (2002), An Overview of Wind Energy-Status 2002, Renewable and Sustainable Energy Reviews 6: 67-127.
2. http://www.gepower.com/prod_serv/products/wind_trubines/en/index.html
3. Burton T, Sharpe D, Jenkins N (2001) Wind Energy Handbook, John Wiley&Sons, Ltd.
4. Kling WL, Slootweg JG (2002) Wind Turbines as Power Plants, Oslo, Norway: in Proceeding of the IEEE/Cigré workshop on Wind Power and the impacts on Power Systems.
5. Xu L, Wei C (1995) Torque and Reactive Power Control of a Doubly Fed Induction Machine by Position Sensorless Scheme. IEEE Trans, Industry Application.
6. Alesina A, Venturini M (1988) Intrinsic Amplitude Limits and Optimum Design of 9 Switches Direct PWM AC-AC converter, Proc. of PESC, 1284-1290.

7. Seyoum D, Grantham C (2003) Terminal Voltage Control of a Wind Turbine Driven Isolated Induction Generator using Stator Oriented Field Control, IEEE Transactions on Industry Applications, 846-852.

8. Davigany A (2007) «Participation aux services système de fermes éoliennes à vitesse variable integrant un stockage inertiel d'énergie,» Thèse de Doctorat, USTL Lille.

3

A Fluid-Based Approach for Modeling Network Activities

Yen-Hung Hu*

Department of Computer Science, Hampton University, Hampton, Virginia 23668, USA

Abstract

Network traffic traces provide valuable information for researchers to study behaviors of normal and malicious network activities. Although traffic traces are enough to reveal packet-level and connection-level details of most network activities, identifying specific malicious network activities is still a huge challenge: many malicious network activities are able to hide themselves behind normal activities with forged packet and connection information. In practice, mechanisms that are able to effectively extract malicious network activities from raw traffic traces are emerging and will benefit network security and other related communities as well. In this paper, a fluid-based approach for modeling simulated normal and malicious flooding-based denial of service network activities is developed. To approach this goal, several raw traffic traces gathered by the Cooperative Association for Internet Data Analysis (CADIA) are analyzed and investigated.

Keywords: Network activities; Denial of service

Introduction

The Internet has merged into our daily life because of its usage and enormous size: it is estimated that at least 8 x 10^8 documents and links covering almost every categories that we need [1]. Since the increasing number of fixed and mobile Internet-enabled devices, economic value of the Internet grows as well. In 2009, the Internet contributed about 3.8 % of the United States (U.S.) Gross Domestic Product (GDP) and the U.S. has led the Internet supply ecosystem [2-4]. Due to its popularity and financial capability, the Internet has become a target of many criminals and terrorists. In the first quarter of year 2012, there were 83 million pieces of malware including 8 thousand mobile malware; more than 1 trillion messaging threats (*e.g.* email spam); more than 4 million messaging botnets; huge number of network threats (*e.g.* Remote Procedure Call (RPC), SQL injection, Browser, cross-site scripting, etc.); and about 8 million websites hosting malicious downloads or browser exploits. The U.S. was almost at the top of every listed attack category [5]. Meanwhile, the number of cyber-attack on U.S. critical infrastructures (*e.g.* dams, energy, water, and cross-sector) increased sharply from 2009 to 2011 (from 9 incidents to 209 incidents) [6]. The report [7] conducted by Ponemon Institute in August 2011 revealed that average financial impact of every victim (private company) due to cyber-crime is in the range from 1.5 million to 3.6 million U.S. dollars and is about 56 percent increase from their last year's report. This report also indicated there is more than 1 successful attack per company per week and such a number is 44 percent increase compared to their last year's report. Paolo Passeri [8] presented monthly reports in cyber-attacks statistics. His observation indicated that Denial of Service (DoS) attack is the top three attack techniques affecting the stability of the Internet. Since flooding-based DoS attack could be launched with very less effort comparing with other attacks, it has been widely adopted to flood resources of victims and cause service disruption. There have been many approaches proposed to reduce Internet threats [9-23].

In this paper, we are interested in flooding-based DoS attack since its simplicity. We develop a fluid-based approach for modeling simulated normal and malicious flooding-based DoS network activities. To achieve objectives of this paper, we first analyze raw traffic traces and calculate statistical data of them. We then, mimic normal and malicious flooding-based DoS network traffic and depict a fluid-based model to study network activities.

Our approach is based on an observation that malicious flooding-based DoS network activities are not isolated, but related as different stages of a series of cyber-attacks. Intuitively, their traces could be caught even though they are carefully hidden behind normal network activities and have forged footprints. For example, the distribution of inter-arrival time of a series of malicious requests on a web-server could be identified even through those malicious requests implemented with forged IP headers. In order to launch a successful flooding-based DoS attack, the hacker has to make large enough requests to overwhelm the target's service capacity. Therefore, such malicious service requests are tended to be intensive and follow best-effort approach.

The remainder of this paper is organized as follows: Section 2 reviews related work. Section 3 covers background of flooding-based DoS attack. Section 4 introduces the simulated normal and malicious traffic. Section 5 describes characteristics of the selected network traffic captured by CADIA. Section 6 explains fluid-based approach on a single congested network. Section 7 discusses performance of our model under the simulated normal and malicious traffic. Section 8 concludes this paper and points out future work.

Related Work

Several literatures have studied and addressed strategies for mitigating cyber-attacks. Lobo et al. [9] studied attacks and countermeasures of the Windows Rootkits: software that is used to hide malicious activities and permit hackers to take control of victims. Several suggestions were issued to the Microsoft and research communities for developing future Windows operating systems. Shafi [10] surveyed security challenges in Cyber-Physical Systems (CPS). Agresti [11] proposed four distinct forces that will shape the future evolution of cybersecurity. Michael et al. [12] emphasized the importance of integrating legal and policy in cyber-preparedness. Eom et al. [13] developed an active cyber-attack model for accessing network vulnerabilities. Yu et al. [14] discussed models and countermeasures for

***Corresponding author:** Yen-Hung Hu, Department of Computer Science, Hampton University, Hampton, Virginia 23668, USA, E-mail: yenhung.hu@hamptonu.edu

attacks that aim at Internet threat monitors. Wang et al. [15] focused on developing a mechanism to gather digital evidences that could be used to defend against cross-site script attack. Tejay et al. [16] analyzed performance of existing information system security countermeasures.

Leland et al. [17] presented a result of Ethernet traffic: "aggregating streams of such traffic typically intensifies the self-similarity instead of smoothing it". Several other researchers adopted the concept of self-similarity as well to propose their approaches for detecting cyber-attacks such as traffic anomaly [18], intrusion [19], spam [20], and Distributed DoS (DDoS) attack [21].

In 1998, Defense Advanced Research Projects Agency (DARPA) and Air Force Research Lab (AFRL) funded a research in MIT Lincoln Laboratory to create large-scale intrusion detection database as the first standard set for measuring performance in terms of false alarm for each intrusion detection system under test. Most intrusion detection systems use signatures of known attacks to detect attacks. Many of these systems suffer high false alarm rates and poor detection of new attacks. Despite its increasing role in intrusion detection system, network traffic analysis approach remains premature: lack of effective malicious patterns and heavy increase of computational overhead [22,23].

Flooding-Based Denial Of Service Attack

An easy way to cause service denial to normal requests is by congesting the target links though high-rate unresponsive malicious flows. Flooding-based DoS attack [24-26] is the most prevalent among all cyber-attacks. It induces attack traffic from a sufficient number of compromised hosts to carry out congestion and cause most packets from normal flows to be dropped at the routers or service stacks. Most of the approaches in literatures for dealing with congestion are dedicated to providing fairness [27-31] to all active flows or rejecting malicious packets before they reach the service stacks [25-29]. Those approaches may not reduce impacts of malicious flows since they are sharing the same bandwidth with normal flows.

The most common way to introduce flooding-based DoS attack is to disrupt connections between victims and legitimate users. For instance, in TCP SYN flooding attack [29], a large number of TCP SYN packets with spoofed source addresses are sending to service ports of the victim to request for establishing new connections. The victim responds those requests with SYN-ACK packets and waits for ACK packets from those requests. Since source addresses in those TCP SYN packets are spoofed and unreachable, these SYN-ACK packets will never reach their destinations. And then the victim is forced to retransmit SYN-ACK packets for each request several times before giving up and could not establish regular connections for legitimate requests.

An alternative way to cause flooding-based DoS attack is to drain the bandwidth of all incident links of the victim to force the nearest router to drop most incoming packets of the victim. Attackers could do this by generating a heavy load of UDP-like unresponsive best-effort traffic (e.g., UDP, ICMP, TCP SYN, etc.) to exhaust bandwidth of the victim. For instance, attackers can broadcast ICMP Echo packets with victim's IP address in the source field [28]. And then huge amount of ICMP Echo replies will be triggered and aim at the victim. These replies would overwhelm the victim's network and consume most of its bandwidth and cause denial of service.

There are several tools (*e.g.*, Shart, TFN, TFN2K, Trinoo, etc.) that could conduct flooding-based DoS attack easily and automatically by using the existing network protocols such as TCP, ICMP, UDP, or mixture of them. Those DoS attacks can either consume all connections or network bandwidth to cause denial of service.

Analyzing Existing Network Traffic Traces

Existing network traffic traces provide clues for researchers to study scenarios and patterns of packets and connections. Researchers can simply derive statistical data regarding to packets, connections, and network resources for conducting complicated simulations.

In our research, we first gather knowledge from existing network traffic traces. 4 network traffic traces (Table 1) provided by the CAIDA (www.caida.org/data) have been analyzed.

They were all captured by the "Equinix San Jose A" monitoring point equipped with OC-192 optical link and dated from year 2009 to 2011. Each of these traffic traces contains 60-second raw network data.

Packet level analysis

We extract packet-level information from those traffic traces listed in Table 1. The packet-level information includes time stamp, source IP address, destination IP address, protocol, packet size (with and without IP header), source port (application), destination port (application), and other information regarding to TCP, etc.

We group packets from every traffic trace into several different streams according to their protocols. In this paper, we differentiate packets into three categories: TCP packet, UDP packet, and Other packet.

The packet-level information of the selected traffic traces is revealed in the Table 2 and 3. We observe that about 78% - 88% of network traffic is made by TCP packets, about 9% - 20% is made by UDP packets, and about 1% - 4% is made by other packets. We also observe that there are more than 85 other protocols (*e.g.*, control messages, peer-to-peer protocol, other special protocols) implemented in the selected traffic traces.

These observations meet our expectation, since the majority of web applications (*e.g.*, HTTP and HTTPS) are implemented upon TCP-related protocols [32,33]. Overall, we observe that TCP and UDP packets make up more than 94% of all packets.

Connection level analysis

To apply connection level analysis, we extract connection information from the first selected traffic trace (labeled 2009-01). In

Traffic Trace	File Name
2009-01	Equinox-sanjose.dirB.20090115-130000.UTC.canon.pcap
2009-02	Equinox-sanjose.dirB.20100121-130000.UTC.canon.pcap
2009-03	Equinox-sanjose.dirB.20110120-130000.UTC.canon.pcap
2009-04	Equinox-sanjose.dirB.20120119-130000.UTC.canon.pcap

Table 1: The Selected Traffic Traces.

	2009-01	2010-01	2011-01	2012-01
TCP Packet	12261116	16456196	24846485	26542956
UDP Packet	1482744	1694519	6396887	3101842
Other Packet	522407	530530	505827	1159979

Table 2: Percentage of Packets in the Selected Traffic Traces.

	2009-01	2010-01	2011-01	2012-01
TCP Packet	85.95%	88.09%	78.26%	86.16%
UDP Packet	10.39%	9.07%	20.15%	10.07%
Other Packet	3.66%	2.84%	1.59%	3.77%

Table 3: Composition of Packets in the Selected Traffic Traces.

here, we use a unique combination of source IP address, destination IP address, and Protocol to represent a "Connection".

Our results show that there are about 589,537 connections in this traffic trace. Among them, 46.62% are TCP connection, 39.95% are UDP connection, and 13.43% are others (Table 4).

We also calculate life (in seconds) and size (in number of packets) of every connection (Table 5 and 6) in this traffic trace. We observe that:

- Average life of TCP connections is longer than that of UDP and Other connections: As shown in the Table 5, there are about 25.3% of TCP connections having life shorter than 1 second. However, the value is 77.93% for UDP connections and 68.53% for Other connections, respectively. Meanwhile, about 50% of TCP connections having life longer than 20 seconds, but only about 6% of UDP connections and 19% of Other connections having life longer than 20 seconds, respectively.
- Average size of TCP connections is larger than that of UDP and Other connections: As shown in Table 6, there are about 76.66% of TCP connections having size smaller than 10 packets, but about 97.67% of UDP connections and 93.4% of other connections having size smaller than 10 packets, respectively.

Overall, we observe that UDP connections are much shorter in size and life than TCP connections. One interesting factor of UDP connections is: about 78% of UDP connections having life shorter than 1 second and more than 97% of them having size less than 10 packets. This could be formed by large amount of short-life streaming video or audio data embedded in webpages.

	All Connection	TCP Connection	UDP Connection	Other Connection
Number	595611	277697	237927	79987
Percentage	100%	46.62%	39.95%	13.43%

Table 4: Connection Number and Percentage in Traffic Trace 2009-01.

X	TCP Connection	UDP Connection	Other Connection
1	25.03%	77.93%	68.53%
10	44.13%	89.88%	76.42%
20	50.19%	93.69%	80.69%
30	54.48%	96.05%	85.13%
40	86.43%	97.51%	90.05%
50	91.91%	98.66%	94.52%

Table 5: Connections with Life ≤ X Seconds in Traffic Trace 2009-01.

X	TCP Connection	UDP Connection	Other Connection
10	76.66%	97.67%	93.40%
100	94.62%	99.65%	99.36%
200	96.90%	99.77%	99.73%
300	97.79%	99.82%	99.85%
400	98.29%	99.85%	99.89%
500	98.59%	99.87%	99.92%

Table 6: Connections with Size ≤ X Packets in Traffic Trace 2009-01.

Traffic Trace	All-stream	TCP-stream	UDP-stream	OTHER-stream
2009-01	0.84	0.84	0.55	0.55
2010-01	0.75	0.76	0.55	0.59
2011-01	0.81	0.81	0.62	0.73
2012-01	0.80	0.80	0.69	0.72

Table 7: Hurst Parameter of Four Different Streams in the Selected Traffic Traces.

Self similarity analysis

Another important factor of network traffic is heavy-tailed distribution [17]. We examine and study this factor of the selected network traffic traces.

Self similarity: Let $X(X_t : t = 0,1,2,3...)$ represent a stationary stochastic process with mean $\bar{X}$, variance S^2 and auto-correction function $r(k), k \geq 0$. Let $X^{(m)} = (X_k^{(m)} : k = 1,2,3...)$ for each m = 1,2,3…, denote the new covariance stationary time series obtained by averaging the original series over non-overlapping blocks of size m. Thus $X_k^{(m)} = \frac{1}{m}(X_{km-m+1} + ... + X_{km}), k \geq 0$. The correction function of $X_k^{(m)}$ is $r^{(m)}(k)$. X is said to be self-similar if $r^{(m)} \rightarrow r(k)$ as $m \rightarrow \infty$.

Hurst parameter: A practical approach to present the scale of self-similarity is to calculate the Hurst parameter H. A time series is self-similar as long as its Hurst parameter is bounded between 0.5 and 1. The larger the value of H indicates the higher the scale of self-similarity. Several approaches have been applied to estimate Hurst parameter. In this paper, we apply re-scaled adjusted range statistic approach (*R/S* statistic) to evaluate Hurst parameter.

R/S statistic:

Let $\frac{R(n)}{S(n)} = \frac{1}{S(n)}[Max(0, W_1, ..., W_t) - \min(0, W_1, ... W_t)]$ with $W_t = (X_1 + ... + X_t) - t\bar{X}$.

Since $\frac{R(n)}{S(n)} \sim a_1 n^H$ as $n \rightarrow \infty$, Hurst parameter H of X could be represented as the slop of $E\left[\frac{R(n)}{S(n)}\right] vs. \log n$.

Hurst parameter of the selected traffic traces: To study self-similar characteristics of the select traffic traces, we apply R/S statistic to data streams that are extracted from them. We extract 4 data streams form each selected traffic trace. These data streams are labeled as: ALL-stream (stream of all packets), TCP-stream (stream of TCP packets only), UDP-stream (stream of UDP packets only), and OTHER-stream (stream of all other packets). We evaluate Hurst parameters of these data streams. The results are in the Table 7.

We observe that TCP-stream is much more self-similar (*i.e.*, Hurst parameter ~ 0.8) than UDP-stream and OTHER-stream. We also observe that the Hurst parameter of ALL-stream is almost equal to the Hurst Parameter of TCP-stream. This is because most packets in ALL-stream are actually TCP packets.

Simulated Normal and Simulated Malicious Traffic

We have learned the following scenarios from the selected traffic traces discussed in the Section III and IV: (1) TCP packets contribute to about 85%, UDP packets contribute to about 10% and the combination of them contribute to about 95% of the network traffic, respectively; (2) TCP-stream is more self-similar than UDP-stream and Other-stream since it tends to be burstiness. (3) Other packets could be treated as UDP-like since characteristics of them are very similar to UDP; (4) Hurst parameter of All-stream is very similar to TCP-stream, since TCP packets make up most of the network traffic.

To simplify this research without loss of generality we build a set of simulated network traffic: simulated normal traffic and simulated malicious traffic.

- Simulated Normal Traffic: It is a combination of TCP and UDP traffic: 85% TCP packets and 15% UDP packets. This simulated normal traffic will behave self-similar with Hurst parameter about 0.85. It is about 50% of TCP flows will last longer than 10 seconds. But only 10% of UDP flows could last longer than 10 seconds. Meanwhile, this traffic is made up by 50% TCP flows and 50% UDP flows.

- Simulated Malicious Traffic: To mimic flooding-based DoS attack, we assume this simulated malicious traffic is made up by a large number of UDP-like flows with short flow life and small flow size. In order to confuse most intrusion detection systems, it is assumed average flow life of the simulated malicious traffic will be shorter than 1 second and average flow size will be smaller than 10 packets. To gain best result, this simulated malicious traffic will not behave self-similar and have Hurst parameter smaller than 0.5.

Fluid-Based Approach for Modeling Network Traffic in a Single Congested Network

To study network behavior without captures of actual network traffic, we develop a fluid-based approach adopting ideas from [24]. We model network traffic as a fluid and use Stochastic Differential Equations (SDE) to model TCP traffic. We also derive differential equations to describe Drop-Tail queuing policy.

As mentioned in the previous sections, TCP and UDP are the two major protocols used in the selected network traffic traces. Therefore, we consider only these two protocols in the fluid-based model. Performance measures used in this paper are throughput, goodput, and drop-rate: throughput represents sending rate (in bits per second) of the source node of a connection, drop-rate represents packet-loss rate (in bits per second) of a connection, and goodput represents receiving rate (in bits per second) of the destination node of a connection.

In our model, we consider network traffic in a link as fluid flows in a pipe. Therefore, for any given connection k at time t with throughput $A_k(t)$ and drop-rate $D_k(t)$, its goodput $GP_k(t)$ can be represented as $A_k(t)$ - $D_k(t)$ That is

$$GP_k(t) = A_k(t) - D_k(t) \tag{1}$$

We first apply our model on a single congested network to study traffic behaviors. We than extend our study to a multi-congested network with complicated network traffic.

Single congested network

In this section, we assume there is only one bottleneck router in the network that causes packet losses. A network G is a directed graph, $G = (V, E)$, where $V=\{v_1, v_2, ..., v_x\}$, denoting a set of routers in G, and $E = \{e_1, e_2, ..., e_y\}$, denoting a set of links in G. We further assume that there are N UDP connections and M TCP connections passing through the only bottleneck v_r in the network G during the monitoring period ∂t.

To study traffic behaviors, we measure throughput of every connection before the bottleneck router v_r and goodput of every connection after the bottleneck router v_r, respectively.

Drop-Tail is the default and one of the most popular queuing management algorithms. In this paper, we implement Drop-Tail in all of our network models.

Throughput of any UDP connection *i*

Since UDP adopts best-effort fashion, throughput of an UDP connection could be constant during its life time. Thus the average inter-arrival time $\overline{\tau}_{udp_i}$ (in seconds) of any UDP connection i could be represented as $\frac{l_{udp_i}}{\overline{\tau}_{udp_i}}$, where l_{udp_i} is packet size (in bits) of UDP flow i. That is

$$A_{udp_i} = \frac{l_{udp_i}}{\overline{\tau} udp_i} \tag{2}$$

Throughput of any TCP connection *j*

To simplify network traffic without losing general characteristics of TCP protocol, we assume TCP implements an Additive Increase Multiplicative Decrease (AIMD) policy: when there is no congestion occurred, the policy of "Additive Increase" will increase the congestion window by 1 for every round trip time; when congestion detected, the policy of "Multiplicative Decrease" will decrease the congestion windows by half. Therefore, for any TCP connection j following the AIMD policy, its dynamic congestion window size can be represented by a SDE listed below.

$$\delta W_j(t) = \frac{1}{R_j(t)}\left(1 - I_j(t-1)\right)\delta t - \frac{W_j(t-1)}{2} I_j(t-1) \tag{3}$$

The first portion denotes that congestion window size W_j (in packets) of TCP connection j will increase by 1 for every round trip time R_j (in seconds) during a non-congestion period (*i.e.*, $I_j(t-1)=0$). But the second portion denotes that congestion window size will become half of the size of its previous state (*i.e.*, $\frac{W_j(t-1)}{2}$), if there is at least one packet of TCP connection j (*i.e.*, $I_j(t-1)=1$) dropped during a congestion period.

In here, I is a packet loss indication function. $I = 1$ while at least a packet loss from TCP connection j is detected. Otherwise, $I = 0$.

At time t, TCP connection j will send out $W_j(t)$ packets. Thus throughput $A_{tcp_j(t)}$ of TCP connection j can be written as $\frac{W_j(t) \times l_{tcp_j}}{R_j(t)}$, where l_{tcp_j} is packet size (in bits) of TCP flow j. That is

$$A_{tcp_j(t)} = \frac{W_j(t) \times l_{tcp_j}}{R_j(t)} \tag{4}$$

Throughput of all connections

Intuitively, aggregate throughput $A_{all}(t)$ of all connections (*i.e.*, N UDP connections + M TCP connections) passing through the bottleneck router v_r having service capacity C_{v_r} and physical queue size $Q_{v_r}^{\max}$, at time t can be represented as $\sum_{i \in N} A_{udp_i}(t) + \sum_{j \in M} A_{tcp_j}(t)$, where $A_{udp_i}(t)$ and $A_{tcp_j}(t)$ can be derived from Equation (2) and Equation (4), respectively. Therefore, we have

$$A_{all}(t) = \sum_{i \in N} \frac{l_{udp_i}}{\overline{\tau}_{udp_i}} + \sum_{j \in M} \frac{W_j(t) \times l_{tcp_j}}{R_j(t)} \tag{5}$$

In order to determine packet loss indication function I in the Equation (3), we introduce a drop probability function P. For any incoming packet z arriving the bottleneck v_r at time t, let's assume its drop probability is $P_z(t)$, where $0 \le P_z(t) \le 1$. We will discuss this $P_z(t)$ in the next section.

Drop probability of any incoming packet *z*

We consider two events that cause packet losses: (1) aggregate throughput $A_{all}(t)$ of all connections during the motoring period is larger than service capacity C_{v_r} of the bottleneck router v_r and (2) instantaneous queue size $Q_{v_r}^t(t)$ at time t of the congested router is larger than its physical queue size $Q_{v_r}^{\max}$.

Since characteristics of TCP connection and UDP connection are different, drop probability $P_z(t)$ of any incoming packet z passing through the bottleneck router v_r with DT policy at time t could be determined by: (1) protocol of this packet (either TCP or UDP); (2)

aggregate throughput $A_{all}(t)$ of all connections; (3) throughput (either $A_{udp_i}(t)$ or $A_{tcp_j}(t)$ of UDP connection *i* or TCP connection *j* where this incoming packet *z* belongs; (4) service capacity C_{v_r} of the bottleneck router v_r; and (5) instantaneous queue size $Q^l_{v_r}(t)$ at time *t* and physical queue size $Q^{\max}_{v_r}$ of the bottleneck router v_r.

To discuss these conditions in detail, we consider the following four cases:

Case 1: $A_{all}(t) \le C_{v_r}$: Since aggregate throughput $A_{all}(t)$ is less than service capacity C_{v_r} of the bottleneck router v_r, there is no packet loss. Therefore, drop probability of incoming packet *z* at time *t* will be 0 (*i.e.*, $P_z(t) = 0$).

Case 2: $A_{all}(t) > C_{v_r}$ and $Q^{\max}_{v_r} - Q^l_{v_r}(t-1) \ge (A_{all}(t) - C_{v_r}) \times \delta t$: Although aggregate throughput $A_{all}(t)$ is larger than service capacity C_{v_r} of the bottleneck router v_r, drop probability $P_z(t)$ of incoming packet *z* at time *t* is still 0 (*i.e.*, $P_z(t) = 0$). This is because available queue space (*i.e.*, $Q^{\max}_{v_r} - Q^l_{v_r}(t-1)$) is still large enough to accommodate this incoming packet.

Case 3: $A_{all}(t) > C_{v_r}$ and $0 < Q^{\max}_{v_r} - Q^l_{v_r}(t-1) < (A_{all}(t) - C_{v_r}) \times \delta t$: In this case, for any incoming packet *z* at time *t*, its drop probability would be proportional to throughput of the connection where this packet belongs. Since *z* could be a TCP or UDP connection, we have to consider both cases. If packet *z* is an UDP packet and belongs to UDP connection *i*, its drop probability is proportional to $\frac{A_{udp_i}(t)}{A_{all}(t)}$. However, if packet *z* is a TCP packet and belongs to TCP connection *j*, its drop probability is proportional to $\frac{A_{tcp_j}(t)}{A_{all}(t)}$. We introduce a random variable *Ran_Var* whose value is between 0 and 1 to determine whether the incoming packet *z* will be dropped or not. We assume $P_z(t) = 1$ if $Ran\_Var \ge (1 - \frac{A_{tcp_j}(t)}{A_{all}(t)})$. Otherwise $P_z(t) = 0$.

Case 4: $A_{all}(t) > C_{v_r}$ and $Q^{\max}_{v_r} = Q^l_{v_r}(t-1)$: Since aggregate throughput $A_{all}(t)$ is larger than service capacity C_{v_r} of the bottleneck router v_r and queue size of this router is full at time *t*-1, the incoming packet *z* will be dropped without doubt. Therefore, $P_z(t) = 1$.

Packet loss indication function of any incoming packet *z* in TCP connection *j*

To determine packet loss indication function *I* of TCP connection *j* demonstrated in the Equation (3), we consider the following four cases for any incoming packet *z* belonging to TCP connection *j*.

Case 1: $A_{all}(t) \le C_{v_r}$: Since $P_z(t) = 0, I_j(t+1) = 0$.

Case 2: $A_{all}(t) > C_{v_r}$ and $Q^{\max}_{v_r} - Q^l_{v_r}(t-1) \ge (A_{all}(t) - C_{v_r}) \times \delta t$: Since $P_z(t) = 0, I_j(t+1) = 0$.

Case 3: $A_{all}(t) > C_{v_r}$ and $0 < Q^{\max}_{v_r} - Q^l_{v_r}(t-1) < (A_{all}(t) - C_{v_r}) \times \delta t$: We have $I_j(t+1) = 1$, if there exists any $P_z(t) = 1$. Otherwise $I_j(t+1) = 0$.

Case 4: $A_{all}(t) > C_{v_r}$ and $Q^{max}_{v_r} = Q^l_{v_r}(t-\ )$: Since $P_z(t) = 1, I_j(t+1) = 1$.

Goodput of any connection *k*

As mentioned in the Equation (1), goodput represents receiving rate of the destination of a connection. For any given connection *k*, its goodput $GP_k(t)$ at time *t* can be represent as $A_k(t) - D_k(t)$, where $A_k(t)$ denotes throughput of connection *k* and $D_k(t)$ denotes drop rate of connection *k*.

In here, throughput $A_k(t)$ could be estimated by using either Equation (2) or Equation (4). Drop-rate $D_k(t)$ of connection *k* could be estimated by determining drop probability of every incoming packet of this connection during time *t*. Therefore, goodput $GP_k(t)$ of connection *k* at time *t* could be measured by using models we addressed in the previous sections.

Modeling Simulated Network Traffic

In this section, we demonstrate various simulated network traffic according to the knowledge learned from the selected raw network traffic traces and models developed from our fluid-based approach.

Modeling simulated normal traffic with single TCP and UDP connection

To understand TCP and UDP involved in the simulated normal and malicious traffic, we design a simple simulation to study their characteristics.

At first, we simulate dynamical change of congestion window size and goodput of a TCP connection. The parameters involve in this simulation are: physical queue size of the congested router is 32,000 bits; minimum window size of TCP is 1 packet; maximum window size of TCP is 80 packets; average TCP round trip time is 20 *ms*; service capacity of the congested router is 10 Mbps; and TCP packet size is 8,000 bits.

As shown in the Figure 1, congestion window size of this TCP connection is fluctuated between 15 and 31 packets after the first packet loss detected. Meanwhile, we also discover goodput of this TCP connection is fluctuated as well (Figure 2). These results indicate that the TCP AIMD policy adopted in our model actively responses to packet losses from this TCP connection.

We then add an UDP connection into the same simulation to study the competition between TCP and UDP. We designate a reserved service capacity of the congested router to this UDP connection and then capture its goodput vs. time. The additional parameters needed for this simulation are: UDP packet size is 1,600 bits and throughput of this UDP connection is fixed to about 15% of the service capacity of the congested router. As we expected, goodput of the TCP connection will be reduced since it responds to network congestion. However, the

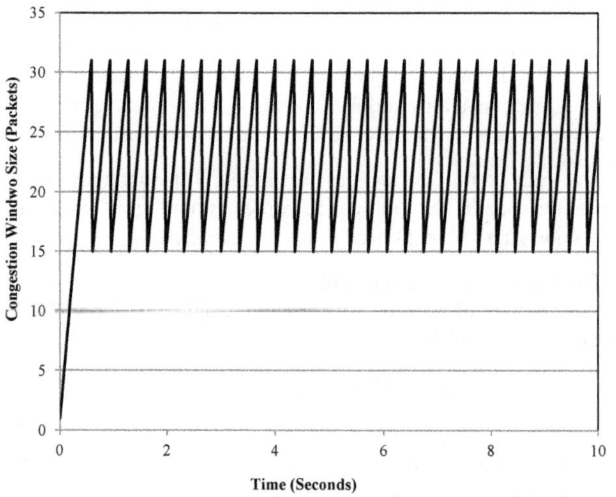

Figure 1: Congestion Window Size of a TCP Connection vs. Time

UDP connection keeps its sending-rate steadily (Figure 3). Therefore, we see a potential that how a single malicious activity can gain largest advantage against normal network activities: using high-rate non-responsiveness packets to flood targeted victims.

Modeling simulated normal traffic with multiple TCP and UDP connections

As we observed in the previous sections, traffic trace 2009-01 has about 590,000 connections within its 60 second monitoring period. Among these connections, 47% are TCP connection, 40% are UDP connection, and 13% are other connections. We also observe that about 51% of them having connection life shorter than 1 second and about 87% of them having connection size less than 10 packets.

These data demonstrate a fact that most connections passing through the monitoring point are very short and fragile and even TCP connection would act like an UDP one and will not perform congestion control as well as it is designed. Therefore, we could see a large amount of burstiness across 60-second monitoring period (Figure 4). This fact explains why TCP SYN attack could bring much more damages than we expected. It could not only hijack services for normal requests, but also deprive them of network bandwidth.

To model multiple connections with various sizes and lives, we introduce three additional variables (start_time, end_time, and size) to

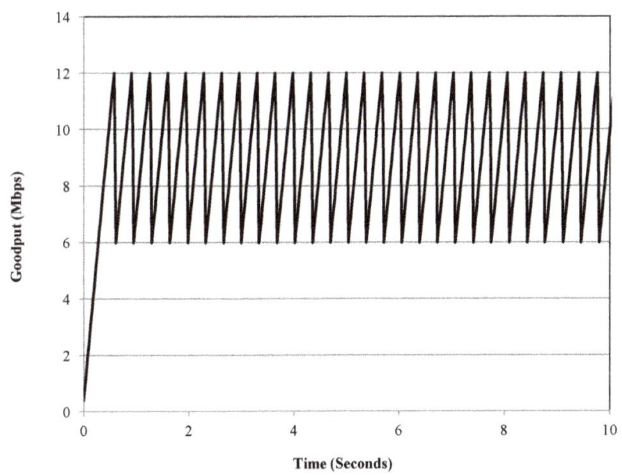

Figure 2: Goodput of a TCP Connection vs. time.

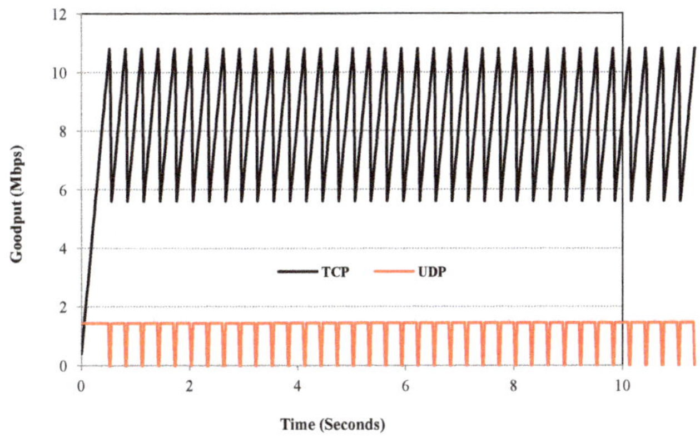

Figure 3: Goodput of a TCP Connection and an UDP Connection vs. Time.

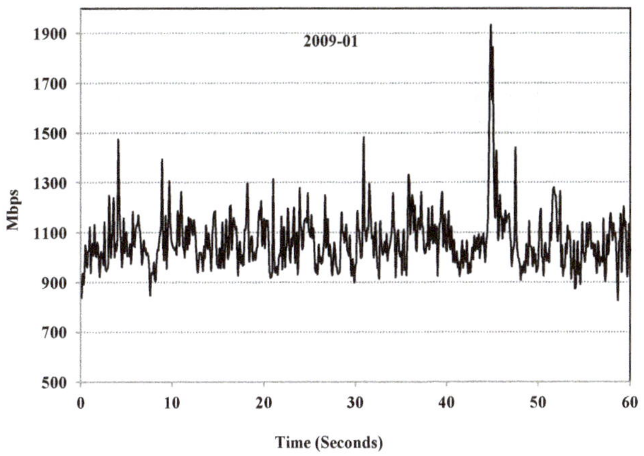

Figure 4: Heavy Burstiness in Traffic Trace 2009-01.

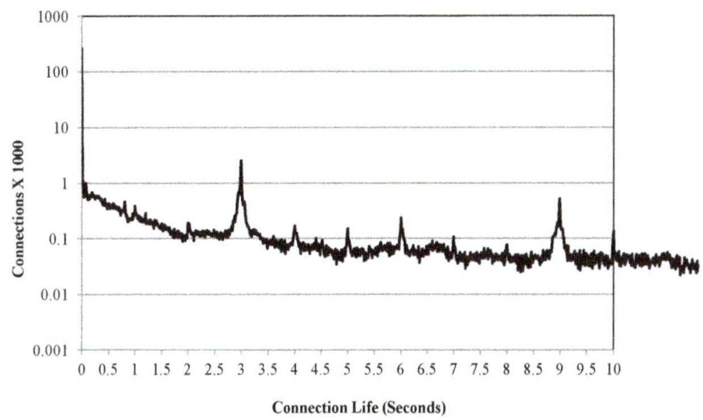

Figure 5: Life Distribution of Traffic Trace 2009-01.

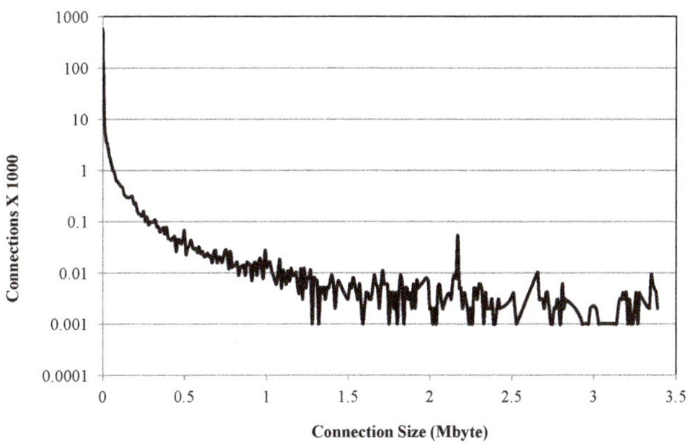

Figure 6: Size Distribution of Traffic Trace 2009-01.

our model to represent start time, end time, and size of a connection. We use life and size distribution learned from traffic trace 2009-01 to mimic start_time, end_time, and size of every connection. As shown in the Figure 5 and 6, life distribution of traffic trace 2009-01 demonstrate a Power Law-like characteristic: exponential decrease with long tail; and size distribution of the same traffic trace demonstrate a Poisson-like characteristic: exponential decrease. Therefore, these two characteristics would be added into our model to produce those three additional parameters for every connection.

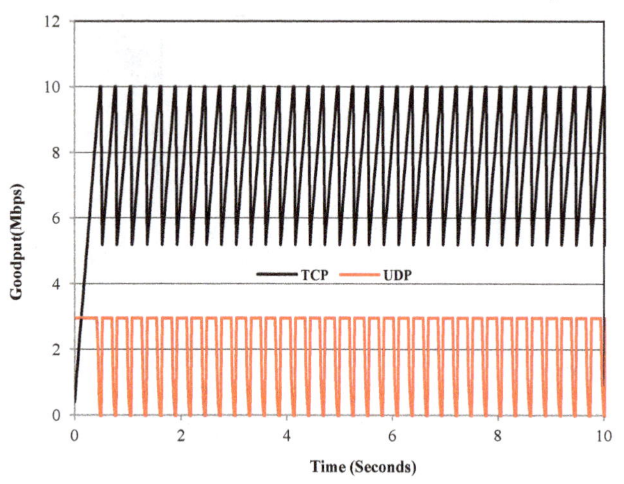

Figure 7: Goodput of a TCP Connection and an Aggregate UDP Connection vs. Time.

Modeling simulated malicious traffic

In this paper, we assume malicious traffic is derived from a certain amount of malicious UDP-like connections with short life and small size and aiming at some predefined network victims. To mimic flooding-based DoS attack, we introduce a simulated malicious traffic which is made up by a number of UDP-like flows with short flow life and small flow size. We assume their average life will be shorter than 1 second and their average size will be smaller than 10 packets as well. To exhaust bandwidth of the victim, this simulated malicious traffic will not behave self-similar and have Hurst parameter smaller than 0.5.

Our approach is based on an observation that malicious flooding-based DoS network activities are not isolated, but related as different stages of a series of malicious attacks. Intuitively, their traces could be caught even though they are carefully hidden behind normal activities and have forged footprints.

To model simulated malicious traffic, we introduce a large amount of UDP-like best-effort packets as an aggregate UDP connection with throughput fixed to about 30% of the service capacity into the monitoring point. Other simulation conditions are the same as modeling simulated normal traffic with single TCP and UDP connection. We observe that goodput of the simulated normal traffic decreased as the number of simulated malicious packet increased (Figure 7).

Conclusion and Future Work

In this paper, we (1) analyze several selected traffic traces; (2) introduce a set of simulated normal traffic and simulated malicious traffic according to the knowledge learned from the selected traffic traces; and (3) develop a fluid-based model to study performance of a single congested network under simulated normal traffic and the simulated malicious traffic. In the future, we will develop more network models (*e.g.*, a network with multiple congestion points) to study performance of the simulated traffic. We will also extend our network model and simulated traffic to study other malicious activities and to evaluate their influences as well.

References

1. Albert R, Jeong H, Barabasi A-L (1999) Diameter of the World-Wide Web. Nature 401: 130-131.
2. Hätönen J (2011) The economic impact of fixed and mobile high-speed networks. EIB Papers 16: 30-59.
3. Greenstein S, McDevitt R (2011) The broadband bonus: estimating broadband Internet's economic value. Telecommunications Policy 35: 617-632.
4. Rausas MP, Manyika J, Hazan E, Bughin J, Chui M, et al. (2011) Internet matters: the net's sweeping impact on growth, jobs, and prosperity. McKinsey Global Institute.
5. McAfee Lab (2012) McAfee Threats Reports: First Quarter 2012.
6. Industrial Control Systems Cyber Emergency Response Team Control Systems Security Program (2011) ICS-CERT incident response summary report 2009-2011.
7. Ponemon Institute (2011) Second annual cost of cyber crime study.
8. Passeri P (2013) Cyber attack statistics.
9. Lobo D, Watters P, Wu X-W, Sun L (2010) Windows rootkits: attacks and countermeasures. Proceeding of 2010 Second Cybercrime and Trustworthy Computing Workshop.
10. Shafi Q (2012) Cyber physical systems security: a brief survey. Proceeding of 12th International Conference on Computational Science and Its Applications.
11. Agresti W (2010) The four forces shaping cybersecurity. Computer 43: 101-104.
12. Michael J, Sarkesain J, Wingfield T, Dementies G, de Sousa GNB (2010) Integrating legal and policy factors in cyberpreparedness. Computer 43: 90-92.
13. Eom JH, Han YJ, Park SH, Chung TM (2008) Active cyber attack model for network system's vulnerability assessment. Proceeding of 2008 International Conference on Information Science and Security.
14. Yu W, Zhang N, Fu X, Battati R, Zhao W (2010) Localization attacks to Internet threat monitors: modeling and countermeasures. IEEE Transactions on Computers 59: 1655-1668.
15. Wang SJ, Chang YH, Chiang WY, Juang WS (2007) Integrations in cross-site script on Web-systems gathering digital evidence against cyber-intrusions. Proceeding of Future Generation Communication and Networking (FGCN 2007).
16. Tejay G, Zadig S (2012) Investigating the effectiveness of IS security countermeasures towards cyber attack deterrence. Proceeding of 45th Hawaii International Conference on System Sciences.
17. Leland W, Taqqu M,Willinger W, Wilson D (1994) On the self-similar nature of Ethernet traffic. IEEE/ACM Transactions on Networking 2: 1-15.
18. Cheng X, Xie K, Wang D (2009) Network traffic anomaly detection based on self-similarity using HHT and wavelet transform. Proceeding of Fifth International Conference of Information Assurance and Security.
19. Allen W, Marin G (2003) On the self-similarity of synthetic traffic for the evaluation of intrusion detection system. Proceedings of 2003 Symposium on Applications and the Internet.
20. Lee J, Jeong HD, McNicke D, Pawlikowshi K (2011) Self-similar properties of spam. Proceeding of Fifth International Conference on Innovative Mobile and Internet Services in Ubiquitous Computing.
21. Zhang S, Zhang Q, Pan X, and Zhu X (2011) Detection of low-rate DDoS attack based on self-similarity. Proceeding of second International Workshop on Education Technology and Computer Science.
22. Lippmann R, Cunningham R (2000) Improving intrusion detection performance using keyword selection and neural networks. Computer Networks 34: 597-603.
23. Peddabachigari S, Abraham A, Grosan C, Thomas J (2007) Modeling intrusion detection system using hybrid intelligent systems. Journal of Network and Computer Applications 30: 114-132.
24. Change RKC (2002) Defending against flooding-based distributed denial-of-service attacks: a tutorial. IEEE Communication Magazine 40: 42-51.
25. Piskozub A (2002) Denial of service and distributed denial of service attacks. Proceedings of the International Conference on Modern Problems of Radio Engineering, Telecommunications and Computer Science.
26. CERT Coordination Center (2001) Denial of service attacks.

27. Floyd, S, Jacobson V (1993) Random early detection gateways for congestion avoidance. IEEE/ACM Transactions on Networking 1: 397-413.

28. Demers A, Keshav S, Shenkar S (1990) Analysis and simulation of a fair queueing algorithm. Proceedings of SIGCOMM '89 on Communications Architectures & Protocols.

29. Shreedhar M, Verghese G (1996) Efficient fair queueing using deficit round robin. IEEE/ACM Transactions on Networking 4: 375-385.

30. Lin D, Morris, R (1997) Dynamics of random early detection. Proceedings of SIGCOMM on Applications, Technologies, Architectures, and Protocols for Computer Communication.

31. Ott TJ, Lakshman TV, Wong, LH (1999) SRED: stabilized RED. Proceedings of IEEE INFOCOM'99.

32. Fielding R, Gettys J, Mogul J, Frystyk H, Masinter L, et al. (1999) Hypertext Transfer Protocol -- HTTP/1.1. Network Working Group, RFC 2616.

33. Rescorla E (2000) HTTP Over TLS. Network Working Group, RFC 2818.

4

Application Analysis of Parameters for Wireless and Wire-Line Network with and without Load Balancer

Raghav Puri*and Navpreet Singh

Department of Electronics and Communication Engineering, Global Institute of Management and Emerging Technologies, Amritsar, Punjab, India

Abstract

The wired computer network provide secure and faster means of connectivity but the need of mobility i.e. anywhere, anytime and anyone access is tilting the network users towards wireless technology. This Paper Analyse the modelling and implementation of Wireless Local Area Network (WLAN) using different factors based on OPNET Modeller. Here OPNET Modeller is used to develop a new model that fits for Academic Site Location. Our model was analysed to measure the performance of factors of the wireless local area network based on such academic site location. Our model was tested adjacent to four applications (FTP, HTTP, Video Conferencing and Database) in four sites and found that other factors also were extremely influenced by the number of users per application with and without load balancer. OPNET Modeller simulation demonstrated the effect of load balancer on wireless and wire-line network for four different types of applications.

Keywords: WLAN; Load balancer; HTTP; FTP; Video conferencing; Database, MAC

Introduction

Wireless local area networks (WLANs) based on the IEEE 802.11 standards are one of the fastest growing wireless access technologies in the world today. These are common place on many academic site locations [1-5]. It have all capabilities of Wired LANs along with additional feature that user terminal do not need to be physically connected to Wired Infrastructure. WLANs bring the user closer to the promise "anything, anytime, anywhere" of future technology.

Wireless resources are inclusive of broadband Internet connection, network printers, data files, and even streaming audio and video. Such resource sharing has become more prevalent as computer users have changed their habits from using single, stand-alone computers to working on networks with multiple computers, each with potentially different operating systems and varying peripheral hardware [2]. Most general technologies are applying for the series of IEEE 802.11 standards. IEEE 802.11b is the well recognized technology to be known using frequency 2.5 GHz range. For this paper we have focused on IEEE 802.11b [4,6-8].

Our Paper focuses to study academic site location scenario. We use the OPNET Modeler [6,9-11] simulation environment, with its detailed models of IEEE 802.11b, TCP/IP, FTP, HTTP, Video Conferencing and Database. We have chosen simulative tool- OPNET Modeler for our research because of the several benefits it offers over the other contemporary tools available. It provides the set of complete tools and a complete user interface for topology design and development. Another advantage of using it is that it is being extensively used and there is wide confidence in the validity of the results it produces. We parameterized the simulation model based on academic site measurements, and validate the model adjacent to WLAN performance metrics using simple FTP, HTTP, Video Conferencing and Database workload models. It was used to investigate the various performance metrics in wireless and wire-line LAN for a balanced and unbalanced network which has been presented.

After briefing the introduction in section I, Section II introduces our model and the scenarios we tested, section III analyses the results and the conclusion is drawn in section IV.

Model Outline

The IEEE 802.11 WLAN architecture is built around a Basic Service Set (BSS). The IEEE 802.11 standard defines a set of wireless LAN protocols that deliver services similar to those found in wired Ethernet LAN environments. A BSS is a set of stations that communicate with one another. When all the stations in the BSS can communicate directly with each other (without a connection to a wired network), the BSS is known as an *ad hoc* WLAN. When a BSS includes a wireless access point (AP) connected to a wired network, the BSS is called an infrastructure network. A simulation model was developed using OPNET Modeller [6,11]. OPNET 802.11b module was used as a standard with default data rate up to 11 Mb/s. IEEE 802.11b Direct Sequence was used as a default Physical Characteristic. In this section we will introduce the two scenarios we tested:

Scenario 1

Here four WLAN Sites (Figure 2) each with 40 Users through 1 access points using DATABASE (10 users), and HTTP (10 users) and FTP (10 Users) and Video Conferencing (10 Users) connected with outside wire-line network without load balance in (Figure 1 and Table 1).

Scenario 2

Here four WLAN Sites (Figure 2) each with 40 Users through 1 access points using DATABASE (10 users), and HTTP (10 users) and FTP (10 Users) and Video Conferencing (10 Users) connected with outside wire-line network with load balance in (Figure 3 and Table 1).

***Corresponding author:** Raghav Puri, Departmet of Electronics and Communication Engineering, Global Institute of Management and Emerging Technologies, Amritsar, Punjab, India, E-mail: dewraghavpuri@gmail.com

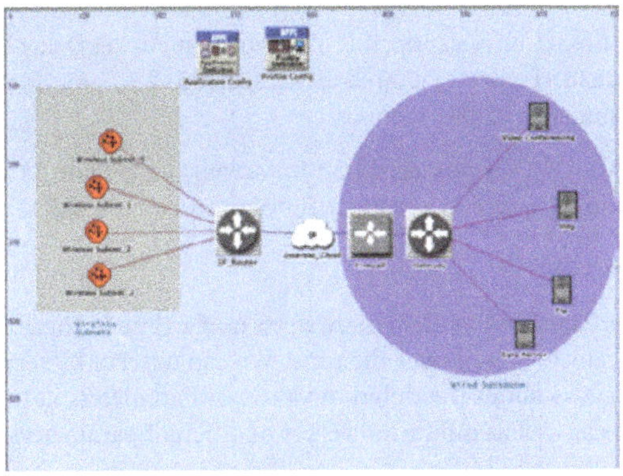

Figure 1: OPNET Model without load balancer.

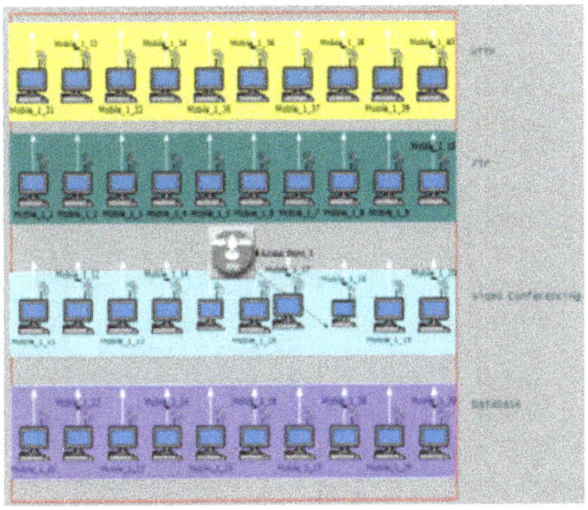

Figure 2: Mix of FTP, HTTP, Video Conferencing and DATABASE clients.

Applications	Attribute
Browsing	HTTP
Transactions (Query/Entry)	Database
File Transfer	FTP
Video Conferencing	Video

Table 1: Application Description.

Atttribute	Value
Start Time Offset (seconds)	uniform (5, 10)
Repeatability	Once at Start Time
Operation Mode	Serial (Random)
Start Time (seconds)	uniform (100,110)
Inter-repetition Time (seconds)	constant (300)
Number of Repetitions	constant (30)
Repetition Pattern	Serial

Table 2: Wireless Lan Traffic Generation Parameters.

Scenario 2 is the duplicate of scenario 1 in terms of number of users and types of application each user accesses. In our model we installed 4 access points (sites) in a Academic site where mix of DATABASE, HTTP, Video Conferencing and FTP clients are there. Simulations have been conceded out for our model to determine performance of the Factors. Tables 1 and 2 indicate the application description and the wireless traffic generation Factors.

Result Analysis along with Web Reports

Six graphs were selected after simulating our model (Figures 4-9). All graphs show a combination of the two scenarios. It has been analysed that DB Query/Entry Response time is less in scenario 2 (Figure 3) i.e. averagely it remains 0.041/0.047 seconds than scenario 1 (Figure 1) i.e. averagely it remains 0.062/0.078 seconds.

For FTP, it is observed that the Download Response Time for

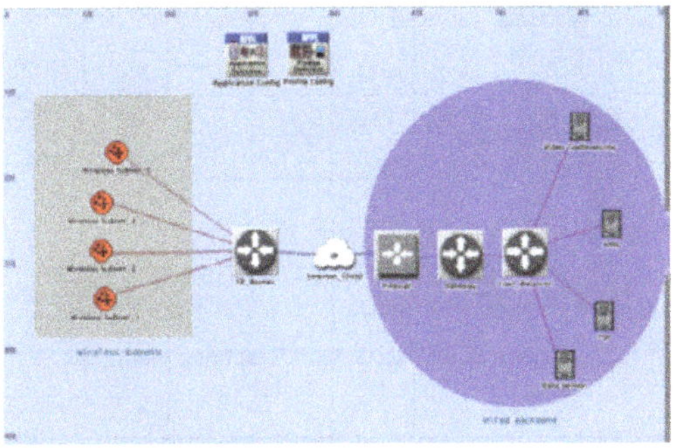

Figure 3: OPNET Model with load balancer.

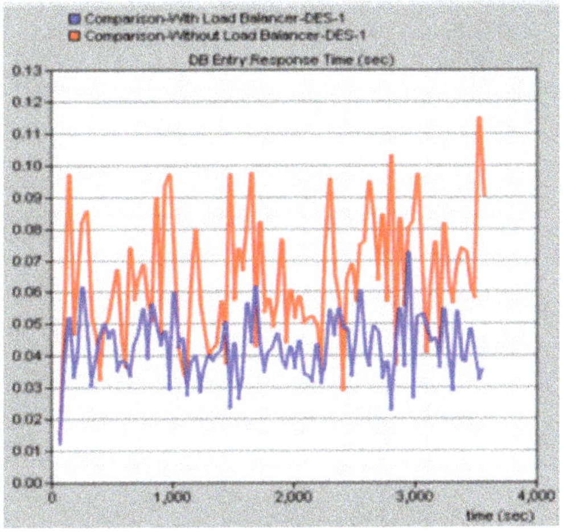

Figure 4: DB Entry Response time (sec).

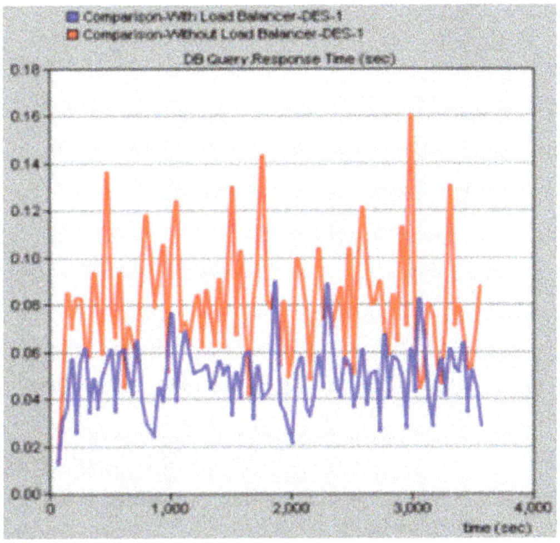

Figure 5: DB Query Response time (sec).

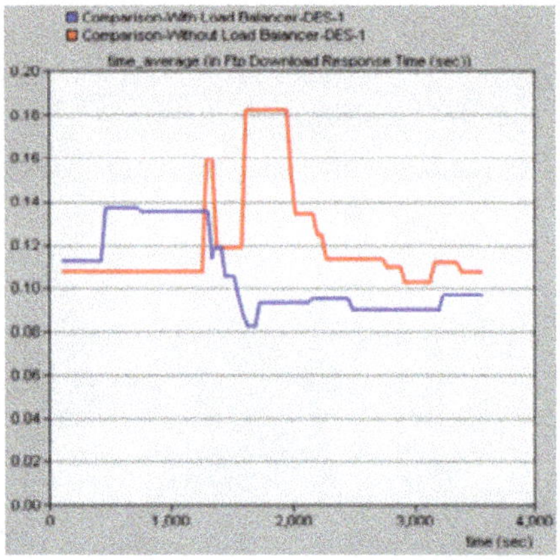

Figure 6: FTP Download Response (sec).

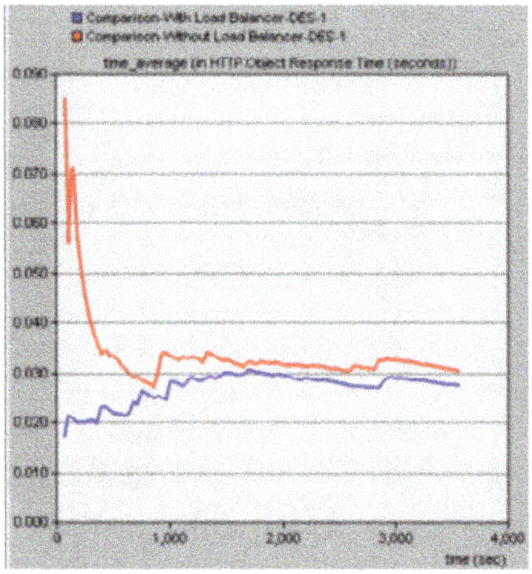

Figure 7: HTTP Page Response Time (sec).

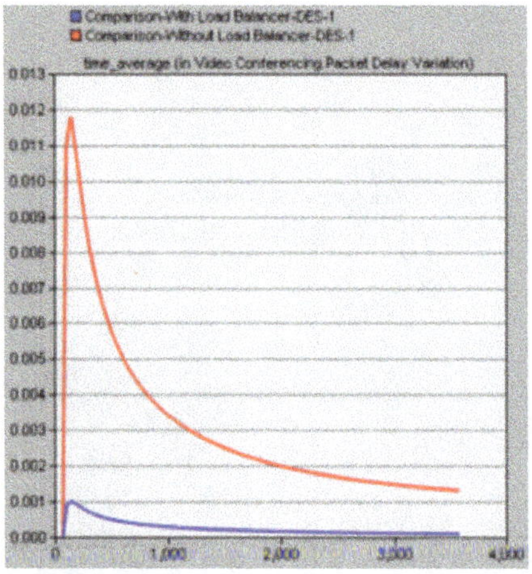

Figure 8: Video Conferencing Packet Delay Variation (sec).

scenario 1 (Figure 1) averagely time consumed is 0.107 seconds where as for scenario 2 (Figure 3) averagely time consumed is 0.097 seconds.

For HTTP, it is observed that the Page Response Time for scenario 1 (Figure 1) averagely time consumed is 0.127 seconds where as for scenario 2 (Figure 3) averagely time consumed is 0.102 seconds.

For Video Conferencing, it is observed that Packet Delay Variation for scenario 1 (Figure 1) as well as for scenario 2 (Figure 3) averagely time consumed is 0.0001 seconds.

For WLAN, it is observed that for scenario 1 (Figure 1) as well as for scenario 2 (Figure 3) averagely time consumed is 0.005 seconds.

Conclusion

In this Paper we have laid more stress on the time factor as all factors or the parameters are as per the time. We can observe by seeing at the Tables 3 and 4 about the different values of Parameters i.e. averagely, maximises as well as minimum values of different parameters.

It is observed from Table 4 that while accessing FTP, HTTP and Database applications there will be less consumption of the time in

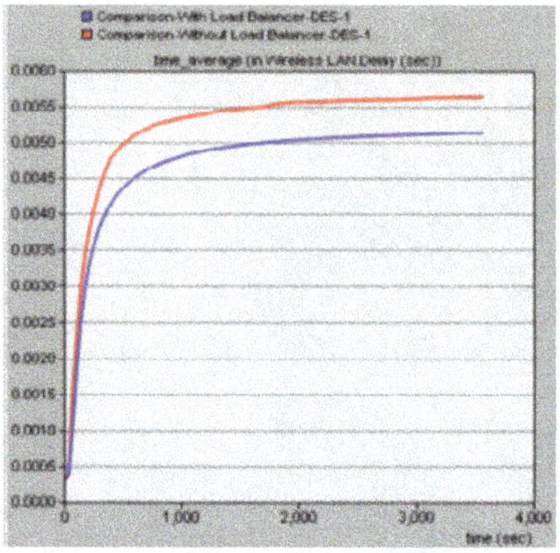

Figure 9: Wireless LAN Delay (sec).

Application	Parameter	Unit
HTTP	Page Response Time Object Response Time	Seconds Seconds
Database (Query/Entry)	Response Time	Seconds
FTP	Download Response Time Upload Response Time	Seconds Seconds
Video Conferencing	Delay Variation End-End Delay	Seconds Seconds
WLAN	Delay Media Access Delay	Seconds Seconds
Ethernet	Delay	Seconds

Table 3: Simulated Parameters.

Parameters	No Load Balancer			Load Balancer		
Statistics	**Avg. Val.**	**Max. Val.**	**Min. Val.**	**Avg. Val.**	**Max. Val.**	**Min. Val.**
DB Entry Response Time (sec.)	0.062	0.115	0.013		0.072	0.012
DB Query Response Time (sec.)	0.078	0.160	0.012	0.047	0.090	0.12
FTP Download Response Time (sec.)	0.107	0.373	0.036	0.097	0.190	0.036
HTTP Page Response Time (sec.)	0.127	0.393	0.039	0.102	0.324	0.03 8
Video Conferencing Packet Delay Variation (sec.)	.0001	.0218	.0000	.0001	.0018	.0000
WLAN Delay (sec.)	0.005	0.006	.0003	0.005	0.005	.0003

Table 4: Web Report.

scenario 2 (Figure 3) as compare to the scenario 1 (Figure 1) but if we see the time consumption at Video Conferencing as well as WLAN there is almost equal time consumed while accessing those applications.

So, it is conclusive that using Load Balancer reduces the time consumption to access different applications as compare to without Load Balancer. Also there is equal Consumption of CPU in both the cases.

References

1. Bennington B, Bartel C (1997) "Wireless Andrew: xperience Building a High Speed, Campus-WideWireless Data Network" Proceedings of ACM MOBICOM Budapest Hungary 55-65.
2. Hansen T, Yalamanchili P, Braun HW (2002) "Wireless Measurement and Analysis on HPWREN" Proceedings of Passive and Active Measurement Workshop Fort Collins Co 222-229.
3. Kotz D, Essein K (2002) "Analysis of a Campus-Wide Wireless Network" Proceedings of ACM MOBICOM Atlanta GA.
4. Tang D, Baker M (2000) "Analysis of a Local-Area Wireless Network" Proceedings of ACM MOBICOM Boston MA 1-10.
5. Sharma M, Kumar M, Sharma AK (2009) "HTTP and FTP Statistics for Wireless and Wire-Line Network with and without Load Balance Based on OPNET" International Journal of Information and Systems Sciences, Institute for Scientific Computing and Information, Canada 5: 112-125.
6. Al-Wabie SA (2002) The New Wireless Local Area Networks (WLAN's) Standard University of Maryland.
7. Gast M (2002) 802.11 Wireless Networks The Definitive Guide O'Reilly and Associates, Inc.
8. Tanenbaum AS (2003) "computer Networks" 4th Edition Prentice-Hall International Inc.
9. Wlan_lab_script_1_2 from http:// www.comnets.unibremen.de/~mms/wlan_lab_script_1_2.pdf
10. Hneiti W (2006) "Performance Enhancement of Wireless Local Area Networks" Amman Arab University for Graduate Studies, Jordan.
11. OPNET Technologies.

5

A Novel Voltage-Mode Lut Using Clock Boosting Technique in Standard CMOS

Sathyavathin S[1]* and Mr Ilanthendral J[2]

[1]*Final Year M.E (VLSI design), Department of ECE, Adhiparasakthi Engineering College, Melmaruvathur, TN, India*
[2]*Assistant Professor, ECE Department, Adhiparasakthi Engineering College, Melmaruvathur, TN, India*

Abstract

In a VLSI circuit, interconnection plays the dominant role in every part of the circuit nearly 70 percent of the area depends on interconnection, 20 percent of area depends on insulation, and remaining 10 percent to devices. The binary logic is limited due to interconnect which occupies large area on a VLSI chip. In this work, the designs of quaternary-valued logic circuits have been explored over multi-valued logic due to the following reasoning. An approach to mitigate the impact of interconnections is to use multiple-valued logic (MVL), hence, more information can be carried in each wire, reducing the routing network. Therefore, a single wire carrying a signal with N logic levels can replace log N having base 2 wires carrying binary signals. Reducing the routing leads to a direct reduction of the line capacitance and the overall circuit area. Therefore, this results in increasing the maximum operation frequency and also reducing the power consumption. The most important characteristics of this method is a voltage-mode structure. Voltage mode structure has the advantages like reduced power consumption implemented in a standard CMOS technology. Our new method overcomes conventional techniques with simple and efficient CMOS structures.

Keywords: Multiple-value logic; Quaternary logics; Look- up tables; FPGAs; Standard CMOS technology

Introduction

Central processing unit power dissipation or CPU power dissipation is the process in which central processing units (CPUs) consume electrical energy, and dissipate this energy both by the action of the switching devices contained in the CPU and by the energy lost in the form of heat due to the impedance of the electronic circuits. Mainly occurs due leakage current and static power dissipation and has formula [1]

$$P_d \propto \mathbf{CV}_{\mathbf{2dd}} \tag{1}$$

Therefore by reducing the capacitance value we can able to reduce the dissipation. One of the important advantage of quaternary logic is that has the reduced noise margin when compared to the conventional binary logic. More over if we use the current mode we have to face the problem for the fabrication process and have the high power consumptions [2].

Binary and quaternary Look-Up tables

In General Look-Up Tables (LUT) are basically memories, which implement a logic Function according to their configuration. Configuration values C=(c0….. ci, ck-1); are initially stored in the look-up table structure, and once inputs are applied to it, the logic value in the addressed position is assigned to the output.

The capacity of a LUT |C| is given by

$$|C| = n_x\, b_k \tag{2}$$

Where n denotes the number of outputs, k denotes the number of inputs and b for the number of logic values. For example, a 4-input binary look-up table with one output is able to store 1 ×24=16 Boolean values.

A binary function implemented by a Binary Look-Up Table (BLUT) is defined as f: Bk → B, over a set of variables X=(x0,…xi,…; xK-1), where each variable xi represents a Boolean value. The total number of different functions |F| that can be implemented in a BLUT with k input variables is given by

$$|F| = b_{|C|} \tag{3}$$

Where b=|B| (b=2 in the binary case). For example, a look-up table with 4 inputs (k=4) can implement one of |F|=65,536 different functions. Quaternary functions are basically generalizations from binary functions. This function implemented by a quaternary look-up table (QLUT) is defined as g: Qk → Q, over a set of quaternary variables Y=(y0,….. yi …….. yk-1), where the values of a variable yi, as the values of the function g(Y), can be in Q= {0,1,2,3}. As in the binary case, the number of possible function in QLUTs is given by (2), where b=4. In this case, the number of functions that can be represented is everywhere 4.3×109 for aQLUT with only two quaternary inputs (k=2), which is much larger than for the BLUT [3].

The quaternary variable y is capable of representing twice as much information as a binary variable x, we note that the cardinality of |Q|=2 × |B| in our experiments. In other words, two binary variables with the same inputs can be grouped in order to represent a quaternary variable. Such procedure is used mainly for reducing both the total number of connections and the number of gates.

Quaternary logic and reference voltages levels

This design was implemented using a standard CMOS technology, a single supply voltage and a clock boosting technique to incorporate a 16 to 1 multiplexer and a dual quaternary decoder. One of the most

***Corresponding author:** Sathyavathin S, Final Year M.E (VLSI design), Department of ECE, Adhiparasakthi Engineering College, Melmaruvathur, Tamilnadu, India, E-mail: nssathyavathi@gmail.com

important feature that was taken into account was the area usage since that, in order to perform more complex functions, this circuit needs to be replicated a millions of times in the FPGA [4].

The circuit depicted in the table below has two quaternary inputs, QA and QB, which are then computed by the dual quaternary decoder into the QLUT's binary control signals, B00-B33. The multiplexer 16-to-1 consists of sixteen NMOS switches enhanced with a clock boosting technique. When one of the control signals is high, the corresponding QLUT's line- switch- is activated connecting the corresponding QLUT's quaternary input to the output. The four voltage levels are represented in Table 1.

A quaternary variable can assume four different logic levels. Assuming a rail-to-rail voltage range and equal noise margins for the four logic levels, three different reference voltage values are required, 1/6*VDD*, 3/6*VDD*, and 5/6*VDD*, to determine a quaternary value.

Value	Voltage value [v]
0	0
1	0.404
2	0.707
3	1.2

Table 1: The four voltage levels.

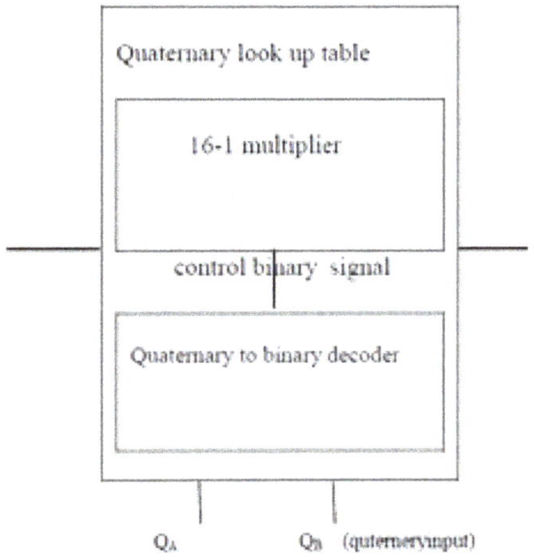

Figure 1: Proposed quaternary LUT.

Decimal	8	4	2	1	Decimal	4	1
0	0	0	0	1	0	0	0
1	0	0	1	0	1	0	1
2	0	0	1	1	2	0	0
3	0	1	0	0	3	0	1
4	0	1	0	0	4	1	0
5	0	1	0	1	5	1	1
6	0	1	1	0	6	1	0
7	0	1	1	1	7	1	1
8	1	0	0	0	8	2	0
9	1	0	0	1	9	2	1
10	1	0	1	0	10	2	0
11	1	0	1	1	11	2	1
12	1	1	0	0	12	3	0
13	1	1	0	1	13	3	1
14	1	1	1	0	14	3	0
15	1	1	1	1	15	3	1

Table 2: Quaternary and binary input table.

Q	Q_0	Q_1	Q_2	Q_3
0_4	1_2	0	0	0
1_4	0	1_2	0	0
2_4	0	0	1_2	0
3_4	0	0	0	1_2

Table 3: The Q-decoder behavior as a function of the quaternary logic value at the input.

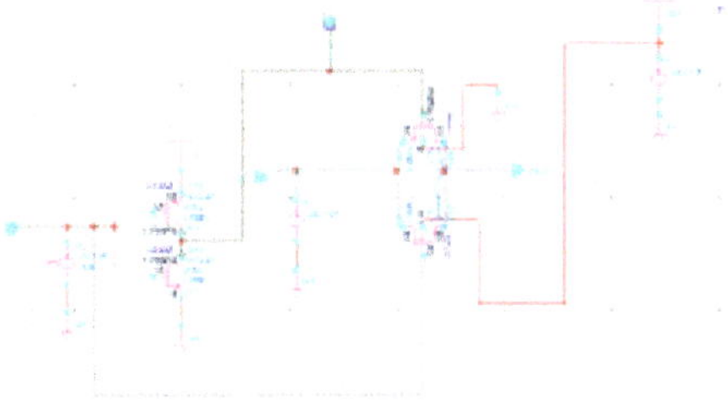

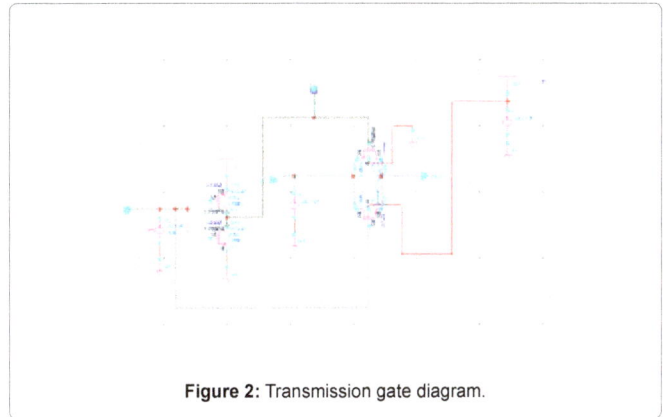

Figure 2: Transmission gate diagram.

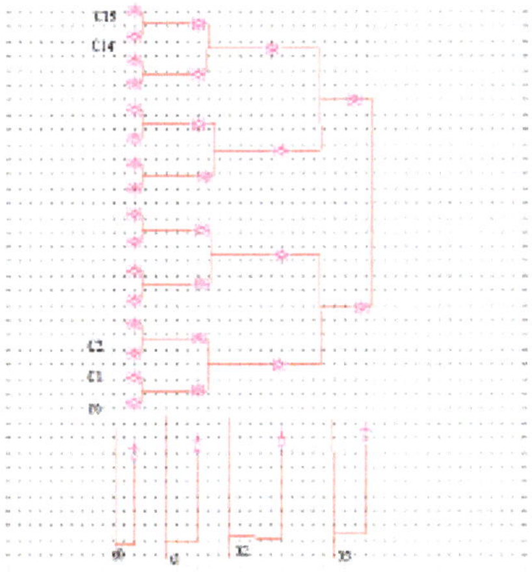

Figure 3: Binary 16-1 mux.

A LUT is an array indexing operator, where the output is mapped by the input, based on the configuration memory. The configuration values are initially stored in the LUT configuration memory, and according to the input, the logic value in the addressed position is assigned to the output (Figure 1).

1V 16-1 MUX: A Multiplexer has many inputs and one output has to be selected. Although, the Use of quaternary logic helps to reduce the number of interconnecting wires, which leads to to a compact layout, with reduced routing capacitance. We used the typical value for a binary FPGA (10 pF), since it maintains same number of wires, we can increase the number of functions in FPGA [5].

When compared to binary quaternary implementation of 16-1 multiplexer had the less number of gates. For the binary implementation nearly 30 transmission gates are used but in case of quaternary only 24 transmission gates are used (Tables 2 and 3) (Figures 2-8).

Quaternary-to-Binary converter

We also implemented the complete binary and quaternary look-

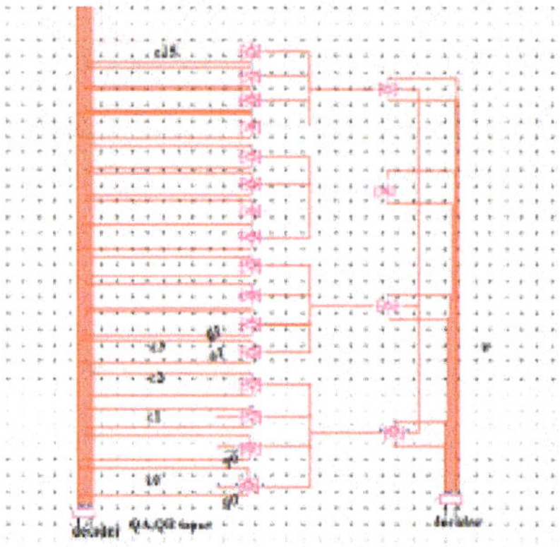

Figure 4: Quaternary 16-1 mux.

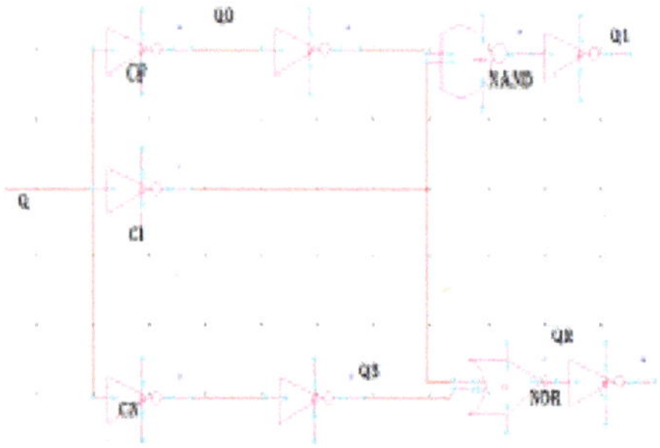

Figure 5: The Q-decoder logic structure.

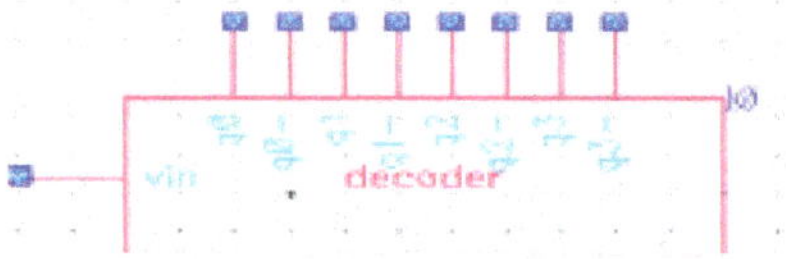

Figure 6: Symbol creation for decoder.

up tables with the UMC 130 nm technology in order to evaluate their performance and power consumption. The development of the binary and quaternary LUTs was performed. Transistor widths were kept to the minimum value in order to have a fair comparison between binary and quaternary versions (Figure 9).

Quaternery mux inputs and outputs waveforms

The quaternary structure proposed in this paper outperforms the binary implementation in both power consumption and propagation delay. These results were obtained through CADENCE Spectre simulation. The propagation delay is simply the largest delay from an input to the output of each LUT.

Conclusion

In this paper, we have reported an innovative QLUT design that can be used for multiple valued combinational logic or as a building

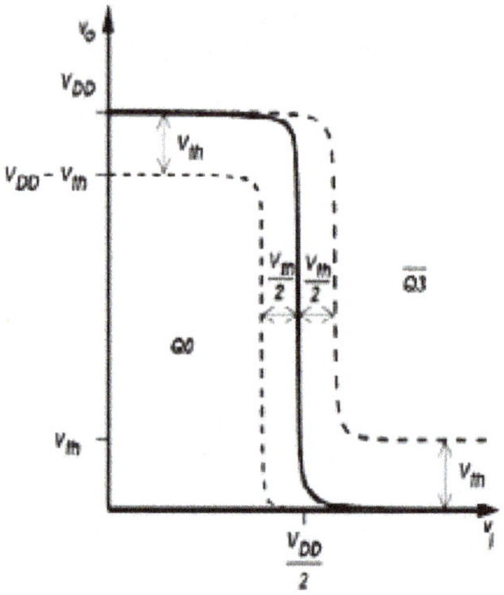

Figure 7: CP and CN transfer functions.

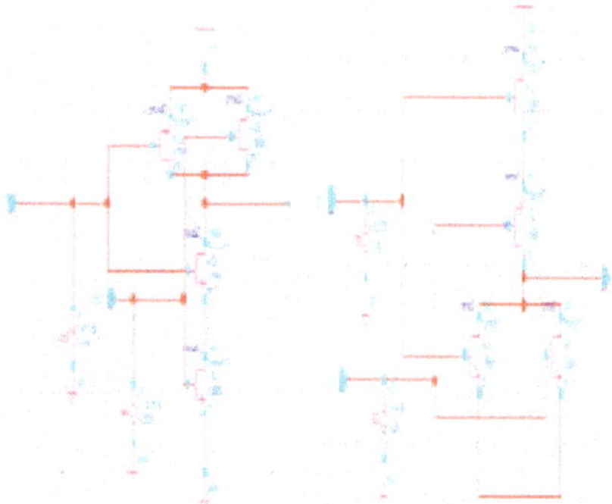

Figure 8: NAND Gate and NOR Gate.

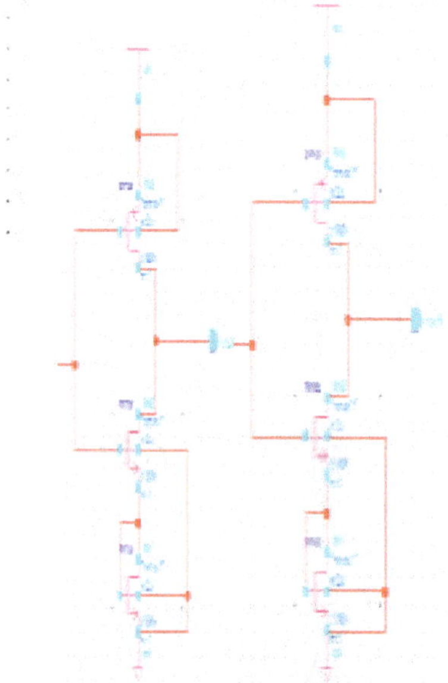

Figure 9: Cp diagram and CN diagram.

block in FPGAs. The QLUT internal functionality is implemented using simple standard CMOS structures. This feature is achieved through quaternary-to-binary decoders that quantize the input signals. This decoder is based on voltage-mode self-referenced comparators that allows the use of a standard CMOS Technology and overcomes previous design drawbacks. Also, a CB technique was used to decrease the switches resistance and increase the operation frequency, while at the same time, achieving low power consumption [6]. Therefore, the presented design is a valid solution to reduce the interconnections impact, without increasing Power consumption or losing performance. Experimental results were performed on an ASIC implementation of a full adder employing the designed QLUT. The obtained results attested the circuit feasibility and its advantages, using a standard CMOS process and its main characteristics (timing and power).

References

1. Shang A, Kaviani S, Bathala K (2002) "Dynamic power consumption in virtex-II FPGA family," in Proc. ACM/SIGDA Int. Symp. Field-Program. Gate Arrays 157-164.
2. Zilic Z, Vranesic Z (1993) "Multiple-valued logic in FPGAs," in Proc. Midwest Symp. Circuits Syst 1553-1556.
3. Ozer E, Sendag R, Gregg D (2006) Multiple-valued logic buses for reducing bus energy in low-power systems," IEE Comput. Digital Tech 153: 270-282.
4. Current K (1994) "Current-mode CMOS multiple-valued logic circuits," IEEE J. Solid-State Circuits 29:-95-107.
5. Kim J. (2010), "An area efficient multiplier using current-mode quaternary logic technique," in Proc. 10th IEEE Int. Solid-State Integr. Circuit Technol 403-405.
6. Rabaey J (2009) Low Power Design Essentials (Integrated Circuits and Systems). New York, NY, USA: Springer-Verlag.

Intrusion Detection System for Wormholes in WSN

Harleen Kaur[1]* and Neetu Gupta[2]

Department of ECE, GIMET, Amritsar, Punjab, India

Abstract

As an increasing number of people are going wireless, reducing the criticism of wireless networks is becoming a top priority. Wormhole attack is a severe threat against ubiquitous sensor networks. In a wormhole attack, the intruder sniffs the packets at one point in the network and forwards them with a less latency and relays them to another point in the network. A strategic placement of the wormhole can result in a significant breakdown in communication across a wireless network. The objective of dissertation addresses the efficient comparing the proposed technique with the previous study. In this paper, we have proposed an algorithm where intrusion detection has been done in a proposed approach to detect the wormhole attacks. The AODV routing protocol is used as the underlying network topology. Data tracker is used for detecting and isolating the malice node i.e. acting as a wormhole. This approach is implemented by using NS-2. The Simulation results are presented to validate the stated goal by comparing various performance metrics.

Keywords: Wormhole; AODV (Ad-Hoc on Demand Distance Vector Protocol Vector); Wireless network; Intrusion detection system; Routing

Introduction

Intrusion Detection System (IDS) in wireless networks has played an important role in network security by providing an additional level of protection to the network topology and applications beyond the traditional security mechanisms such as encryption and authentication. It detects the attacks and isolates the malicious nodes by matching the patterns of known intrusions or discovering the anomalies in the network activities. Its application environments cover almost all wireless networking scenarios such as ad hoc networks, wireless LANs, and sensor network [1]. Wireless Sensor Networks (WSNs) have been applied in more and more applications; however in sensor network sensor nodes are responsible not only for-the monitoring of the environment but also for forwarding the data packets toward base station on behalf of other sensor nodes. The sensors must be able to trace the routes to the base station and aware of their neighbours. An attacker can easily access of this, and may try to control the routes and to monitor the data packets that are sent along these routes [2]. One way to achieve this is to set up a wormhole in the network. A wormhole is a specialized man in- the-middle attack in which the adversary connects two otherwise distant regions of the network. We proposed a scheme for intrusion detection in WSN. They proposed distributed and cooperative framework to detect the attack. Every node in the WSN participates in the process of intrusion detection. It detects the sign of intrusion locally and independently and also propagates this information to other nodes in the network. Intrusion Detection is a security technology that attempts to identify individuals who are trying to break into and misuse a system without authorization and those who have legitimate access to the system and are abusing their privileges. The system protected is used to denote an information system being monitored by the Intrusion Detection system. Routing protocols [3] like table-driven/proactive, demand-driven/reactive or hybrid variants are subjected to routing attacks resulting in compromised confidentiality, integrity and message authentication.

Outline of the paper

The rest of this paper is organized as follows. A brief survey of related work is given in the next section. Section II describes the wormhole attack model and its types for implementing the wormhole attack. Section III shows the related work and its solution. In Section IV proposed algorithm for detection and isolation method is described. In Section V an analysis of the results is presented. Finally conclude the paper in section VI.

Wormhole Attack

In the wormhole attack, an attacker tunnels messages received in one part of the network over a low latency link and replays them in a different part. The simplest instance of this attack was a single node situated between two other nodes forwarding messages between the two of them as shown in Figure 1. However, wormhole attacks more commonly involve two distant malice nodes colluding to understate their distance from each other by relaying packets along an out-of-

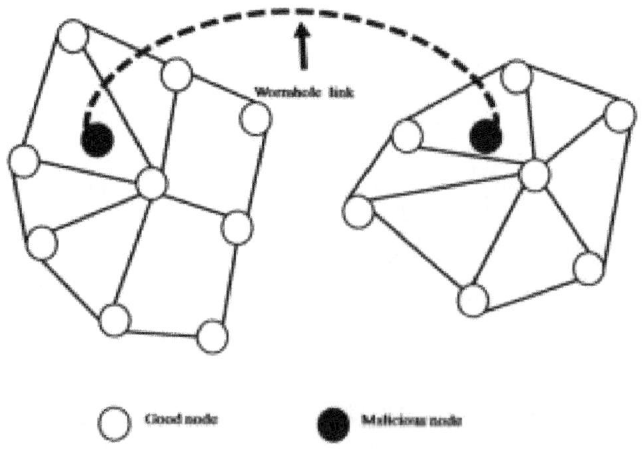

Figure 1: Wormhole attack.

***Corresponding author:** Harleen Kaur, Department of ECE, GIMET, Amritsar, Punjab, India, E-mail: harleen.kaur15@yahoo.com

bound channel available only to the attacker. An attacker is situated close to a base station can easily disrupt routing by creating a well-placed wormhole and convince the nodes that they were only one or two hops away via the wormhole. This can create a sinkhole: that is the attacker on the other side of the wormhole draws all the traffic if alternate routes are less attractive and provides artificially a high-quality route to the base station. This will most likely always be the case when the endpoint of the wormhole was relatively far from a base station [4].

Types of wormhole attack

Wormhole attacks are divided on the basis of implementation technique used for launching it and the number of nodes involved in establishing wormhole into the following types:

Encapsulation of the packet: Wormhole attacks are the disaster against many ad-hoc routing protocols, such as the two ad-hoc on-demands routing protocols DSR and AODV, and the sensor Tiny OS beaconing routing protocol. In DSR, RREQ floods the traffic in the network, when node S needs to discover a route to a destination, say D. Any node that receives the request, adds its identity to the source route, and rebroadcasts it. Each node broadcasts only the first route request it receives and drops any further copies of the same request. D generates a route reply when it receives each route request and sends it back to S and selects with the shortest number of hops or the path associated with the first arrived reply. This protocol will fail in a malice environment. When a malicious node at one part of the network receives the route request packet, it send to a second colluding party at a distant location near the destination, then the two colluding nodes will be said to have a wormhole . This prevents nodes from discovering true paths that are more than two hops away.

Consider Figure 2 in which nodes A and B try to maintain the shortest path between two spiteful nodes X and Y. Node A broadcasts a route request (RREQ), X gets the REQ and encapsulates it in a packet destined to Y through the path that exists between X and Y (U-V-W-Z). Node Y de marshals' the packet, and replays it again, which reaches B. The hop count does not increase during the traversal through U-VW-Z due to packet encapsulation. From A to B request travels through C-D-E. There are two routes for node B, the first is four hops long (A-C-D-E-B), and the second is apparently three hops long (A-X-Y-B). Node B will choose the second route since it appears to be the shortest while in reality it is seven hops long. So X and Y succeed in involving themselves in the route between A and B. A simple way of countering this mode of attack is a by-product of the secure routing protocol ARAN, which chooses the fastest route reply rather than the one which claims the shortest number of hops.

Out-of-band channel: In this mode, the wormhole attack was

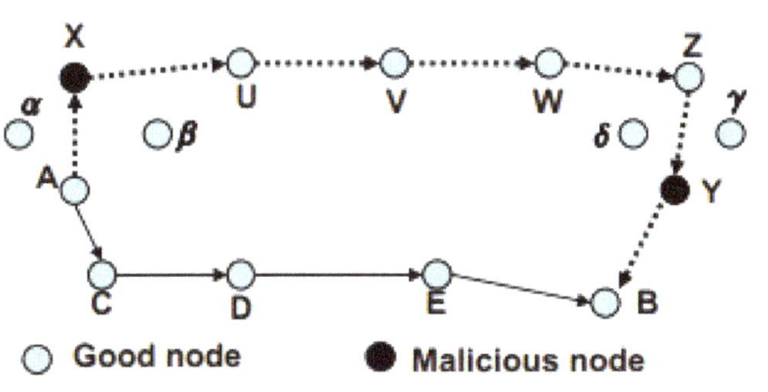

Figure 2: Packet Encapsulation.

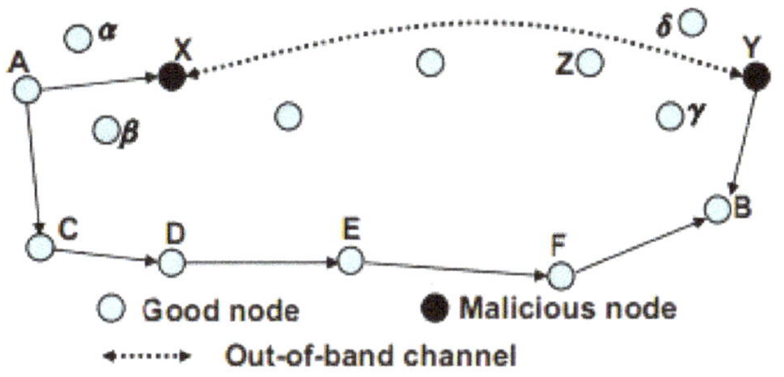

Figure 3: Out of band channel.

launched by having a high-quality, single-hop, out-of-band link (called tunnel) between the malicious nodes. By using a long-range directional wireless link this tunnel is achieved. This mode of attack was more difficult to launch than the packet encapsulation method since it needs specialized hardware capability [5].

Consider the scenario depicted in Figure 3. Node A is sending a route request to node B, nodes X and Y are malicious having an out-of-band channel. Node X tunnels the route request to Y, which is a true neighbour of B and node Y broadcasts the packet to its neighbours. Node B gets two route requests-A-X-Y-B and A C-D-E-F-B. Node B chooses the first route which is shorter and faster than the second and thus wormhole is established between X and Y in the route between A and B.

High-power transmission capability: In this type of wormhole attack, only one malice node with high-power transmission can communicate with other normal nodes from a long distance. When a malice node receives an RREQ, it broadcasts the request at a high-power level and rebroadcasts the RREQ towards the destination so that another node hears. By this method, the spiteful node increases its chance to be in the routes established between the source and the destination even without the participation of another malicious node. This attack can be minimized if each sensor node measures the received signal strength accurately [4].

Packet relay: Packet-relay-based wormhole attacks can be launched by one or more malicious nodes. In this attack, a malice node relays data packets of two distant sensor nodes to convince them that they were neighbour. This kind of attack was also known as "replay-based attack".

Protocol distortion: In this mode of wormhole attack, one malicious node tries to sniffs the network traffic by distorting the routing protocol. Routing protocols that were based on the 'shortest delay' instead of the 'smallest hop count' was at the risk of wormhole attacks and also called as "rushing attack".

Related Work

Detection of wormhole attack has been an active area of research and many mechanisms have been proposed so far luring the various behaviours of wormhole attack.

Kashyap Patel and Manoranjitham [6] addresses several types of sensor nodes and many network layer attacks can be perform on the network. Wormhole Attack was one of them which were most destructive routing attack for wireless sensor network and can be implemented by using Mint route protocol. In wormhole attack two or more node creates a Virtual tunnel in that network which transfer data packet. This virtual tunnel creates the shorter link in wireless sensor network. This paper presents the high level security and detection of wormhole attack by using Simulation results.

Issa Khalil et al. [7] represent the multihop wireless systems to relay each other's packets expose them to a wide range of security attacks. A particularly severe threat was known as the wormhole attack, where spiteful nodes records and control the data traffic at one location and tunnels it to another node, which replays it locally. For such sensor networks. A lightweight countermeasure for the wormhole attack, called LITEWORP was suitable for resource-constrained. Simulation results show that every wormhole was detected and isolated within a very short period of time. The results also show that the fraction of packets lost is less when LITEWORP was applied.

Yih-Chun Hu et al. [8] represents wormhole attack, a severe attack in the network that was particularly challenging to defend against and was possible even if the attacker has not compromised any hosts and even if all communication provides authenticity and confidentiality. The wormhole attack can form a serious threat in wireless networks, against many network routing protocols and location-based wireless security systems. A general mechanism, called packet leashes, for detecting and defending against wormhole attacks, and a specific protocol, called TIK that implements leashes. Topology-based wormhole detection was discussed, and shows that it was impossible for these approaches to detect some wormhole topologies.

Bintu Kadhiwala and Harsh Shah [9] propose the security emerges as a central requirement as mobile ad hoc network applications were deployed. Wormhole attacks enable an attacker with limited resources and no cryptographic material to wreak havoc on wireless networks. It is possible even if the attacker has not compromised any hosts and all communication provides authenticity and confidently. Wormhole attacks can form a serious threat in wireless networks.

Devendra Singh et al. [10] addresses the multiple -hop Mobile ad hoc networks which establish the routes involving with each node acting as a host and router. The wormhole attack was considered to be a serious security attack in multi-hop ad hoc networks. A simple technique to effectively detect wormhole attacks without the need for special hardware and/or strict location or synchronization requirements was proposed. The base of dissertation is to find alternative path from source to second hop and calculate the number of hops to detect the wormhole.

Saurabh Gupta et al. [11] represent specific attack called Wormhole attack that enables an attacker to record packets at one location in the network, send them to another location, and retransmits them into the network. After the route discovery, source node initiates wormhole detection process in the established path which count shop difference between the neighbours of the one hop away nodes in the route. If the hop difference between neighbours of the nodes exceeds the acceptable level then the destination node detects the wormhole. Our simulation results show that the WHOP is quite excellent in detecting wormhole of large tunnel lengths.

Proposed Work

Brief overview of proposed algorithm

- Divide the network in number of zones information of number of nodes and packet routing.
- Select the leader for the respective zones giving the information of each node.
- Assign the data tracker for each zone keeping the track of data send and received by the destination.
- Mismatch between data sent by source and received by destination will lead to the detection of the wormhole in the network.
- If the number of received packet - number of forwarded packets was more.
- Isolate the wormhole nodes from the network by sending alert messages to the nodes.
- Nodes after receiving the alert message will not communicate with the wormhole.

Network consists of number of nodes. To start the simulation, wormhole attack is created with animated rate of 5 ms and movement of the nodes are started at 0.1 sec in the network. At 2.0 sec all the mobile nodes are placed in the area of 40 m. Now divide the network into number of clusters. Each cluster has its own leader i.e. its own cluster head. The cluster head is elected on the basis of energy the node having the highest energy is elected as the leader. To each cluster data tracker is assigned. Data tracker contain all the information about the number of nodes in the cluster, number of packets forwarded and receive and communication path. When CBR is attached with UDP, communication between the nodes is started. After 14 sec tunnel is created between the two nodes i.e. wormhole is present. Wormhole is detected if threshold value (number of forwarded packets – number of received packets) is more and to isolate the scheme two nodes are created in the path through which data packets are send i.e.by sending the alert message to the nodes. Nodes after receiving the alert message will not communicate with the wormhole

Simulation Results

The results are shown below deals with the comparison of the routing protocol with mobile nodes. The work focuses on traceability of the wormhole attack. The result of this work is derived by using NS2 simulator.

NS-2 was used to verify the performance of the previous and proposed wormhole mechanisms. The network topology illustrates that there were fifty six random movable regular nodes with maximum speed in 5m/s randomly distributed in an area of 1100 m×1100 m, and AODV was performed for regular routing. A pair of wormhole nodes is developed wherein two tunnel nodes were applied to play the secret tunnel for wormhole attack. A connection with UDP-CBR is set up for communication (Figures 4-7).

The above results show the comparison between AODV and DSDV for the wormhole attack on the basis of performance metrics. As the results of DSDV are based on the routing table and number of hop counts and in AODV routing protocol, routes are established dynamically at intermediate nodes. Each node maintains sequence numbers to determine freshness of routing information and avoid routing loops. The impact of the above mentioned results for AODV will check out Wormhole attack when cluster head does not receive any data from the mobile nodes. The data tracker informs all the nodes and cluster head to trace the wormhole attack. The proposed AODV and base DSDV routing protocol produce result for tracing and isolation of wormhole attack by calculating its throughput and overhead. From the Figure 7 of throughput it shows that throughput of AODV routing protocol is more as compared with routing protocol DSDV. From the

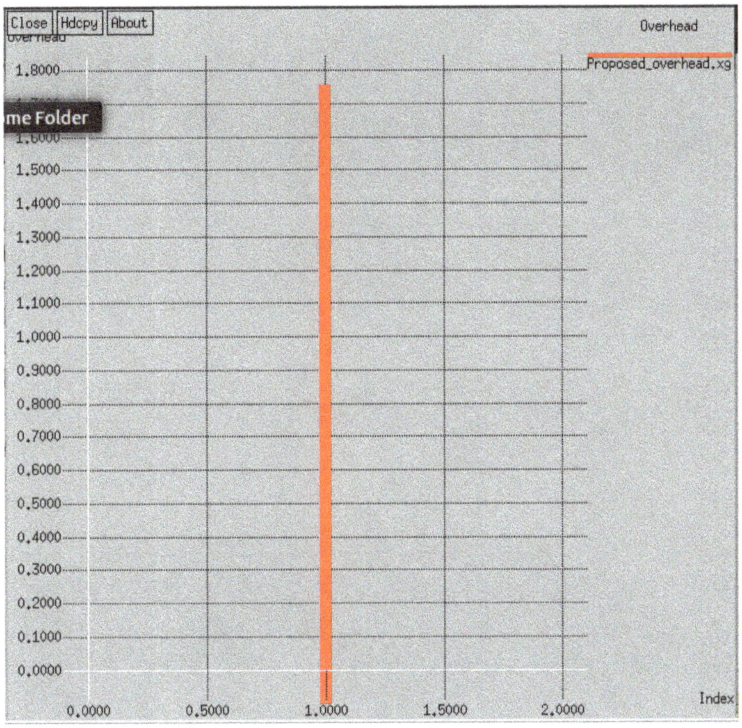

Figure 4: Proposed Overhead.

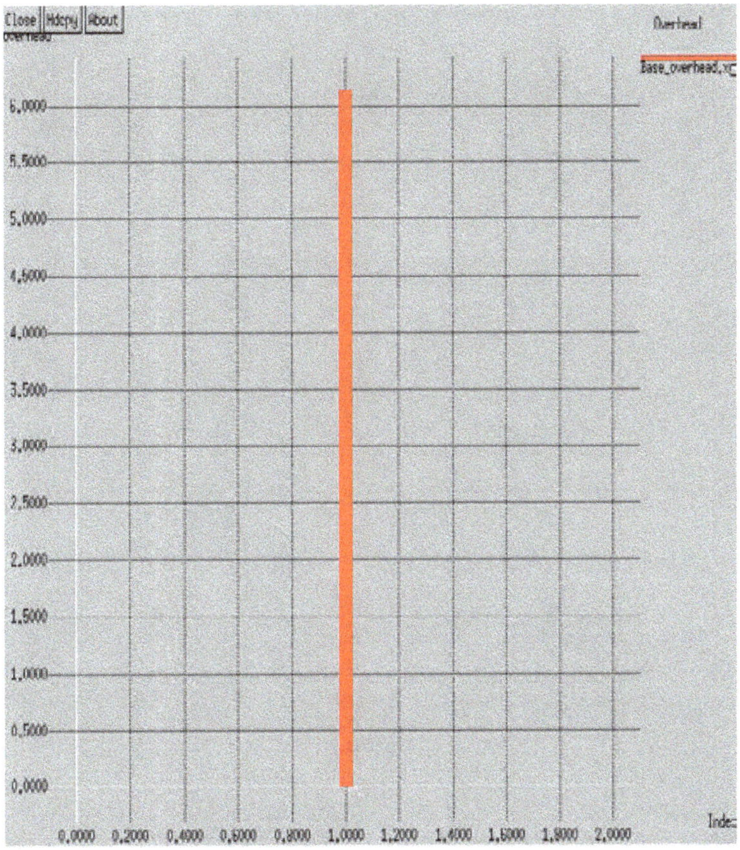

Figure 5: Base Overhead.

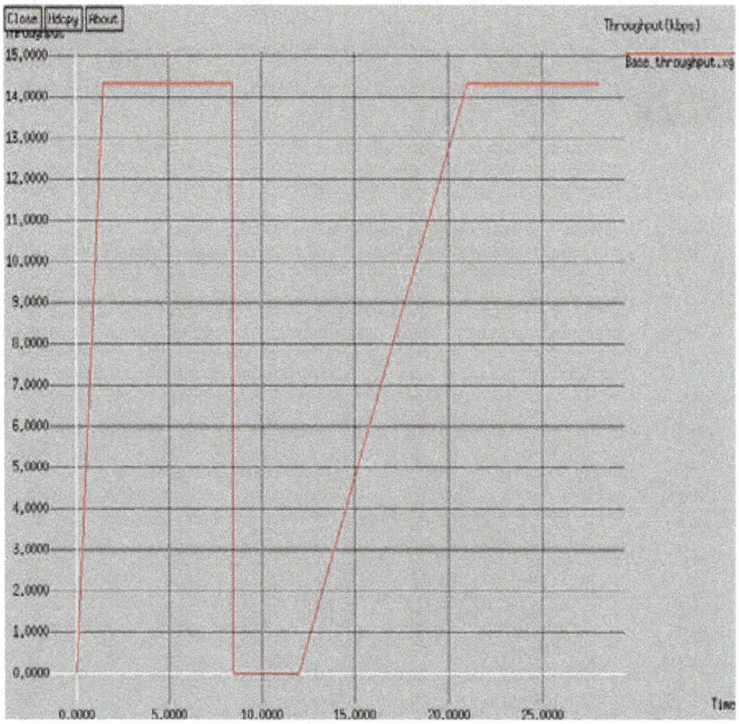

Figure 6: Base Throughput.

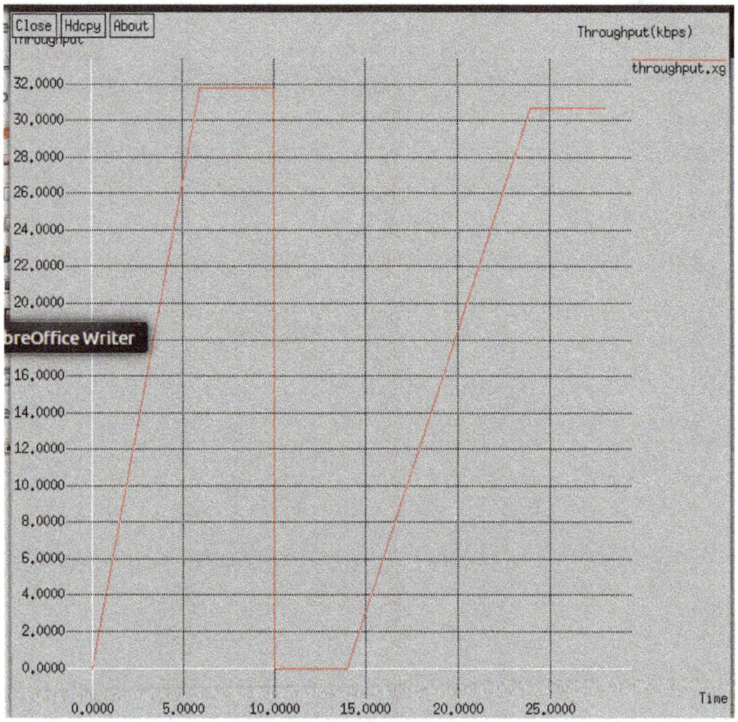

Figure 7: Proposed Throughput.

Figure 4, it shows that overhead of AODV is less as compared with DSDV i.e. overhead of 1.75 is observed which signifies the amount of routing required to transmit the data in the network is 1.75 times the data packets in AODV.

Conclusion and Future Work

The deployment of mobile nodes in an attended environment makes the network vulnerable. This paper gives a bird eye over WSN and intrusion of malicious nodes may cause serious impairment to security.

Wormhole presents an illusion of shortest path and tries to attack all the traffic over the network. The objectives listed have been carried out. In the presented work, we have discussed the routing protocol AODV with their working. With the results of AWK programming and trace graph, we can conclude that in the case of simple AODV increases in throughput by isolating the wormholes and decrease in the packet overhead by 1.75 and in future, work out with the unprotected protocols different types of attacks including group attacks and their relations can be studied and to study the robustness of Wireless Ad Hoc Networks for all types of protocols.

References

1. Wang W, Lu A (2006) Interactive Wormhole Detection in Large Scale Wireless Networks, IEEE Symposium on Visual Analytics Science and Technology, Baltimore, MD, USA.
2. Butty´an L, D´ora L, Vajda I (2005) Statistical Wormhole Detection in Sensor Networks. Lecture Notes in Computer Science Springer-Verlag Berlin Heidelberg ESAS 3813: 128-141.
3. Stephen Glass NICTA, Vallipuram Muthukkumurasamy, Marius Portmann (2013) MLDW- a Multilayered Detection mechanism for Wormhole attack in AODV based MANET, International Journal of Security, Privacy and Trust Management.
4. Sharma N, Singh U (2014) Various Approaches to Detect Wormhole Attack in Wireless Sensor Networks. International Journal of Computer Science and Mobile Computing 3: 29-33.
5. Issa Khalil (2007) Mitigation of control and data traffic attacks in wireless Ad-hoc and sensor networks, Purdue University.
6. Patel K, Manoranjitham T (2013) Detection of wormhole attack in wireless sensor network, International Journal of Engineering Research & Technology (IJERT).
7. Khalil I, Bagchi S, Shroff NB (2005) LITEWORP: A Lightweight Countermeasure for the Wormhole Attack in Multihop Wireless Networks, Proceedings of the 2005 International Conference on Dependable Systems and Networks.
8. Hu Y-C, Perrig A, Johnson DB (2006) Wormhole attacks in wireless networks. IEEE Journal on Selected Areas in Communications 24: 370-380.
9. Kadhiwala B, Shah H (2012) Exploration of Wormhole Attack with its Detection and Prevention Techniques in Wireless Ad-hoc Networks, International Conference in Recent Trends in Information Technology and Computer Science (ICRTITCS-2012).
10. Singh D, Khare KA, Rana JL (2013) Improved Trustful Routing Protocol to Detect Wormhole Attack in MANET. International Journal of Computer Applications (0975-8887) 62: 21-25.
11. Gupta S, Kar S, Dharmaraja S (2011) WHOP: Wormhole Attack Detection Protocol using Hound Packet, IEEE International Conference on Innovations in Information Technology.

Expansion of Power System Corridors Using Tier-1 Technique for Reactive Power Compensation

Ezennaya SO[1]*, Ezechukwu OA[1], Anierobi CC[1] and Akpe VA[2]

[1]*Department of Electrical Engineering, Nnamdi Azikiwe University, Awka, Nigeria*

[2]*Transmission Company of Nigeria (TCN)*

Abstract

This paper develops a novel strategy for the expansion of the power system corridors for the release of the embedded transmission capacity. Both theoretical and practical network models are presented with a focus on power flow studies which concentrates on the steady state or static behavior of electrical power system. The methodology involves the power flow analysis revalidation of the existing standard IEEE 14 bus system and simulation using Newton-Raphson method in both MATLAB and Powerworld simulator (PWS) environment. This paper therefore establishes that an original designed network could be modified to take more loads without building new generators or transmission lines. The expansion of the existing IEEE 14 bus network to accommodatemore load involves the use of static compensators incorporated at the transmission lines. This technique is then analyzed extensively when distributed along the lines through the use of a distributed capacitors compensators, (DCC). DCC can affect significant change in power line impedance to improve the power transfer capacity of an interconnected power system. The application of the DCC on the line is the tier-1 technique. The results obtained show that by applyingthe tier-1 techniques to the transmission line, the system's capacity will remarkably improve and the transmission line will accept extra loading.

Keywords: Power corridors, Newton-Raphson, Static compensators, DCC, Tier-1 compensation

Overview

Electrical energy efficiency is of prime importance to industrial and commercial companies operating in today's competitive markets. Optimum use of power system components is one main concern that needs to be balanced with energy efficiency, for both economic and environmental reasons. Electricity plays a fundamental role in the economic development of any country. Every country seeks to ensure supply of electricity that is affordable, reliable and secure in order to sustain modern ways of living. The availability of electricity greatly facilitates industrialization. This is because, electricity is a convenient way to transport energy in which they are also converted into transmission, distribution, and consumption [1]. Investigations are done in this paper to see how capacitors distributed along the transmission lines can expand the transmission line corridor by the release of embedded transmission capacity.

During the past two decades, the increase in electrical energy demand has presented higher requirements from the power industry. In interconnected power systems, it has become important to fully utilize the existing transmission facilities in preference to building new power plants and transmission lines that are costly to implement and involving long construction times. This necessitated the need for alternative technology through the use of solid state electronic devices with fast response characteristics [2]. The requirement was fuelled by worldwide restructuring of electric utilities, increased environmental and efficiency regulations and difficulty in getting permit and right of ways for the construction of overhead power transmission lines. Different approaches such as reactive power compensation and phase shifting have been applied to increase the capacity, stability and security of the power system. This need in conjunction with the development of semiconductor thyristor switch opened the door for the development of flexible alternating current transmission system (FACTS) controllers [3]. FACTS controllers make it possible to control the voltage magnitude of a bus, active and reactive power flows through transmission line of a system.

Power systems control

Reactive power control service should satisfy the following system requirements [4];

1. Satisfy overall system and customer requirements for reactive energy on a continuous basis;

2. Maintaining system voltages within acceptable limits;

3. Provide a reserve to cover the changed reactive requirements caused by contingencies, against which the system is normally secured, and satisfy certain quality criteria in relation to speed of response;

4. Optimize system losses.

Three tiers could be established in reactive power control. These are tier-1, tier-2, and tier-3 controls. However, two or more of the three tiers can simultaneously be applied to form a hybrid tier control. A description of the three tiers of reactive power control could be made;

a. **Tier-1 control** co-ordinates the action of voltage and reactive power control devices within the transmission zones of the network in order to maintain the requisite voltage level at a certain node points in the system.

b. **Tier-2 control** involves a process of load optimization by improving load power factors which influence the distribution of

***Corresponding author:** Ezennaya SO, Department of Electrical Engineering, Nnamdi Azikiwe University, Awka, Nigeria, E-mail: ezennayasamuel@gmail.com

reactive power, where the system load is high, and the operator must be certain that, in case of a loss of generation, the remaining facilities will be able to deliver enough reactive power to keep the voltage within the required range. The same applies to the converse situation, where the system load is low and reactive power needs to be absorbed.

c. **Tier-3 control** is the generator control.

d. **Hybrid-Tier control** is the simultaneous application of both the tier-1, and tier-2 or tier-3. It can also involve the control at the three tier controls to the power system at the same time.

Sources of reactive power

Reactive power is produced or absorbed by all major components of a power system [4];

1. Generators
2. Power transfer components
3. Loads
4. Reactive power compensation devices

Power Systems Reactive Power Compensation

Reactive power compensation otherwise called reactive var compensation is the management of reactive power to improve the performance of AC power systems, maximizing stability by increasing flow of active power. Compensation can be carried out in series or in parallel (shunt). Series and shunt var compensation are used to modify the natural electrical characteristic of AC transmission or distribution system parameters as well as changes the equivalent impedance of the load.

1. devices for reactive power compensation
2. synchronous condensers
3. Flexible alternating current transmission system (FACTS) controllers.
4. the distributed capacitor compensation (DCC)

The distributed capacitor compensation (Dcc) basis

There are many different methods used for compensation in power systems. Some of these methods include reducing generator and transformer reactance, increasing the number of parallel lines used, using shunt capacitor compensators, or using series capacitor compensators [5].

DCC can be used in series or in parallel on a transmission line. The addition of DCC in series serves multiple purposes, the most important being the improvement in stability along the entire line. Its addition in parallel (shunt compensation) is used to support voltage at certain point on the line as opposed to the entire line and also inject or absorb reactive power to the loads. Series and Shunt compensation have been in use since the early part of the 20th century. The first application of shunt compensation was in 1914 and has been used ever since becoming the most common method of capacitive compensation. Series compensation was first used in the United States for NY Power & Light in 1928, but didn't become popular until the 1950's when the voltage levels that could be handled began increasing. By 1968, a 550 kV application had been implemented and today there are applications approaching 800 kV [6].

The principal applications of DCC are;

- Improves voltage regulation
- Expand power transmission corridor of the transmission line
- Improves system stability

The applications previously mentioned are merely a selected few of the uses that DCC devices can provide. These applications and others are used throughout the world to improve the system as a whole. One common location where DCC devices are used heavily is on long transmission lines fed from hydroelectric generating plants. Many of the lines use the DCC devices to improve voltage regulation because the main load area is commonly several hundred kilometers from the generating station, allowing for large voltage decay.

DCC circuit

Capacitor compensator circuit is made up of the capacitor module and its protective scheme. The protective scheme shown in Figure 1 consists of [7];

1. A metal oxide varistor (MOV)
2. Current limiting damping equipment (CLDE)
3. Fast protective Device (FPD) and
4. By-pass switch (B)

The MOV has been designed to withstand the energy from external faults; faults appearing outside the series compensated circuit, without by-passing the DCC. The DCC module may be by-passed for any internal fault, (faults in the same circuit where the DCC is located). Each DCC is connected and disconnected from the line by means of two isolating disconnectors and one by-pass disconnector. The by-pass switch is of Sf_6 type, with a spring operating mechanism.

The CLDE consists of a current limiting reactor, a resistor and a varistor in parallel with the reactor. The purpose of the resistor is to add damping to the capacitor discharge current, and thus quickly reduce the voltage across the capacitor after a by-pass operation. The varistor help to avoid fundamental frequency losses in the damping resistor during steady state operation.

The FPD scheme is based on a hermetically sealed and very fast high power switch, which replaces conventional spark gaps. The FPD works in combination with the MOV, and allows by-passing in a very controlled way in order to reduce the energy dissipated in the MOV.

The Mathematical Model of Tier-1 Compensation

Electrical power is transmitted through the transmission line from the sending-end of the line to the receiving-end of the line.

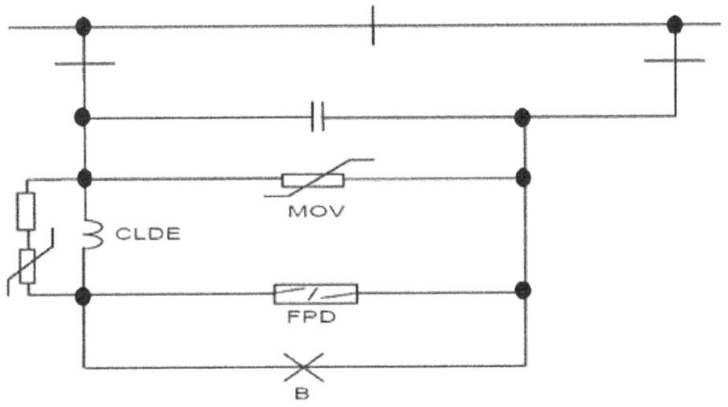

Figure 1: The single line diagram of a one-capacitor compensator in series.

This can be analyzed through parameterization and modeling of the transmission line with passive components such as resistors, capacitors and inductors. The quantities of these parameters depend mostly on the line conductors and the physical or geometrical configuration of the lines. These conductors will have certain characteristics such as resistance and reactance both in series (from sending to receiving-ends of the line) and shunts (from the line to ground) associated with them.

Basic principle of power in transmission

Loads are more often expressed in terms of real (watts/KW) and reactive (vars/Kvars) power. It is convenient to deal with transmission line equations for the sending and receiving-end complex power and voltages [8,9].

For a two-bus system shown in Figure 2, the sending and receiving-end voltages are represented by the bus voltages while the sending end voltage leads the receiving end voltage by an angle, δ. This angle is called the torque angle. The complex power leaving the receiving end and entering the sending-end of the transmission line can be expressed as

$$S_j = P_j + jQ_j = V_j I_j^* \text{ and } S_i = P_i + jQ_i = V_i I_i^* \tag{1}$$

Where

$$P_j = \frac{|V_j||V_i|}{|Z|}\cos(\beta-\delta) - \frac{|V_j|^2}{|Z|}\cos(\beta) \text{ And}$$

$$Q_j = \frac{|V_j||V_i|}{|Z|}\sin(\beta-\delta) - \frac{|V_j|^2}{|Z|}\sin(\beta) \tag{2}$$

Similarly,

$$P_i = \left|\frac{1}{Z}\right||V_i|^2\cos(\beta) - \left|\frac{|V_i||V_j|}{|Z|}\right|\cos(\beta+\delta) \tag{3}$$

$$Q_i = \left|\frac{1}{Z}\right||V_i|^2\sin(\beta) - \left|\frac{|V_i||V_j|}{|Z|}\right|\sin(\beta+\delta) \tag{4}$$

At $\delta = \beta$, the maximum power delivered at the load will be;

$$P_j = \frac{|V_j||V_i|}{|Z|} - \frac{|V_j|^2}{|Z|}\cos(\beta); \tag{5}$$

$$\text{If, } \beta = \theta, \text{then } cos\theta = \frac{R}{|Z|} \tag{6}$$

But the resistance R of a transmission line is very small compared to its reactance, so that;

$$\theta = \tan^{-1}\left(\frac{X}{R}\right) \approx 90^0 \tag{7}$$

Where $Z = R + jX$ and $\theta = \delta$.

Therefore the receiving-end power (P_j) becomes;

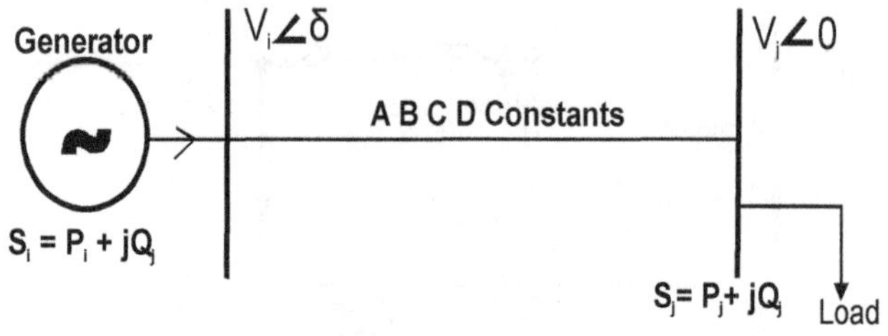

Figure 2: A two bus system.

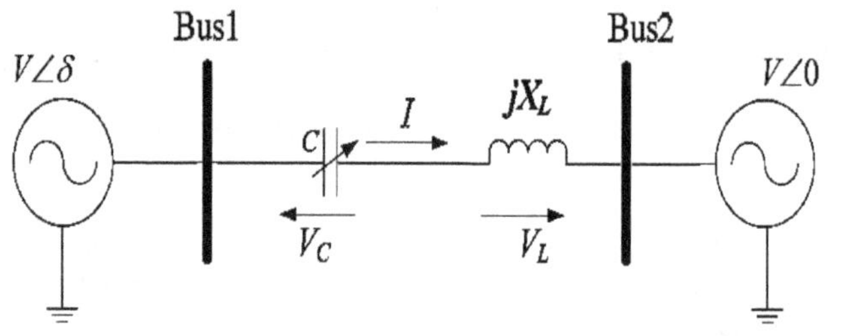

Figure 3: A simplified model of transmission system with series compensation.

$$P_j = \frac{|V_j||V_i|}{|X|}\sin(\delta) \text{ and } Q_j = \frac{|V_j||V_i|}{|X|}\cos(\delta) - \frac{|V_j|^2}{|X|} \tag{8}$$

Hence $Z \approx jX$.

For a very small value of δ, $cos\ \delta$=1 thus;

$$Q_j = \frac{|V_j|}{|X|}\left(|V_i| - |V_j|\right); \tag{9}$$

$$\text{where } \left(|V_i| - |V_j|\right) = |\Delta V| \tag{10}$$

$|\Delta V|$ is called the magnitude of voltage drop across the transmission line.

Therefore;

$$Q_j = \frac{|V_j|}{|X|}|\Delta V| \tag{11}$$

Reactive Power compensation of transmission lines

Equations (8) through (11) indicate that the active and reactive power/current flow can be regulated by controlling the voltages, phase angles and line impedances of the transmission system. It has been shown above that the active power flow will reach the maximum when angle δ is 90^0.

Series Compensation of A Transmission Line: A series-connected capacitor adds a voltage in opposition to the transmission line voltage drop, therefore reducing the series line impedance.

Figure 3 show a simplified model of a transmission system with series compensation. The voltage magnitude of the sending-end is assumed equal as $|V|$, and the phase angle between them is δ. The transmission line is assumed lossless and represented by the reactance X_L. A control capacitor is series-connected in the transmission line with voltage addition V_{inj}.

The Degree Of Series Compensation (Ks): The degree of series compensation or percentage compensation (K_s) is used to analyze a transmission line with the required addition of series capacitor. It is defined as the fraction of X_c, which refers to the total capacitive reactance of series compensators and X_L, which refers to the total inductive reactance of the line, as defined in equation 12;

$$K_s = \frac{X_c}{X_L} \tag{12}$$

Therefore, the capacitance, C as a portion of the line react

$X_c^{(n)} = X_c^{(n-1)} + \left(\frac{\blacklozenge X_c}{X_c}\right)^{(n)} X_c^{(n-1)}$ ance can be obtained from

$$X_c = K_s X_L \tag{13}$$

and

$$C = \frac{1}{2\pi f X_c} \tag{14}$$

The overall series reactance, X of the transmission line is;

$$X = X_L - X_c = (1 - K_s) X_L \tag{15}$$

Thus the active power transmitted becomes;

$$P_i = P_j = \frac{|V|^2}{(1-K_s)X_L} sin\delta \tag{16}$$

The reactive power supplied by the capacitor is calculated as;

$$Q_c = 2\frac{|V|^2 K_s}{X_L (1-K_s)^2}(1 - cos\delta) \tag{17}$$

From the above equation, it can be seen that transmitted active power increases with Ks [10].

Effective line reactance with and without dcc device

Figure 4 shows a simple transmission line without a compensating device. Equation (18) is the effective line reactance in matrix form.

$$X_{eff} = \begin{bmatrix} A & B \\ C & D \end{bmatrix} \tag{18}$$

Where X_{eff} is the effective reactance of the line.

The power flow equation becomes

$$\begin{bmatrix} V_i \\ I_i \end{bmatrix} = \begin{bmatrix} A & B \\ C & D \end{bmatrix} \begin{bmatrix} V_j \\ I_j \end{bmatrix} \tag{19}$$

Inserting a single series capacitor device on the line as in Figure 4 changes the ABCD parameters and the effective reactance of the line becomes (Figure 5)

$$X_{eff} = \begin{bmatrix} A_{1/2} & B_{1/2} \\ C_{1/2} & D_{1/2} \end{bmatrix} * \begin{bmatrix} 1 & X_c \\ 0 & 1 \end{bmatrix} * \begin{bmatrix} A_{1/2} & B_{1/2} \\ C_{1/2} & D_{1/2} \end{bmatrix} \tag{20}$$

As the power flow equation changes to;

$$\begin{bmatrix} V_i \\ I_i \end{bmatrix} = \begin{bmatrix} A_{1/2} & B_{1/2} \\ C_{1/2} & D_{1/2} \end{bmatrix} * \begin{bmatrix} 1 & X_c \\ 0 & 1 \end{bmatrix} * \begin{bmatrix} A_{1/2} & B_{1/2} \\ C_{1/2} & D_{1/2} \end{bmatrix} \begin{bmatrix} V_j \\ I_j \end{bmatrix} \tag{21}$$

The ABCD parameters are halved because the DCC is place at exactly midpoint (Figure 6) to the length of the line hence one DCC device is used.

Inserting several series capacitor devices on the line will change

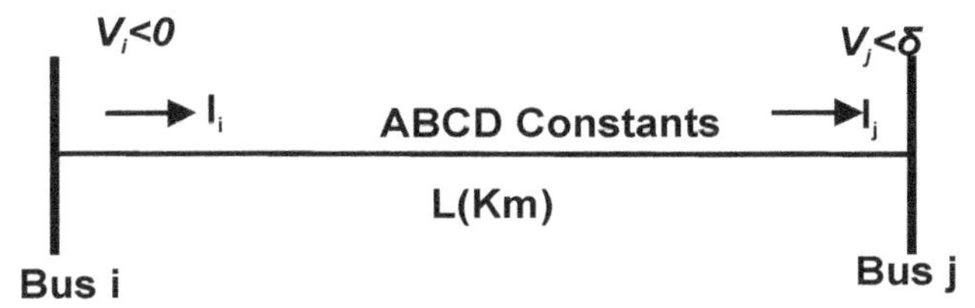

Figure 4: A simple transmission line without compensation.

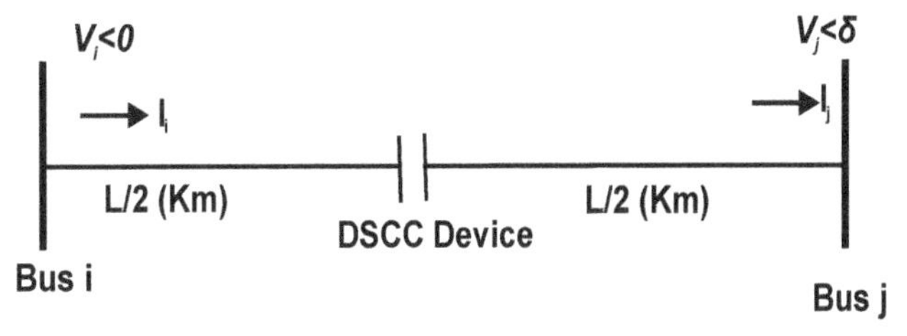

Figure 5: A transmission line with single DCC device (compensated line).

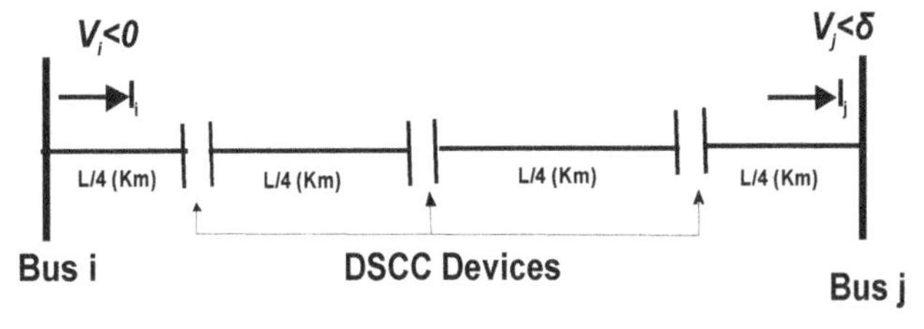

Figure 6: A transmission line with multiple DCC devices.

the ABCD parameters hence the more the capacitors on the line are distributed, the better the performance. Figure 6 shows a transmission line with multiple series capacitor devices and equation 21 changes to;

$$X_{eff} = \begin{bmatrix} A_{1/4} & B_{1/4} \\ C_{1/4} & D_{1/4} \end{bmatrix} * \begin{bmatrix} 1 & X_c \\ 0 & 1 \end{bmatrix} * \begin{bmatrix} A_{1/4} & B_{1/4} \\ C_{1/4} & D_{1/4} \end{bmatrix} * \begin{bmatrix} 1 & X_c \\ 0 & 1 \end{bmatrix} * \begin{bmatrix} A_{1/4} & B_{1/4} \\ C_{1/4} & D_{1/4} \end{bmatrix} * \begin{bmatrix} 1 & X_c \\ 0 & 1 \end{bmatrix} * \begin{bmatrix} A_{1/4} & B_{1/4} \\ C_{1/4} & D_{1/4} \end{bmatrix} * \begin{bmatrix} 1 & X_c \\ 0 & 1 \end{bmatrix} \tag{22}$$

The ABCD constants are divided by four (Figure 6) when the DCC is placed at quarter of the line hence three Capacitors are used and placed at every quarter of the line.

Power flow including dcc in matrix forms

From equation (21), the transfer admittance matrix of the DCC is given by [11];

$$\begin{bmatrix} I_i \\ I_j \end{bmatrix} = \begin{bmatrix} jB_{ii} & jB_{ij} \\ jB_{ji} & jB_{jj} \end{bmatrix} * \begin{bmatrix} I_i \\ I_j \end{bmatrix} \tag{23}$$

Where

$$B_{ii} = B_{jj} = \frac{-1}{X_c}, B_{ji} = B_{ij} = \frac{1}{X_c} \tag{24}$$

Equation (23) holds for inductive operation while for capacitive operation, the sign are reversed. The active and reactive power equations at bus ***j*** are as in equations (25) and (26) below;

$$P_j = V_j V_i B_{ij} \sin(\delta_j - \delta_i) \tag{25}$$

$$Q_i = -V_j^2 B_{jj} - V_j V_i B_{ij} \cos(\delta_j - \delta_i) \tag{26}$$

In Newton-Raphson solutions, these equations are linearized with respect to the series reactance. For the condition shown in Figure 3 where series reactance regulates the amount of active power flowing from bus ***i*** to ***j*** at a value ***P***, [11] the set of linearized power equation is,

$$\begin{bmatrix} \Delta P_i \\ \Delta P_j \\ \Delta Q_j \\ \Delta P^{Xc}{}_{ij} \end{bmatrix} = \begin{bmatrix} \frac{\partial P_i}{\partial \theta_i} & \frac{\partial P_i}{\partial \theta_i} & \frac{\partial P_i V_i}{\partial V_i} & \frac{\partial P_i V_j}{\partial V_j} & \frac{\partial P_i X_c}{\partial V_i} \\ & & & & \frac{\partial P_i X_c}{\partial X_c} \\ \frac{\partial P_j}{\partial \delta_i} & \frac{\partial P_j}{\partial \delta_j} & \frac{\partial P_j V_i}{\partial V_i} & \frac{\partial P_j V_j}{\partial V_i} & \frac{\partial Q_i X_c}{\partial X_c} \\ \frac{\partial Q_i}{\partial \delta_i} & \frac{\partial Q_i}{\partial \delta_j} & \frac{\partial Q_i V_i}{\partial V_i} & \frac{\partial Q_j V_j}{\partial V_j} & \frac{\partial Q_j X_c}{\partial X_C} \\ \frac{\partial P_{ij}^{Xc}}{\partial \delta_i} & \frac{\partial P_{ij}^{Xc}}{\partial \delta_i} & \frac{\partial P_{ij}^{Xc}}{\partial V_i} & \frac{\partial P_{ij}^{Xc}}{\partial V_j} & \frac{\partial P_{ij}^{Xc}}{\partial \delta_i} \end{bmatrix} \begin{bmatrix} \Delta\delta_i \\ \Delta\delta_j \\ \frac{\Delta V_i}{V_i} \\ \frac{\Delta V_j}{Vj} \\ \frac{\Delta X_c}{X_c} \end{bmatrix} \tag{27}$$

$$\Delta P_{ij}^{X_c} = P_{ij}^{reg.} - P_{ij}^{X_c\,cal.} \quad (28)$$

Where, $\Delta P_{ij}^{X_c}$ is the active power flow mismatch for the series reactance calculated;

$$\Delta X_c = X_c^{(n)} - X_c^{(n-1)} \quad (29)$$

ΔX_c is the incremental change in series reactance; and $P_{ij}^{X_c\,cal.}$ is the calculated power given by equation (25). The state variable X_c of the DCC controller is updated at the end of each iterative step according to equation (30);

$$X_c^{(n)} = X_c^{(n-1)} + \left(\frac{\blacklozenge X_c}{X_c}\right)^{(n)} X_c^{(n-1)} \quad (30)$$

The Standard IEEE 14 Bus Test Systems (Revalidation)

One of the international load flow test systems is the IEEE-14 bus system. Load flow analysis is carried out in IEEE 14 bus test system. Figure 7 show the standard IEEE 14 bus network simulated in Powerworld platform. The run mode of Power world simulator enable the simulation of the existing IEEE 14 bus test system model using N-R iterative method to obtain the bus voltages, phase angles, line losses, real and reactive power flows. The system topology consists of 14 buses, 20 transmission lines or branches, 2 online generators, 3 online synchronous compensators used only for reactive power support, and 11 loads totaling 259 MW and 78.7 Mvar.

The simulated result of the test system in Power world shown in Table 1 gives a very close result when compared with the MATLAB results of Table 2. It was therefore confirmed that the result obtained when DCC is applied on the IEEE 14 bus network using only Power world simulation software due to its flexibility and simplicity.

Using codes written in MATLAB and system information exported from Power world simulator, the standard IEEE 14 bus network is revalidated and reconfirmed.

Simulation result

The revalidated Standard IEEE 14 bus network shows that the total real and reactive power loss of the system are 15.31 MW and 36.77 Mvar respectively with the systems maximum current rating totaling 2948.91 Amps. As a result, the system's maximum MVA loading becomes 696.604 MVA. These results confirmed and agreed with the standard performance of the standard IEEE 14 bus system as shown in Tables 1 and 2 (Power world simulator tool) and Table 2 (MATLAB simulator tool). All bus voltages were also confirmed to fall within the recommended limit ($0.9 \leq V \leq 1.1\ p.u$).

Total power loss in Powerworld simulator: 15.31 MW and 36.77 Mvar

The Table 3 confirmed that the standard IEEE 14 bus system total power loss in Powerworld simulator is 15.31 MW and 36.77 Mvar and that of MATLAB is 15.031 MW and 35.348 Mvar. This is a clear indication that the software tool used for this model is validated.

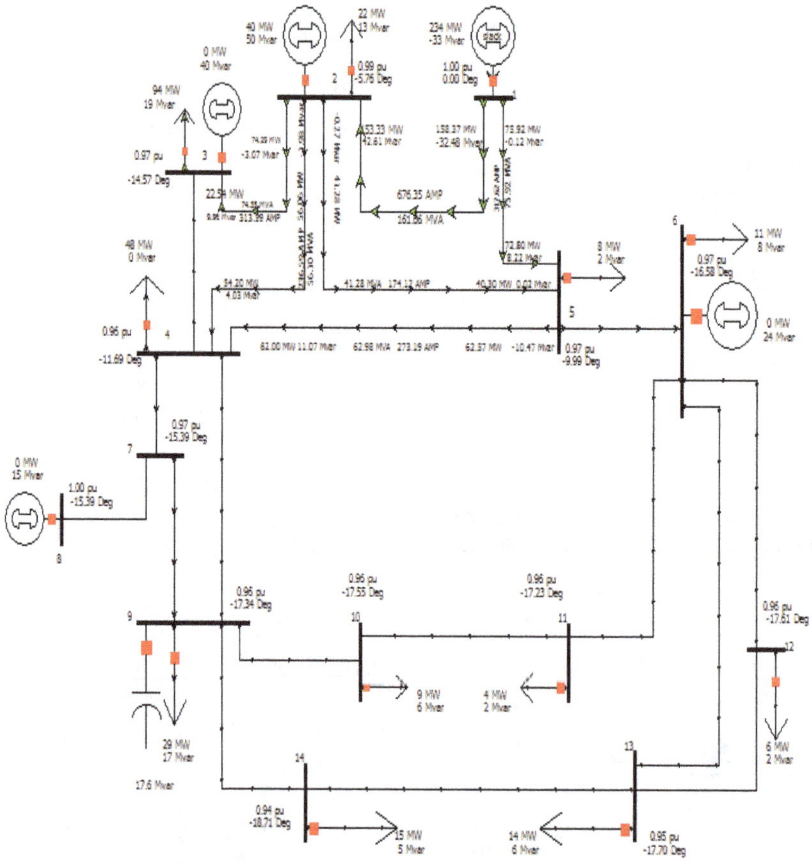

Figure 7: Standard IEEE 14 bus test system in Powerworld simulator environment.

Line records		MW From	MW To	Max MW	Mvar From	Mvar To	Max Mvar	MVA From	MVA To	Max MVA	Amps From	Amps To	Max Amps	MW Loss	Mvar Loss
From NO	To NO														
1	2	158.4	153.333	158.366	-32.5	42.612	42.612	161.7	159.144	161.633	676.35	671.202	676.35	5.03	10.13
1	5	75.9	-72.8	75.917	-0.1	8.218	8.212	75.9	73.263	75.917	317.615	316.494	317.615	3.12	8.1
2	3	74.3	-71.654	74.289	-3.1	9.958	9.958	74.4	72.342	74.353	313.588	311.998	313.588	2.64	6.89
2	4	56.1	-54.205	56.061	-2.0	4.028	4.028	56.1	54.354	56.096	236.587	235.762	236.587	1.86	2.05
2	5	41.3	-40.295	41.283	-0.3	0.015	0.268	41.3	40.295	41.283	174.116	174.074	174.116	0.99	-0.25
3	4	-22.5	23.019	23.019	11.0	-13.064	13.064	25.1	26.468	26.468	105.256	114.804	114.804	0.48	-2.02
4	5	-62	62.574	62.574	11.1	-10.468	11.074	63	63.443	63.443	273.192	274.074	274.074	0.57	0.61
4	7	29	-28.962	28.962	-3.1	5.033	5.033	29.1	29.396	29.396	126.351	126.351	126.351	0.00	1.91
4	9	16.4	-16.431	16.431	1.1	0.534	1.087	16.5	16.439	16.466	71.424	71.424	71.424	0.00	1.62
5	6	42.9	-42.922	42.922	0.6	4.317	4.317	42.9	43.139	43.139	185.444	185.444	185.444	0.00	4.95
6	11	6.6	-6.584	6.636	2.8	-2.707	2.816	7.2	7.118	7.209	30.988	30.988	30.998	0.05	0.11
6	12	7.7	-7.604	7.688	2.5	-2.278	2.454	8.1	7.938	8.07	34.693	34.693	34.693	0.08	0.18
6	13	17.4	-17.156	17.401	6.9	-6.429	6.911	18.7	18.322	18.724	80.488	80.488	80.488	0.24	0.48
7	8	0	-0.006	0.006	-14.7	15.132	15.132	14.7	15.132	15.132	63.308	63.308	63.308	0.00	0.4
7	9	29	-28.956	28.956	9.7	-8.613	9.695	30.5	30.21	30.536	131.253	131.253	131.253	0.00	1.08
9	10	5.9	-5.922	5.943	5.0	-4.911	4.966	7.7	7.694	7.745	33.649	33.649	33.649	0.02	0.05
9	14	10	-9.973	9.953	4.1	-3.8	4.139	10.8	10.505	10.779	46.831	13.996	13.996	0.16	0.34
10	11	-3.1	3.084	3.084	-0.9	0.907	0.907	3.2	3.215	3.215	13.996	7.215	7.251	0.01	0.02
12	13	1.5	-1.498	1.504	0.7	-0.673	0.679	1.7	1.642	1.651	7.251	23.336	23.366	0.01	0.01
13	14	5.2	-5.103	5.156	1.3	-1.197	1.305	5.3	5.241	5.319	23.336	23.336	23.336	0.05	0.11
Total		490	474.547	666.151	0.1	36.614	148.663	693.3	685.3	696.604	2948.71	2947.414	2956.14	15.31	36.77

Table 1: Line records of the standard IEEE 14 bus system in Power world simulator.15.31-0.25.

Number	Name	Nom kV	PU Volt	Volt (kV)	Angle (Deg)	Load MW	Load Mvar	Gen MW	Gen Mvar	s.sht Mvar
1	1	138	1.00000	138.00	0.00			234.28	-32.6	
2	2	138	0.99197	136.892	-5.76	21.7	12.7	40	50	
3	3	138	0.97006	133.869	-14.57	94.2	19	0	40	
4	4	138	0.96454	133.107	-11.69	47.8	0			
5	5	138	0.96845	133.646	-9.99	7.6	1.6			
6	6	138	0.97323	134.306	-16.58	11.2	7.5	0	24	
7	7	138	0.97334	134.322	-15.39					
8	8	138	1.00000	138.000	-15.39			0	15.13	
9	9	138	0.96294	132.886	-17.34	29.5	16.6			17.62
10	10	138	0.95663	132.015	-17.55	9	5.8			
11	11	138	0.96106	132.626	-17.23	3.5	1.8			
12	12	138	0.95722	132.097	-17.61	6.1	1.6			
13	13	138	0.95233	132.422	-17.7	13.5	5.8			
14	14	138	0.93845	129.507	-18.71	14.9	5			

Table 2: Bus records of IEEE 14 bus system in Powerworld simulator.

Case 1: Modification of the ieee 14 bus network (addition of an excess load to the network)

To illustrate that an already saturated network can be expanded by the use of capacitors distributed along the lines at strategic places, an existing load of a selected Company in Nigeria was used - the General Steel Mills (GSM), Asaba. The Company's total maximum active and reactive power demand are **17.80 MW** and **25.71 Mvar** respectively [12]. These loads were added to bus 6 of the standard IEEE 14 bus system which modified the revalidated results of the system. The active and reactive power losses increased from the normal operating performance of the 14 bus standard network to **59.85 MW** and **226.84 Mvar** respectively. The system's maximum amperage was **5658.287 Amps** as all the bus voltages dropped below **0.9 p.u** except the slack bus-bus 1 (Table 4). The MVA maximum loading also increased to **1062.225 MVA (Overloading)**. For these reasons, it is enough to say

Bus No.	Bus Voltage		Load		Generation		Injected Mvar
	Magnitude	Angle	MW	MVAR	MW	MVAR	
1	1.050	0	0	0	234.029	61.659	0
2	1.000	-4.664	21.7	12.7	40	-11.307	0
3	0.970	-13.283	94.2	19	0	36.233	0
4	0.964	-10.372	47.8	0	0	0	0
5	0.970	-8.79	7.6	1.6	0	0	0
6	1.000	-15.109	11.2	7.5	0	14.105	0
7	0.979	-13.768	0	0	0	0	0
8	1.000	-13.768	0	0	0	12.064	0
9	0.963	-15.594	29.5	16.6	0	0	0
10	0.962	-15.835	9	5.8	0	0	0
11	0.977	-15.608	3.5	1.8	0	0	0
12	0.982	-16.085	6.1	1.6	0	0	0
13	0.976	-16.134	13.5	5.8	0	0	0
14	0.949	-17.014	14.9	5	0	0	0

Table 3: The bus records of the IEEE 14 bus system in MATLAB tool.

Bus records		Nom kV	PU Volt	Volt (kV)	Angle (Deg)	Load MW	Load MVar	Gen MW	Gen MVar	s.shnts Mvar
NO	Name									
1	1	138	1	138	0			256.71	371.82	
2	2	138	0.82572	113.95	-4.21	21.7	12.7	40	50	
3	3	138	0.62455	83.188	-17.9	91.53	19.72	0	0	
4	4	138	0.59434	82.018	-10.26	45.16	113.38			
5	5	138	0.6584	90.859	-8.27	7.53	1.59			
6	6	138	0.48283	66.63	-24.44	13.9	20.07	0	0	
7	7	138	0.51624	71.241	-19.52					
8	8	138	0.51624	71.241	-25.52			0	0	
9	9	138	0.48187	66.497	-26.41	22.98	12.53			0
10	10	138	0.4696	66.804	-25.93	6.8	4.38			
11	11	138	0.47045	64.923	-27.46	2.65	1.36			
12	12	138	0.45884	63.32	-27.67	4.48	1.18			
13	13	138	0.45279	62.486	-27.67	9.75	4.19			
14	14	138	0.43901	6.583	-30.11	10.35	3.47			

Table 4: Bus records of the modified standard IEEE 14 bus network (with **17.80 MW** and **25.71 Mvar** extra load addition).

Line Records		MW From	MW To	Max MW	Mvar from	Mvar To	Max Mvar	MVA From	MVA To	Max MVA	Amps From	Amps To	Max Amps	MW loss	Mvar loss
From NO	To NO														
1	2	180.8	163.425	180.84	236.4	187.704	236.434	297.7	248.879	297.665	1234.34	1260.995	1260.995	17.41	48.73
1	5	75.9	-62.494	75.872	135.4	-83.694	135.39	155.2	104.452	155.2	649.309	663.723	663.723	13.38	51.7
2	3	78.9	-70.943	78.86	71.1	-40.085	71.092	106.2	81.485	106.74	537.953	545.84	545.84	7.92	32.01
2	4	59.1	-49.186	59.135	89.1	-60.896	89.147	107	78.279	106.978	542.024	551.028	551.028	9.95	28.25
2	5	43.7	-38.501	43.728	64.8	-50.701	64.764	78.1	63.662	78.144	395.932	404.532	404.532	5.23	14.06
3	4	-20.6	22.075	22.075	20.4	-17.848	20.36	29	28.338	28.953	193.949	199.83	199.83	1.49	2.51
4	5	-55.2	58.335	58.335	-72.6	81.951	81.951	91.2	100.593	100.593	641.907	639.204	641.907	3.13	9.37
4	7	23.6	23.608	23.608	24.1	-17.367	24.107	33.7	29.308	33.741	237.514	237.514	237.514	0.00	6.74
4	9	13.5	-13.546	13.546	13.8	-7.931	13.832	19.4	15.697	19.36	137.283	136.283	136.283	0.00	5.9
5	6	35.1	-35.126	35.126	50.9	-28.647	50.857	61.8	45.326	61.108	392.751	392.751	392.751	0.00	22.21
6	11	3.6	-3.543	3.603	1.3	-1.196	1.322	3.8	3.74	3.838	33.258	33.258	33.258	0.06	0.13
6	12	5.5	-5.339	5.521	2	-1.617	1.995	5.9	5.579	5.87	50.866	50.866	50.866	0.18	0.38
6	13	12.1	-11.607	12.101	5.3	-4.28	5.252	13.2	12.371	13.192	114.305	114.305	114.305	0.49	0.97
7	8	0	0	0	0	0	0	0	0	0	0	0	0	0	0
7	9	23.6	-23.608	23.608	17.4	-13.822	17.367	29.3	27.356	29.308	237.514	237.514	237.514	0.00	3.55
9	10	6	-5.914	5.995	4.8	-4.558	4.772	7.7	7.467	7.662	66.525	66.525	66.525	0.08	0.21
9	14	8.2	-7.718	8.174	4	-3.078	4.047	9.1	8.31	9.121	79.189	79.189	79.189	0.45	0.97
10	11	-0.9	-0.892	0.892	0.2	-0.167	0.175	0.9	0.908	0.908	8.071	8.071	8.071	0.00	0.01
12	13	0.9	-0.847	0.856	0.4	-0.433	0.442	1	0.951	0.964	8.786	8.786	8.786	0.01	0.01
13	14	2.7	-2.633	2.696	0.5	-0.393	0.521	2.7	2.662	2.746	25.37	25.37	25.37	0.06	0.13
Total		496.5	436.736	654.571	669.3	442.466	823.827	1052.9	865.413	1062.25	5596.846	5655.584	5658.287	59.85	226.84

Table 5: Line records of the modified standard IEEE 14 bus network (with **17.80 MW** and **25.71 Mvar** extra load addition).

Bus records		Nom kV	PU Volt	Volt (kV)	Angle (Deg)	Load MW	Load MVar	Gen MW	Gen MVar	s.shnts Mvar
NO	Name									
1	1	138	1.0000	138	0.01			243.29	48.73	
2	2	138	0.96558	133.25	-1.46	21.7	12.7	40	50	
3	3	138	0.9115	125.787	-3.67	94.2	19.0	0	0	
4	4	138	0.92276	127.34	-2.72	47.8	0.00			
5	5	138	0.93304	128.76	-2.25	7.6	1.6			
6	6	138	0.90345	124.676	-4.59	17.8	25.71	0	0	
7	7	138	0.90974	125.544	-4.06					
8	8	138	0.90974	125.544	-4.06			0	0	
9	9	138	0.90309	124.627	-4.78	29.5	16.6			0
10	10	138	0.89828	123.962	-4.78	9	5.8			
11	11	138	0.89837	123.975	-4.7	3.5	1.8			
12	12	138	0.8919	123.028	-4.8	6.1	1.6			
13	13	138	0.88911	122.697	-4.77	13.5	5.8			
14	14	138	0.88218	121.741	-5.01	14.9	5.0			

Table 6: Bus records of the Modified standard IEEE 14 bus network with tier-1 compensation.

Line Records		MW From	MW To	Max MW	Mvar from	Mvar To	Max Mvar	MVA From	MVA To	Max MVA	Amps From	Amps To	Max Amps	MW loss	Mvar loss
From NO	To NO														
1	2	161.1	-156.037	161.15	22.3	-11.831	22.333	162.7	156.49	162.69	677.898	680.23	680.23	5.11	10.5
1	5	82.1	-77.844	82.141	37.9	-24.778	37.889	90.5	81.693	90.459	308.555	375.837	375.837	4.3	13.11
2	3	78	-74.856	78.019	29.9	-19.244	29.875	83.5	77.019	83.544	354.832	350.199	350.199	3.44	10.63
2	4	55.6	-53.473	55.59	17.7	-14.656	17.742	58.4	55.445	58.352	250.09	254.31	254.31	2.12	3.09
2	5	40.7	-39.561	40.278	15.4	-14.892	15.393	43.5	42.271	43.54	186.856	191.631	191.631	1.17	0.5
3	4	-19.6	19.963	19.963	5.7	-5.885	5.885	20.4	20.812	20.812	93.387	94.364	94.364	0.34	0.21
4	5	-63.7	64.349	64.349	-6.9	8.912	8.912	64.1	64.963	64.963	290.858	290.868	290.868	0.65	2.04
4	7	31.4	-31.398	31.398	21.9	-18.51	21.866	38.3	36.448	36.262	167.615	167.615	167.615	0.00	3.36
4	9	18	-18.016	18.016	13.3	-10.317	13.256	22.4	20.761	22.367	96.176	96.176	96.176	0.00	2.94
5	6	45.5	-45.456	45.456	44.5	-34.435	44.476	63.6	57.027	63.595	264.079	264.079	264.079	0.00	10.04
6	11	4.2	-4.222	4.244	1	-0.905	0.951	4.3	4.317	4.349	20.122	20.12	20.12	0.02	0.05
6	12	7.2	-7.106	7.191	2.2	-2.308	2.214	7.5	7.393	7.524	34.677	34.677	34.677	0.8	0.18
6	13	16.2	-15.985	16.244	6	-5.558	6.03	17.3	16.923	17.309	79.633	79.633	79.633	0.24	0.47
7	8	0	0	0	0	0	0	0	0	0	0	0	0	0	0
7	9	31.4	-31.398	31.398	19.7	-17.98	19.745	37.1	36.982	37.091	167.615	167.615	167.615	0.00	1.77
9	10	8.3	-8.277	8.321	6.8	-6.68	6.799	10.7	10.636	10.746	49.539	49.539	49.539	0.04	0.12
9	14	11.7	-11.434	11.686	5.4	-4.821	5.356	12.9	12.409	12.855	58.85	58.85	58.85	0.25	0.53
10	11	-0.7	0.722	0.722	0.9	-0.879	0.882	1.1	1.138	1.139	5.298	5.298	5.298	0.00	0.00
12	13	1.00	-1.033	1.007	0.4	-0.437	0.44	1.1	1.094	1.099	5.15	5.15	5.15	0.00	0.00
13	14	3.5	-3.463	3.489	0.2	-0.179	0.233	3.5	3.468	3.497	16.455	16.455	16.455	0.03	0.05
Total		511.9	494.215	681.092	244.3	185.113	260.278	742.9	706.48	744.193	3187.685	3212.958	3212.958	3212.958	59.17

Table 7: Line Records of the the Modified standard IEEE 14 bus network with tier-1 compensation.

that the modified IEEE 14 bus system got overloaded and cannot accept this extra load (Table 5).

Case 2: Application of tier-1 compensation to the modified standard IEEE 14 bus network

The distributed capacitor technology applied on the transmission lines were used in this case to know how much the lines can be freed of their carriage even when the loads were operating. This was verified by placing capacitors on all the lines (interline action) with degree of compensation Ks allowed to operate by 0.7 or 70% of the original line reactance value (Table 6). The bus and line results were compared, from which the total active power loss reduced from 59.85 to 17.79 MW (70.28% reduction) while the total reactive power loss reduced from 226.84 to 59.17 Mvar (73.92% reduction). By this margin, the system's MVA loading was released from 1062.225 to 744.193 MVA (29.94% released) (Figure 8). This is resulted from the reduction in system's current from 5658.287 to 3212.958 Amps (43.94% reduction) with all lines still operating within their normal limits. Bus voltages were also restored appreciably (Table 7).

Figures 9A-9D show the graphical plots of line numbers of the Modified standard IEEE 14 bus network (with tier-1 compensation) against the MW, Mvar, MVA loadings and bus p.u voltages with and without compensation.

The percentage reduction in MVA loading and savings determines how much the systems corridors have been expanded by the release of the embedded system capacity on which the system can be available for extra loadings (Table 8).

Bus records		Nom kV	The modified IEEE 14 bus system (overloaded system)		Use of tier-1 compensator	
NO	Name		PU Volt	Volt (kV)	PU Volt	Volt (kV)
1	1	138	1	138	1.0000	138
2	2	138	0.82572	113.95	0.96558	133.25
3	3	138	0.62455	86.188	0.9115	125.787
4	4	138	0.59434	82.018	0.92276	127.34
5	5	138	0.6584	90.859	0.93304	128.76
6	6	138	0.48283	66.63	0.90345	124.676
7	7	138	0.51624	71.241	0.90974	125.544
8	8	138	0.51624	71.241	0.90974	12.544
9	9	138	0.48187	66.497	0.90309	124.627
10	10	138	0.4696	64.804	0.89828	123.962
11	11	138	0.47045	64.923	0.89837	123.975
12	12	138	0.45884	63.32	0.8919	123.082
13	13	138	0.45279	62.486	0.88911	122.697
14	14	138	0.43901	60.583	0.88218	121.741

Table 8: Comparison of the modified IEEE 14 bus system voltage performance against operation results of the tier-1 voltage result as a mean of validating the proposed compensation strategy for the bus records.

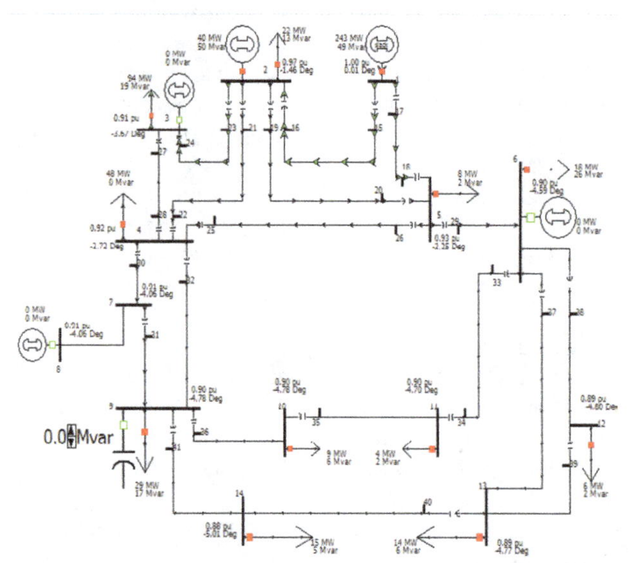

Figure 8: Modified standard IEEE 14 bus network with tier-1 compensation.

Conclusion

The use of DCC creates more loops in the transmission system by providing more active power routes without having to build new generating stations, new transmission stations or dealing with right-of-way issues. Application of the tier-1 compensation was able to accommodate the added load in the existing 14 bus system by releasing the system up to 29.94% MVAof its overloading which remarkably reduced the system's active losses by 70.28% and reactive losses by 73.92%. Recommendations are made to simultaneously apply also the teir-2 and 3 to the network in other to ensure maximum restoration of the power (up to 100% restoration). This efficient control method can salvage the power system from total collapse and as well serves as a quick way to respond to consumers power satisfaction quest.

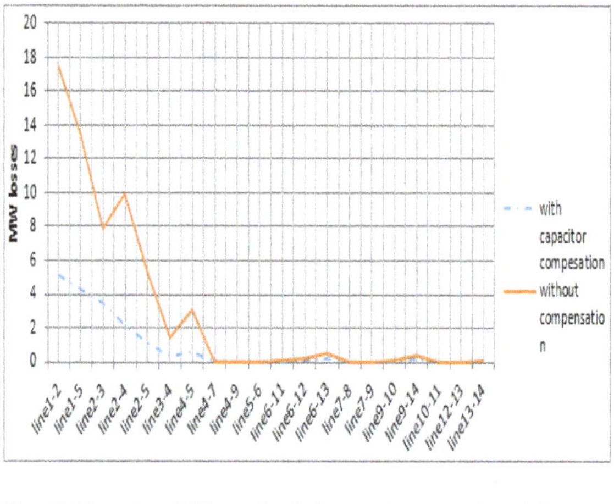

Figure 9A: Line number vs MW losses with and without capacitor compensation on the lines

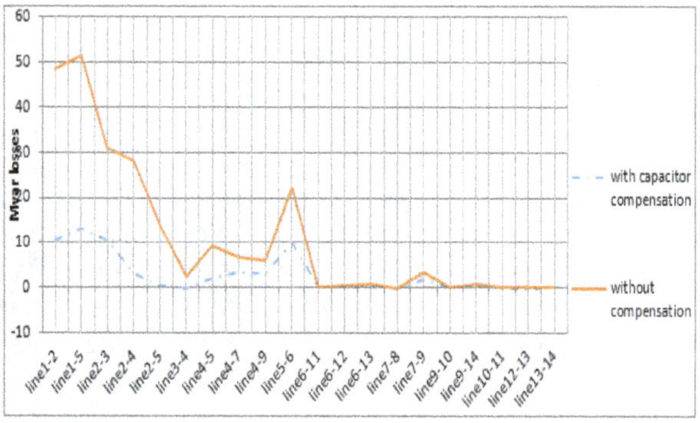

Figure 9B: Line Number vs Mvar Losses with and without capacitor compensation on the lines

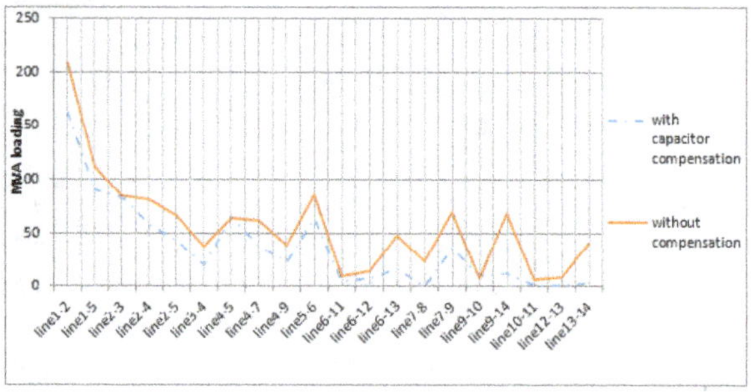

Figure 9C: Line numbers vs MVA Loadings with and without capacitor compensation on the lines

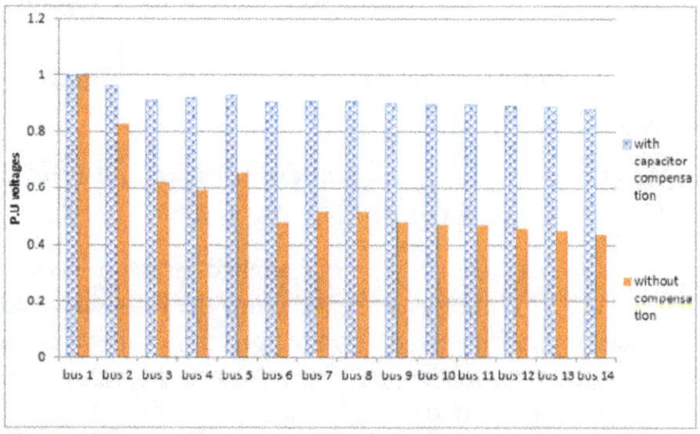

Figure 9D: Bus numbers vs p.u voltages with and without capacitor compensation on the lines

Figures 9A-D: Show the graphical plots of line numbers of the Modified standard IEEE 14 bus network (with tier-1 compensation) against the MW, Mvar, MVA loadings and bus p.u voltages with and without compensation.

References

1. Nagesh HB, Puttas Wamy PS (2009) Power flow Model of Static VAR Compensator and Enhancement of Voltage Stability, IJAET.
2. Adepoju GA, Komolafe OA, Aborisade DO (2011) Power Flow Analysis of the Nigerian Transmission System Incorporating Facts Controllers. International Journal of Applied Science and Technology 1: 186-200.
3. Biswas MM, Das KK (2011) Voltage level Improvement by Using Static Var Compensator (SVC), GJRE J 11: 12-18.
4. Staniulis R (2001) Reactive Power Valuation, Lund University.
5. Rogers KM, Overbye TJ (2008) Some Applications of Distributed AC Transmission System (D-FACTS) devices in power systems.
6. Kiran IK, Laximi JA (2011) Shunt Versus Series Compensation in the improvement of power system performance, Int J Applied Eng Res, Dindigul.
7. Gruenbum R, Rasmussen J (2010) Series capacitors for increased power transmission capability of a 500 Kv Grid intertile, ABB AB, Vasteras, Sweden.
8. Saadat H (1999) Power system Analysis, McGraw Hill Singapore.
9. Nagrath DP (2008) Power System Engineering, Tata McGraw Hill publishing company Ltd, New Delhi India.
10. Deng Y (2010) Reactive Power compensation of Transmission Lines, Concordia University.
11. Adebayo IG, Adejumobi IA, Olajire OS (2013) Power Flow Analysis and Voltage Stability Enhancement Using Thyristor Controlled Series Capacitor (TCSC) Facts Controller. IJEAT 2: 100-104.
12. Peter U (2013) Managerial Electrical Maintenance Report, Unpublished technical Report, Electrical Department, General Steel Mill.

Online Tuning of Power System Stabilizers using Fuzzy Logic Network with Fuzzy C-Means Clustering

Hajizade Kanafgorabi M* and Dr. Karami A

Department of Electrical Engineering, Faculty of Engineering, University of Guilan, Rasht, Iran

Abstract

Power system stabilizers (PSS) have been widely used to enhance damping due to the electromechanical low frequency oscillations occurrence in power systems. In this paper, a new method is used for the online tuning of parameters of conventional power system stabilizers (CPSS) using fuzzy logic. Fuzzy logic enables mathematical modeling and computation of some nonlinear parameters of the system, which are usually, derived empirically by utilization of expert knowledge rules. Various literatures has shown that fuzzy logic controller is one of the most useful methods for expert knowledge utilization. This type of controller is adaptive in nature and can be used successfully as a power system stabilizer. The design of fuzzy logic controllers is mainly based on fuzzy rules and input/output membership functions. Simple and efficient clustering algorithms allow data classification in distinct groups using distance and/or similarity functions. In the present paper, the optimum generation of fuzzy rules base using Fuzzy C-means (FCM) clustering technique is used. In fact, data are classified and the number of fuzzy rules which depends on convergence radius is determined. Finally, the performance of proposed FCM controller is compared with that of conventional controller. The active power, reactive power and bus voltages used as inputs to the fuzzy logic network based power system stabilizer and the parameters of the optimum stabilizer , i.e. gain factor as well as time constants of the lead/lag compensator, are the outputs of the proposed system. The design method has been successfully implemented on a single machine power system connected to an infinite bus over various operating conditions.

Keywords: Dynamic stabilizer; Power system stabilizers; online tuning of parameters; Fuzzy C-Means clustering Prediction

Introduction

Power systems are complex and nonlinear. In these systems, electromechanical low frequency oscillations are produced. The electromechanical oscillations between connected synchronous generators are an inherent phenomenon. To overcome these oscillations, power system stabilizer is widely used. Recent research is related to lead/lag compensators [1-3].

Electromechanical low frequency oscillations in transmission networks are an important issue in power systems which its study will contribute largely to stability problem [1,2]. The performance of excitation systems and high gain AVR in terms of transient stability improvement and normal performance of the system is very desirable, but these excitation systems with high gain and fast action also can cause system instability. This type of instability known as low frequency oscillation (LFO) in the range of 0.2-3 Hz reveals negative impact of excitation systems on utilization of a power system. Depending frequency oscillation, these oscillations are classified into three types of local (Local Mode), Inter area (Inter Area Mode) and intra area (Intra Plant Mode) which among them, local mode is considered here. In other words, this type of instability can be harmful to system safety and also can limit the maximum transmittable power by the system [4]. Low frequency oscillations (LFO) occurrence is due to in adequacy of inherent damping of the system. To modify and improve the dynamic stability and enhance the damping of low frequency oscillations, various solutions has been proposed and applied. These solutions include fast acting governor, system topology alteration, modification in synchronous machine design, protection devices exploitation, FACTS devices exploitation, characteristics modification of voltage regulators and excitation systems, and installation of power system stabilizer in production units. These methods aren't economic and since the designed systems for generators are often old, their structural replacement isn't possible. The most economic and efficient way to overcome oscillations problem is providing sufficient damping for rotor oscillations. This is done by effective exploitation of the power system stabilizer. In fact, PSS applies a supplement control signal to the generator excitation system in order to quickly damp the power system oscillations following disturbance. The aim of a power system stabilizer design is to provide generator with additional damping torque in critical oscillating frequencies without influencing synchronizing torque. As mentioned earlier, the input also can be a signal of frequency error, speed error, electric power, and/or a combination of these signals and the output signal of the stabilizer is applied to the generator excitation system.

Today, Most of the generators available in power systems are equipped with Automatic Voltage Regulator (AVR) to automatically regulate the terminal voltage of generator [5]. In power systems study and control, power system stabilizers are used to generate the supplement control signal for the synchronous generator excitation system in order to damp low frequency oscillations. The design method of conventional power system stabilizers (CPSS) is based on the application of compensation theory and lead/lag compensators in the frequency domain. In this method, CPSS parameters are calculated based on the line arized model of power systems considering a system operating point, and then these parameters are assumed constant for all system operating points. In order to enable the conventional power

***Corresponding author:** Hajizade Kanafgorabi M, Electrical Engineering Department, Faculty of Engineering, University of Guilan, Rasht, P.O. Box 3756, Iran, E-mail: M.hajizade87@gmail.com

system stabilizer to provide appropriate damping over a wide range of the system operating points, the stabilizer parameters are recalculated and then retuned with respect to the current system operating point.

In order to improve the performance of power system stabilizers, various design methods has been presented for them, including varying structure PSSs, neural network based PSSs, and fuzzy logic based PSSs [6-12]. The application of fuzzy logic with Fuzzy C-Means clustering (FCM) [13] is of interest in this paper. The methods already presented for this purpose are more based on Multi Layer Perceptron (MLP) Neural Network, Radial Basis Function (RBF), and Fuzzy Logic Network which each of them has problems such as unavailability of the number of hidden layer neurons for MLP, long training time for RBF, and optimal retuning of fuzzy network parameters with computational intelligence methods for Fuzzy Logic Network. The method presented here uses fuzzy logic network with Fuzzy C-Means clustering (FCM) which leads to reduction of the number of data base rules and automatic training and specification of membership parameters at the minimum possible time, causing proper fitting of membership functions to input and output for a given class center [13]. Inputs/outputs has obtained by the design method of conventional lead/lag controllers and are used for training of FCM based fuzzy logic network [13].

In this paper, we have used a single machine system connected to an infinite bus (SMIB) and in order to damp the low frequency oscillations in this system, a PSS is exploited which its parameters tuned online with respect to each system operating point [14-20].

The power systems stability is also affected by these poorly damped oscillations and can lead to the system instability. The paper presents a design PSSs based on μ-controller to enhance power systems stability and improve power transfer capability. MATLAB dynamic model was developed for a power system and lead-lag PSS structure is considered in the model. Damping torque technique is applied to tune the PSS parameters. The results of this technique have been verified by eigen value analysis and time-domain simulations. The optimal sampling time was determined for transferring the s-domain of PSS model to digital (z-domain) model and then it was implemented on μ-controller chip. The peripheral interface controller (PIC) μ-controller type was used and the developed MATLAB model was interfaced with the μ-controller. The simulations results show that the system time responses under different operating conditions are well damped with the designed PSS. Moreover, the proposed PSS based μ-controller is relatively simple and suitable for real-time applications in the future smart power grid where the stabilizing signals to the PSSs will be provided by wide-area measurement signals using the new technology of synchrophasors [20-23].

Single Machine System Connected to an Infinite Bus

In this paper, we use a simplified dynamic model of a power system, i.e. single machine system connected to an infinite bus (SMIB) as shown in Figure 1 [20]. This system includes a synchronous generator with a fast acting excitation system which its rated parameters are given in [20-22]. This single machine system is connected to an infinite bus through an external reactance X_e and external resistance R_e.

Considering typical representations, the dynamic equations governing this system are as:

$$\dot{E}'_q = -\frac{1}{T'_{do}}(E'_q + (X_d - X'_d)I_d - E_{fd}) \quad (1)$$

$$\dot{\delta} = \omega - \omega_s \quad (2)$$

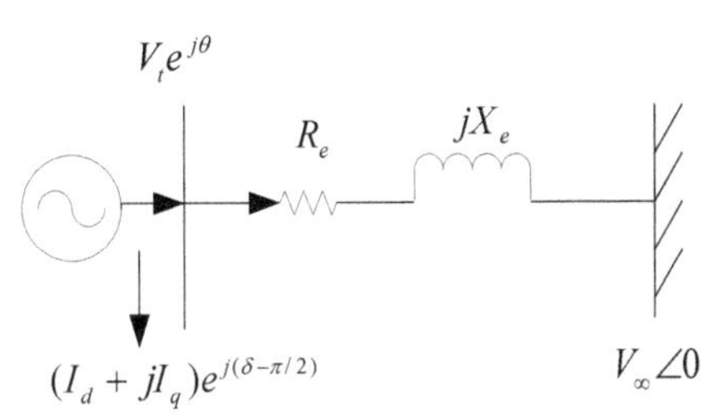

Figure 1: Single machine system connected to an infinite bus.

$$\dot{\omega} = \frac{\omega_s}{2H}[T_M - (E'_q I_q + (X_q - X'_d)I_d I_q + D(\omega - \omega_s))] \quad (3)$$

$$T_A \dot{E}_{fd} = -E_{fd} + K_A(V_{ref} - V_t) \quad (4)$$

Also ignoring the stator resistance of the synchronous machine, the algebraic equations governing stator can be written as:

$$X_q I_q - V_d = 0 \quad (5)$$

$$E'_q - V_q - X'_d I_d = 0 \quad (6)$$

In addition, the equations governing this network can be written as:

$$R_e I_d - X_e I_q = V_d - V_\infty \sin\delta \quad (7)$$

$$X_e I_d + R_e I_q = V_q - V_\infty \cos\delta \quad (8)$$

Since small disturbances are considered in dynamic stability studies, the linearized system equations can be used.

In this paper, we assume that besides the variation in the system operating point, the transmission line reactance X_e also changes. Here the variation of X_e implies the variation in configuration and/or structure of the system. Since X_e is not measurable in practice, it cannot be used as an input to fuzzy logic networks. As we know, the variation in transmission line reactance actually makes the reactive power production of generator (Q) change so that X_a can be substituted by Q. Therefore, we assume that in this system, the active power production of generator (P), generator terminal voltage (V_t), and even transmission line reactance (X_e) are variable and the other system parameters including the amplitude of infinite bus voltage are constant. Now ignoring the transmission line resistance (, i.e. R_e=0), the following equation can be written from Figure 1:

$$V_\infty \angle 0 = V_t - jX_e\left(\frac{P - jQ}{V_t^*}\right) \quad (9)$$

Using eq. (9) and considering the definition, $V_t = V_{td} + jV_{tq}$ we get:

$$V_\infty(V_{td} - jV_{tq}) = V_t^2 - jX_e(P - jQ) \quad (10)$$

Separating real and imaginary parts of eq. (10) and considering $V_{tq} = \sqrt{V_t^2 - V_{td}^2}$, gives two following equations:

$$V_\infty V_{td} = V_t^2 - X_e Q \quad (11)$$

$$V_\infty \sqrt{V_t^2 - V_{td}^2} = X_e P, \; V_{tq} = \sqrt{V_t^2 - V_{td}^2} \quad (12)$$

Now with the knowledge of V_t (terminal voltage) and P (active power) as well as the infinite bus voltage and reactance X_e, we only have two unknowns in eqs. (11) and (12) which are Q and V_{td}. Equations.

(11) and (12) are nonlinear equations which their solution involves using a mathematical method for the calculation of Q and V_{dt}. We have used Broyden method [21] for solving above equations. The important feature of Broyden method is that it doesn't require Jacobian matrix. In fact, it can be said that with the variation in system operating points, i.e. P, V_t, and the transmission line reactance X_e, the active power production of generator Q can be calculated from (11) and (12), and then P, Q and V_t are used as inputs to fuzzy logic network. The only remaining issue is that in eqs. (1) to (8), two variables, δ and θ, is needed which their computation is as follows.

Ignoring the stator resistance as well as the transmission line reactance, the active power production of generator is given by:

$$P = \text{Real}\{V_t I^*\} = \text{Real}\left\{V_t e^{j\theta} \times \left(\frac{V_t e^{j\theta} - V_\infty}{jX_e}\right)^*\right\} \tag{13}$$

Solving the equation above, we get:

$$P = \frac{V_\infty V_t}{X_e}\sin\theta \tag{14}$$

Therefore, with the knowledge of P, V_t, X_e and also the infinite bus voltage, the angle of generator terminal voltage θ can be simply obtained by using eq. (14). Moreover, the angle of synchronous machine internal voltage δ can be calculated as follows:

$$\delta = \angle\{V_t e^{j\theta} + jX_q I\} \tag{15}$$

Where, $I = \frac{V_t e^{j\theta} - V_\infty}{jX_e}$

Conventional Power System Stabilizers Design

As we know, currently the usage of conventional power system stabilizers (CPSS) is the most economic way for enhancing the power systems damping due to the low frequency oscillations occurrence. This stabilizer actually creates a damping torque in phase with generator speed variations. The CPSS input can be the speed deviation of a generator from synchronous speed, frequency variations of the system and/or accelerating torque designed by the phase compensation method [20-22]. The block diagram of such a PSS along with the linearized equations of a single machine system connected to an infinite bus presented in the previous section is given in Figure 2.

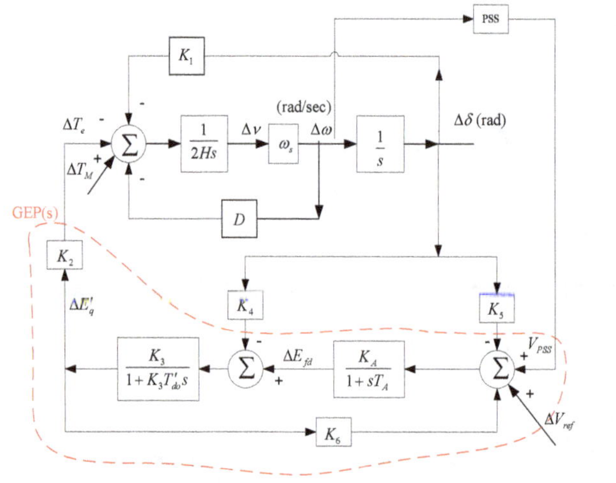

Figure 2: Linearized model of a single machine system.

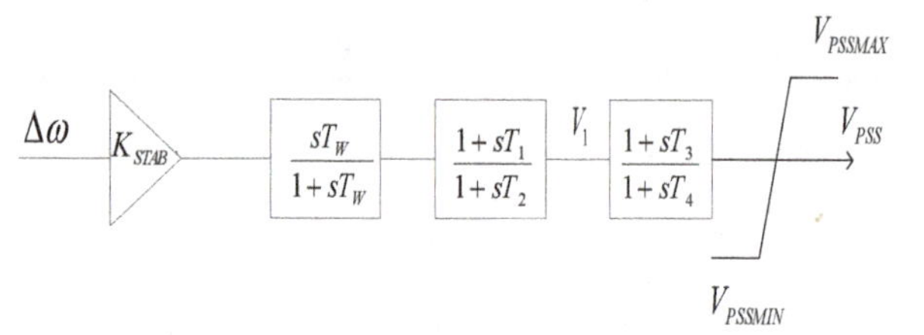

Figure 3: Block diagram of a conventional power system stabilizer.

where $= \frac{\omega}{\omega_s}$

The block diagram of a CPSS involving a two stages lead/lag compensator block, a washout block and an amplification factor K_{STAB} is shown in Figure 3.

The aim of lead/lag blocks in the model of a CPSS is cancellation of lagging caused by blocks following PSS which have been shown in Figure 2 with dashed lines as the transfer function GEP(s). Also it can be shown that ignoring constant K_4, the transfer function GEP(s) is calculated as follows [20-22]:

$$GEP(s) = \frac{K_2 K_A K_3}{K_A K_3 K_6 + (1 + sT'_{do}K_3)(1 + sT_A)} \tag{16}$$

As mentioned before, the aim of lead/lag blocks in a CPSS is cancellation of lagging caused by the transfer function GEP(s) during the low frequency oscillations of a synchronous machine. Ignoring the inherent damping of synchronous machine which implies D=0 in the machine model, the frequency of these oscillations is [20-22]:

$$\omega_n = \sqrt{\frac{K_1\omega_s}{2H}} \tag{17}$$

Now if we denote the created phase angle by a block GEP(s) of frequency $s=j\omega_n$ with β, then considering a lead/lag block with m similar stages and also assuming time constants T_1 and T_2 for these blocks, these constants are given by:

$$\beta = \angle GEP(j\omega_n)$$

$$\alpha = \frac{1-\sin(\beta/\text{m})}{1+\sin(\beta/\text{m})}$$

$$T_1 = \frac{1}{\omega_n\sqrt{\alpha}}$$

$$T_2 = \alpha T_1 \tag{18}$$

As said, here we have used a two stages lead/lag block (m=2) in a CPSS. Now assuming a value for the damping coefficient ξ, the stabilizer gain factor (K_{STAB}) can be calculated as follows [20-22]:

$$K_{STAB} = \frac{4\xi\omega_n H}{|GEP(j\omega_n)||G_1(j\omega_n)|} \tag{19}$$

Where $G_1(s)$ is the transfer function of a m stages lead/lag block which can be represented by:

$$G_1(s) = \left(\frac{1+sT_1}{1+sT_2}\right)^{\text{m}} \tag{20}$$

Also, we have assumed the value of ξ=0.7 for all system operating points. Finally, considering the CPSS block involving a gain K_{STAB} and a two stages lead/lag block corresponding to T_1 and T_2 time constants, the linearized equations of the system can be written as the following state equations from Figure 2 (because of papers limitation, we ignore the computational details):

$$\dot{x} = Ax + B_1 \Delta V_{ref} + B_2 \Delta T_M \qquad (21)$$
$$y = Cx$$

where

$$x = \begin{bmatrix} \Delta E'_q & \Delta\delta & \Delta v & \Delta E_{fd} & V_1 & V_{PSS} \end{bmatrix}^t$$

$$y = \begin{bmatrix} \Delta V_t & \Delta v & \Delta P_e \end{bmatrix}^t$$

in the expressions above, V_1 and V_{PSS} are two additional state variables which created by the CPSS block in the system and have been shown in Figure 3. Also A, B_1, B_2 and C are as:

$$A = \begin{bmatrix} \frac{-1}{K_3 T'_{do}} & \frac{-K_4}{T'_{do}} & 0 & \frac{1}{T'_{do}} & 0 & 0 \\ 0 & 0 & \omega_s & 0 & 0 & 0 \\ \frac{-K_2}{2H} & \frac{-K_1}{2H} & -\frac{D\omega_s}{2H} & 0 & 0 & 0 \\ \frac{-K_A K_6}{T_A} & -\frac{K_A K_5}{T_A} & 0 & \frac{-1}{T_A} & \frac{K_A}{T_A} & 0 \\ -\frac{T_1 K_2}{2HT_2} & -\frac{T_1 K_1}{2HT_2} & \frac{1}{T_2} & 0 & -\frac{1}{T_2} & 0 \\ -\frac{K_{STAB} K_2 T_1^2}{2HT_2^2} & -\frac{K_{STAB} K_1 T_1^2}{2HT_2^2} & \frac{K_{STAB} T_1}{T_2^2} & 0 & \frac{K_{STAB}}{T_2}(1-\frac{T_1}{T_2}) & -\frac{1}{T_2} \end{bmatrix}$$

$$B_1 = [0 \quad 0 \quad 0 \quad \frac{K_A}{T_A} \quad 0 \quad 0]^t$$

$$B_2 = [0 \quad 0 \quad \frac{1}{2H} \quad 0 \quad 0 \quad 0]^t$$

$$C = \begin{bmatrix} K_6 & K_5 & 0 & 0 & 0 & 0 \\ 0 & 0 & 1.0 & 0 & 0 & 0 \\ K_2 & K_1 & 0 & 0 & 0 & 0 \end{bmatrix}$$

Here we initially want to show which shape does the root locus of the system take with and without PSS (Figures 4 and 5).

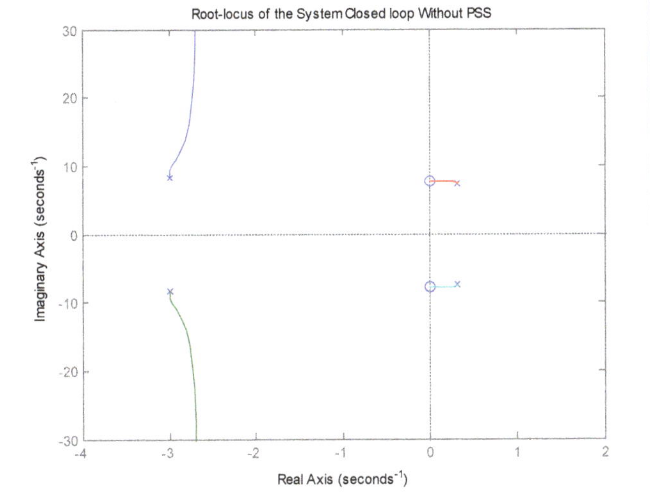

Figure 4: Root locus of the system without stabilizer.

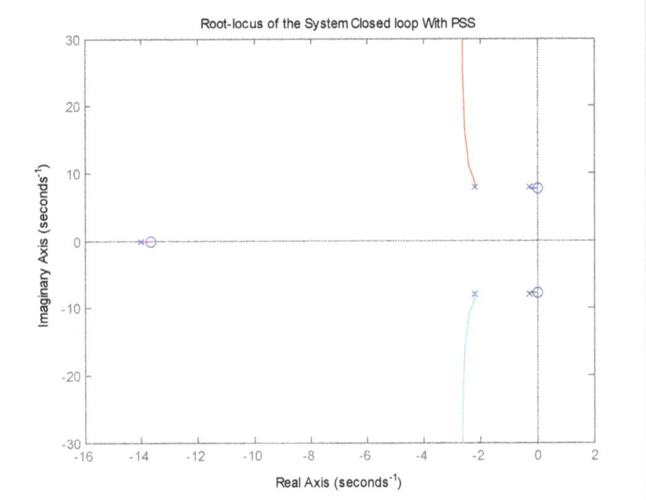

Figure 5: Root locus of the system with sta.

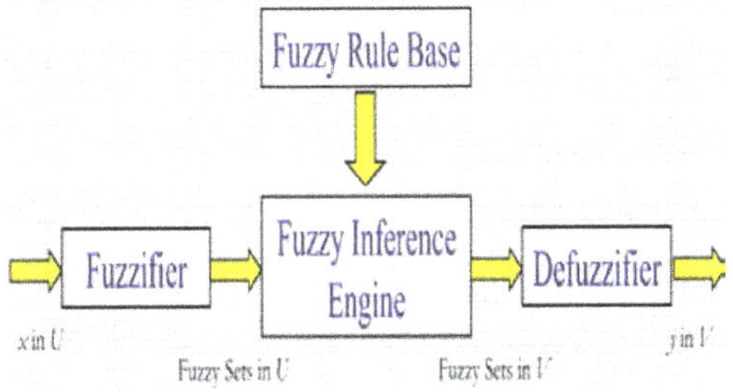

Figure 6: General form of a fuzzy system.

Fuzzy Logic Network, Data Base rules and the Usage Method

Fuzzy logic controller structure

The conventional stabilizer is tuned for one operating point, but using fuzzy logic network and training with clustering method, it can be used for a wide range of the system operating points. Fuzzy systems are nonlinear and based on human knowledge. The core of these systems is a rules base which composes of fuzzy IF-THEN rules. A fuzzy IF-THEN rule is an if-then expression which some of its words specify by means of the fuzzy membership functions. As said, utilizing human experience in the form of formula is difficult. Fuzzy logic provides a simple means for such an application. As shown in Figure 6, the basic structure of a fuzzy logic controller includes three following parts [19].

1. Fuzzifiers- The inputs are read or measured and then we change the measured values to the form corresponding to the linguistic variable (the values proportional to membership function values).

2. Fuzzy inference engine-includes rules that relate the input membership functions to the output ones.

3. defuzzifiers- includes functions which transform the fuzzy output in the form of membership functions to a specified and acceptable point for the general output of the system [1,4,7,18].

There are three kinds of fuzzy systems:

1. Pure fuzzy system
2. Takagi-Sugeno-Kang fuzzy systems
3. Systems with fuzzifiers and defuzzifiers

The basic structure of fuzzy inference systems is a model which maps input properties to input membership functions, input membership functions to rules, rules to a set of output properties, output properties to output membership functions and ultimately output membership functions to a unique output value and/or a decision. Fuzzy inference is only used for modeling systems which their related rules were already determined by your interpretation of application. Figure. 6 represent the simple model of a fuzzy system. Sometimes in system modeling you cannot specify membership function by means of data investigation. Although the parameters associated with membership function can be arbitrarily determined, it is to be noted that the appropriate selection of these parameters would have a substantial effect on the system performance. In such cases, the design techniques of fuzzy systems can be used which one of them is the clustering method where an optimal tuning of the membership functions' parameters is achieved by finding the class center and designating the membership functions appropriate to the class center [1,4,7].

Types of fuzzy system design methods are:

1. First order estimator
2. Second order estimator
3. Look-up table
4. Gradient Descent
5. Clustering method
6. Fuzzy C-means clustering analysis

Explanation of the final method, i.e. FCM clustering analysis, is the aim of this paper [13,14,18].

Fuzzy C-Means (FCM) clustering analysis

There isn't a systematic procedure for specifying the number of fuzzy system rules in the previous methods. In the gradient descent method, the number of rules was fixed before training. In the table and estimator methods, the number of fuzzy sets is determined and then this number initially specifies the number of rules (Figure 7).

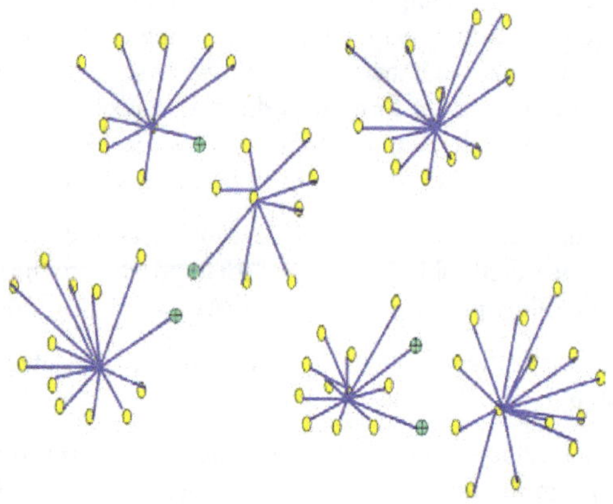

Figure 7: Clustering and centers mean.

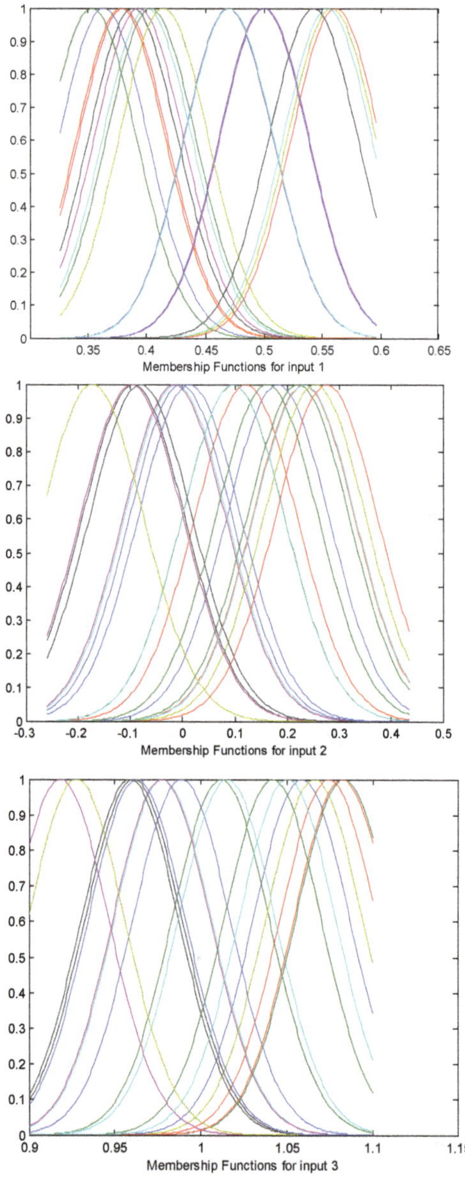

Figure 8: Trained membership functions with Fuzzy C-Means clustering method.

Clustering method includes an algorithm for selecting the number of rules [5]. The input/output data sets divides into some clusters and a rule is expressed for each cluster.

First, a proper algorithm is selected for the limited number of input/output pairs, and then the nearest neighbour method is applied for clusters selection, and finally the optimum fuzzy system is designed for data fitting [13].

Assume an input/output pair $(\chi_0^k, y_0^k), 1 < l < N$ where N is small. Our task is developing a fuzzy system that is able to match all N input/output pairs with an arbitrary given accuracy. This requires that $\left|f(\chi_0^{'}) - y_0^l\right| < \varepsilon, \forall \xi > 0$.

To summarize, the training steps are given in the appendix. And now we assume that the trained fuzzy logic network of interest is ready

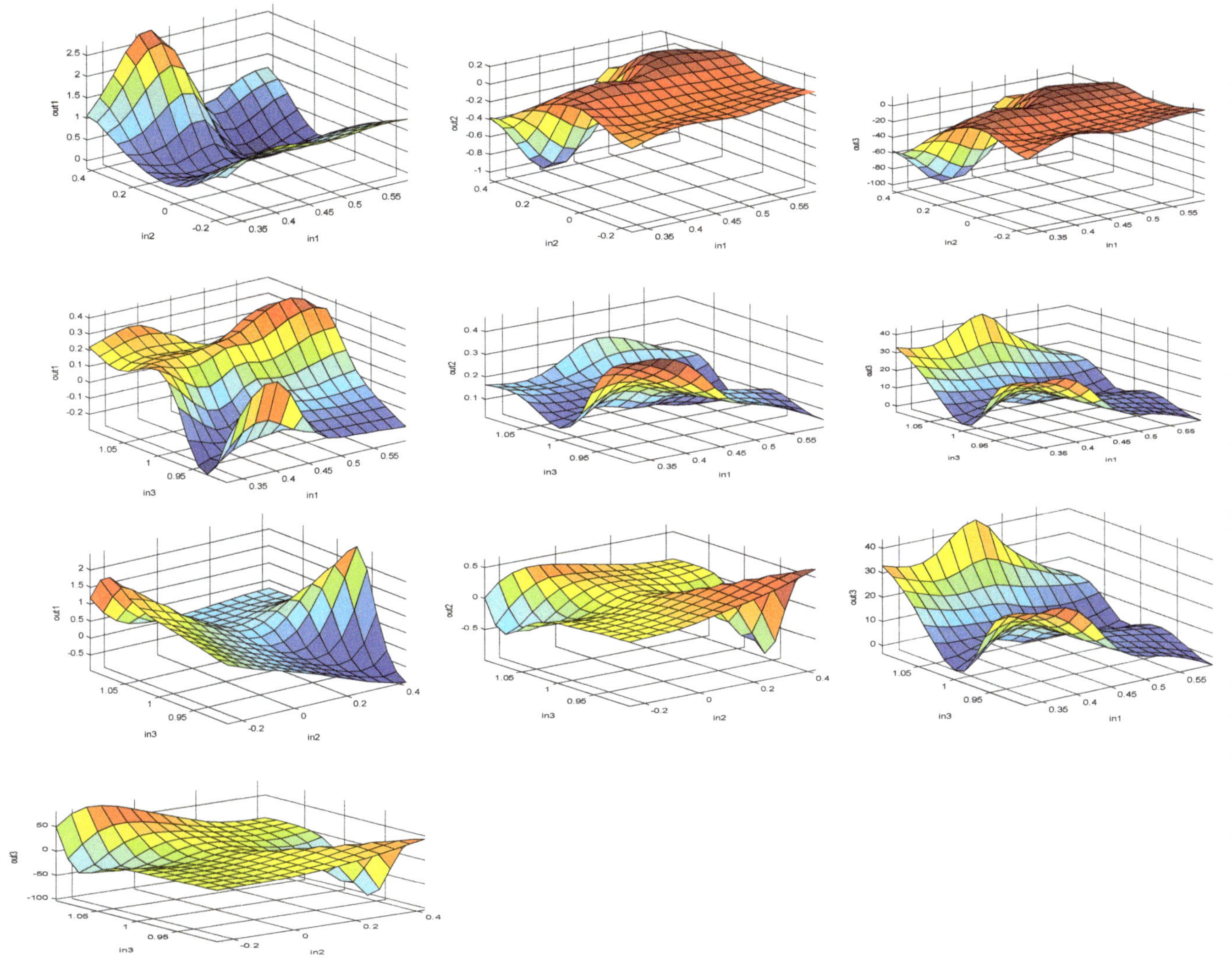

Figure 9: Decision-making borders.

to use (Figure 9).

The shape of input membership functions after training is shown in Figure 8.

Simulation Results

In order to obtain the training data for fuzzy logic network, we assumed that the active power production of generator (P) is 0.5 to 1.1 times the rated active power production of generator. We also assumed that the amplitude of generator voltage is 0.95 to 1.05 times the nominal terminal voltage. The transmission line reactance was assumed constant. To produce training vectors, each time a random number with uniform distribution is independently selected for P and V_t in the range of associated variations and then solving eqs. (11) and (12) presented in section 2 gives the amount of reactive power production of generator (Q). Finally, we calculate the optimum values of stabilizer gain (K_{STAB}) and time constants T_1 and T_2 with the phase compensation method. Here we have used a stabilizer with a two stages lead/lag block for the random operating points. Therefore, a training pattern (vector) for fuzzy logic network actually includes P, X_e=cte, and V as inputs and K_{STAB}, T_1 and T_2 as outputs.

Using the method explained above, we generated 2000 training patterns and trained the fuzzy logic network with FCM clustering method. After some simulations and training, we investigated the network performance considering a neighboring radius of 0.5 for Gaussian membership functions. Test results show that Root Mean-squared Error (RMSE) for 3 desirable outputs, K_{STAB}, T_1 and T_2 and the values obtained by trained fuzzy logic network is 0.7278,0.0138, and 0.0053, respectively which implies the ability of trained fuzzy logic network in good estimation of the PSS parameters. Estimations of the fuzzy logic network for 50 test data are shown in Figure 10.

As said earlier, one of the main problems with conventional power system stabilizers is that they are only tuned for only one system operating point. The optimum performance of these stabilizers requires retuning of these parameters as the system operating points change. To investigate this and also the advantages of online tuning of CPSS parameters using fuzzy logic network, we studied the performance of a single machine system connected to an infinite bus with applying two types of disturbance to the system. For the first disturbance, the variations in the generator terminal voltages (V_t) are plotted for a 0.1 p.u. step change in the generator reference excitation voltage (V_{ref}). For

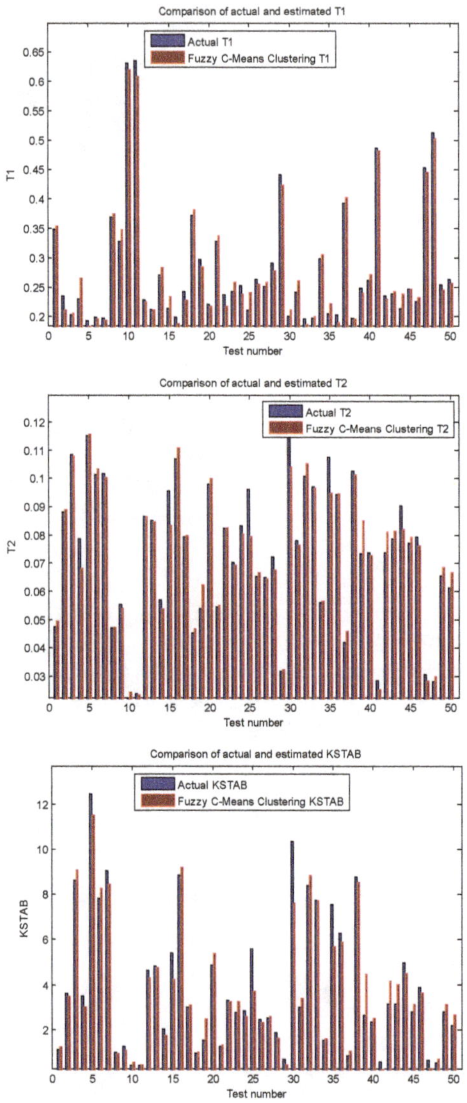

Figure 10: Estimations of fuzzy logic network for 50 test datas.

the second disturbance, the variations in generator speed (ω) for a 0.5 p.u. step change in the machine input mechanical power is plotted. This is done for two operating points and we considered these three situations:

1. Single machine system without PSS
2. Single machine system with constant parameters PSS
3. Single machine system with PSS whose parameters retuned online for each operating point of the trained fuzzy logic network

The comparison for the first operating point is done among all 3 situations and for the second one, it is only done between (b) and (c). In the case of constant parameters PSS, the values of PSS parameters for both operating points were obtained considering the nominal values of the single machine system parameters and then the obtained values assumed constant for all system operating points. For the first operating point, the diagrams of the generator terminal voltage variations and generator speed variations are shown in Figures 11 and 12, respectively. For the second operating point, these diagrams are shown in Figures 13 and 14, respectively. These diagrams are associated with a random system operating point which is much different from the nominal operating point. As can be seen from these two figures, PSS with parameters tuned by FCM practically achieved damping of created oscillations, and also the system without PSS was unsuccessful in damping these oscillations and hence the system has become unstable.

Operating points are as:

- Operating point-1:

Active Power (P)= 0.6;

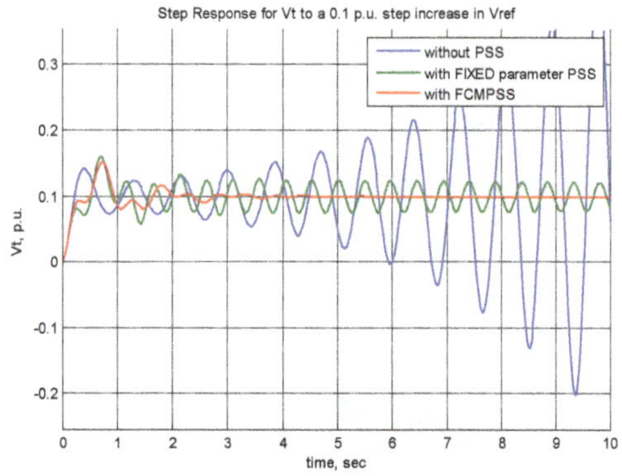

Figure 11: Diagram of generator terminal voltage variations.

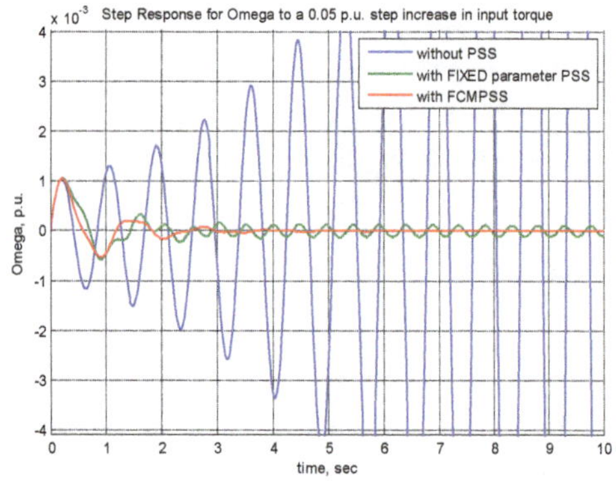

Figure 12: Diagram of generator speed variations.

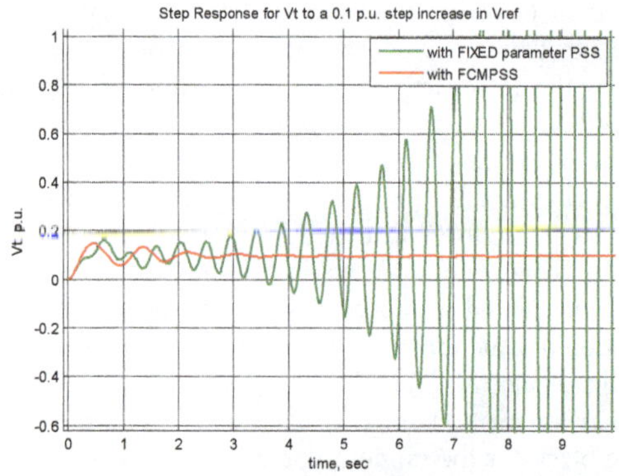

Figure 13: Diagram of generator terminal voltage variations.

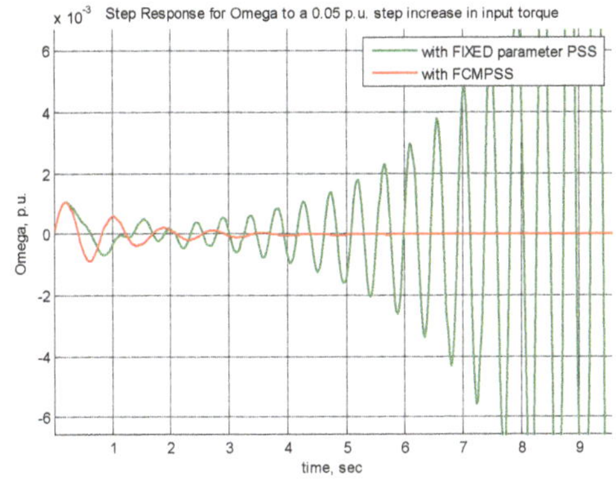

Figure 14: Diagram of generator speed variations.

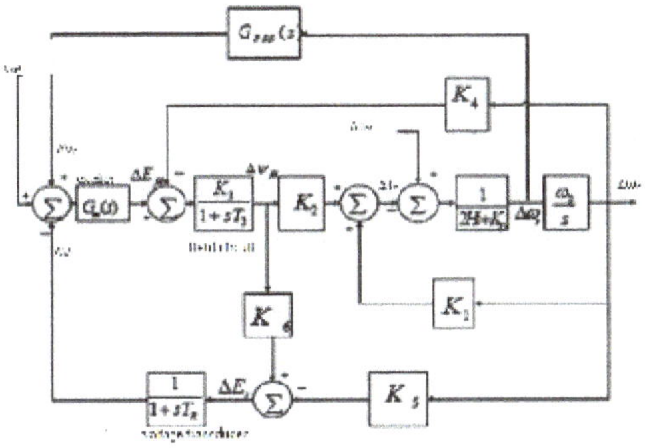

Figure 15: Power distributer.

Type	RMSE
Fuzzy C-Means	0.0053
MLP	0.5168
RBF	0.7183

Table 1: The comparison of stabilizers for T_2.

- Operating point-2: with 0.4 p.u. change in the transmitted power and increasing it to 1 p.u.

Conclusion

In this paper, an effective method as been presented for the online tuning of parameters of conventional power system stabilizers (CPSS) using fuzzy logic network (Fuzzy C-Means Clustering). The inputs are quantities like the active power production which are directly measurable, and hence after training network, the optimum stabilizer values corresponding to the current system operating point can be obtained. The required equations for designing CPSS with the phase compensation method have been completely explained. Considering some disturbances, the performance of the dynamic system equipped with PSS with parameters tuned by fuzzy logic network (Fuzzy C-means) was studied. The simulation results verified good performance of the designed PSS. Also FCM based PSS is able to create good damping over a wide range of the system operating points. More damping means the generator can be utilized stably at its maximum capacity, leading to economic saving of cost. Table 1 shows a comparison of stabilizers based on three networks, i.e. MLP, RBF and Fuzzy C-means.

In this paper, a hybrid approach for tuning and placement of power system stabilizers (PSS) in multi-machine power systems is provided with the aim of reducing low-frequency oscillations (LFO) and improving power system dynamic stability with wide range of changes in system parameters and operating point of the system. PSS parameters are adjusted using Particle Swarm Optimization algorithm (PSO), and using Takagi-Sugeno (TS) fuzzy, optimal location for the PSS is determined. The employed fuzzy system has two inputs, the real part and the damping coefficient of network eigen values. The results of test on a grid four machine-two area sample, show optimal performance of the proposed method in improving system stability and reducing low-frequency oscillations of local and inter-area modes Figure 15.

It is clear from Table 1 that the neural network estimation has more error. Moreover, training time with clustering method was very low within 1 to 2 s, but training neural network has done in 39 iterations in a very large time.

Appendix

Case Study of the System:

XE=0.5, V=1, Tw=2, f=60, Tdo=9.6, xd=2.5, xpd=0.39, xq=2.1, H=3.2,

KA=400, TA=0.2, Ra=0, Re=0, Eb=1.05

The steps of Fuzzy C-Means network training are as follows.

Step 1: starting from input/output (χ_0^1, y_0^1), we create a cluster with the center χ_c^1 at χ_0^1 and we put:

$$A^1(1) = y_0^1 \qquad B^1(1) = 1$$

Now we select a value for the radius, r.

Step 2: introducing the *k*th input/output pair (χ_0^k, y_0^k), we have M clusters with centers χ_c^1 .... χ_c^m. We calculate the distance of χ_0^k from the center of these M clusters as:

$$|\chi_0^k - \chi_c^l| \qquad 1 < l < M$$

and we take the lowest distance, i.e. $|\chi_0^k - \chi_c^{l_k}|$.

a) if $|\chi_0^k - \chi_c^{l_k}| > r$, then χ_0^k is a new class center and $\chi_c^{m+1} = \chi_0^k$. We put:

$$A^{m+1}(k) = y_0^k \;,\; B^{m+1}(k) = 1$$

$$A^l(k) = A^l(k-1) \qquad B^l(k) = B^l(k-1)$$

b) if $|\chi_0^k - \chi_c^{l_k}| \le r$, then

$$A^{l_k}(k) = A^{l_k}(k-1) + y_0^k$$

$$B^{l_k}(k) = B^{l_k}(k-1) + 1$$

$$A^l(k) = A^l(k-1)$$

c) $B^l(k) = B^l(k-1)$

If χ_0^k doesn't create a new cluster, then according to k input/output pairs (χ_0^l, y_0^l), we get:

$$f(x) = \frac{\sum_{l=1}^{m} A^l(k)\exp(-(\frac{\chi_i - \bar{X}_i^l}{\sigma})^2)}{\sum_{l=1}^{m} B^l(k)\exp(-(\frac{\chi_i - \bar{X}_i^l}{\sigma})^2)}$$

And if χ_0^k is a new cluster, M is substituted by M+1 in the new formula.

Step 3: step 2 with k=k+1

In the above relations, $B^{l}(k)$ is the number of input/output pairs in the lth cluster and $A^{l}(k)$ is the sum of output values of its input/output pairs. The radius r determines the degree of complexity of the designed fuzzy system.

Since $A^{l}(k)$ and $B^{l}(k)$ are calculated using recursive equations, a forgetfulness factor can be simply used which is useful for modeling systems with variable structures. Using this factor, we get:

$$A^{l_k}(k)=\frac{\tau-1}{\tau}A^{l_k}(k-1)+\frac{1}{\tau}y_0^k \quad b_1$$

$$B^{l_k}(k)=\frac{\tau-1}{\tau}B^{l_k}(k-1)+\frac{1}{\tau} \quad b_2$$

$$A^{l_k}(k)=\frac{\tau-1}{\tau}A^{l_k}(k-1) \quad b_3$$

$$B^{l_k}(k)=\frac{\tau-1}{\tau}B^{l_k}(k-1) \quad b_4$$

Where τ is time constant of the decreasing exponential function. In practice, $B^{l}(k)$ must have a threshold so that this cluster eliminates if $B^{l}(k)$ goes below this threshold.

References

1. Radaideh SM, Nejdawi IM, Mushtaha MH (2012) "Design of power system stabilizers using two level fuzzy and adaptive neuro-fuzzy inference systems ", Electrical Power and Energy Systems 35: 47-56.
2. Hardiansyah, Furuye S, Irisawa J (2006) A robust power system stabilizer design using reduced-order models. Electr Power Energy System 28: 21-8.
3. Tse CT, Tso SK (1993) Refinement of conventional PSS design in multimachine system by modal analysis. IEEE Trans Power Syst 8(2).
4. Hussein T, Saad MS, Elshafei AL, Bahgat A (2009) ,"Robust adaptive fuzzy logic power system stabilizer", Expert Systems with Applications 36: 12104-12112.
5. Larsen EV, Swann DA (1981) "Applying power system stabilizers, Part I, II, III", IEEE Transaction on Power Apparatus and Systems (PAS) 100: 3017-3041.
6. Lee SS, Park JK (1998) "Design of reduced-order observer-based variable structure power system stabilizer for unmeasurable state variables", in: IEE Proceedings of the Generation, Transmission and Distribution 145: 525-530.
7. El-Metwally KA, Hancock GC, Malik OP (1996) "Implementation of a fuzzy logic PSS using a micro-controller and experimental test results", IEEE Transaction on Energy Conversion 11: 91-96.
8. Hsu YY, Chen CL (1991) "Tuning of power system stabilizers using an artificial neural network", IEEE Transaction on Energy Conversion 6: 612-619.
9. Park Y, Choi M, Lee KY (1996) "A neural network-based power system stabilizer using power flow characteristics", IEEE Transactions on Energy Conversion 11: 435-441.
10. Zhang Y, Malik OP, Chen GP (1995) "Artificial neural network power system stabilizers in multi-machine power systemenvironment", IEEE Transactions on Energy Conversion 10: 147-155.
11. Changaroon B, Srivastava SC, Thukaram D (2000) "A neural network based power system stabilizer suitable for on-line training—a practical case study for EGAT system", IEEE Transactions on Energy Conversion 15: 103-109.
12. Bouchama Z, Harmas MN (2012) "Optimal robust adaptive fuzzy synergetic power system stabilizer design" 83: 170-175.
13. Sudha KR, ButchiRaju Y, Sekhar AC (2012) "Fuzzy C-Means clustering for robust decentralized load frequency controlof interconnected power system with Generation Rate Constraint" ,Electrical Power and Energy Systems 37: 58-66
14. Zadeh NH, Kalam A (2002) "An indirect adaptive fuzzy-logic power system stabilizer "Electrical Power and Energy Systems. 24: 837-842
15. Talaat HEA, Abdennour A, Al-Sulaiman AA (2010) "Design and experimental investigation of a decentralized GA-optimized neuro-fuzzy power system stabilizer" Electrical Power and Energy Systems 32: 751-759.
16. Bouchama Z, Harmas MN (2012) "Optimal robust adaptive fuzzy synergetic power system stabilizer design", Electric Power Systems Research 83: 170-175.
17. Sambariya DK, Prasad R (2012) "Robust Power System Stabilizer Design for Single Machine Infinite Bus System with Different Membership Functions for Fuzzy Logic Controller", IEEE.
18. Corcau JI, Stoenescu E (2007) "Fuzzy logic controller as a power system stabilizer ",Int J Circuits, Systems and Signal Processing 1: 266-273.
19. Dr. Kumar J, Kumar PP, Mahesh A, Shrivastava A (2011) Power System Stabilizer Based On Artificial Neural Network, Department of Electrical Engineering PEC University of Technology, Chandigarh, IEEE.
20. Sauer PW, Pai MA (1998) Power System Dynamics and Stability, Prentice-Hall, Inc., New Jersey.
21. Lindfield G, Penny J (1995) Numerical Methods using MATLAB, Ellis Horwood Limited.
22. Anderson PM, Foaud AA (1977) Power System Control and Stability, Ames: Iowa State Univ. Press.
23. Padiyar KR, Power System Dynamics.

Optimization and Cost Benefit Analysis of a Large PV Installation in Delhi

Shaheen H* and Nawaz I

Department of Mechanical Engineering, Jamia Millia Islamia, Delhi, India

Abstract

Acute power crisis in Delhi is stress on the people. Delhi dwellers suffer many problems due to power crisis. Dwellers spent a dark summer in Delhi. Apart from these problem the temperature in summer steadily rising to make their life harder. Presently Delhi has multiple problems, one is already discuss (acute electricity) and second that is storm, more than 10 tower and lines in different parts of the capital got damages. According to centre for science and environment (CSE) says that the domestic sector is the biggest guzzler of electricity in Delhi.

According to newly released report of the Central Electricity Authority (CEA) on load Generation Balance Report 2015-16, Delhi consumes more electricity than the states of others. The household electricity consumption per capita is about 43 units per month against a national average of 25 CEA project Delhi, speak will cross 6,300 MW this year and 12,000 MW by 2012.

While the Delhi govt. plan to reduce its dependence on the NTPC by producing its own power. There are an alternative sources of power generation in short, mid and long term basis. If solar panels be the alternative short term power source, then per unit production cost is a huge barrier in making this technology popular among the mass. But when the people installs roof top, ground solar power plant the cost of installation initially is high. The installation of roof top to get a new electricity connection power utilities. In such a situation, a solar energy rate has been proposed that will cover the solar electricity production cost, alternately called the cost recovery scheme. This rate is near about the maximum per unit rate of energy the government purchases from quick rental plants. Also, different types of solar modules have been compared. This can be a technical support for the city dwellers to who wants to buy solar panels.

Keywords: Solar panel; Battery efficiency; Economic impact; Emission; Environmentally friendly; Roof top photovoltaic; Tilt angle

Introduction

Delhi is situated in northern India with a land area of 1,484 with a population of 25,753,235. The second most populous city and second most populous urban agglomeration in India. The actual power requirement of Delhi is 29,231 MU but the availability of energy in Delhi 29,106 MU. So the deficit of energy is 0.4% [1,2].

Delhi is a state which suffering from significant energy poverty and pervasive electricity deficit. In recent years, Delhi's energy consumption has been increasing at a relative fast rate due to population growth, Metro and economic development. As per the estimates made in the Integrated Energy Policy Report of Planning Commission of India, 2006, if the country is to progress on the path of this sustained GDP growth rate during the next 25 years, it would imply quadrupling of its energy needs over 2003-04 levels with a six-fold increase in the requirement of electricity and a quadrupling in the requirement of crude oil. Delhi at this time suffering many problem such as transportation, pollution, water, residential problem and crime problem etc., (Table 1).

As per Figure 1, Delhi subjected to acute power so for this situation solar PV system is a better option to produce electricity for Delhi dwellers. In this research work, several proposals have been made to keep the PV energy price affordable to the city dwellers [3].

Technical as well as financial have been done on the installation of solar system on the roof-top of high rise building to produce electricity for different type of solar modules and comparing their result, the best one from them have been proposed. The production/installation of solar are high so that we have proposed a solar system that will cover a production cost of solar electricity. Although green energy is costlier compared to utility energy rate, it is not so compared to a small diesel generator frequently used during the time of load shedding [4-6].

Solar Power Calculation

A building has been selected in a highly dense populated area of Delhi city and a solar system is designed to be installed on the unused roof top and parking place. The total number of module used is calculated. The average output and price of the electricity generation by the solar is calculated [7,8].

Site area calculation

Available space area is calculated: The area available for installed a solar system has been calculated as follows:

Available open space for system = The area of stair rooms + Machine room + Meeting room +Parking place +Void area + South wall

Loss area = 5% of available open space, so that area can be calculated for installing solar panel = Available open space - Loose area

Our calculated area was 2787.1 m^2 and effective area of installation is 2,648 m^2.

Tilt angle calculation

When we adjust the tilt angle in such a way that it get the highest amount of solar radiation. So that panel efficiency is optimize. We can

***Corresponding author:** Shaheen H, Department of Mechanical Engineering, Jamia Millia Islamia, Delhi, India, E-mail: Shasan39@gmail.com

Delhi	Apr-14	May-14	Jun-14	Jul-14	Aug-14	Sep-14	Oct-14	Nov-14	Dec-14	Jan-15	Feb-15	Mar-15	2014-15
Peak Demand(MW)	4418	5358	5533	6006	5589	4882	4570	3408	4271	4405	3847	3589	6006
Peak Availability(MW)	4418	5338	5533	5925	5507	4882	4570	3408	4271	4405	3847	3589	5925
Surplus(+)/Deficit(-) (MW)	0	-20	0	-81	-82	0	0	0	0	0	0	0	-81
(%)	0.0	-0.4	0.0	-1.3	-1.5	0.0	-1.8	0.0	-0.1	0.0	0.0	0.0	-1.3

Table 1: Month-wise power supply position of Delhi during the year 2014-15(in terms of peak demand).

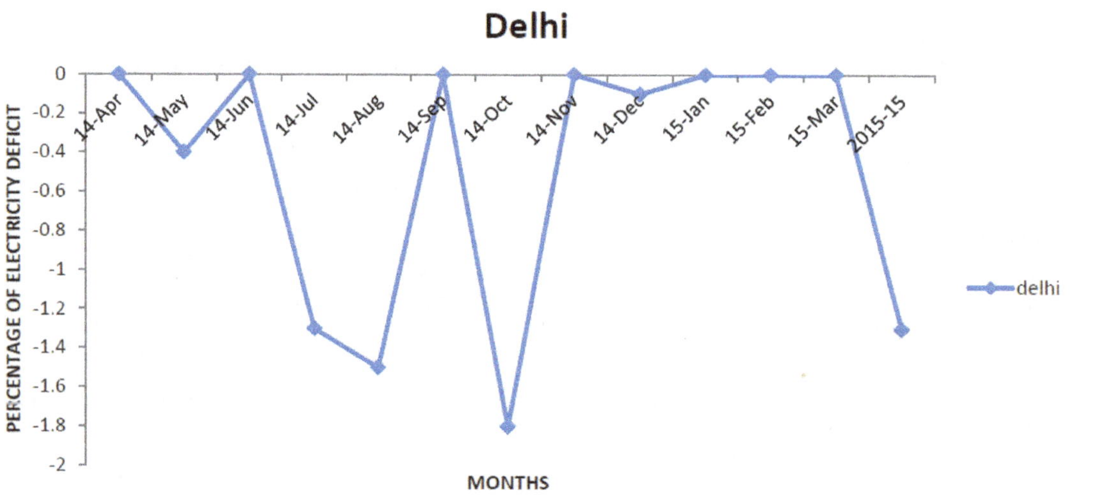

Figure 1: Shows Electricity Reduction in Delhi.

adjust the tilt angle as per requirement or each seasonal. But is a very complicated and costly process. So to reduce this problem we adjust the fix angle to get most energy the whole year. Formula [2] can be used for calculating the tilt angle,

$$\beta = 0.76 \times \phi + 3.1^{\circ} \quad (1)$$

Where ϕ = geographic latitude (Delhi is situated). So the optimum tilt angle for this region is

$$\beta = 0.76 \times 28.38^{\circ} + 3.1 = 24.67^{\circ}$$

Row distance calculation

One of the boundary conditions for the installation and performance of PV modules is to determine the correct distance between two consecutive arrays. To avoid excessive shadowing, the arrays have to be spaced apart by a distance, d in relation to the module width, a [2]:

$$d/a = \cos\beta + \sin\beta/\tan\varepsilon \quad (2)$$

$$\text{and, } \varepsilon = 90^{\circ} - \delta - \phi \quad (3)$$

Where, ε = shadowing angle, and δ = ecliptic angle = 23.5° [4].

From eq. (1), (2) and (3), ε = 41.8°

Determining the total load of building

Design solar photovoltaic system

SPV design in a 3 steps

Step 1: Load Estimate, power, converter Rating, system voltage decision.

The available area for install the solar system in the building is 2,648 m^2 (Figure 2 and Table 2).

Total load =503.36 Kw

Assume we take 4 hr, energy supplied.

Load in kWh= 503.36 × 4 = 2013.44 kWh

Note: Total number of load does not run continuous, if we take 70% use load. This term 70% is called load factor.

Actual load = 2013.44 kWh × 0.7 = 1409.408 kWh.

Energy supplied to the Inverter every day

The available efficiency of the inverter is in the range of 0.90 to 0.97%. If we choose the inverter efficiency 0.94%, then the energy supplied to the input end of the inverter should be adjust accordingly.

Input energy of the inverter = 1409.408 kWh/0.94

=1499.37 kWh.

System voltage determination: The photo voltaic output voltage is specified as 15-16 V at standard conditions, actual voltage at operationing conditions and after all other voltage drops becomes 12 V. Therefore, we say solar panels are available for 12 V and in multiplies of 12, (12 V, 24 V, 36 V, 48 V etc.,). I study on for all 4 cases of voltage: 12 V, 24 V, 36 V, 48 V.

Steps 2: Sizing the batteries:

How many batteries and what A-hr. capacity, we have to consider battery parameters like,

- Depth of Discharge (DOD)
- Operating voltage of battery and it"s A-hr. capacity
- Number of day of autonomy (how many cloudy days without sunshine you want to consider may be 2,3,4) Normally batteries with

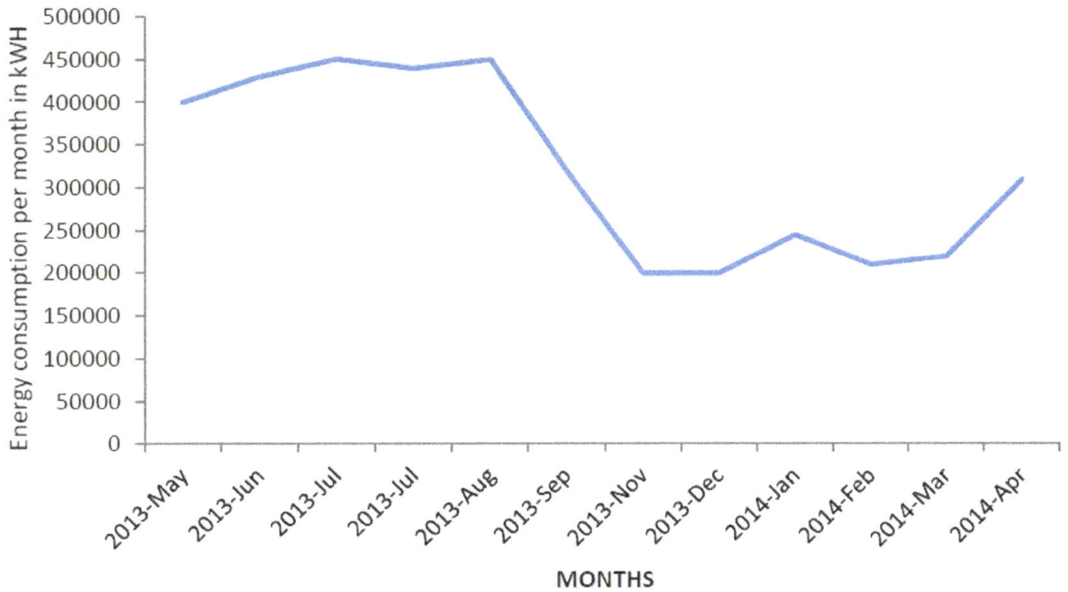

Figure 2: Energy consumption.

No.of equipment	W(Watt)	%
5 PAC (packed air condition)	289.5(TR)	57.5
82 lighting (CFL/LED/Tube light)	40.5	8.04
971 Celling fans	72.833	14.46
50 Chiller	37.5	7.49
50 Water pump	18.75	3.72
20 Exhaust fan	15	2.97
20 Fridge (310 Ltrs)	7.98	1.58
23 Fridge (165 Ltrs)	2.3	0.456
50 Computer	15	2.98
35 Monitors	3.2	0.695
20 Printer	0.5	0.0993
Total	503.36	100%

Table 2: Total load The available area for install the solar system in the building is 2,648m^2.

DOD 60-80% are available for the PV system. Batteries are available for 12 V and different A-hr., 50 A-hr., 100 A-hr., 150 A-hr., etc. for my calculation I choose 12 V - 150 A-H-hr. batteries.

If we take DOD value as 0.7

Required charge capacity= 1499.37 kWh/12 V=124,947.5 A-hr. (For 12 V) (4)

No of batteries required =124,947/150 0.7=1190

- For 3 days autonomy (no sunshine):
- No. of batteries =1190 × 3 = 3570

Required charge capacity =1499370 W-hr. / 24 V = 62,473.75 A-hr. (For…24 V) (5)

- No. of batteries required = 62,473.75 A-hr. / 150 × 0.7 = 595 batteries
- Actual no. of batteries required (as battery voltage is 12 V) =1190 batteries
- For 3 days autonomy

No. of batteries = 1190 × 3 = 3570

Required charge capacity =1499370 W-hr. / 36 V= 31,236.875 A-hr. (For 36 V) (6)

No. of batteries required 31,236.875 / 150 × 0.7 = 397 batteries

Actual no. of batteries required (As battery voltage is 36 V) = 397 × 3 = 1191 batteries.

For 3 days autonomy:

No. of batteries = 1191 × 3 = 3573

Required charge capacity = 1499370 W-hr. / 48 V = 31,236.875 A-hr. (For 48 V) (7)

No. of batteries required =31,236.875 / 150 × 0.7 = 298 batteries

Actual no. of batteries required (as batteries voltage is 12 V) = 1190 batteries

For 3 days autonomy (no sunshine):1190 × 3 = 3570

Steps 3: Sizing the solar array

The panel with specification 220 W, at 16 V standard conditions. This panel produce 220/16 = 13.75 Amps.

Actual operating conditions of these solar panels are: 12 V, 13.5 Amps.

The PV panels are connected to the battery. The efficiency of the battery depends on the type of the battery; normally 0.80-0.90.

We take battery efficiency as 0.85. And the efficiency of the controller circuit (of the battery as 0.90).

Then the solar array has to generate = 1499370 = W-hr./0.85; 0.90=1,959,960 W-hr. every day.

Again, we will calculated for 4 option of system voltage: 12 V, 24 V, 36 V and 48 V

If the array voltage is 12 V, it needs to generated 1959.960 kWh/12 V=163,330 A-hr.

Assuming good sunshine of 6 hr. most of the days, the solar array has to generate:

163,330 A-hr. / 6 hr = 27,221.67 Amps.

27,221 Amps / 13.75 Amps = 1980 solar panel (as each produces 13.75 Amps.)

For 12 V PV system No. of panels required =1980

For 12 V PV system No. of batteries required = 3570

If the array voltage is 24 V (two panels connected in series),

It needs to generated 1,959,960 W-hr. / 24 V = 81665 A-hr.

Assuming good sunshine of 6 hr. most of the days, the solar array has to generate;

81665 A-hr 8/ 6 hr. =13610.8 Amps.

13610.8 Amps / 13.75 = 990 solar panels (as each produces 13.75 Amps.)

For 24 V PV system No. of panels required = 990 × 2 = 1980.

For 24 V PV system No. of batteries required = 3570

If array voltage is 36 V (three panels connected in series connection),

It needs to generated 1,959,960 W-hr. / 36 V = 54,443 A-hr.

Assuming good voltage of 6 hr. most of the days, the solar array has to generate;

54,443 A-hr / 6 hr. = 9,074 Amps

9,074 Amps / 13.75 Amps = 660 solar panels (as each produces 13.75 Amps.)

For 36 V PV system No. of panels required = 660 × 3 = 1980

For 36 V PV system No. of batteries required = 3573

If the array voltage is 48 V (four panels connected in series)

It needs to generate 1,959,960 W-hr. / 48 V = 40,832.25 A-hr.

Assuming good voltage of 6 hr. most of the days, the solar array has to generate:

40,832.5 A-hr / 6 hr = 6,805.4 Amps

6,805 Amps / 13.75 Amps = 495 solar panels (as each produces 13.75 Amps.)

For 48 V PV system no of panels required = 495 × 4 = 1980

For 48 V PV system no. of batteries required = 3570

Solar panel area required for this load

Area of each panel 4.5 × 2 = 9

1980 × 9 = 17820

The best design in the study chosen is SPV system with 12 V 220 Watt solar panel and 3570 No. of 12 V -150 A-hr. batteries.

Solar panel system

Specification

- Peak power generation per panel; 220 Wp

Types of Solar Panel	η (%)	Average Output(kW)	Annual Energy Production(kWh)
Monocrystalline	15.6	6.112	26,747
Polycrystalline	13.5	5.501	24,072
Thin Film	11.1	4.676	20,461

Table 3: Calculated Efficiencies and Average Output Power for Different Types of Solar Panels.

- Number of panel needs 1980
- Dimension of each panel: 2 feet feet
- Life 25 years
- % degradation every 10 yr.: 8-9%

Batteries

Specifications

- Total capacity: 12 V - 100 Ah
- Number of batteries: 3570
- Life: 6-7 years
- Warranty; 5 years with 8 years maintenance contract
- Cost: 27-41%

7. Inverter

Specifications

- Sine wave in inverter 0.9 kW
- Cost 50%

Conclusion

By the calculation and study of SPV (solar photovoltaic system) in building we find that we reduce the power load which take from any other energy resources and use renewable and non-polluted energy (Table 3). It is available in an abundantly amount. By using solar system we consumed an environmental and eco –friendly energy. As per we know that in a Delhi no hydraulic system and geo- thermal system and thermal energy produce a more quantity of carbon dioxide, hydrogen sulphide, and other component of nitrogen oxide. We already know that Delhi is a high dense city so electricity consumption in a Delhi is high and Government does not fulfil criteria of energy demand. So it is very important to initiate the renewable energy used. So by study of solar photovoltaic system we conclude that installing charge of SPV is high and it is not available 24 hr so by using batteries we use this energy in autonomy day and in the night. In SPV field more research is necessary which reduce the starting cost of SPV. When cost is reduced then it is a very useful because it is more in quantity. It is easily available and environmental friendly. It is not harmful for worker and human. It does not smoke and not required any extra site. It installed in roof, open area car parking, etc.,

Reference

1. Roy P, Arafat Y, Upama MB, Hoque A (2012) Technical and Financial Aspects of Solar PV System for City Dwellers of Bangladesh Where Green Energy Installation is Mandatory to Get Utility Power Supply, 7th International Conference on Electrical and Computer Engineering, Bangladesh.
2. Berger AG, Hoffmann V (2005) Photovoltaic Solar Energy Generation, Springer-Verlag Berlin Germany.
3. Aki H (2007) The penetration of micro CHP in residential dwellings in Japan, in Proc. IEEE-PES Gen. Meet: 1-4.

4. Nyeng P, Østergaard J (2011) Information and communications systems for control-by-price of distributed energy resources and flexible demand, IEEE Trans. Smart Grid 2: 334-341.

5. (2011) Electro Solar Power Ltd. website on PV Module.

6. (2011) HOMER, Software for Micro Power Optimization Modeling.

7. (2011) The GREENZU homepage on Solar Inverter.

8. Aziz S, Chowdhury SA, Al-Hammad H (2009) Marketing and Financing of Solar Home Systems in Bangladesh: Assessment of Success, International Conference on the developments in Renewable Energy (ICDRET): 34-37.

Pulsed Inductive Discharge as New Method for Gas Lasers Pumping

Razhev AM[1,2], Churkin DS[1,2] and Kargapol'tsev ES[1]

[1]*Institute of Laser Physics SB RAS, Lavrentyeva ave. 13/3, Novosibirsk, 630090, Russia*
[2]*Novosibirsk State University, Pirogova st. 2, Novosibirsk, 630090, Russia*

Abstract

Results of experimental investigations into the possibility of a pulsed inductive cylindrical discharge as a new method of pumping gas lasers operating at different transitions of atoms and molecules with different mechanisms of formation of inversion population are presented. The excitation systems of a pulsed inductive cylindrical discharge (pulsed inductively coupled plasma) in the gases are developed and experimentally investigated. For the first time five pulsed inductive lasers on the different transitions of atoms and molecules are created. Characteristic feature of the emission of pulsed inductive lasers is ring-shaped laser beam with low divergence and pulse-to-pulse instability is within 1%.

Keywords: Pulsed inductive discharge; Gas lasers pumping; Pulsed inductively coupled plasma

Introduction

The RF induction excitation of continuous-wave lasing was reported in [1-3]. Continuous-wave lasing on transitions in atomic argon ions in the green spectral range under excitation by a longitudinal inductive RF discharge was obtained in [1,2]. Lasing on vibrational-rotational transitions in CO_2 molecules in a wavelength range of 10.6 μm in an expanding nitrogen flow heated by an inductive discharge after the addition of cold CO_2 to it was reported in [3].

In this work, a method for exciting gas laser active media by a pulsed inductive cylindrical discharge is proposed and experimentally implemented in order to obtain lasing on electron transitions in atoms and molecules and vibrational-rotational transitions in molecules. It is important that the pulse repetition rate has to be several hertz or higher; i.e., all processes of discharge formation, creation of population inversion, amplification, absorption, and quenching must occur during each pulse irrespective of the past history of the preceding pulse. In contrast to conventional pulsed longitudinal and transverse electric discharges, a pulsed inductive cylindrical discharge is formed due to the magnetic field induction produced by the pumping system without any electrodes in the active medium. An appropriate choice of the tube material may ensure the purity of the active medium and considerable endurance of lasers. The formation of such a discharge is not accompanied by the appearance of cathode spots on the surface of the electrodes, which are responsible for the instability and contraction of the discharge, deterioration of the homogeneity of the discharge, contamination of the gas mixture, quenching of lasing, and limitation of the pulse repetition rate. The application of the pulsed inductive discharge for excitation is a promising method for pumping not only gas lasers, but also metal vapor and solid state lasers. In addition, this method can be used to produce the plasma for obtaining radiation (including induced radiation) in any spectral range, especially that extending from 100 nm to THz, which is of considerable interest for microelectronics, photolithography and biomedicine.

Apparatus

In our measurements, the spontaneous emission spectra of the inductive discharge in gases and the lasing spectrum were recorded with a Ocean Optics HR 2000 spectrometer, S-150 Solar LS spectrometer with a resolution of 0.66 nm in the spectral range from 200 to 1100 nm and a SpectraPro-500 Acton Research Corp. spectrograph with a resolution of 0.025 nm in the spectral range from 180 to 700 nm with different photodiodes and photomultipliers. The output laser energy was measured with a PE50-BB Ophir pyroelectric pulse energy meter (Ophir Optronics Ltd). The temporal parameters of electric pulses were recorded with high-voltage P6015A probes and a 200-MHz TDS-2024 Tektronix oscilloscope. The accuracy of measurements was 5%. The temporal parameters of optical pulses were recorded with a PhEC-22 and PhC-15 coaxial photocells with a temporal resolution of 10^{-10} s and infrared photodetector. The spatial distribution of the laser radiation intensity over the tube cross section and the light beam profile were analyzed by using a WinCamD-UCM digital video camera (Data Ray Inc).

Experimental Setup

In our experiments we used two excitation systems. The first one is shown in Figure 1. The excitation system operated in the following

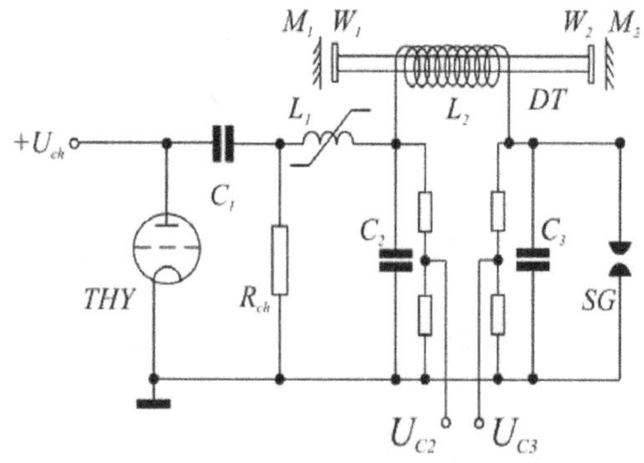

Figure 1: Electric circuit of the excitation system of pulsed inductive lasers. ***THY*** - thyratron ***TPI***1 - 10*k*/50, C_1 = 80 nF, C_2 = 8 nF, C_3 = 18 nF, L_1 - choking-coil, L_2 - inductor, ***DT*** - discharge tube, ***SG*** - self-triggered gas-filled spark gap.

***Corresponding author:** Dmitry Churkin, Institute of Laser Physics SB RAS, Lavrentyeva ave. 13/3, Novosibirsk, 630090, Russia, E-mail: churkin@laser.nsc.ru

way. The capacitor C_1=80 nF was charged from an ALE 152A Lambda EMI pulsed power supply up to the voltage 20-27 kV of the positive polarity. When the voltage across capacitor C_1 achieved the maximum, a triggering pulse was fed to a high-voltage switch (TPI1-10*k*/50 thyratron). After the thyratron switched on, the capacitor C_1 began to discharge, a negative voltage appeared across charging choking-coil L_1, and energy was transferred to capacitors C_2 and C_3. Both capacitors were charged during the time 1.5-2.0 µs up to the breakdown voltage of a spark gap. The capacitor C_3 and the spark gap represented a low-inductive circuit in which the capacitor C_3 began to discharge after the actuation of the spark gap slightly earlier than the capacitor C_2. During the discharge of the latter, a time-varying electric current passed through the inductor L_2 placed on Discharge Tube *DT* and creates varying magnetic field around it, which induces azimuthally electric current in gases, leading to breakdown and formation of a pulsed inductive discharge (pulsed inductively coupled plasma).

This excitation system provided comparatively efficient energy transfer from peaking capacitors C_2 and C_3 to the active medium [4]. However this system contained nonlinear element L_1 which reduced total scheme efficiency. Moreover additional self-triggered spark gap *SG* has not allowed to achieve high pulse-to-pulse stability and to operate with high pulse repetition rate.

To avoid these problems a new excitation system was developed (Figure 2). This system was based on well known scheme of Blumlein-type. This excitation scheme had a very simple design and provided high stable operating conditions. Using this systems, inductive lasers had high pulse-to-pulse stability: amplitude differences were not more than 0.5%.

The ignition moment of the inductive discharge was determined by the appearance of spontaneous emission of gases in the discharge tube. Under pressures below 1 Torr, inductive discharge filled the discharge tube completely, uniformly and had high intensity. As the mixture pressure increased above several Torr the intensity of inductive discharge in the center of the tube rapidly decreased. The discharge started to assume cylindrical form and concentrated near the inner tube wall surface. According to our observations discharge thickness decreased as a function of pressure. It was found in [4-8] that the output energy of the inductive lasers was proportional to the discharge tube diameter, i.e. the lasing efficiency and output energy increased with increasing the tube diameter. Because of this, we used in our experiments a ceramic discharge tube with maximal inner diameter we had (42 mm, external diameter was 50 mm). The tube was sealed by means of plane-parallel windows of MgF_2 or KCl oriented perpendicular to the tube axis (Figure 1). The optical resonator was

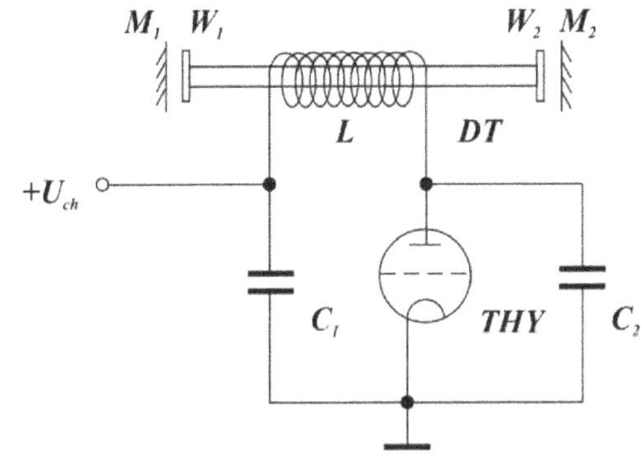

Figure 2: Electric circuit of the Blumlein-type excitation system. ***THY*** - thyratron ***TPI*1** - 10*k*/50, C_1 = 18 - 30 nF, C_2 = 30 - 55 nF, ***L*** - inductor, ***DT*** - discharge tube.

formed by external plane dielectric mirrors. The rear dielectric mirror had the reflectance 99% in the selected spectral region. The reflectance of the output mirror was optimized during experiments to obtain the maximum output energy. Mirrors with reflectance from 8% to 93% were used. The inductor L_2 consisted of separate sections representing solenoids made of a cable wire of cross section 1.5-6 mm^2. The results presented in this paper were obtained by using the inductor containing 30 sections, each of them consisting of four coils. The solenoids were connected in parallel, and the total length (~ 68 cm) of the inductor determined the length of the active medium of the gas laser. Gases or its mixtures were admitted from a gas system into the tube up to pressures 0.1-300 Torr. Gases flowed longitudinally during experiments.

Results

Red laser on the electronic transitions of atomic fluorine

In our first experiments to demonstrate the possibility of the creation of the pulsed inductive gas laser, we chose transitions in neutral fluorine atoms which works by the excitation of the pulsed cylindrical inductive discharge in a He:F_2(NF_3) mixture, because the population inversion in these transitions is reached at comparatively low excitation levels in a wide pressure range [9-13]. In this case, a high gain is achieved, which ensures super luminescence regime in a low-Q resonator, and lasing takes place in the red spectral region and the transition from spontaneous emission to the lasing mode is easily detected.

In this work the developed laser on the electron excited transition of Fluorine (FI) atoms pumped by a pulsed inductive cylindrical discharge is described. Lasing at 8 wavelengths in range of 624-755 nm is obtained by exciting He-F_2 (NF_3) gas mixtures in a pressure range from 20 to 300 Torr. The energy of laser radiation depended from the ratio of the mixture components and total pressure. The optimal composition for our pumping conditions was He:F_2 (80:1) and total pressure 40-50 Torr. When F_2 was replaced by NF_3, lasing was also obtained, but with a lower intensity. For this reason, subsequent experiments were made on the He:F_2 (80:1) mixtures. Maximum energy of FI laser in these experiments was achieved 2.6 mJ at pulse power 30 kW and durations 80 ns. The laser beam divergence was 0.4 mrad.

Ultraviolet laser on the self-limiting electronic transitions of the N_2 molecules

The purpose of this part of paper is to study the ability of the pulse periodic inductive cylindrical discharge to be a new method for pumping laser on self-limited electronic excited transitions of molecules and atoms (such as N_2, H_2, He, metal vapor). The aim of this paper is to show that a pulsed inductive discharge is an efficient alternative tool for exciting UV nitrogen lasers, offering a number of advantages compared to transverse and longitudinal electric discharges.

Experiments with the pulsed inductive discharge in nitrogen showed that the pressure range in which the inductively coupled plasma can exist is quite narrow, from 0.1 to 10 Torr. Visual observations with the use of different optical filters showed that the inductive discharge was homogeneous in the entire pressure range. No sparks and streamers were observed in the inductive discharge in pure nitrogen and its mixtures. At pressures 0.2-3.0 Torr, the two most intense lines were observed at 337.1 nm and 357.7 nm (0–0 band 0–1 band of $C^3\Pi_u \rightarrow B^3\Pi_g$ transition respectively). After the mounting of dielectric mirrors and alignment of the optical resonator, we obtained UV lasing at these transitions of molecular nitrogen excited by the pulsed inductive discharge in the pressure range from 0.3 to 3 Torr. The maximum of the lasing efficiency was achieved at pressures 0.5-0.8 Torr. The intensity of the 337.1 nm line exceeded that of the 357.7 nm line more than by two orders of magnitude. It is interesting that simultaneous lasing at these two lines in pure nitrogen at low pressures (1-2 Torr) was previously reported only in one paper [14]. The maximum output energy 6 mJ is achieved for resonator with front mirror with R_2=60 %. The laser pulse FWHM was 15 ± 1 ns. This corresponds to the laser pulse power of 500 kW. As the reflectance of the output mirror was further increased, the output energy decreased. However, the laser pulse duration increased from 13 ± 1 ns for R_2=16% to 18 ± 1 ns for R_2 = 93 %. This is larger than in electric-discharge lasers, where the pulse duration does not exceed 5–10 ns [15,16]. The duration of UV laser pulses (at the base level) in the dense resonator exceeded 35 ns. It is important to note that such a high pulsed power was never achieved before in a nitrogen laser at low pressures 1 Torr. This result demonstrates the specific features of the operation of the inductive nitrogen laser such as its emission spectrum containing many vibrational lines, which is obtained upon such excitation, and a long duration of pulses with a comparatively flat leading edge (about 7 ns). According to our measurements, the pulse-to-pulse instability of the lasing amplitude was 0.5%. We performed experiments with the inductive nitrogen laser operating in the pulse periodic regime. The pulse repetition rate was varied from 1 to 30 Hz. We found that the average output power increased linearly with increasing pulse repetition rate. For the repetition rate of 30 Hz and output power of the nitrogen laser of 6 mJ, the average output power was 180 mW. The output energy of a single pulse was independent of the pulse repetition rate because the active medium was cooled at the ceramic tube wall directly adjacent to the lasing region.

Near infrared laser on the electronic transitions of the H_2 molecules

There are several works on lasing on the electronic transitions in H_2 molecules in near IR spectra region [17,18]. Lasing was observed on 7 lines. Maximum peak power 1.5 kW was obtained. In our experiments lasing on the electronic transitions of H_2 molecules in near IR spectra region with excitation by pulsed inductive discharge is achieved. The generation is observed on 4 lines. The wavelength λ_1 =0.835 μm (band (2,1) rotational line *P*(2), λ_2 =0.89 μm (band (1,0) rotational line *P*(2), λ_3 =1.116 μm (band (0,0) rotational line *P*(4), the λ_4 =1.122 μm (band (0,0) rotational line *P*(2) that corresponds to $2s\sigma^1\Sigma^+_g(E) \rightarrow (2ps)^{21}\Sigma^+_g(B)$ transition. Maximum of laser emission peak power was at two lines: 7 kW for wavelength λ_2 =0.89 μm and 5 kW for wavelength λ_4 =1.122 μm, respectively. Peak power of the lines λ_1 =0.835 μm and λ_3 =1.116 μm was much weaker. Pulse duration was 18-20 ns (FWHM). Laser is working both on one wavelength and on two wavelengths simultaneously with competition between these transitions. The active medium was hydrogen at optimal pressure 0.5-0.8 Torr.

Mid and far infrared lasers on the vibrational-rotational transitions of the HF and CO_2 molecules

The purpose of this part of paper is to study the ability of the pulsed inductive cylindrical discharge to be a new method for pumping laser media and for creating the population inversion at vibrational-rotational transitions of molecules (such as CO_2, HF or DF) in the ground electronic state.

In the experiments we studied the effect of gas mixture composition on the lasing power characteristics of the inductive plasma HF laser. As a fluorine donor one used F_2, NF_3 and SF_6, as hydrogen donor- H_2. The highest laser power has been obtained with SF_6 as the fluorine donor. The proportion of the components has been H_2:SF_6 - 1:2. For the mixture of H_2-NF_3 lasing has also obtained, but its parameters were not studied due to dusting of the resonator's windows with the products of discharge. For H_2-F_2 we have not been able to obtain the infrared lasing. Comparing with electric discharge HF lasers, the pulsed inductive HF laser reaches its maximum of generation energy when Ne has added as a buffer gas in the mixture, but not He. During the experiments we have obtained 10 mJ for H_2-SF_6-Ne mixture, which is 10% higher than lasing in H_2-SF_6-He mixture.

To create the inductive CO_2 laser in our experiments we used CO_2:N_2:He gas mixture as an active laser medium. The main purpose of our experiments was to investigate the temporal, energy, and spatial characteristics of pulsed inductive CO_2 laser for gas mixtures with a different ratio of components as a function of the total pressure and excitation parameters.

Under total pressures of 10–15 Torr of the gas mixture CO_2:N_2:He - 1:4:12 lasing on vibrational-rotational transitions of CO_2 molecules at a wavelength of 10.6 lm in a pulsed inductive discharge has been observed. The maximum generation energy 152 mJ is obtained in a ceramic tube with an inner diameter of 42 mm in the CO_2:N_2:He - 1:4:12 gas mixture at a pressure of 15 Torr. The maximum efficiency in these first experiments did not exceed 0.3%. We expect that it is possible to increase the inductive CO_2 laser efficiency by using pulsed RF inductive discharge. The development of such excitation system is one of the goals of our further investigations. The measured divergence of CO_2 laser radiation is 8 mrad. The possibility of the inductive discharge CO_2 laser operation in the pulse-periodic regime was analyzed in the experiments. The pulse repetition rate varied from 1 to 50 Hz. It was revealed that the average radiation power increases linearly with the pulse repetition rate. Thus an average power of 3.0 W is obtained at a repetition rate of 50 Hz and generation energy of 60 mJ.

Characteristics of the inductive discharge laser emission

Characteristic feature of the emission of pulsed inductive lasers is ring-shaped laser beam. The thickness of the ring depends on active gas medium and pumping conditions. The study of the beam profile showed that the laser radiation intensity at the external boundary of the ring is lowest (Figure 3). The lasing intensity increases towards the ring

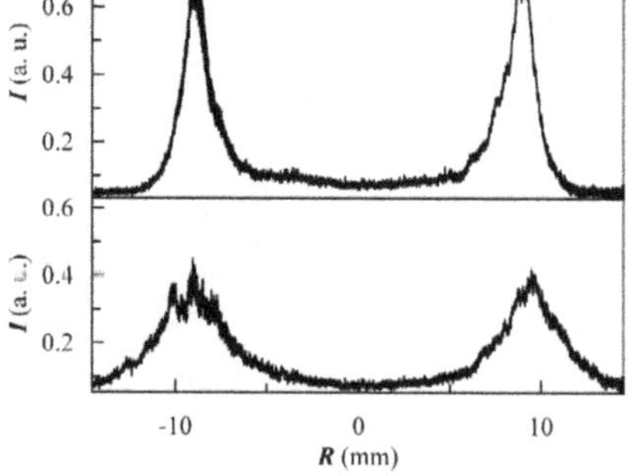

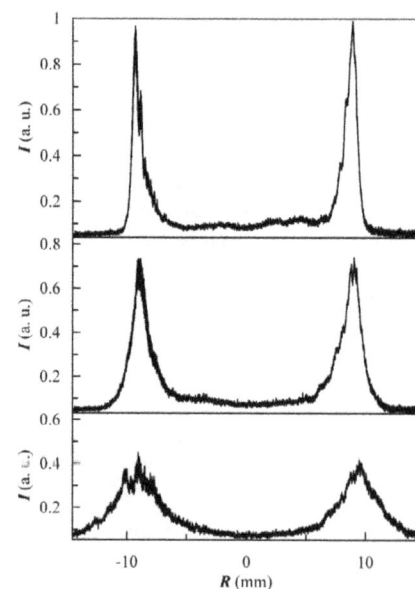

Figure 3: The profile of the laser beam of inductive discharge nitrogen laser at the distance 0.1 m, 4 m and 8 m from output window. Because of small aperture of beam profiler, experiments were performed with discharge tube with inner diameter of 20 mm.

centre and achieves the maximum at a distance of about 1 mm from the external boundary. Then, the radiation intensity decreases almost to zero at a distance of 4 mm from the boundary. This value can be treated as the ring width. The laser radiation divergence, which is determined by measuring the size of the laser radiation ring at different distances from the laser 0.1 m to 8 m and is 0,3÷8 mrad in dependence of the laser wavelength. The circular structure of the beam cross section is a specific feature of pulsed inductive lasers with a cylindrical inductive discharge. Such beams in the case of a low radiation divergence offer certain advantages because they can be focused to produce the radiation intensity distribution similarly to Bessel beams.

Conclusion

A pulsed inductive cylindrical discharge as a new method of pumping gas lasers operating at different transitions of atoms and molecules with different mechanisms of formation of inversion population is proposed and experimentally realized. The excitation systems of a pulsed inductive cylindrical discharge (pulsed inductively coupled plasma) in the gases are developed and experimentally investigated. For the first time five pulsed inductive lasers on the different transitions of atoms and molecules were created. Characteristic feature of the emission of pulsed inductive lasers is ring-shaped laser beam with low divergence and pulse-to-pulse instability is within 0.5%.

References

1. Bell WE (1965) Ring discharge excitation of gas ion lasers. Appl Phys Lett 7: 190-191.
2. Goldborough JP, Hodges EB, Bell WE (1966) RF induction excitation of CW visible laser transition in ionized gases. Appl Phys Lett 8: 137-139.
3. Kiselevskii LI, Skutov DK, Sokolov SA (1974) Primenenie visokochastotnogo indukzionnogo razrjada dla poluchenija lasernoi generazii v neprerivnom regime. Zh Prikl Spektroskopiya 21: 951-955.
4. Razhev AM, Churkin DS, Zhupikov AA (2009) Study of the UV emission of an inductive nitrogen laser. Quantum Electronics 39: 901-905.
5. Razhev AM, Mkhitaryan VM, Churkin DS (2005) 703-to 731-nm FI laser excited by a transverse inductive discharge. JETP Letters 82: 259-262.
6. Razhev AM, Churkin DS, Zavyalov AS (2009) Pulsed inductive discharge molecular hydrogen laser. Vestnik NSU Seria Fizika 4: 12-19.
7. Razhev AM, Churkin DS (2009) Pulsed inductive discharge CO_2 laser. Opt Commun 282: 1354-1357.
8. Razhev AM, Churkin DS (2007) Inductive ultraviolet nitrogen laser. JETP Lett 86: 420-423.
9. Hocker LO, Phi TB (1976) Pressure dependence of the atomic fluorine laser transition intensities. Appl Phys Lett 29: 493-494.
10. Loree TR, Sze RC (1977) The atomic fluorine laser: spectral pressure dependence. Opt Commun 21: 255-257.
11. Lisitsin VN, Razhev AM (1977) High-power, high-pressure laser based on red fluorine lines. Pisma v Zhurnal Tekhnischeskoi Fiziki 3: 862-864.
12. Lawler JE, Parker JW, Anderson LW, Fitzsimmons WA (1979) Experimental investigation of the atomic fluorine laser. IEEE J Quantum Electron 15: 609-613.
13. Zaeferani MS, Parvin P, Sadighi R (1996) Pressure dependence of the spectral lines of a high power, high pressure atomic fluorine laser pumped by a charge transfer from He^{+}_{2}. Opt Laser Technol 28: 203-205.
14. Kaslin VM, Petrash GG (1966) Rotational structure of ultraviolet generation of molecular nitrogen. JETP Letters 3: 88-92.
15. Shipman JD (1967) Traveling wave excitation of high power gas lasers. Appl Phys Lett 10: 3-4.
16. Wang CP (1976) Simple fast-discharge device for high-power pulsed lasers. Rev Sci Instrum 47: 92-95.
17. Bazhulin PA, Knyazev IN, Petrash GG (1965) Stimulated radiation from hydrogen and deuterium molecules in the near infrared region. JETP 49: 16-23.
18. Bockasten K, Lundholm T, Andrade O (1966) Laser lines in atomic and molecular hydrogen. J Opt Soc Am 56: 1260-1261.

Development of Low Power Dynamic Threshold PCS System

Pawan Whig[1]* and Syed Naseem Ahmad[2]

[1]Department of Electronics and Communication, Bhagwan Parshuram College of Engineering and Technology Rohini, Delhi, India
[2]Head of Department, Department of Electronics and Communication Engineering, Jamia Millia Islamia, New Delhi 110025, India

Abstract

Real-time reception of sensor signals is one of the most significant operations in analog signal processing. Driven by low-power and low-voltage requirements for integrated mixed signal portable applications, this paper propose a novel low power Dynamic Threshold Photo Catalytic Sensor (DTMOS) system to work efficiently below lower bound of power supply independent of substrate bias effect. Another important feature of DTMOS is that the fabrication of circuits based on DTPCS is less complex and more economical. The SPICE simulations were performed with 120 nm technology parameters and results verify the performance of the circuit. The proposed system has high linearity and simple structure hence it is suitable for high-performance and low-power analog VLSI applications. The main reason of employing a readout circuit to PCS circuitry, is the fact that the fluctuation of O_2 influences the threshold voltage, which is internal parameter of the FET and can manifest itself as a voltage signal at output but as a function of the trans-conductance gain. The trans- conductance is a passive parameter and in order to derive voltage or current signal from its fluctuations the sensor has to be attached to readout circuit. This circuit provides high sensitivity to the changes in percentage of O_2 in the solution.

Keywords: Sensor; Dynamic Threshold Photo catalytic Sensor (DTMOS); Low power; VLSI

Introduction

Water is indispensable source of our life. According to Central Pollution Control Board, 90% of the water supplied in India to the town and cities is polluted, out of which only 1.6% gets treated. Therefore, water quality management is fundamental for the human welfare. The statistical regression analysis has been found to be a highly useful tool for correlating different parameters. Correlation analysis measures the closeness of the relationship between chosen independent and dependent variables. If the correlation coefficient is nearer to +1 or -1, it shows the probability of linear relationship between the variables x and y . This way analysis attempts to establish the nature of the relationship between the variables and thereby provides a mechanism for prediction or forecasting.

In the field of artificial waste water analysis and COD determination, TiO_2 sensors have large number of potential applications due to their highly efficient photo activity, steadiness and least cost. From the earlier times TiO_2 is used as white pigment which is safe for human use. The results achieved by the use of TiO_2 Photo Catalytic Sensor are in good agreement with those from the conventional Dichromate method. Very complex and bulky set up is needed for the conventional method used for Photo Catalytic Sensor applications and consumes a plenty of time for computation. To provide a solution to these problems and make the application faster, Whig and Ahmad developed a Simulation Program with Integrated Circuit Emphasis (SPICE) model for Photo Catalytic Sensor [1]. With the advancement in semiconductor technology development of sensors has become easy due to advantages of low power, high speed signal processing capabilities. CAD tools are an added advantage as they provide a method for simulation and synthesis of semiconductor sensors. By using the SPICE model, the size and power of the overall system can be minimized thus increasing the reliability of the system.

In the Modern electronics there is a great demand for low power and high performance digital system [2]. The Low power concept in semiconductor industries relies on a basic concept of power supply scaling. Since reduction in power supply below $3V_t$ will degrade the speed of circuit. Hence scaling of power supply should be done along with threshold voltage reduction. The threshold voltage cannot be reduced by lower bound a certain lower limit. The lower limit of threshold voltage can be set by keeping the value of offset leakage current that can be withstand by the circuit. To increase the flexibility of operation of device to work efficiently below lower bound of power supply a new technique DTPCS with highest V_t at zero bias and lower limit at $V_{gs} = V_{dd}$ is proposed in the rest of the paper.

Photocatalysis Process

The process of photocatalysis is a proficient method for degrading organic compounds. Various literatures are available on the different mechanisms and equations involved in the process for gaining a better knowledge [3]. The semiconductor material consists of two bands which are valence band and conduction band. The energy gap between these two bands is known as band gap given by E_g. The electrons from the valence band jump to conduction band which may be empty when a light of energy higher than band gap energy falls on the semiconductor material. Holes are left behind in the valence band due to excitation of electrons to higher energy band. These holes on reaching the surface of the organic molecule reacts with water to give OH^- radicals for oxidizing the organic pollutants. The dissolved oxygen in the molecular form acts as a scavenger of the photogenerated electrons and forms a superoxide radical ion. Titanium oxide has the ability to cause photo-oxidative destruction of the organic pollutants and is non-corrosive in nature due to which it is used as a catalyst in the process [4]. The oxygen content in any given sample can be determined by observing the change in dissolved oxygen concentration during the process of photocatalysis.

***Corresponding author:** Pawan Whig, Assistant Professor, Department of Electronics and Communication Engineering, Bhagwan Parashuram Institute of Technology, New Delhi-110025, India, E-mail: pawanwhig@gmail.com

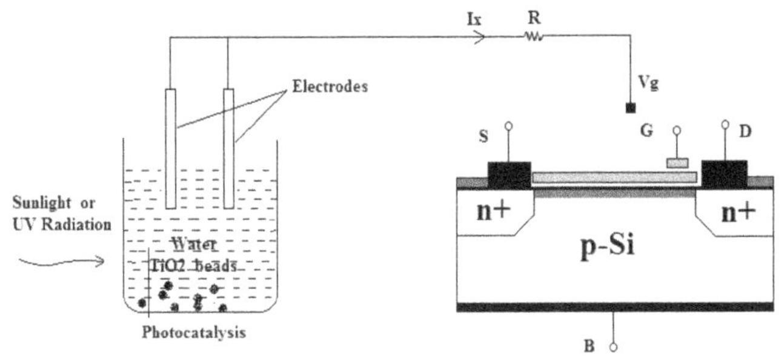

Figure 1: Photo catalysis process.

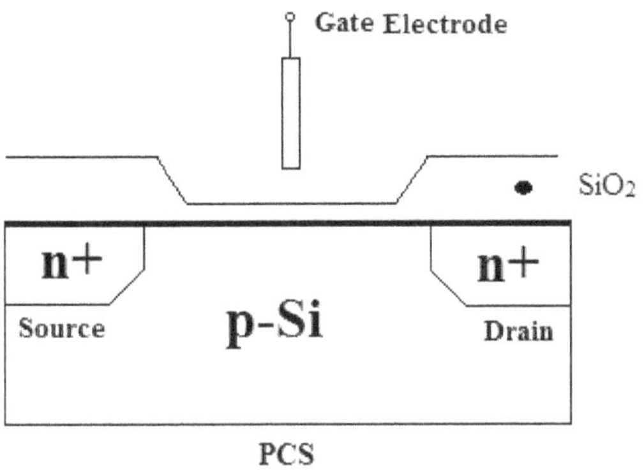

Figure 2: Cross-section of PCS.

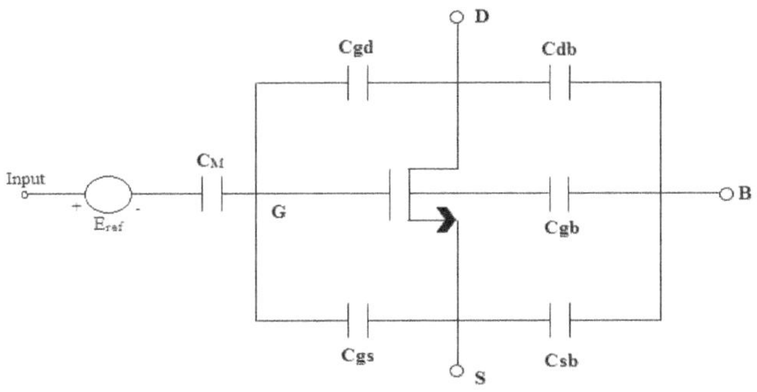

Figure 3: Circuit for PCS.

The Photo Catalytic Sensor (PCS) senses the changes in the oxygen concentration and its voltage levels change as an indication. The cross section of the PCS is shown in Figure 1.

PCS (Photo catalytic sensor)

The SPICE model for PCS is given in [5]. It is basically a MOSFET having structural difference in which the gate terminal is kept inside the solution and diffusion and quantum capacitances are added to explain the effect of Helmholtz and diffusion layer [6]. The cross section of PCS is shown in Figure 2.

The threshold voltage equation for the PCS model is given in equation 1:

$$V_{th}(\text{PCS}) = \text{E}_{\text{Ref}} - \Psi_{sol} + \chi^{sol} + \frac{-\Phi_s}{q} - \frac{Q_{ox} + Q_{ss} + Q_B}{C_{ox}} + 2\Phi_f \quad (1)$$

Where Ψ_{sol} is an input parameter of the equation which is dependent on the concentration of O_2 in the solution and surface dipole potential (χ^{sol}). Here E_{Ref} is the constant reference electrode potential. For different concentrations of O_2, different V-I curves for PCS can be plotted. Ψ_{sol} is a function of O_2 and as the saturation cut-off current I_{ds} increases the value of the oxygen concentration level decreases. The circuit for PCS as given in is shown in Figure 3.

Here C_M is the resultant of C_{ox} and C_q which are oxide and quantum capacitances respectively. The equivalent capacitance C_M is given in equation 2.

$$\frac{1}{C_M} = \frac{1}{C_q} + \frac{1}{C_{ox}} \quad (2)$$

The drain current equation in non-saturation mode for PCS is given in equation 3.

$$I_{ds} = C_{ox}\mu\frac{W}{L}\left[\left(V_{gs} - V_t\right)V_{ds} - \frac{1}{2}V_{ds}^2\right] \quad (3)$$

where

C_{ox} = Oxide capacitance per unit area,

μ = Electron mobility of the channel,

W = Channel width

L = Length of the Channel

The curves between I_{ds} and V_{ds} is shown in Figure 4 for different values of oxygen contet.

Device Description and Analysis

The PCS generates potential proportional to activity of detected oxygen ion. Potential in PCS is measured against the reference electrode. The Potentiometric method previously used had a serious limitation that for multiple sensors network multiple reference electrodes are needed [7] also the system is suffer from body bias effect. To measure the change in the concentration of dissolved oxygen through a corresponding shift in the device threshold voltage independent of substrate bias effect using DTMOS PCS is shown in Figure 5.

The threshold voltage of DTMOS transistor is given as:

$$V_{th} = V_{to} - \gamma\left(\sqrt{2\phi_F + V_{SB}} - \sqrt{2\phi_F}\right) - \eta V_{DS} \quad (4)$$

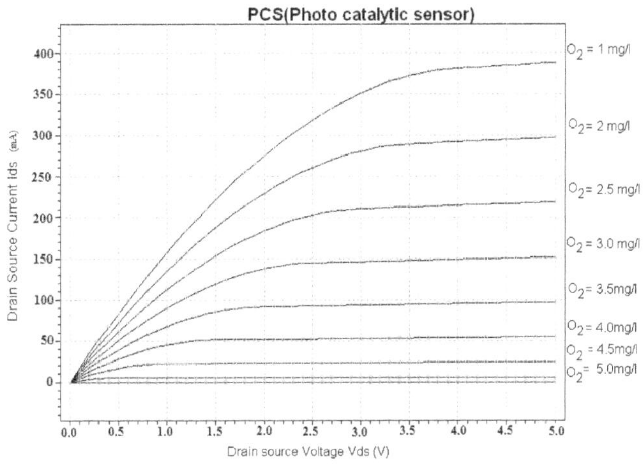

Figure 4: Characteristic curve for PCS.

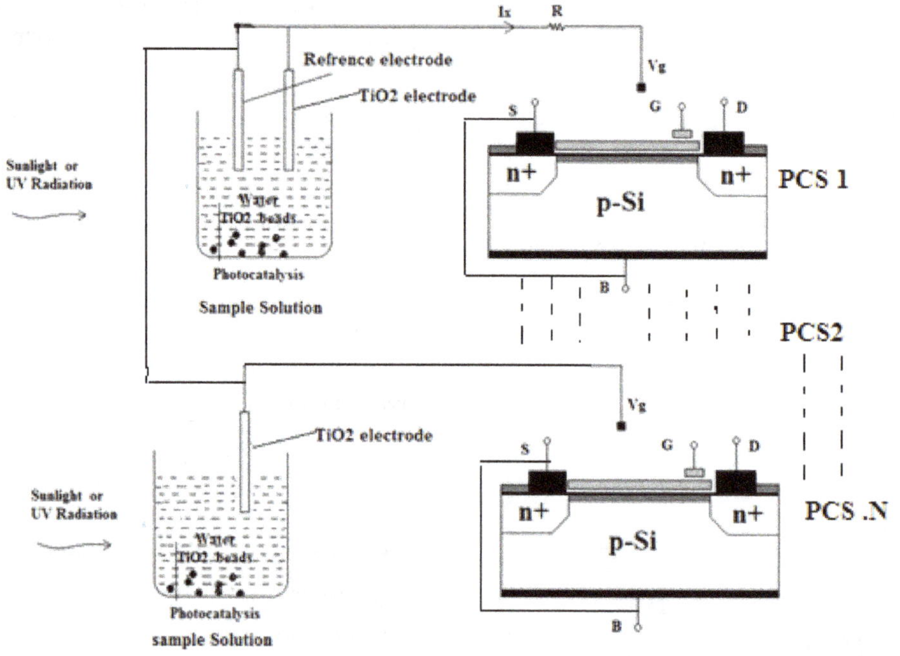

Figure 5: Multiple sensor network using single reference electrode.

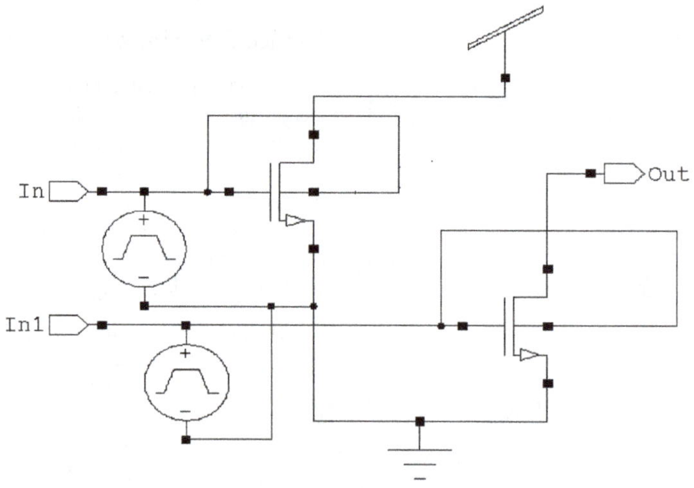

Figure 6: Circuit diagram of DTMOS PCS.

Where V_{to} threshold voltage is when substrate bias voltage is zero γ is a body bias coefficient

ϕ_F is surface potential.

V_{SB} is source to body voltage.

ηV_{DS} represent the effect of drain induced barrier lowering (DIBL)

η is DIBL coefficient and has a typical value 0.02-0.1

When forward bias is applied to the junction then the depletion region decrease which in turn reduces the threshold voltage. On the other hand when reverse bias is applied this will increase the depletion width and hence increases body charges which in turn increases the threshold voltage. Hence during the Forward Bias mode the DTMOS PCS have higher driving capabilities and during reverse bias mode it will have low leakage current hence the threshold voltage of the DTMOS PCS can be change dynamically according to the inputs applied at the gate. Also, DTMOS PCS have larger effective gate capacitance which causes the system to operate with less delay then regular MOS Circuit.

We know that saturation current equation in a MOSFET is given as:

$$I_d = \frac{k\left(V_{gs} - V_{th}\right)^2}{2} \tag{5}$$

Where K is constant and its value depends upon $\left(\frac{W}{L}\right)$ ratio and W is channel width and L is Channel Length.

In a given circuit the two PCS are connected to the V_{gs1} and V_{gs2} as shown in Figure 6.

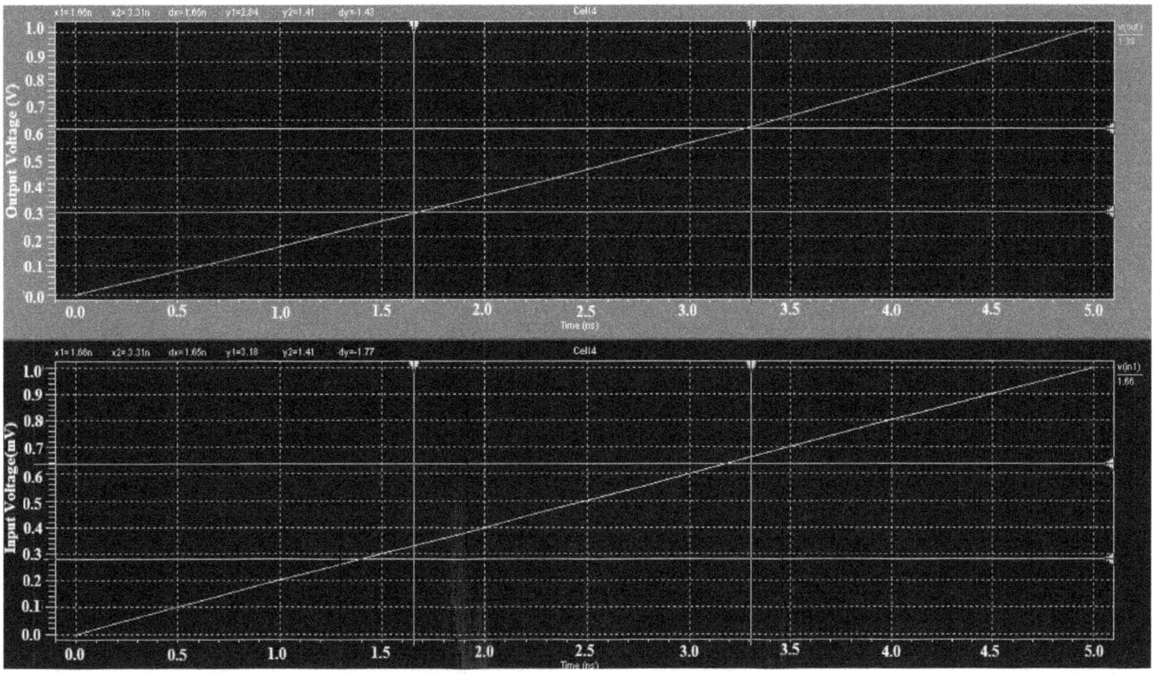

Figure 7: Transient analysis of the device.

Potentiometric Circuit Integrated Noise Analysis	
Onoise_total	0.00876p
Inoise_total	0.00880p

Table 1: Integrated Noise Analysis.

The sum of the gate source voltage is kept constant i.e.

$$V_g = V_{gs1} + V_{gs2} \tag{6}$$

And using eq. 1 and eq. 2 two drain current can be given by

$$I_{d1}\alpha\left[V_{gs1}-\left(V_{to}-\gamma\left(\sqrt{2\phi_F+V_{SB}-\sqrt{2\phi_F}}\right)-\eta V_{DS}\right)\right]^2 \tag{7}$$

Similarly

$$I_{d2}\alpha\left[V_{gs2}-\left(V_{to}-\gamma\left(\sqrt{2\phi_F+V_{SB}-\sqrt{2\phi_F}}\right)-\eta V_{DS}\right)\right]^2 \tag{8}$$

It is observed that the difference I_{d1}-I_{d2} in the drain currents will depends upon V_{gs1} or V_{gs2} only, i.e. the device is free from substrate bias effect.

Transient analysis

The transient analysis of the device is done on Tanner Tool version 15 and shown in Figure 7 and it is observed that the response is highly linear indicating that the device is stable. On plotting a linear trend line between V_{out} and V_{in} the coefficient of determination R^2 is found to be 99.7% with standard error of 0.02. The coefficient of determination R^2 is useful because it gives the proportion of the variance (fluctuation) of one variable that is predictable from the other variable. It is a measure that allows us to determine how, certain one, can be in making predictions from a certain model. The coefficient of determination is a measure of how well the regression line represents the data. If the regression line passes exactly through every point on the scatter plot, it would be easy to explain all the variations.

Noise analysis

Noise is electrical or electromagnetic energy that reduces the quality of a signal. Noise affects digital, analog and all communications systems. For noise analysis a noise model of the circuit, using noise models of each resistor and semiconductor device is obtained. It calculates the noise contribution of each component and propagates it to the output of the circuit sweeping through the frequency range specified in the analysis dialog box. Noise analysis calculates the noise contribution from each resistor and semiconductor device at the specified output node. Each resistor and semiconductor device is considered a noise generator. Each noise generator's contribution is calculated and propagated by the appropriate transfer function to the output of the circuit. The total output noise" at the output node is the root mean square (RMS) sum of the individual noise contribution. The result is then divided by the gain from input source to the output source to get the equivalent input noise". This is the amount of noise which, if injected at the input source into a noiseless circuit, would cause the previously calculated amount of noise at the output. The total output noise" voltage can be referenced to ground or it may be referenced to another node in the circuit. In this case, the total output noise is taken across these two nodes. The onoise and inoise for the given device is shown in Table 1.

The noise figure is used to specify exactly how noisy a device is. For a transistor, noise figure is simply a measure of how much noise the transistor adds to the signal during the amplification process. In a circuit network, the noise figure is used as a "Figure-of-merit" to compare the noise in a network with the noise in an ideal or noiseless network. It is a measure of the degradation in signal-to-noise ratio (SNR) between the input and output ports of a network. When calculating the noise figure of a circuit design, Noise Factor (F) must also be determined. This is the numerical ratio of noise figure, where noise figure is expressed in dB. Thus,

1	Fourier analysis for V(Probe					
2	DC component:	-0.8368				
3	No. Harmonics:	9				
4	THD:	188.676 %				
5	Grid size:	256				
6	Interpolation Degree:	1				
7						
8	Harmonic	Frequency	Magnitude	Phase	Norm. Mag	Norm. Phase
9	1	1000	5.55329e-008	-173.4	1	0
10	2	2000	2.79115e-008	-170.07	0.502611	3.33058
11	3	3000	1.9046e-008	-161.07	0.342967	12.3328
12	4	4000	1.46309e-008	-155.11	0.263464	18.2914
13	5	5000	1.20534e-008	-149.38	0.21705	24.0258
14	6	6000	9.62329e-008	157.492	1.7329	330.895
15	7	7000	9.2298e-009	-138.73	0.166204	34.6733
16	8	8000	8.40249e-009	-133.8	0.151307	39.6016
17	9	9000	7.78371e-009	-129.15	0.140164	44.2567
18						

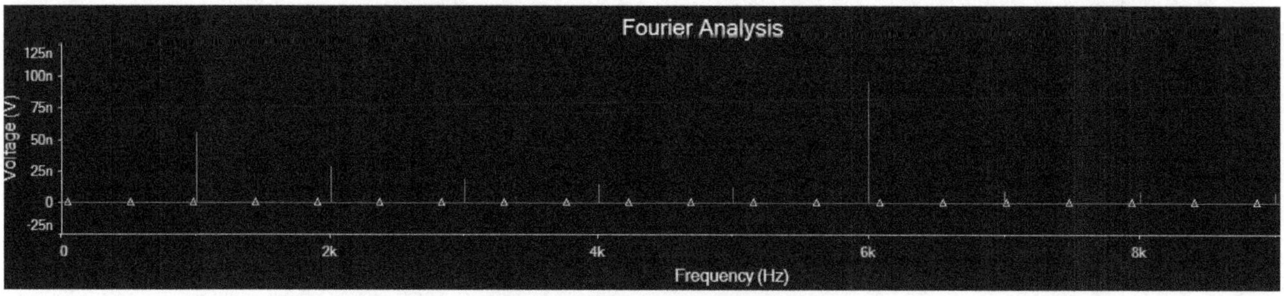

Figure 8: Fourier analysis of the device.

→ ↓ Parameters	Results obtained from Spice Model	Results obtained from FIA analysis
Multiple R	0.983	0.958
R^2	0.966	0.918
Standard Error	0.026	0.040
Complexities	Less complex	More complex
Cost	Inexpensive	Expensive
Accuracy	More accurate	Less accurate
Behavior	Fairly linear	Non linear

Table 2: Result comparison.

$$\text{Noise Figure} = 10\log_{10} F$$

$$F = \frac{\text{Input SNR}}{\text{Output SNR}}$$

The noise figure analysis of the device is observed to be 0.0399db.

Fourier analysis

Fourier analysis is a method of analysing complex periodic waveforms. It permits any non-sinusoidal period function to be resolved into sine or cosine waves and a DC component. This permits further analysis and allows you to determine the effect of combining the waveform with other signals. Each frequency component of the response is produced by the corresponding harmonic of the periodic waveform. Each term is considered a separate source. According to the principle of superposition, the total response is the sum of the responses produced by each term. It is observed that, amplitude of the harmonics decreases progressively as the order of the harmonics increases. This indicates that comparatively few terms yield a good approximation. Fourier analysis of the device is shown in Figure 8.

The comparison between Spice Model and the FIA analysis readings has been shown in Table 2.

Inference from Table 2:

a. The value of R^2 in case of Spice Model which shows the direction of a linear relationship between peak height decrease in current (ΔI) and dissolved oxygen concentration decrease (ΔO_2) is greater compared to FIA model.

b. The value of standard error in Spice Model is found to be smaller which shows better accuracy of the spice model.

c. Spice Model is designed on CAD tools hence it is less complex, inexpensive, fairly linear and more accurate as compared to FIA Model.

Result and Conclusion

Various analyses on the given device reveal that the device has fairly good performance. Power analysis on Tanner Tool shows that the device consumes very low power in order of 7.1 mW. The slew rate of the device is good. The output observed in Figure 7 is highly linear, indicating that the device is stable. Coefficient of determination R^2 is found to be 99.7% with standard error of 0.02. A significant advantage of the proposed design is that with the use of only few active components and using grounded reference electrode can overcome the problem of using multiple reference electrodes as inputs in an array of sensors. The device has a simple architecture, and hence it is very suitable for water quality monitoring applications. This study may be extended for further improvements in terms of power and size, besides the wiring and layout characteristics level.

References

1. Whig P, Ahmad SN (2014) Development of Economical ASIC For PCS for Water Quality Monitoring. JCSC.
2. Whig P, Ahmad SN (2012) Performance analysis of various readout circuits for monitoring quality of water using analog integrated circuits. International Journal of Intelligent Systems and Applications 11: 91-98.
3. Kim YC, Sasaki S, Yano K, Ikebukuro K, Hashimoto K, et al. (2001) Photocatalytic sensor for the determination of chemical oxygen demand using flow injection analysis. Analytica Chimica Acta 432: 59-66.
4. Whig P, Ahmad SN (2013) Simulation of linear dynamic macro model of photo catalytic sensor in SPICE. Compel the Int. J. Comput. Math. Electric Electron Eng 33: 611–629.
5. Brown WD, Grannemann WW (1978) C-V characteristics of metal titanium dioxide-silicon capacitors. Solid-State Electron 21: 837-846.
6. Campbell SA, Gilmer DC, Wang XC, Hsieh MT, Kim HS, et al. (1997) MOSFET transistors fabricated with high permitivity TiO2 dielectrics. IEEE Transactions on Electron Devices 44: 104-109.
7. Duffy JA (1990) Bonding Energy Levels and Bands in Inorganic Solids, Wiley, New York, NY.

12

Factors Affecting Solar Photovoltaic Power Output at Particular Location and Cost Estimation

Nakkela H*

Andhra University College of Engineering, Visakhapatnam, Andhra Pradesh, India

Abstract

Fossil fuels like coal, gas, diesel etc., used for generate electrical energy are in exhaust stage and causes pollution. This necessitates alternative sources of fuels which are renewable and non-pollutant like wind, solar, tidal to generate electrical energy. These renewable fuels cannot be supplied to generate power as required like conventional fuels, as they are naturally available and depends on environmental conditions. Hence they have to be efficiently utilized when they are available. Solar energy is the electromagnetic waves in the form of light which contains packets of energy known as photons. These photons produce electrical energy by the process known photovoltaic effect. In this paper studies are made to utilization of solar energy to produce electrical energy, for this sun-earth angular relation was found using declination angle of earth based on day number, azimuth angle, altitude angle, angle of incidence for particular location i.e., Andhra university, Visakhapatnam with latitude 17.680° N and longitude 83.320° E. Collecting radiation data for the consider site and calculating actual radiation on tilt surface by applying correction to the data collected. Simulation of solar photovoltaic panel using five parameter model of cell considering top Indian manufactured Solar Panels and estimating output DC power by using location Radiation, temperature and wind as input. Comparing output of simulation with on-line simulators like PV watts and off-line software's SAM, PVs yst. Second part of this paper is estimation of installation cost of solar power plant, cost of generations, Performance ratio, Capacity utilization Factor, LCOE, return of investment, comparing with conventional utility rates.

Keywords: Altitude angle; Azimuth angle; Declination angle; Latitude; LCOE; Longitude; PR; PVSYST; PV watts; Radiation data; CUF

Introduction

Solar Energy is a free source of energy which is very important to all living beings. This intensity of Solar Energy is not continuously available in a particular location as Earth moves around itself and round the Sun in elliptical orbit that varies distance between sun and earth. The axis of rotation Earth is not parallel to the axis of rotation of Sun, because of this the rays of sun sometimes need to travel more distance i.e., need to travel through atmosphere which attenuates the irradiation energy. According to World Radiation Centre (WRC) solar radiation is spectral distribution which was divided in to three sub categories which are Ultraviolet radiation $\lambda \leq 0.38$ μm the visible radiation $0{:}38 \leq \lambda \leq 0{:}78$ μm and the infrared radiation $\lambda \geq 0{:}78$ μm where λ is wavelength. The Radiation above the Earth's atmosphere is known as Extra-terrestrial radiations with solar constant 1367 W/ m2. These solar constants are not constant for the whole year as we send above that distance between sun and earth is not constant [1].

Solar Photovoltaic Power output estimation using Matlab programme requires accurate inputs i.e., radiations data, temperature, wind speed, Solar cell parameters. In this paper the factors studied are proper orientations of solar panel, radiation data collected should be recalculated to get radiation on tilted surface, accurate measurement of parameters of solar cell. Proper orientation of solar photovoltaic surface to make it perpendicular to incident radiation is important to maximize conversion efficiency. For this Sun-Earth angular relations are calculated using different mathematical equations. Normally radiation measurements are made with sensors in a horizontal position as in the case of pyranometer, pyrradiometer or pyrgeometer and at normal incidence when direct solar irradiances are required. But for solar energy application the received surfaces are normally kept in a tilted position to maximize capture of solar energy incident on it. Hence it is necessary to derive the values of radiant energy falling on inclined surface from the values of direct and global solar irradiance. Modelling of solar cells in Matlab requires accurate values of series resistance, shunt resistance, saturation current, ideality factor, photo current. Series resistance of the solar cell consists of several components. Of these components, the emitter and top grid dominates overall series resistance. Shunt resistance of solar cell plays important role in current output. It's value determines recombination's of electron hole pair without passing through the load [2,3].

Electrical Energy is known to be most efficient form of energy which plays important role in industrialization and development. To generate this there are conventional means like thermal power, nuclear power, gas, diesel, etc., which are fossil fuels based power plant and more over they cause pollution and which are in exhaust stage because of this we are in search of alternate energy which are not exhaust and not polluting known as unconventional energy generation like wind, solar, tidal etc., In this paper Solar Energy is considered as source for electrical power generations. Solar photovoltaic panels convert solar energy directly in to direct current electrical energy. The solar photovoltaic panels are made of semiconductor materials i.e., silicon, germanium which when doped with impurities i.e., fifth and third group elements converts into N-type and P-type semiconductor materials. These N-type and P-type materials when sandwiched forms PN junction which when placed in sun light due to photovoltaic effect i.e., when Photon strike the

***Corresponding author:** Nakkela H, Andhra University College of Engineering, Visakhapatnam, Andhra Pradesh, India, E-mail: hari.nakkella@gmail.com

Photovoltaic module surface creates electron hole pair if we connect two conductors on either side of P-N junction and connected to load. These electron hole pairs due electric field at the junction pass through load and thus generate electrical energy. The factor need to be considered is the energy of photon is sufficient to generate electron hole pair. The energy of photon is dependent on wave length of incident radiation. Silicon has affinity of wave length $\lambda \leq 1{:}15$ µm. It has energy sufficient to overcome band gap energy of silicon i.e., 1:121 eV. We need to make the surface to orient perpendicular to solar radiation so that the output of photovoltaic panel will be more.

Solar photovoltaic power plant initial cost is higher than any other conventional and non-conventional energy power plants. In India for the year 2015-16 benchmark cost of 1 MW solar photovoltaic plant is 605.85 Lakhs [4]. In this overall cost 55% cost is on Photovoltaic Modules. Before actually install photovoltaic plant one should first know load profile, retail price of utility, annual amount paid to the utility, location radiation data and environmental conditions of the location, using these one can calculate solar photovoltaic power output at considered locations and cost of generation per of energy, which can be compared with retail price of conventional utility. Solar photovoltaic power plant installed at considered location some useful factors like performance Ratio, capacity utilization factor and levelized Cost of Energy plays important role in estimating feasibility of solar power plant at considered location [5]. For this there some on-line and off-line software's like PVwatts, PVsyst, SAM etc., are available to find and compare above said parameters. In this paper Matlab programme was written to estimate DC power output at considered location by considering modules of Indian

manufactures, location radiation, environmental conditions like temperature and wind speed and multiplying this DC output with overall efficiency other balance of system like cables, mppt tracker, inverter, transformer, soiling losses etc., to know useful AC output energy of the installed plant. Performance Ratio gives insights how efficient the available solar energy is converted into electrical energy. Knowing highest and lowest temperature of location, one can design no of modules of particular rating should be connected in parallel and no. of modules connected in series. As highest temperatures leads to increase of current and lowest temperature leads to increase of voltage.

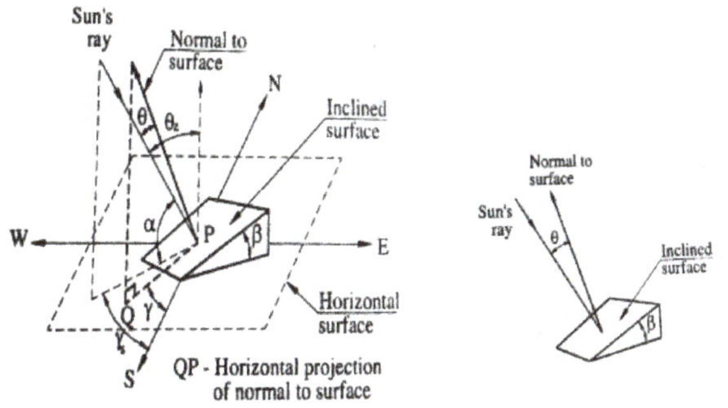

Figure 1: Sun-Earth angular relation.

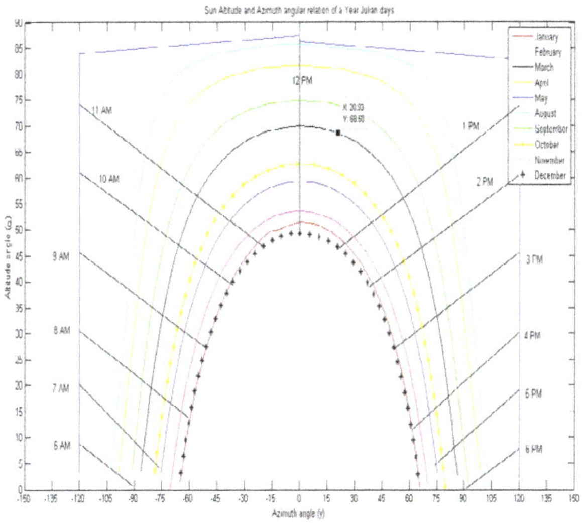

Figure 2: Azimuth angle vs Altitude angle.

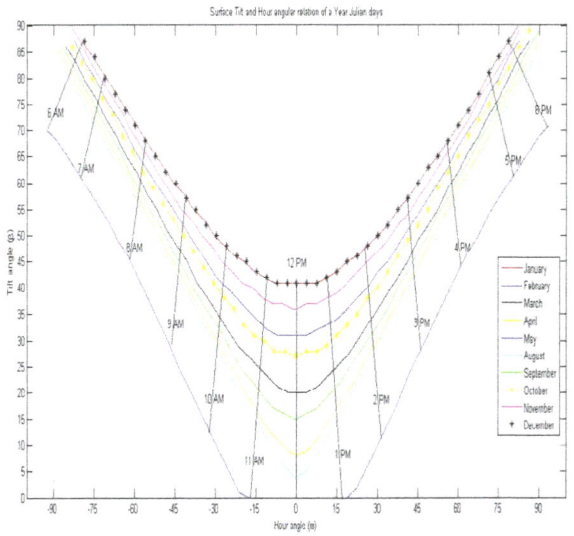

Figure 3: Hour angle vs Slope of surface.

Sun and Earth Relations

The Sun position at the location considered i.e., Andhra university, Visakhapatnam, various angles and the relation between incident radiation and surface is explained with help of Figure 1. In order to maximize output from installed photovoltaic power plant, the surface needs to orient perpendicular to rays of sun. This can be possible by proper tracking thesun. To track the sun we need to know two angles Azimuth angle, altitude angle represent in Figures 2 and 3. To find these angles, latitude angle, declination angle, Hour angle, slope of the considered plane, azimuth angle of plane should be known. Latitude angle is location latitude. In this paper location considered is Andhra University, Visakhapatnam (=17:68) and Figure 4 gives required inclination and azimuth angle of surface. Declination angle changes every day and this can be calculated from the equation of Cooper and other equations are given below section A [1,6].

Sun-Earth angular relations

The angular relation of Sun-Earth is given by Figure 4.

Declination: (δ)

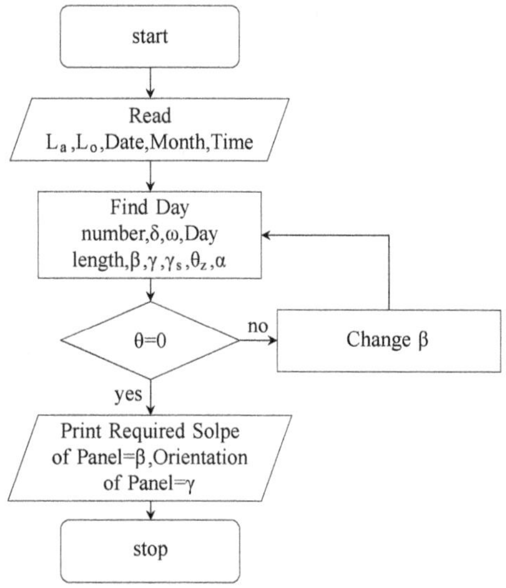

Figure 4: Flow chart to find slope of surface and surface azimuth angle.

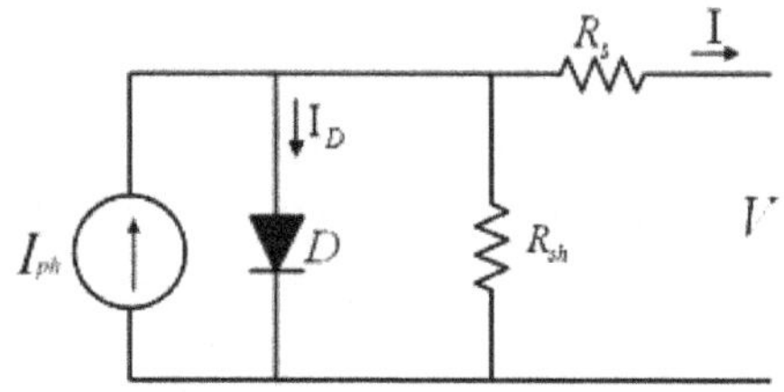

Figure 5: Electrical equivalent circuit of solar cell.

$$\delta = 23{:}45 \sin(360(284 + n){=}365) \quad (1)$$

Where n is day number.

Hour Angle ω The angular displacement of Sun east or west of local meridian as shown in Figure 5 this is due rotation of earth around the sun, 150 of angular displacement per hour, hour angle due east negative and due west is positive. Day Length is given by the equation (2)

$$DL = \frac{2}{15}\cos^{-1}(-\tan\phi\tan\delta) \quad (2)$$

Altitude Angle: (α)

The actual location of sun in the sky at particular place and time is given by equation (3)

$$\alpha = \sin^{-1}(\sin\phi\sin\delta + \cos\phi\cos\delta\cos\omega) \quad (3)$$

Azimuth angle: (γ)

The surface azimuth of plane should track sun azimuth angle which is given by the equation (4)

$$\gamma = \sin^{-1}(\cos\delta\sin\omega / \sin\alpha) \quad (4)$$

Incident Angle (θ)

The angle of incidence of solar rays on plane of surface considered should be perpendicular to it and should be parallel to perpendicular drawn to the considered surface. This incidence angle is given by (5)

$$\theta = \cos^{-1}(\cos\theta_z\cos\beta + \sin\theta_z\sin\beta(\gamma_s - \gamma)) \quad (5)$$

Where

θ_z is Zenith angle

γ_s is Sun Azimuth angle

γ is Surface Azimuth angle

Zenith angle (θ_z)

The of Sun rays subtend an angle with respect to the Polar axis of earth is known as Zenith angle. It can be from the below equation (6)

$$\theta_z = \cos^{-1}(\cos\phi\cos\delta\cos\omega + \sin\phi\sin\delta) \quad (6)$$

Necessary corrections to available radiation data of considered location

Solar Radiation data which is divided into three major groups i.e., direct or beam radiation, diffuse radiation, Total solar Radiation. Total solar radiation is sum of direct, diffuse, terrestrial radiation. Solar Radiation data plays a vital role in estimating electrical power output from the installed capacity if we want to simulate by using software, the radiation data which can be obtained from IMD on payment basis which is horizontal data need to be corrected if the considered plane is tilt with some angle. Here considered IMD's relations and

Mani's relations are consider to find total solar radiation which are obtained from the report Solar Radiant Energy over India [7].

Relationships given by IMD: The Total Solar Radiation G_T falling on inclined surface has three components. They are direct solar radiation or Beam G_B, diffuse solar radiation G_D and reflected solar radiation G_R which are measured in W/ m^2 i.e., Power, if we integrate for specific time then in J/m^2 i.e., Energy.

$$G_T = G_B + G_D\downarrow + G_R\uparrow \quad (7)$$

Beam Radiation G_B of Inclined surface is given by G_{BH} which is beam radiation on horizontal surface.

$$G_B = G_{BH}(\cos\theta / \cos\theta_z)$$

Mani's Relation Ship between global solar radiation on horizontal surface and global solar radiation on horizontal surface. Tilt Factor: K_I

$$K_I = (G_T \downarrow / G\downarrow) \quad (8)$$

$$K_I = \cos\beta + \frac{\sin\beta\cos(\gamma_s - \gamma)}{\tan\alpha} \quad (9)$$

Where G is Global Solar Radiation on Horizontal Surface G_T is Total Solar Radiation on tilted Surface G_D is diffuse solar radiation on horizontal surface β is inclination angle of surface

Extraction of Electrical Parameters of Solar Cell

The solar modules are direct current electrical generators in solar photovoltaic power plant. To estimate electrical output from the considered solar photovoltaic power plant using software's as the direct installation is costly, as manufacturing cost is more. So, before installing solar photovoltaic power plant in a particular location one must be aware of meteorological data of the location which plays very important role in electrical output from the installed plant and the required meteorological data is Radiation data, Wind speed, Temperature which affects the electrical parameters of solar modules. The electrical parameters of solar modules are photo

Current I_{ph}, Dark saturations current I_o, Ideality Constant A, Series resistance R_s, Shunt Resistance R_p [2,8]. These electrical parameters which are not present in the data sheet given by manufacturer so we need find these parameters in order to estimate electrical output from the consider module at particular location i.e., Radiation, Temperature and Wind speed.

Solar photovoltaic module: Data sheet terminology

The solar photovoltaic module is made of Ns cells connected in series and Np series strings connected in parallel. Thus formed solar photovoltaic module is Electrical DC generator using solar radiation as input is current source. The current is obtained by the photons of light which when fall on the module creates electron-hole pairs this effect is known as photovoltaic effect, which are allowed to flow through the load connected across the module. The flow of electrons through load is possible because electric field created across the junction of each cell because each photovoltaic cell is a diode which forms by combining P-type and N-type material, each diode has breakdown voltage of around 0.6-0.8 (V). Thus voltage of each cell when combines in series in module adds up to form module voltage. From the data sheet of considered photovoltaic module we get Maximum power P_{mp}, open circuit voltage V_{oc}, short circuit I_{sc}, maximum power point voltage V_{mp}, maximum power point current I_{mp}, efficiency of module η_m, at STC conditions i.e., $G_{n,T}$ =at 1000 W/m², T_a=T_c=298 K, air mass 1.5, wind speed W_i=1 m/s, and as well as at NOCT condition i.e., at $G_{T,NOCT}$ =800 W/m², T_{NOCT} =293 K, air mass 1.5, wind speed W_i=1 m/s. Temperature coefficients of Short circuit current κI_{SC}, Open circuit Voltage κV_{OC}, Maximum Power Point κP_{mp}, No. of Cells in series and parallel, area of cell and module, weight of the module. By using these values we need to find the above said electrical parameters of considered module to accurately estimate electrical output from the module at particular location and climatic conditions.

Mathematical equations

Electrical parameters of considered solar module as mention above are combine to form basic equation as show below which are obtained from equivalent electrical circuit of solar cell and applying KCL to the circuit

$$I_o = I_{o,n}(\frac{T_c}{T_a})^3 \exp\left[\frac{qE_g}{A_k}(\frac{1}{T_a}-\frac{1}{T_c})\right] \quad I = I_{ph} - I_0\left(\exp\left(\frac{V+IR_S}{AV_t}\right)-1\right) - \frac{V+IR_S}{R_p} \quad (10)$$

$$V_t = N_S kT_{c/q}$$

$$I_{ph} = \frac{G_T}{G_{n,T}}[I_{sc} + \kappa I_{SC}(T_c - T_a)] \quad (11)$$

$$I_{o,n} = \frac{I_{SC}}{\exp(\frac{V_{oc}}{AV_{T,n}})-1}$$

$$I_o = I_{o,n}(\frac{T_c}{T_a})^3 \exp\left[\frac{qE_g}{A_k}(\frac{1}{T_a}-\frac{1}{T_c})\right] \quad (12)$$

Thermal Voltage at particular cell temperature

After seeing these equations we came to know out off five unknown variables we can find two variables I_{ph},I_0 and other three variables need to be calculated those are R_s, R_p, A. Out of these three A which is ideality constant of diode is taken the value between $1 \le A \le 2$. To get R_s and R_p we need substitute points (V_{oc},0) open circuit condition, (0,I_{sc}) short circuit condition, (V_{mp}, I_{mp}) maximum power point condition which we can obtain from data sheet and solve the equation numerically to get unknown variable and this process is trial and error because here A is assumed, final values of unknown variables can be obtained by observing V $_{vs}$I and V $_{vs}$P characteristics of the module using Matlab program. The minimum and maximum values of R_s and R_p are obtained respectively using below equations [9].

$$R_{s,\max} = \frac{V_{oc} - V_{mp}}{I_{mp}}$$

$$R_{p,\min} = \frac{V_{mp}}{I_{SC} - I_{mp}} - \frac{V_{oc} - V_{mp}}{I_{mp}}$$

Finding cell temperature of module which is different from ambient temperature which plays important role in simulating electrical output from the photovoltaic module which was designed in Matlab/Simulink. The relation between cell and ambient temperature is given by the below equation.

$$T_c = T_a + \frac{0.32}{(8.91 + 2W_i)} G_T$$

Or

$$T_c = T_a + \left(1 - \frac{\eta_m}{\Gamma\psi}\right) G_T \left(\frac{\Gamma\psi}{\upsilon_L}\right)$$

where Γ is transmittance of cover over cell ψ is fraction of radiation incident on the surface to the observed υ_L is loss coefficient i.e., by convection and radiation from the surface

$$\frac{\Gamma\psi}{\upsilon_L} = \left(T_{c,NOCT} - T_a\right) / G_{T,NOCT}$$

Data sheet values of different solar modules

From the below data sheet values given in Tables 1 and 2 and numerical analysis using Matlab program given by Figure 6 and comparing with I-V, P-V characteristics of data sheet and plot obtained from program written in Matlab accurate values of A, R_s and R_p are tabulated below Table 3 - A2 as I_{ph} and I_o varies with radiation, Temperature of the cell which depends on ambient temperature and wind speed [10].

Meteorological data and electrical energy output estimation

Meteorological data which plays important role in estimating electrical output from solar photovoltaic power plant which was not installed but simulated using Matlab/Simulink software. In this paper

Parameters	Tata Solar	Titan Solar	Vikram Solar	Surana solar
Pmp W	250	250	250	250
Voc V	37.	37.80	37.45	37.8
Iac A	8.71	8.63	8.70	8. 63
Vmp V	30.2	30.72	30.58	30.72
Imp V	8.3	8.14	8.18	8.14
nm	0.15	-	0.153	-
Ns	60	60	60	60
ĶPmp	-1.095	-1.075	-1.025	-1.025
ĶV oc	-0.123	-0.120	-0.116	-0.120
ĶIsc	0.0055	0.00345	0.00504	0.00345
Area m²	1.460	1.460	-	1.460
TcNOCT K	320	318	318	-
Bipass Diodes	-	3	3	3

Table 1: Data sheet values of different manufactures at stc.

Parameters	Photon Solar	Jupiter solar	EMMVEE solar
Pmp W	250	250	250
Voc V	43.06	37.0	37.62
lac A	7.74	8.87	8.76
Vmp V	36.07	30.5	29.76
Imp V	6.93	8.3	8.40
nm	0.129	0.16	0.154
Ns	72	60	60
ĶPmp	-1.075	-1.05	-1
ĶV oc	-0.142	-0.1073	-0.127
ĶIsc	0.00448	0.00354	0.00438
Area m^2	1.752	-	1.46
TcNOCT K	-	318.5	319
Bipass Diodes	-	-	3

Table 2: Data sheet values of different manufactures at stc.

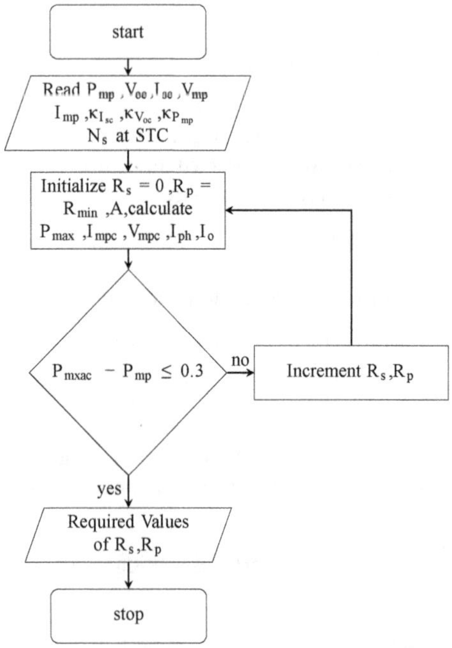

Figure 6: Flow chart to find five parameters of solar cell.

Manufacturer	Idelity constant A	RsΩ	RpΩ
TATA Solar	1.3	0.1165	1444.9
Vikram Solar	1.154	0.2	810
Titan Solar	1.14	0.23	1290
Surana Solar	1.138	0.23	1176
Photon Solar	1.64	0	919.905
Jupiter Solar	1.52	0.3	875
EMMVEE Solar	1.2	0.23	1449

Table 3: Calculated electrical parameters of different modules.

author considered data of Radiation on tilt surface with tilt angel of $\beta = 17.7$ because it gives maximum output for fixed mounting of solar photovoltaic module. This data consists of monthly average daily hourly average instantaneous radiations, temperature, wind speed from year 1986-2000 [7] and comparing with meteorological data collected for month April 2015 [11] at considered location and comparing simulation results electrical energy output for month April and total year using above data from installed 110.25 KW solar photovoltaic power plant. The yearly electrical energy output at particular location is compared with other software's like PVsyst, SAM, PVwatts and with program developed in this paper values given in Tables 4-6 [12].

Cost Estimation

Solar Photovoltaic Power Plant installation is more costly compare to conventional power plants like Thermal, nuclear, Gas and Oil. But operating cost of solar photovoltaic power is negligible compare to conventional power plants. More over there is no pollution and hence known as green power. By proper sizing and selecting location with care can make comparison cost of energy generation from solar plant with conventional power plants. Electrical Energy generated from the solar power plant is not continuous and thus limits to replace conventional power plants. From Table 7 it was known that considered load consumes 16,57,614 units of electrical energy per year, hourly demand of load is not known as the load is educational institute it works for 10 hours a day and with working days around 240 working days, per day energy consumption on an average is around 6906.72 units, per hour around 690.67 units. To generate 690.67 units per hours solar photovoltaic power plant capacity should be around 1.5 MW, by considering energy output from 1 KW is around 4.8-5.13 units per day at considered location i.e., Andhra university, Visakhapatnam. Initial cost is more and more over educational institutes will get 30% subsidy on 100 KW solar photovoltaic power plant from government of India [13,14]. To estimate cost of electrical energy per unit we need to know solar photovoltaic power plant equipment used and there cost. According to Government of India bench mark cost for 1 MW solar photovoltaic power plant for the year 2015-2016 is Rs 605.85 Lakhs. Which include Photovoltaic Modules, Land cost, civil and general works, mounting structures, Power conditioning unit, Cables and Transformers, Preliminary and Pre-operative Expenses. Photovoltaic panel costs around 55% of total cost so one must be careful in choosing

Source	Electrical Energy Output
Matlab data (1986-2000)	20395 KWh
Matlab data (April 2015)	17605 KWh
PVwatts	17444 KWh
PVsyst	17130 KWh
SAM	18320 KWh

Table 4: Electrical energy produced from 110.25 kw for the month April of considered location.

Source	Electrical Energy Output
Matlab	159818 KWh
PVwatts	175266.32 KWh
PVsyst	179770 KWh
SAM	195724 KWh

Table 5: Electrical energy produced from 110.25 kw for year of considered location.

Year	Rs/KVAh	Rs/KVA	Bill paid (Rs)
2013-14	6.90	350	-
2014-15	6.90	350	1,52,58,163
2015-16	7.31	371	-

Table 6: Tariff per kvah paid to conventional utility.

Year	MD(KVA)	Units(KVAh)	Rs/KVAh
2014-15	650.46 (700)	1657614	9.204

Table 7: Yearly electrical energy consumed by considered load.

KVAh generated	Total Cost(Rs)	Rs/KVAh
3579930	11754295	3.8

Table 8: Life time energy output from installed 110.25 kw Solar Photovoltaic power plant.

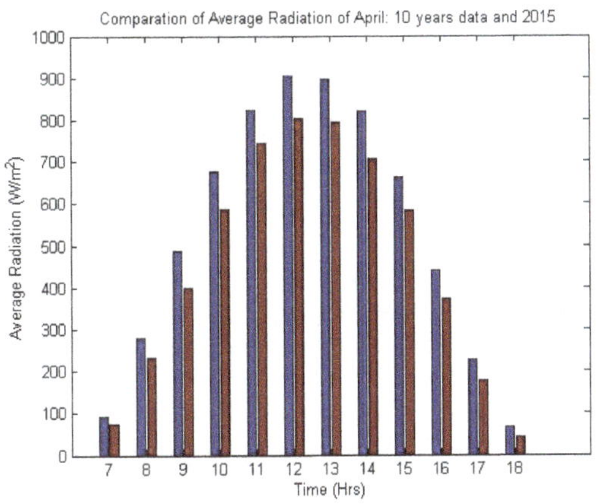

Figure 7: Average radiation data of month April.

photovoltaic panes. From the above benchmark cost for 1 MW we can estimate cost for 110: 25 KW i.e., around Rs 58.733 lakhs. Now in this paper, considered load College of Engineering, Andhra University which consumes electrical energy as shown in Table 8 [15-17].

Factors Determining Solar Plant Feasibility

Performance Ratio, Capacity Utilization Factor, LCOE are important factors to compare photovoltaic power plant with conventional power plants [18]. Performance Ratio is the ratio of energy generated or measured to energy modelled. This parameter is used to compare the actual plant performance which used to rectify fault which decrease energy output from the installed system.

$$PR = \frac{\text{EnergyMeasured}}{\text{EnergyModelled}} \tag{14}$$

Capacity Utilization Factor is the ratio energy output from installed system to the energy it would have generated if the system is available for 24 hours and 365 days. $LCOE = \frac{\text{Toatal Life cycle Energy generated}}{\text{Initialcost} + \text{O \& Mcost}}$

$$CUF = \frac{\text{Energy Measured}}{\text{Capacity of Plant} * 8760} \tag{15}$$

Levelized Cost of Electricity is used to compare cost per unit of non-conventional system with conventional system. It is the ratio of total life cycle cost to total lifetime energy production. Total life cost includes initial cost, operation and maintenance cost.

$$LCOE = \frac{\text{Toatal Life cycle Energy generated}}{\text{Initialcost} + \text{O \& Mcost}} \tag{16}$$

From Table 7 it is known how much money paid by the consumer to the utility and units consumed by considered Load. If for the considered load 110.25 KW solar photovoltaic power is installed, in this paper comparison was made between retailer price and renewable energy source unit price [19]. To calculate cost of generation of one unit electrical energy from the installed load and payback period, Performance ratio and capacity utilization factor of installed solar photovoltaic plant. Solar photovoltaic power life time is considered 25 years in which it yields 90% of electrical energy for first 10 years and then up to 25 years it yields about 80% comparing with first years

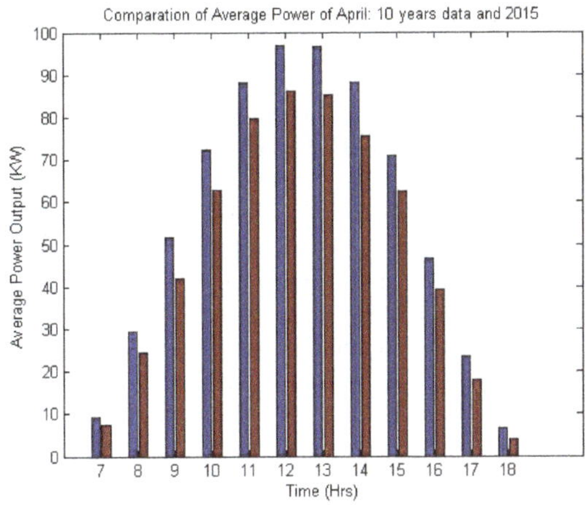

Figure 8: Comparison of power output for month April using 1986-2000 data and 2015 April data.

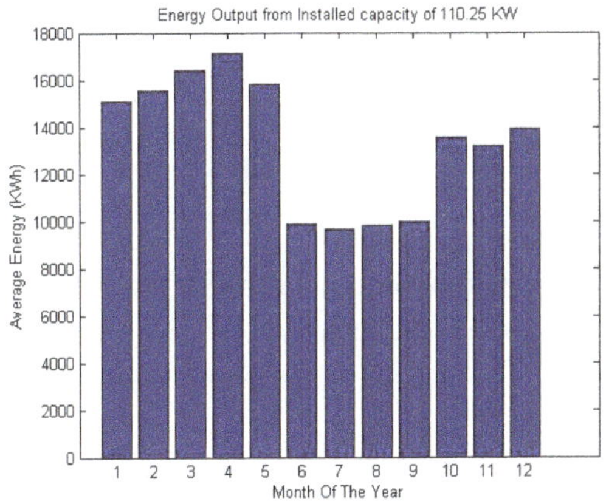

Figure 9: Estimating electrical energy output from 110.25 kw solar PV system using 1986-2000 data.

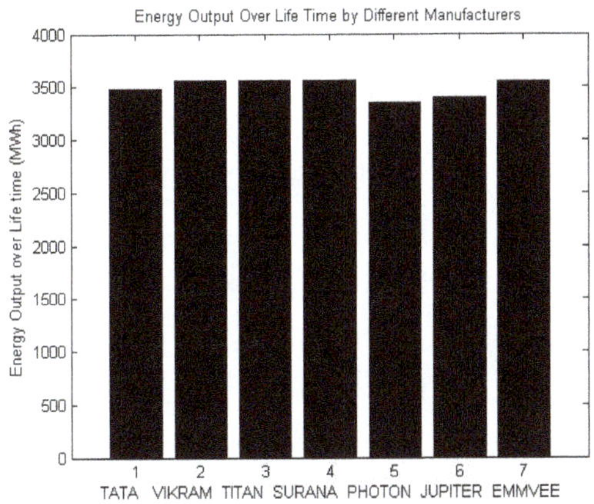

Figure 10: Energy output by using modules of different manufacturers.

output. Operation and maintenance cost will be around Rs 1.3 lakhs [20,21].

Conclusion

Estimation of electrical power output from Solar Photovoltaic Power Plant at particular location using Matlab programme requires accurate calculation of five parameters of solar cell, calculation of radiation data on tilted surface and knowing location temperature and wind speed (Figures 7-10). In this paper 110.25 KW solar photovoltaic power plant is considered and using radiation data of Visakhapatnam from year (1986-2000) and radiation data of the month April 2015 to calculate AC energy output and compare these results with PVwatts, PVsyst, SAM. The calculated values of LCOE is 3.8 Rs/KVAh given in Table 5 if plant operates for 25 years and first year energy Production was 159818 KVAh and for 25 years the energy output is 3579930 KVAh.

Reference

1. Duffie JA, Beckman WA (2013) Solar Engineering of Thermal Processes (2nd edn). New York.
2. Sulaiman SA, Hussain HH, NikLeh NSH, Razali MSI (2011) Effects of Dust on the Performance of PV Panels. World Academy of Science, Engineering and Technology 5: 2028-2033.
3. Skoplaki E, Boudouvis AG, Palyvos JA (2008) A simple correlation for the operating temperature of photovoltaic modules of arbitrary mounting. Solar Energy Materials and Solar Cells 92:1393-1402.
4. Green MA (1982) Solar cells: operating principles, technology and system applications. Prentice-Hall, Inc., Englewood Cliffs, NJ, USA.
5. Cambell M (2008) The Drivers of the Levelized Cost of Electricity for Utility-Scale Photovoltaics. Sun Power.
6. Solar Radiation Monitoring Laboratory, University of Oregon.
7. Tyagi AP (1988) Solar Radiant Energy Over India. Indian Meteorological Department, India.
8. Shongwe S, Hanif M (2015) Comparative Analysis of Different Single-Diode PV Modeling Methods. IEEE Journal of Photovoltaics 5: 938-946.
9. Villalva MG, Gazoli JR, Filho ER (2009) Comprehensive Approach to Modeling and Simulation of Photovoltaic Arrays. IEEE Transactions On Power Electronics 24: 1198-1208.
10. Sera D, Teodorescu R, Rodriguez P (2007) PV Panel model based on datasheet values. IEEE 2392-2396.
11. http://www.imdaws.com/ViewRadiationData.aspx.
12. Kandasamy CP, Prabu P, Niruba K (2013) Solar Potential Assessment Using PVSYST Software. IEEE 667-672.
13. Solar (2012) Government of India, Ministry of New and Renewable Energy.
14. New & Renewable Energy Development Corporation of Andhra Pradesh (NREDCAP).
15. Central Electricity Regulatory Commission (2015) CERC, New Delhi.
16. Gan CK, Tan PH, Khalid S (2013) System Performance Comparison Between Crystalline and Thin-Film Technologies under Different Installation Conditions. IEEE Conference on Clean Energy and Technology 362-367.
17. Ahmed MM (2003) Design and Proper Sizing of Solar Energy Schemes for Electricity Production in Malaysia. National Power and Energy Conference (PECon) Proceedings 268-271.
18. Singh VP, Ravindra B, Vijay V, Bhatt MS (2014) A comparative performance analysis of C-Si and A-Si PV based rooftop grid tied solar photovoltaic systems in Jodhpur. 3rd International Conference On Renewable Energy Research And Applications 250-255.
19. Endo E, Kurokawa K (1994) Sizing Procedure For Photovoltaic Systems. IEEE 1:1196-1199.
20. Zolkapli M, Al-Junid SAM, Othman Z, ManutA, MohdZulkifli MA (2013) High-Efficiency Dual-Axis Solar Tracking Developement using Arduino. International Conference on Technology, Informatics, Management, Engineering and Environment 23-26.
21. Bajpai P, Kumar S (2011) Design, Development and Performance Test of an Automatic Two-Axis Solar Tracker System. India Conference (INDICON), 2011 Annual IEEE 1-6.

13

Reactive Power Management using Firefly and Spiral Optimization under Static and Dynamic Loading Conditions

Ripunjoy Phukan*

Indian Institute of Technology, Guwahati, India

Abstract

Power System planning encompasses the concept of minimization of transmission losses keeping in mind the voltage stability and system reliability. Voltage profile decides the state of a system and its control is dependent on Generator source voltage, shunt/series injection, transformer taps etc. Optimal parameter setting in system level is needed for managing the available resources economically. This paper presents the use of Firefly and Spiral optimization as novel schemes for minimizing the active power loss along with partial compensation of inter bus voltage drop. The objective function has been evaluated under both static and dynamic loading conditions. The control variables being generator bus voltage, capacitor shunts and transformer taps. These methods were employed in an IEEE 6-bus system and the results were tabulated.

Keywords: Compensation; Dynamic loading; Firefly; Spiral optimization; Transmission loss

Introduction

Reactive power optimization problem is a non-linear combinatorial optimization problem. This problem came to focus on account of the majority of loads being reactive in nature. The modern power system is usually controlled by a monopoly resulting in freedom of action and better controllability of generator voltage levels, shunt capacitors and transformer taps (vertical integration). Under stable running conditions the devices are operational at rated values and have minimal stress. However, a static condition is expected to stay for a maximum duration of 10 minutes. During dynamic loading, the optimal parameters must be incorporated into the system to curtail minimum losses and prevent a power outage.

Shunt injection using reactive sources, such as a capacitor, creates phase advancement in line voltage. Assuming lossless lines, the resistance is assumed zero, creating a purely inductive line. Transformer taps also cause the necessary stepping of voltages near stressed lines (Lines where the thermal limits are easily attained on account of loading). Research into accurate location of such transformers has been shown in the study of Kargarian and Raoofat [1]. The positioning in this paper is assumed fixed, i.e. the system is functional since a long time period, and relocation may add to costs. One advantage for using compensators is the absence of fuel costs, while generator scheduling adds to fuel costs as well. However in this paper the economic aspect of VAR Compensation is neglected. The efficiency of Shunt VAR Compensators, Thyristor Switched Capacitors, Thyristor Controlled Reactors have proved the validity in the use of FC-TCR elements in reactive voltage control. STATCOM devices utilize switching strategies for controlling reactive power flow through a bridged circuit, as proposed in the study of Singh and Elmoursi [2,3].

The conventional methods of LaGrange's Multipliers prove incompatible with these applications due to stagnation and localization of optimum solutions [4]. Some of these techniques include linear programming, non-linear programming, mixed integer method, decomposition method. In recent years some Artificial Intelligence methods such as expert systems, neural networks and simulated annealing have been developed. The modern optimization techniques fall under the genre of Artificial Intelligence, where the natural mimicry is postulated in the form of equations. Several such techniques have been stated in [5]. Previous works on Reactive Power Planning include the application of Genetic Algorithms [6], followed by Particle Swarm Optimization [7,8]. Genetic-Simulated annealing and interior point method [9] along with adaptive PSO [10] based on optimal control principle were a few improvisations. Other approaches incorporate the usage of load forecasting using Radial Basis functions [11] and successive quadratic programming methods [12]. These methods have proven to be effective and are useful in their own versions. Al-alawi, et al. [13] have presented an ANN based technique for tuning SVCs in a power system. Significant amount of literature survey indicates that ANN and deregulated conditions have been enforced in [14-16]. Hybridized market and congestion management have been invoked in [17-19]. Voltage control area reserves and wind power volatility have been dealt with in papers [20] and [21].

The Firefly Optimization invokes the luminescent behavior of fireflies [22]. Earlier works with fireflies include the Economic Load Dispatch problem [23]. Spiral optimization is a latest technique introduced by Kenichi Tamura [24]. It interprets the spiral phenomenon in nature such as hurricanes and whirling currents. These techniques have been applied to a static IEEE-6 bus system with partial compensation. The system architecture is similar to [7]. The concept of compensation has been additionally introduced in the fitness function. Further calculations were done for a dynamic system as well [25,26]. A comparison between both the algorithms has been given and the success of spiral optimization as a potent Reactive Power Optimization tool has been justified. The following two sections (4 and 5) deal with the algorithms employed in detail. This is followed by the VAR optimization method in section 6 with results and tabulation in 7. Finally, the Conclusion is derived in Section 8.

***Corresponding author:** Ripunjoy Phukan, Indian Institute of Technology, Guwahati, India, E-mail: ripun000@yahoo.co.in

Firefly Optimization

Fireflies

This optimization algorithm can be used for solving real valued multi-modal optimization problems. It was first proposed by D. Ghose in 2006. If there are no fireflies brighter than a given firefly, it will move randomly. Fireflies use the concept of path identification based on light intensity from another firefly. Being a stochastic algorithm, it exploits all possible regions of the search space for information. This algorithm was inspired by the motion of fireflies and glow-worms in search for a mate, based on their luminescence behavior. Naturally, a firefly would move in the direction of maximum brightness indicating the most feasible mate. The light intensity (brightness) is given by equation (1). It varies exponentially with the distance between two fireflies and is irrespective of their sex. As the attractiveness of fireflies is proportional to the light intensity, similar variations are found between attractiveness and distance (2). In practical problems the light intensity is determined by the landscape of the objective function.

$I = I_0.e^{-\gamma r}$ (1)

$\beta = \beta_0.e^{-\gamma r}$ (2)

The movement of fireflies towards one another is governed by equation (3). For most cases in our implementation we take γ as 1.0 and β_0 as 1.0.

$x_i = x_i + \beta.e^{-\gamma r^2}.(x_j - x_i) + \alpha.\epsilon_i$ (3)

where, x_i is the current position in 1 Dimension and x_j is the previous best position of the fireflies. The following update is carried over all the dimensions. The parameter $\alpha.\epsilon_i$ is a user defined bias control to make corrections in deviations from the optimal value on account of external disturbances. The algorithm for Fireflies is expressed in Figure 1.

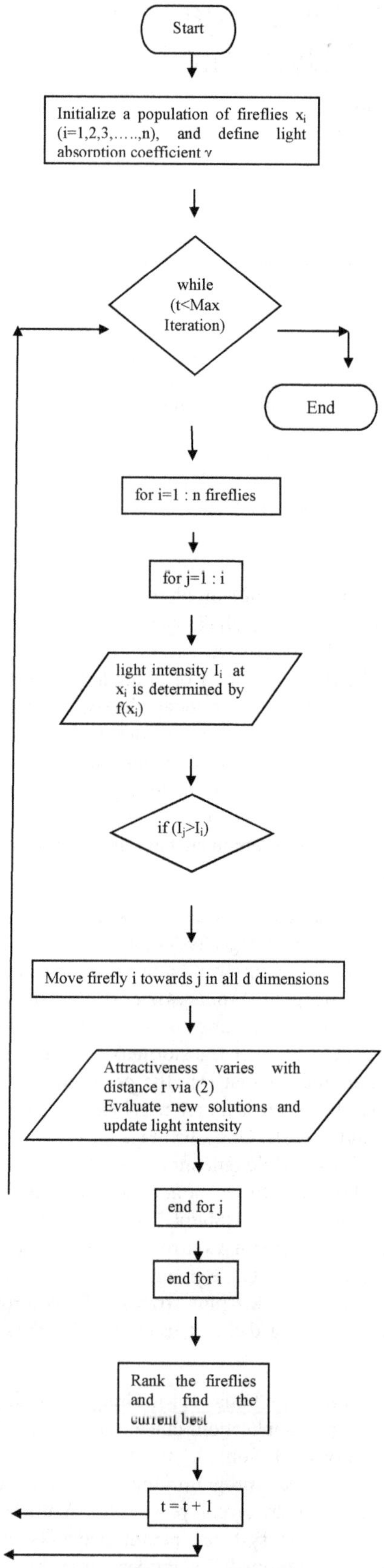

Figure 1: Firefly Algorithm.

Spiral Optimization

Spiral phenomenon

The focused spiral phenomenon is approximated to logarithmic spirals which frequently appear in nature. For example, the Nautilus shell, hurricanes and whirlpools. These spirals are associated with gradually decreasing radius as the vector rotates with reference to the previous points. A 2-dimensional model uses the rotation vector as given in equation (4). The movement of particles is described by equation (6). For a greater degree of search space exploitation, the angle, θ, can be increased while for concentration over a smaller region it can be decreased. Similar variations can be done with the radius as well.

$$x' = R_{1,2}^{(2)}(\theta)x \tag{4}$$

where, rotation matrix is given by

$$R_{1,2}^{(2)}(\theta) = \begin{bmatrix} \cos\theta & -\sin\theta \\ \sin\theta & \cos\theta \end{bmatrix}$$

For an 'n' dimensional system the rotation matrix is modified as shown by (5),

$$R^{(n)}(\theta_{1,2},\theta_{1,3},\theta_{1,4},\theta_{n-1,n}) = R_{n-1,n}^{(n)}(\theta_{n-1,n}) \times R_{n-2,n}^{(n)}(\theta_{n-2,n}) \times R_{1,n}^{(n)}(\theta_{1,n}) \times R_{1,3}^{(n)}(\theta_{1,3}) \times R_{1,2}^{(n)}(\theta_{1,2})$$

$$= \Pi_{i=1}^{n-1}(\Pi_{j=1}^{i} R_{n-i,n+1-j}^{(n)}(\theta_{n-i,n+1-j})) \tag{5}$$

where, $0 < \theta_{i,j} < 2\pi$ are rotation angles each plane around the origin at every k, $0 < r < 1$ is the convergence rate between any point and the optimal solution. The positional update is given by equation (6).

$x(k+1)=S_n(r,\theta)x(k)-(S_n(r,\theta)-I_n)x^*$ (6)

where, I_n is the identity matrix of order (n x n) and x^* is the previous best global solution.

Algorithm in N dimensions

Step 0: [Preparation]: Select the number of search points m ≥ 2, the parameters $0 < \theta < 2\pi$, $0 < r < 1$, of $S_n(r,\theta)$, and the maximum number of iterations k_{max} with $k = 0$.

Step 1: [Initialization]: Set the initial points $x_i(0) \in R^n$, $i=1,2,3,\ldots,m$ in the feasibility region at random and the center x^* as $x^* = x_{ig}(0)$, $ig = \arg\min f(x_i(0))$, $i = 1,2,\ldots,m$.

Step 2: [Updating x_i]: X is updated as per equation (6).

Step 3: [Updating x^*]: $x^* = x_{ig}(k+1)$, $ig = \arg\min f(x_i(k+1))$, $i = 1,2,\ldots,m$.

Step 4: [Checking termination criteria]: If $k = k_{max}$ then terminate. Otherwise set $k = k + 1$, and return to Step 2.

The spiral phenomenon from equation (4) for $r = 0.95$

and $\theta = \pi/4$ is as shown in Figure 2. As observed both these algorithms have the potential for attaining a minimum, with parameter optimization. This phenomenon is also termed as combinatorial optimization. Further, it will be shown that these algorithms have the capability to converge within lesser iteration count. So, for dynamic processes these algorithms prove efficient.

Problem Formulation

Reactive power scheduling is usually carried out every ten minutes to one hour [13]. It consists of the optimization at each discrete time instant k of a power system characterized by operational conditions $r(k)$, which represent the scheduled load demand, active power generation pattern, and network topology at instant k.The main objective of the Reactive Power Optimization problem is to minimize the active power loss by means of controlling the Generator real power outputs, Shunt capacitor banks and the transformer tap positions. An-online simulation during dynamic conditions can prove to be ineffective due to time constraint. So, offline studies are done to mitigate security problems in future avoiding grid collapse or blackouts. A repeated simulation for various

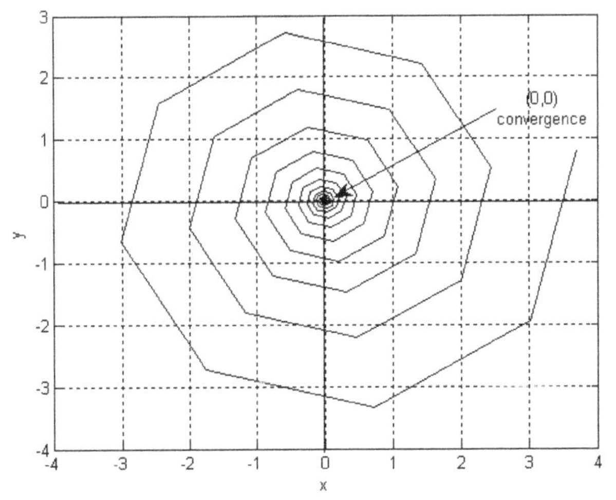

Figure 2: A 2-dimensional spiral towards (0,0).

loading conditions can be done to create an adaptive look up table that eases the process of decision making. For the sake of convenience we have assumed the generator and load buses to be loaded to 1.2 times of their base values. These parameters are the positional co-ordinates in our optimization algorithm and are updated after every iteration. The bus voltages are calculated by means of a suitable load flow analysis. The scope of the problem is enhanced when we introduce several operating constraints in the problem. The transformer taps, voltage magnitudes and the reactive shunt compensation must be within the prescribed limits. The equality constraints are automatically satisfied using load flow analysis, while the inequality constraints need to be satisfied using randomization in optimization algorithms. Another possible solution would be to increase the step size in parameter movement.

Static loading conditions

The static loading condition occurs in a system when it is operating under normal state. The active real power losses in this system are given by equation (7).

$$P_{Loss} = \sum_{k=1}^{ntl} G_k(V_i^2 + V_j^2 - 2V_iV_j Cos(\delta_i - \delta_j)) \quad (8)$$

The constraints involved are shown below:

(1) Load flow constraint:

$$\begin{bmatrix} \Delta P_i \\ \Delta Q_i \end{bmatrix} = \begin{bmatrix} P_i(V,\theta)-(P_{gi}-P_{di}) \\ Q_i(V,\theta)-(Q_{gi}-Q_{di}) \end{bmatrix} = 0$$

where,

$$P_i(V,\theta) = \sum_{j=1}^{nbus} V_iV_j(G_{ij}\cos\theta_{ij} + B_{ij}\sin\theta_{ij})$$

and

$$Q_i(V,\theta) = \sum_{j=1}^{nbus} V_iV_j(G_{ij}\sin\theta_{ij} + B_{ij}\cos\theta_{ij})$$

$i \in n$, where set of buses except the swing bus.

(2) Bus voltage magnitude constraints:

$V_{i,min} \le V_i \le V_{i,max}$ (9)

$i \in$ number of buses

(3) Generator bus reactive power constraints:

$Q_{Gi\text{-}min} < Q_{Gi} < Q_{Gi\text{-}max}$ (10)

$i \in$ number of PV buses

(4) Reactive power source capacity constraints:

$Q_{sh\text{-}min} < Q_{sh\text{-}i} < Q_{sh\text{-}max}$ (11)

$i \in$ number of buses with shunt capacitors

(5) Transformer taps position constraints:

$T_{i\text{-}min} \le T_i \le T_{i\text{-}max}$ (12)

$i \in$ number of transmission lines with OLTC transformers

Dynamic loading conditions

During dynamic loading conditions the power system network is subjected to intermittent load changes resulting in an acute necessity to govern the reactive power. While doing this, the need for short reaction times is increasingly felt. The usage of participation factors is considered for calculating the dynamic real and reactive power of generator, for a time instant (say T+ΔT). The percentage of load increase is modeled by $\lambda.K_{gi}$ and $\lambda.K_{di}$, where K_{gi} and K_{di} are the distribution factors for generators, and load buses respectively. The parameter λ specifies the direction and factor by which the load increases and the generation must increase. Rather than simulating in a dynamic environment with frequent loading and unloading conditions, we have tested the algorithms on a smaller version involving step changes in load. This can also be referred to as a perturbation equation.

$$P_{gi}(loaded) = (1 + K_{gi}.\lambda.\sum_{i=1}^{Ng} P_{gi}^0).P_{gi}^0 \quad (13)$$

$$P_{di}(loaded) = (1 + K_{di}.\lambda.\sum_{i=1}^{Ng} P_{di}^0).P_{di}^0 \quad (14)$$

$$K_{gi} = \frac{P_{gi}}{\sum_{i=1}^{Ng} P_{gi}^0} \quad (15)$$

$$K_{di} = \frac{P_{di}}{\sum_{i=1}^{Ng} P_{di}^0} \quad (16)$$

Compensation using OLTC transformers

In general a tap changing transformer is placed between any two buses for reducing the voltage drop and hence, the losses. This method has gained massive potential due to its equivalence with an auto-transformer. This arrangement works best if the voltages at the two ends are equal. However, a practical condition will always show different voltages at the two ends. So the system is said to be partially compensated. With this in mind, equation (17) holds true. V_1 and V_2 are obtained from the optimization algorithm and the transformer tap positions are checked directly from equation (17). The resulting losses are given by equation (18) (Figure 3).

$$t_r^2.\left(1 - \frac{P_i.R + X_iQ}{V_iV_j}\right) = \frac{V_i}{V_j} \quad (17)$$

$$P_{Loss} = \frac{\left(t_r.V_i - \frac{V_j}{t_r}\right)^2}{Z_{ij}} \quad (18)$$

V_1 and V_2 are the nominal voltages at the ends of the line and the actual voltages being t_sV_1 and t_rV_2. The tap changing ratios are used to partially compensate the system. As t_st_r is made equal to unity it ensures that the overall voltage level remains in the same order. The impedance of the line is *R+jX*.

Parameters:

P_{loss} = System loss (Fitness Function)

V_i = Voltage at bus 'i'

V_j = Voltage at bus 'j'

δ_i = Phase angle of V_i

δ_i = Phase angle of V_j

P_{Gi} = Bus 'i' real power supply

Q_{Gi} = Bus 'i' reactive power supply

P_{Di} = Bus 'i' real power load

Q_{Di} = Bus 'i' reactive power load

G_{ij} = Mutual conductance between bus i and j

B_{ij} = Mutual susceptance between bus i and j

θ_{ij} = Phase angle difference between buses 'i' and 'j'

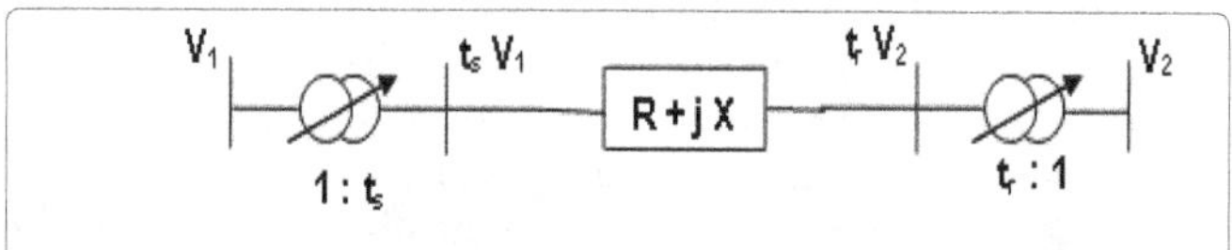

Figure 3: Transmission Line with tap setting transformer.

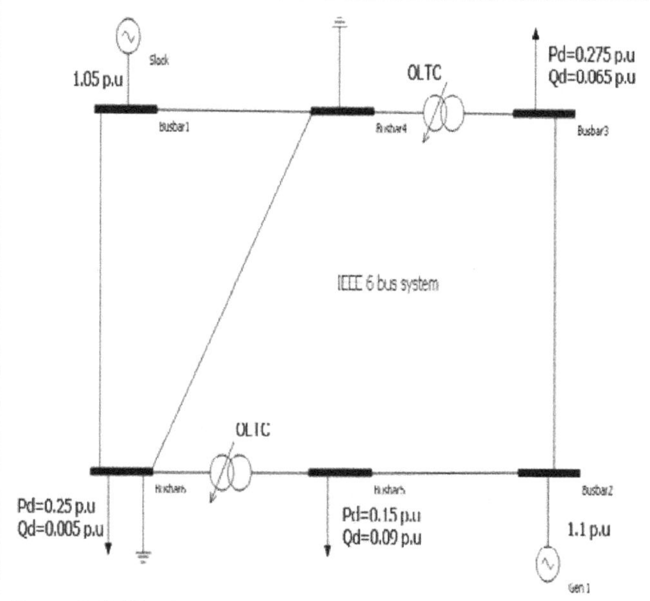

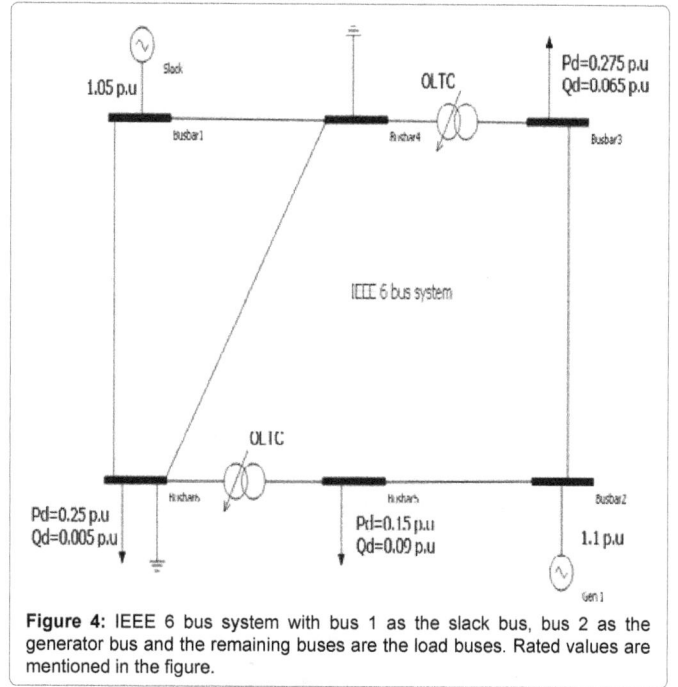

Figure 4: IEEE 6 bus system with bus 1 as the slack bus, bus 2 as the generator bus and the remaining buses are the load buses. Rated values are mentioned in the figure.

$V_{i,min}$, $V_{i,max}$ = Bus 'i' voltage limits

Q_{Gi-min}, Q_{Gi-max} = Bus 'i' reactive power limits

Q_{sh-min}, Q_{sh-max} = Shunt reactive power limits

T_{i-min}, T_{i-max} = Transformer 'i' tap position limits

T_i = transformer tap position

$Q_{sh,i}$ = Bus 'i' shunt reactive power

With compensation and dynamic loading the overall fitness function is given by:

$$P_{Loss} = f(V_i, Q_{sh-i}, T_i) = \sum_{k=1}^{ntl} G_k(V_i^2 + V_j^2 - 2V_iV_jCos(\delta_i - \delta_j)) + \frac{\left(t_r.V_i - \frac{V_j}{t_r}\right)^2}{Z_{ij}} \quad (19)$$

In the above equation, the real power losses between nodes '*i*' and '*j*' must be neglected and reproduced in the form of the second term in the expression.

Results and Discussion

An IEEE 6 bus system is considered and the optimal value for generation and transformer taps is tabulated. The minimum loss is obtained from the fitness function (19). The IEEE 6 bus system is shown in Figure 4. The standard IEEE 6 bus system data was taken from the work of Tamura and Yasuda [24] (Tables 1 and 2). The figure was made in Power System Simulator for clarity.

The optimization parameter results for a static and a dynamic system have been tabulated in Tables 3 and 4 respectively. The optimization parameter results for a static and a dynamic compensated system have been tabulated in Tables 5 and 6 respectively.

The load flow study was conducted using Gauss Siedal method for 5 iterations. β_0 was taken as any random variable between (0 and 1). Spiral and F.A. have been run for 100 iterations each. λ has been taken as 0.2. The net power loss is tabulated for all the cases considered in Table 7.

The convergence characteristic of these algorithms is studied

LSB	LTB	R (pu)	X (pu)	OLTC
1	6	0.123	0.518	1.0
1	4	0.08	0.370	1.0
4	6	0.097	0.407	1.0
6	5	0.000	0.300	1.025
5	2	0.282	0.320	1.0
2	3	0.723	1.050	1.0
4	3	0.000	0.133	1.1

Table 1: System Line Data.

Bus No.	Vm (pu)	P_d (pu)	Q_d (pu)	Q_g min (pu)	Q_g max (pu)	Q_{sh} (pu)
1	1.05	0	0	0	0	0
2	1.1	0	0	-0.4	0.5	0
3	1.0	0.275	0.065	0	0	0
4	1.0	0	0	0	0	0
5	1.0	0.15	0.09	0	0	0
6	1.0	0.25	0.005	0	0	0

Table 2: System Bus Data.

Variable	Lower limit	Upper limit	Firefly (pu)	Spiral (pu)
V_{G1}	1.000	1.100	1.0054	1.0157
V_{G2}	1.100	1.150	1.1173	1.1224
Q_{SH1}	0.000	0.050	0.035	0.0183
Q_{SH2}	0.000	0.055	0.0044	0.0069

Table 3: Optimization Results.

Variable	Lower limit	Upper limit	Firefly (pu)	Spiral (pu)
V_{G1}	1.000	1.100	1.0064	1.0005
V_{G2}	1.100	1.150	1.1023	1.1088
Q_{SH1}	0.000	0.050	0.0009	0.0248
Q_{SH2}	0.000	0.055	0.0131	0.0345

Table 4: Optimization Results.

Variable	Lower limit	Upper limit	Firefly (pu)	Spiral (pu)
V_{G1}	1.000	1.100	1.0979	1.0921
V_{G2}	1.100	1.150	1.1472	1.1130
Q_{SH1}	0.000	0.050	0.0361	0.0475
Q_{SH2}	0.000	0.055	0.0033	0.0013
T_{43}	0.9	1.1	1.0216	1.0229
T_{56}	0.9	1.1	1.0353	1.0342

Table 5: Optimization Results.

Variable	Lower limit	Upper limit	Firefly (pu)	Spiral (pu)
V_{G1}	1.000	1.100	1.0026	1.0903
V_{G2}	1.100	1.150	1.1244	1.1487
Q_{SH1}	0.000	0.050	0.0440	0.0548
Q_{SH2}	0.000	0.055	0.0069	0.0042
T_{43}	0.9	1.1	1.0229	1.0239
T_{56}	0.9	1.1	1.0359	1.0351

Table 6: Optimization Results.

and their fitness is plotted with respect to the iteration count. Firefly algorithm is simulated by including a random walk module in the original algorithm (Figure 5). The Spiral algorithm is run for 100 iterations under uncompensated static conditions Figure 6 and a comparison between F.A and Spiral has been made in Figure 7. The N.R. load flow is done for the base case [6] as well as the optimal parameters for static compensated system. The results are shown in Tables 7 and 8.

Test Case	Firefly (MW)	Spiral (MW)
1. Static system without compensation	4.56	4.53
2. Static system with compensation	4.52	4.43
3. Dynamic system without compensation	5.56	5.49
4. Dynamic system with compensation	5.4	5.36

Table 7: Optimum System Loss.

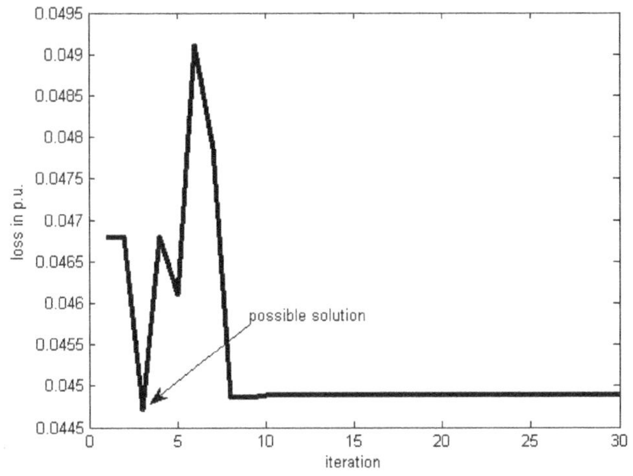

Figure 5: Loss in p.u vs. iteration number using Firefly Algorithm and random walk.

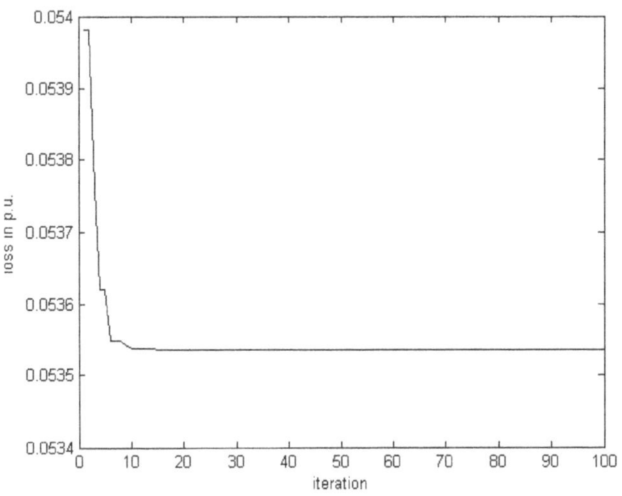

Figure 6: Loss in p.u. vs iteration using spiral optimization algorithm for 100 iterations.

The convergence criterion is taken as 0.000001. Acceptable tolerance is to be greater than 0.000001. All the results have been simulated using MATLAB Software (version 7.0). Separate coding has been done for N.R, Gauss Siedal, Spiral and Firefly algorithms with all the test cases considered.

The discrepancy between the values of Tables 7 and 9 is indicative of the fact that the Gauss Siedal method computes the active losses with greater precision for a 6 bus system. However, as the size of the network increases, N.R. method is more favorable.

Both the techniques adopted have successfully allocated the

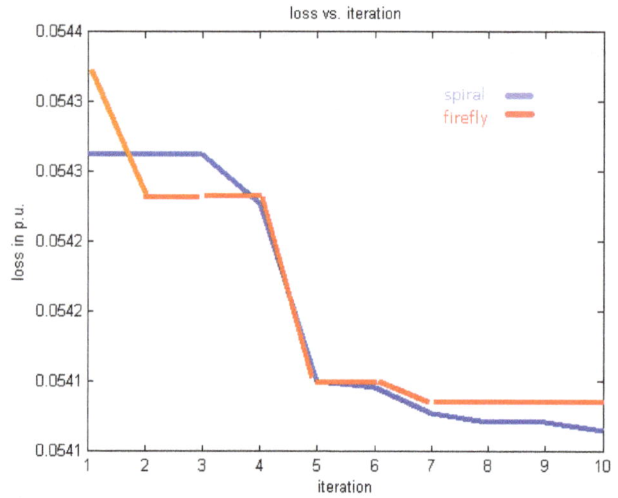

Figure 7: Comparison between spiral and firefly optimization for 10 iterations under static uncompensated system.

Normal (MW)	Firefly (MW)	Spiral (MW)
5.358	5.264	4.89

Table 8: Optimum System Loss.

From Bus	To Bus	P (MW)	Q (MVAR)	Real Power Loss	Reactive Power loss
1	6	10.99	2.21	0.13	0.546
1	4	16.39	-34.5	0.979	4.527
2	3	14.6	-10	0.183	2.655
2	5	30.4	-15.39	2.643	0.6
3	4	-13.08	6.43	0	0.19
4	6	2.33	37.63	0.958	4.021
6	5	-12.75	9.07	0	0.62
Net Loss				4.89	27.752

Table 9: Line Flow Data for Spiral Algorithm.

parameters within the optimal conditions (Tables 3-6). As observed from Table 7, Spiral algorithm provides better optimal values than F.A. under all the test cases considered. As shown in Figure 7, Spiral curve is going under the firefly graph during the later course of the iterations. As expected, Compensation reduces the losses by an appreciable 0.1 MW in most of the cases Table 7. Eventually for huge power systems this value can be in the order of tens or hundreds. One special feature observed is that both the algorithms attain fast convergence between 9 and 10 iterations. This means that for discrete time step of 0.005 seconds the scheduling will occur in 0.05 seconds itself (considering 10 iterations). Moreover, F.A. could be enhanced by using random walk simulation (Figure 5) to increase the possibility of a better solution. But, that feature is not used as it violates the inequality constraints and increases the computation time. A further study is done using Newton Raphson method, for a static compensated system (Table 9). Spiral Optimization proved better in that regard as well giving a minimal real power loss of 4.89 MW. From the line flow data it is shown that real power loss between buses 3-4 and 5-6 have been compensated due to the presence of transformer taps at those locations (Figure 4).

Conclusion

Reactive power should be properly controlled and applied to maintain system balance. It is the responsibility of the utilities to effectively control and plan the reactive power losses along with the financial expenses. Since the concept here involves the operation of a centralized TSO, effective communication channels for processing data is mandatory. Spiral optimization is a promising meta heuristic technique based on random sampling of search space. It is comparable to other evolutionary algorithms such as Firefly algorithm with even better performance. The ability of spiral to converge to the desired location with logarithmic convergence is suitable for Reactive Power Planning. Furthermore, Gauss Siedal analysis proved to be a better option for load flow compared to N.R. method for a 6 bus system. Partial compensation with tap setting transformers can be successfully implemented in the power system to avoid voltage drop between buses where critical loads have been stationed. This will reduce thermal stress and avoid systemic outages. The ability of both these algorithms to react within short epoch duration justifies their significance in a dynamic simulation platform.

References

1. Kargarian A, Raoofat M (2011) Stochastic Reactive Power Market with Volatility of Wind Power Considering Voltage Security. Energy 36: 2565-2571.
2. Singh B, Murthy SS (2004) Analysis and Design of STATCOM based Voltage Regulator for self excited Induction Generators. IEEE T Energy Conver 19: 783-790.
3. Elmoursi MS, Sharaf AM (2006) Novel STATCOM Controllers for Voltage Stabilization of Stand Alone hybrid (Wind/Small Hydro) Schemes. International Journal of Emerging Electric Power Systems 7: 1-25.
4. Opoku G (1990) Optimal Power System VAr Planning. IEEE T Power Syst 5: 53-60.
5. Bansal RC, Bhatti TS, Kothari DP (2003) Artificial Intelligence technique for reactive power/voltage control in power systems: A review. IEEE International Journal of Power and Energy systems 23: 81-89.
6. Zhang H (1998) Reactive Power Optimization based on Genetic Algorithm. International Conference on Power System Technology 2: 1448-1453.
7. Sharma NK, Babu DS (2012) Application of Particle Swarm Optimization for Reactive Power Optimization. ICAESM.
8. Arya LD, Titare LS, Kothari DP (2010) Improved particle swarm optimization applied to reactive power reserve maximization. International Journal of Electrical Power and Energy Systems 32: 368-374.
9. Liya G (2010) A Combination strategy for Reactive Power Optimization based on model of soft constraint considered interior point method and genetic simulated annealing algorithm. ISME.
10. Liu HS (2010) Reactive Power Optimization in Power system based on Adaptive Focussing PSO. ICECE.
11. Da ZB (2012) Control strategy for optimization of voltage and reactive power in substation based on load forecasting. IEEE transactions.
12. Grudinin N (1998) Reactive Power Optimization using successive quadratic programming method. IEEE T Power Syst 13: 1219-1225.
13. Al-Alawi SM, Ellithy KA (2000) Tuning of SVC Damping Controllers over a wide range of Load Models using an Artificial Neural Network. International Journal of Electrical Power and Energy Systems 22: 405-420.
14. Kargarian A, Ebrahimi F, Sarikhani M (2011) An economic based approach for optimal and secure reactive power provision in deregulated environments. Modern Applied Science 5: 111-118.
15. Kargarian A, Khorammi RS (2011) A Fast Algorithm for Reactive Power Market Management Using Artificial Neural Networks. International Journal of Applied Science and Technology 1: 59-65.
16. Kargarian A, Falahati B, Fu Y (2012) Stochastic active and reactive power dispatch in electricity markets with wind power volatility. IEEE Power and Energy Society General Meeting.
17. Kargarian A, Raoofat M, Mohammadi M (2012) Probabilistic reactive power procurement in hybrid electricity markets with uncertain loads. Electr Pow Syst Res 82: 68-70.

18. Kargarian A, Raoofat M, Mohammadi M (2011) Reactive Power Provision in Electricity Markets Considering Voltage Stability and Transmission Congestion. Electric Power Components and Systems 39: 1212-1226.

19. El-Samahy I, Bhattacharya K, Canizares C, Anjos MF, Pan J (2008) A procurement market model for reactive power services considering system security. IEEE T Power Syst 23: 137-149.

20. Martins N, Macedo NJP, Lima LTG, Pinto HJCP (1993) Control strategies for multiple static VAr compensators in long distance voltage supported transmission systems. IEEE T Power Syst 8: 1107-1117.

21. Kargarian A, Raoofat M, Mohammadi M (2011) Reactive Power Market Management Considering Voltage Control Area Reserve and System Security. Appl Energ 88: 3832-3840.

22. Yang XS (2010) Firefly Algorithm, Stochastic Test functions and design optimization. International Journal for Bio-inspired Computation 2: 78-84.

23. Sulaiman MS, Mustafa MW, Zakhari ZN (2012) Firefly algorithm technique for solving Economic load dispatch problem. PEOCO.

24. Tamura K, Yasuda K (2011) Spiral Optimization: A New multipoint search method. IEEE International Conference on Systems, Man, and Cybernetics.

25. Mozafari B, Amran T (2007) Particle Swarm Optimization Method for optimal reactive Power Procurement considering voltage stability. Scientia Iranica 14: 534-545.

26. Okuwari Y, Aoki H (2012) Voltage and Reactive Power Control with load change. Power and Energy Society General Meeting.

14

Non-Parametric Estimation of a Single Inflection Point in Noisy Observed Signal

Nezamoddin N. Kachouie[1]* and Armin Schwartzman[2]

[1]*Department of Mathematical Sciences, Florida Institute of Technology, USA*
[2]*Department of Statistics North Carolina State University Raleigh, NC*

Abstract

Inflection point detection is an important yet challenging problem in science and engineering. This paper addresses the estimation of a single inflection point location in noisy observations using non-parametric polynomial regression. To address the bandwidth problem, a constrained approach is proposed to ensure having a single inflection point, thereby reducing the uncertainty in the inflection point location whereas being flexible on the shape of the underlying signal. The performance of the proposed method is evaluated through simulations. It is shown that the proposed method can effectively estimate the inflection point under high noise conditions.

Keywords: Inflection point estimation; Local polynomial regression; Bandwidth selection; Cross validation

Introduction

Detection of abrupt changes, including inflection points, in the observed noisy signals is an important and challenging problem in engineering and science. Several methods have been proposed to solve this problem and proved to be successful for specific applications. Yet, there is demand for new and general approaches that are applicable to broader range of applications. The change point detection problem can be divided into two categories; *i)* sequential or on-line, and *ii)* batch or off-line. In on-line methods, the observed data are inspected sequentially to catch a change as soon as it occurs. These methods are mainly used in quality control, real time surveillance & vision systems, and real time fault detection in computer networks [1-6]. In off-line methods the entire data set is observed and can be processed at once. The off-line problems are growing fast and have attracted many researchers in computational biology and biostatistics as well as off-line computer vision applications [7-14]. In previous work [15,16], often Cross Validation (CV) is used to select the bandwidth to locate the discontinuities in the observed signal including inflection points, change points, jumps and valleys. Discontinuities in [15], assuming an unknown number, are located by detecting the zero crossing of the second derivative and the local maxima of the first derivative while CV is used to estimate bandwidth parameters. Bootstrap is used in [16] for designing a nested CV method for detecting the change points by finding the local maxima of the first derivative while considering the number of change points is unknown.

For smooth underlying signals, a change point may be defined as an inflection point, i.e. a point where the second derivative of the signal changes sign. In this work, we focus on the problem of estimating the location of an inflection point when it is known that only one such point exists in the underlying signal. When considering smoothing methods for estimating the underlying signal, standard methods for bandwidth selection, such as cross validation, are designed to optimally estimate the underlying function, and can thus produce many inflection points; creating uncertainty in the location of the true inflection point. Our goal is to design a fast and simple, yet effective method to address the inflection point detection. The interest here is to detect a single significant inflection point in either entire range or a fragment of observed signal. A non-parametric inflection point detection method is proposed in which we smooth the noisy observations and locate the inflection point using the estimated zero crossing of the second derivative of the smoothed curve. Spatial curve fitting in our case is accomplished by applying local polynomial regression to smooth the noisy data. We propose a constrained method for bandwidth selection intended to allow a single inflection point, thereby it increases the accuracy of the estimated inflection point location, whereas it permits fitting a flexible function and avoids over-smoothing. Using non-parametric regression allows estimating the standard error of the located inflection point to evaluate the accuracy of the estimation. The performance of the method is evaluated through simulations using a sigmoid function that is corrupted by adding different levels of Gaussian noise.

The proposed method is discussed in the next section by describing local polynomial regression, inflection point detection, and optimal bandwidth selection. The results are then demonstrated, compared with cross validation and discussed at the end.

Methods

Nonparametric polynomial regression

Suppose that n pairs of observations

$$(s_i, Y_i), i = 1,2,\ldots,n \tag{1}$$

consist of a response variable and a location and are related by the signal plus noise model

$$Y_i = r(s_i) + \varepsilon_i, \varepsilon_i : N(0,\sigma_\varepsilon), i \in [1,n] \tag{2}$$

where $Y_i, i \in [1,n]$ are observed noisy samples, ε_i is Gaussian noise, $N(0,\sigma_\varepsilon)$, $s \in \mathrm{R}$, and r is an unknown underlying regression function. The regression function r can be locally approximated at the point s by a polynomial of order p

***Corresponding author:** Nezamoddin N. Kachouie, Department of Mathematical Sciences, Florida Institute of Technology, USA
E-mail: nezamoddin@fit.edu

$$g(z;A) = A(s)+a'(s)(z-s)+a''(s)(z-s)^2/2!+\ldots+a^{(p)}(s)(z-s)^p/p! \quad (3)$$

where z is a point in a neighborhood of s. We use least squares to estimate the polynomial coefficient vector $A(s)=(a(s),a'(s),a''(s)\ldots,a^{(p)}(s))^T$. The estimated polynomial coefficient vector is

$$\hat{A}(s) = (X_s^T W_s X_s)^{-1} X_s^T W_s Y = LY \quad (4)$$

where $\hat{A}(s) = (\hat{a}(s),\hat{a}'(s),\hat{a}''(s),\ldots\hat{a}^{(p)}(s))^T$ is obtained by minimization of the least square problem

$$f(A(s)) = \sum_{i=1}^{n} K\gamma(s_i - s)\left(Y_i - \sum_{p=0}^{P} a^{(p)}(s)(s_i - s)^p/p!\right)^2 = (Y - X_s A)^T W_s (Y - X_s A) \quad (5)$$

where s_i is a point in a neighborhood of s, $L = (X_s^T W_s X_s)^{-1} X_s^T W_s$ is the smoothing matrix, X_s is a $n\times(P+1)$ matrix.

$$X_s = \begin{pmatrix} 1 & s_1 - s & \cdots & \frac{(s_1-s)^P}{p!} \\ 1 & s_2 - s & \cdots & \frac{(s_2-s)^P}{p!} \\ \vdots & \vdots & \ddots & \vdots \\ 1 & s_n - s & \cdots & \frac{(s_n-s)^P}{p!} \end{pmatrix} \quad (6)$$

and W_s is an $n\times n$ diagonal matrix.

$$W_s = \begin{pmatrix} K_\gamma(s_1 - s) & 0 & \cdots & 0 \\ 0 & K_\gamma(s_2 - s) & \cdots & 0 \\ \vdots & \vdots & \ddots & \vdots \\ 0 & 0 & \cdots & K_\gamma(s_n - s) \end{pmatrix} \quad (7)$$

and $K_\gamma(s)$ is a kernel with bandwidth γ

$$K_\gamma(s) = \frac{1}{\gamma}K\left(\frac{s}{\gamma}\right), \int K(s)ds = 1 \quad (8)$$

To estimate *r(s)=a(s)*, the inner product of the first row of with is computed by L with Y is computed by

$$\hat{r}(s) = ([1 \quad 0 \quad 0 \quad \cdots \quad 0]\times L)Y = e^1 LY = \sum_{1}^{n} l_{\gamma,i}(s)Y_i \quad (9)$$

where $e^1 L = l_\gamma(x) = (l_{\gamma,1}(s), l_{\gamma,2}(s),\cdots,l_{\gamma,n}(s))^T$. The variance of this estimator for independent noise that is of interest here is

$$\mathrm{var}(\hat{r}(s)) = \sigma_\varepsilon^2 \|l(s)\|^2 \quad (10)$$

Similarly, the derivatives $r'(s), r''(s),\cdots,r^{(p)}(s)$ can be estimated by the inner product of Y with the second row, third row, and $(P+1)^{th}$ row of L respectively.

Inflection point detection

An inflection point is a point (*D*) on the estimated curve where the curvature changes sign. Here we discuss the positive inflection point where the curvature changes from being concave upward to being concave downward, i.e., $r''(s)$ crosses the zero line from being positive ($r''(s) > 0$) to being negative ($r''(s) < 0$) and the first derivative has a local maximum ($r'(s) > 0$) there. In a similar way to (9), the estimated 2nd derivative ($\hat{r}'(s)$) is

$$\hat{r}''(s) = ([0,\ 0,\ 1\ \cdots\ 0]\times L)Y = \sum_{1}^{n} l_i''(s)Y_i \quad (11)$$

and its associated standard error can be computed by

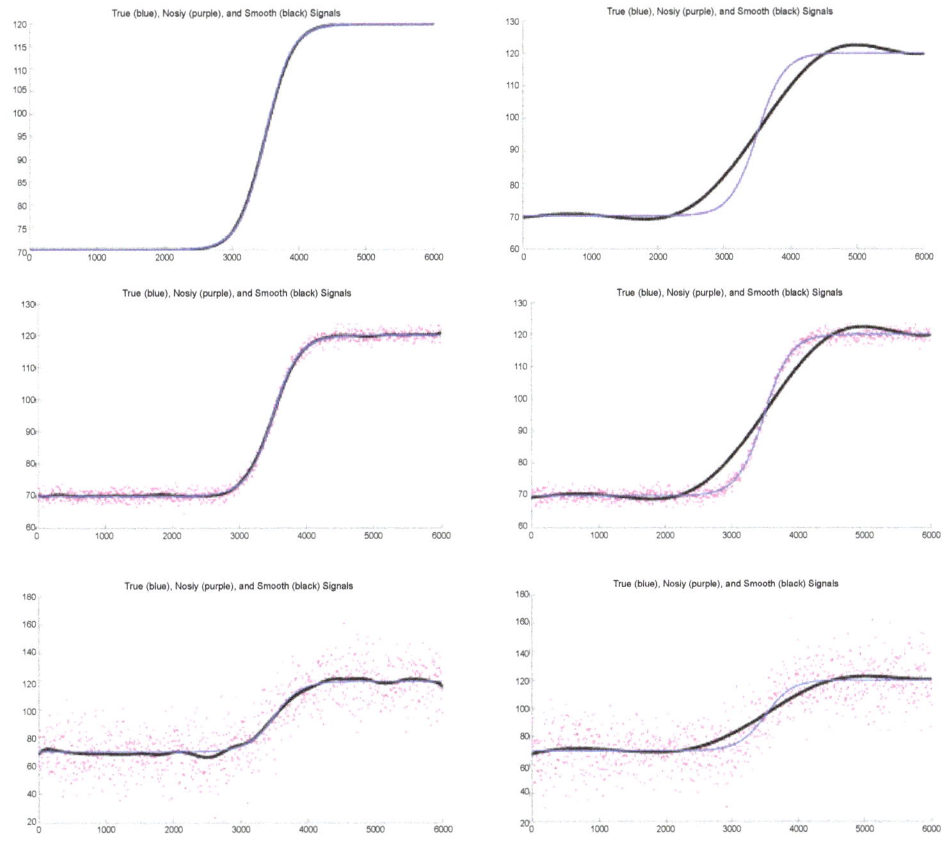

Figure 1: Sigmoid function corrupted with different noise levels: no noise (first row), low noise (second row), and high noise (third row). True sigmoid (blue), noisy observations (purple dots), and estimated sigmoid (black) applying a small bandwidth (left) and applying a large bandwidth (right).

$$\mathrm{var}(\hat{r}''(s)) = \sigma_\varepsilon^2 \left\| l''(s) \right\|^2 \quad (12)$$

The negative inflection point can be identified in a similar way where the second derivative changes sign from negative to positive. As illustrated in Fig. 1, the standard error of the estimated inflection point location is approximated by

$$Se(D) \cong \frac{Se(r''(D))}{r'''(D)} \quad (13)$$

where D is the identified inflection point location, $Se(r''(D)) = \sqrt{\mathrm{var}(\hat{r}''(s))}$ is the standard error of the estimated second derivative at the inflection point based on (12), and $r'''(D)$ is the estimated third derivative at the inflection point, computed in a similar way as (9) and (11).

Optimal bandwidth selection

Conventionally, the optimal bandwidth is chosen using the leave-one-out cross validation score

$$CV = \hat{R}(\gamma) = \frac{1}{n}\sum_{i=1}^{n}(Y_i - \hat{r}_{(-i)}(s_i))^2 \quad (14)$$

where $\hat{r}_{(-i)}$ is estimated by excluding i^{th} pair (s_i, Y_i). The optimal bandwidth chosen by cross validation may produce many inflection points. As an illustration, a sigmoid function is corrupted with different noise levels and is estimated by small, large (Figure 1), and optimal bandwidth chosen by cross validation (Figure 2). Applying the selected optimal bandwidth obtained by cross validation, we may identify zero, one, or several inflection points (Figure 3 and 4) and it is difficult to identify the main one as the estimation of the true inflection point of the original unknown curve (sigmoid function here).

The new proposed method applies a constraint for bandwidth selection to ensure that the smoothed curve (by the selected bandwidth) has only one inflection point. The bandwidth γ is the smallest bandwidth producing the smoothed curve

$$\hat{r}_\gamma(s) = \sum_{1}^{n} l_{\gamma,i}(s) Y_i \quad (15)$$

that satisfies the constraint

$$|D| = 1 \quad (16)$$

where

$$D = \left\{ s : \begin{matrix} \hat{r}''_\gamma(s) = 0 \\ \hat{r}'_\gamma(s) > 0 \end{matrix} \right\} \quad (17)$$

is the set of zero down-crossings of the second derivative. Equation 17 ensures that the number of positive inflection points of the smoothed curve applying bandwidth γ is one.

Results and Discussion

The proposed method is applied to a simulated sigmoid function which is corrupted by adding different levels of independent Gaussian noise. For comparison, first we show the application of the regression-based inflection point detection where the bandwidth is selected by cross validation. The results obtained by applying the proposed constrained bandwidth selection are then presented. Finally the performance is quantified by running several hundreds

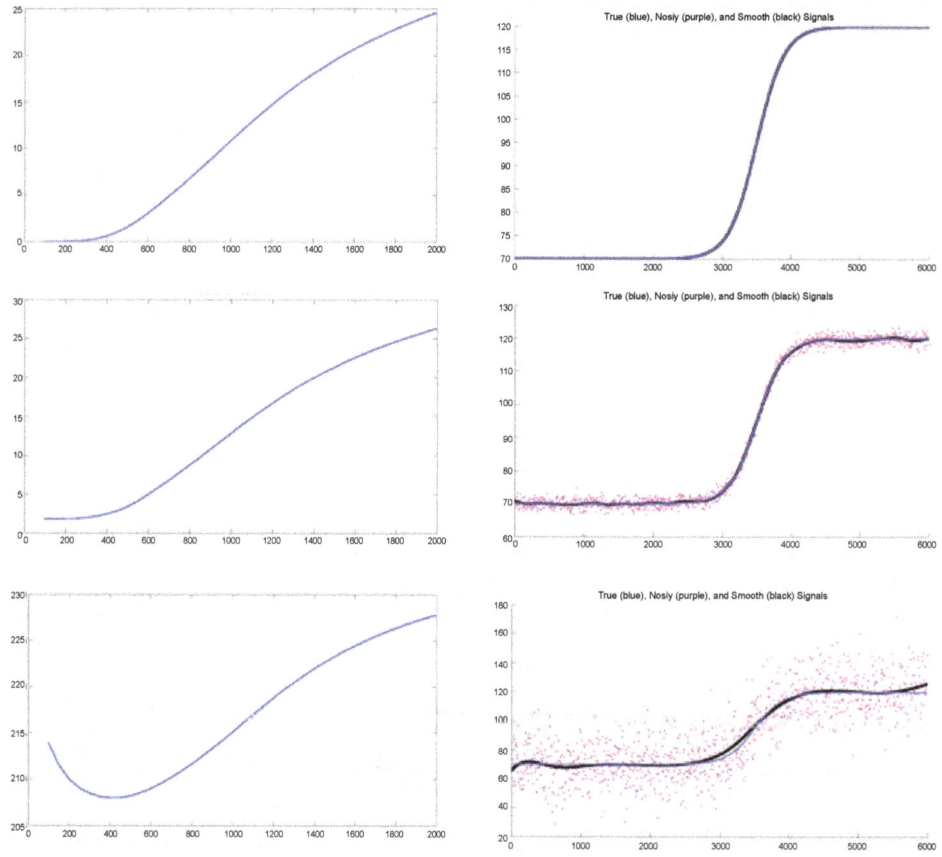

Figure 2: (Left column) Cross validation curve for different noise levels: no noise (first row), low noise (second row), and high noise (third row). (Right column) True sigmoid (blue), noisy observations (purple dots), and estimated sigmoid applying the optimal bandwidth chosen by cross validation (black).

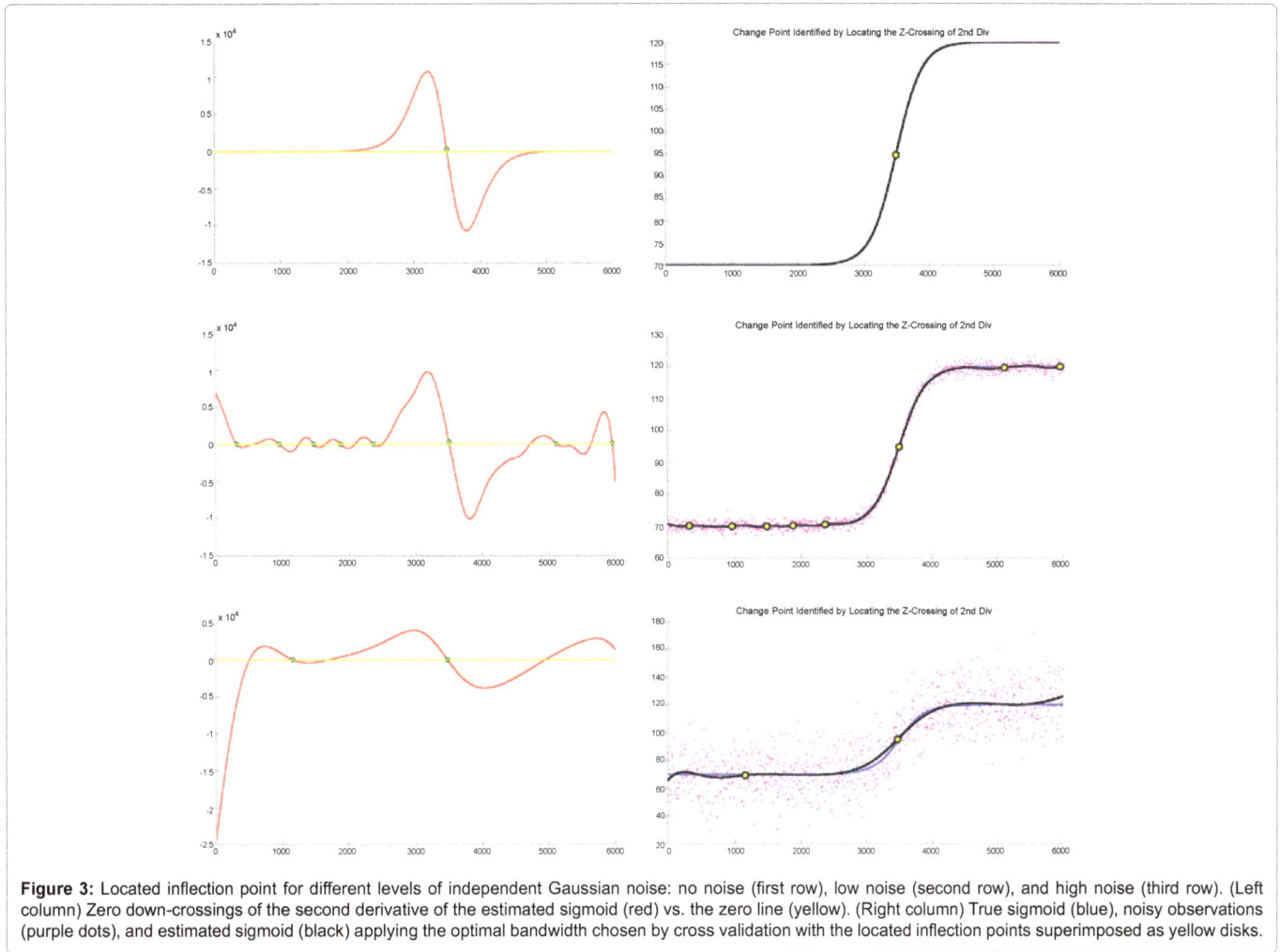

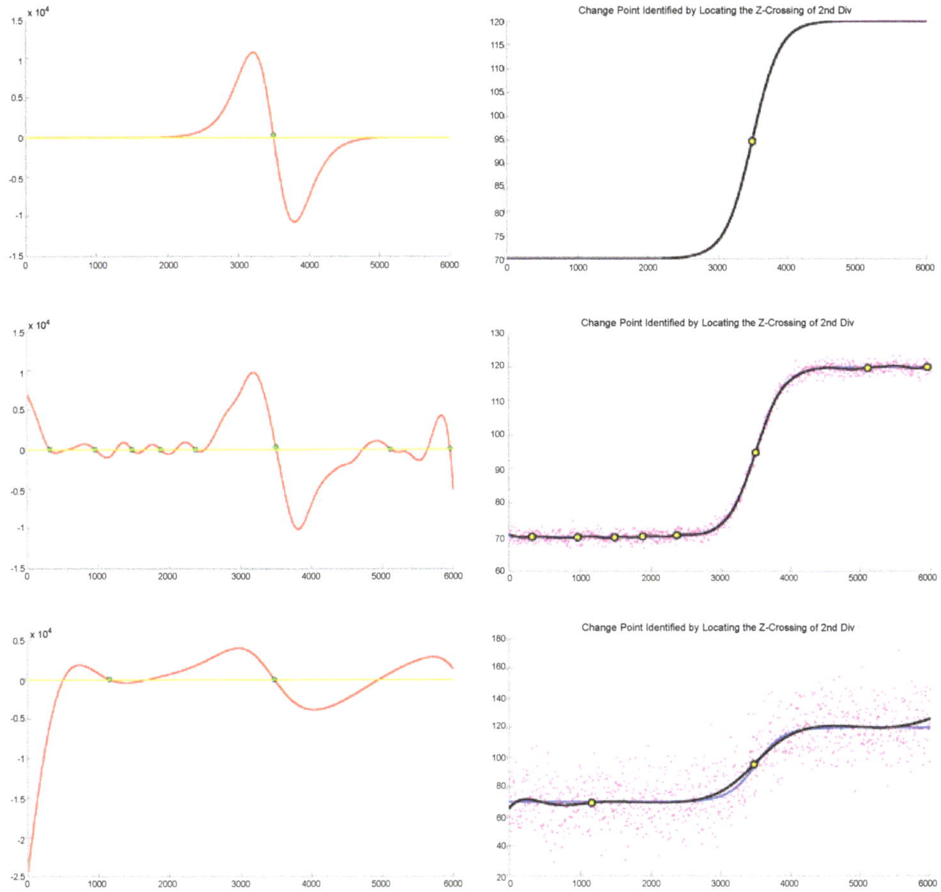

Figure 3: Located inflection point for different levels of independent Gaussian noise: no noise (first row), low noise (second row), and high noise (third row). (Left column) Zero down-crossings of the second derivative of the estimated sigmoid (red) vs. the zero line (yellow). (Right column) True sigmoid (blue), noisy observations (purple dots), and estimated sigmoid (black) applying the optimal bandwidth chosen by cross validation with the located inflection points superimposed as yellow disks.

of simulation instances.

Bandwidth selection by cross validation

A sigmoid function was simulated with $s = [0,6000]$

$$r(s) = 70 + \frac{50}{1+e^{-5\left(\frac{s-3500}{1000}\right)}} \tag{18}$$

and corrupted with $\sigma_\varepsilon^2 = 0$ (no noise), $\sigma_\varepsilon^2 = 1.85$ (low noise with $3\sigma_\varepsilon = 8.16\%$ of the sigmoid height), and $\sigma_\varepsilon^2 = 185$ (high noise with $3\sigma_\varepsilon = 81.6\%$ of the sigmoid height). The estimated smooth curves for small ($\gamma = 200$) and large ($\gamma = 1000$) bandwidths for different noise levels are depicted in Figure 1. The cross validation curve and the smoothed curve applying the selected bandwidth identified by minimum cross validation score for different noise levels is shown in Figure 2. The optimal bandwidth (γ_{opt}) is 100 for no noise, 150 for low noise and 400 for high noise. The number of detected inflection points is 1, 8, and 2 for different noise levels respectively (Figure 3). Although the inflection point closest to the true one is among the detected inflection points, further processing is needed to identify it. The identification of the main inflection point is even more challenging and prone to error when the detected candidate inflection points have close values of their estimated first derivative ($\hat{r}'_\gamma$).

Next, the sigmoid $r(s) = \frac{1}{1+e^{-4(s-5)}}$ was simulated and independent Gaussian noise with $\sigma_\varepsilon^2 = 0.01$ (medium noise with $3\sigma = 30\%$ of the sigmoid height) was added. Five hundred independent simulation instances were performed, inflection points were located, and the number of inflection points was counted for each simulation. The cross validation curve and the smoothed curve applying the optimal cross validation bandwidth for a typical simulation (γ_{opt}) is shown in Figure 4 (First row). The location of inflection points and the number of detected inflection points for the five hundred runs are depicted in Figure 4 (Middle row). The distribution (histogram) of inflection point location and the number of detected inflection points are shown in Figure 4 (Bottom row). Applying the optimal cross validation bandwidth, up to seven inflection points was identified (Figure 4).

Bandwidth selection by the proposed method

To avoid detecting multiple false inflection points, the proposed method was applied to select the optimal bandwidth while ensuring to have only one inflection point in the smoothed curve. Independent Gaussian noise with different levels $\sigma_\varepsilon^2 = 0$ (no noise), $\sigma_\varepsilon^2 = 1.85$ (low noise), $\sigma_\varepsilon^2 = 185$ (high noise), and $\sigma_\varepsilon^2 = 1665$ (very high noise) was added to corrupt the signal. The noisy sigmoid was then estimated using the non-parametric regression where the bandwidth was selected by the proposed method (Figure 5). Regardless of the noise level, the significant inflection point was closely estimated (Figure 5 (Right column)).

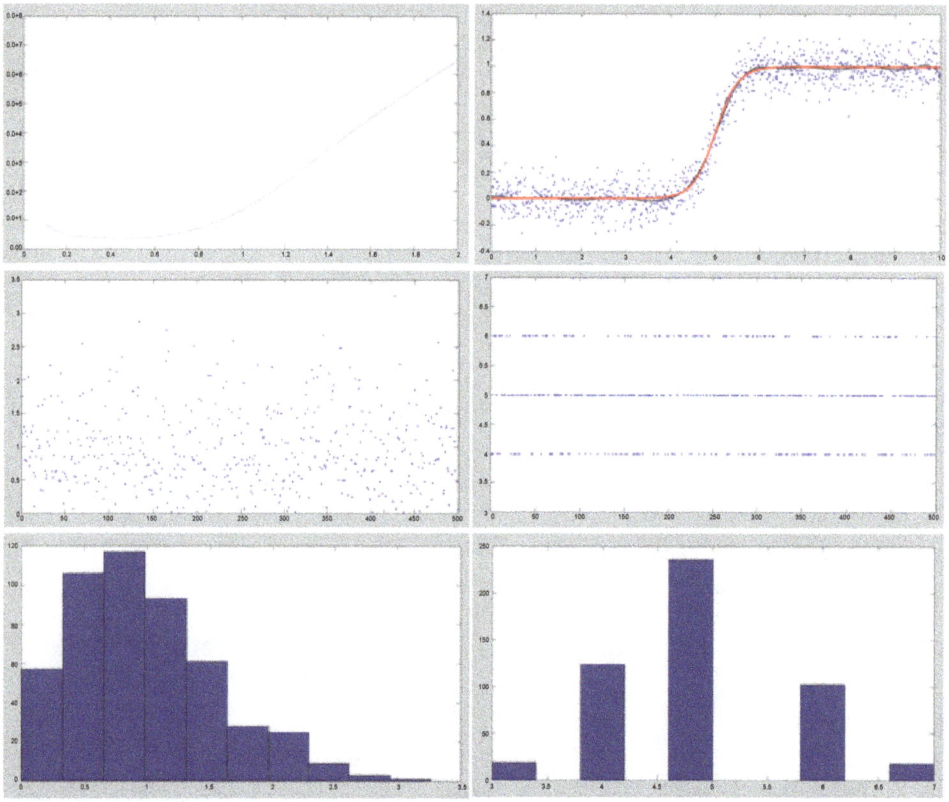

Figure 4: (Top row) Cross validation curve (left) and the estimated sigmoid applying the optimal bandwidth chosen by cross validation (right) for a typical realization. (Middle row) Located inflection point for 500 simulations applying the optimal bandwidth chosen by cross validation (left) and the number of detected inflection points for each simulation (right). (Bottom row) The distribution of the locations of the inflection points for 500 simulations (left) and the distribution of the number of inflection points for the same 500 simulations.

Figure 6 further illustrates the inflection point detection and the estimation error under high noise conditions. The confidence interval was computed by $conf = 1.96 * Se(\hat{r}(s))$ and is superimposed in red where the standard error ($(Se = \sqrt{\text{var}(\hat{r}(s))})$ of the smoothed curve was computed using (10). For high level of Gaussian noise with $\sigma_\varepsilon^2 = 272.25$ or equivalently $3\sigma_\varepsilon = 50$ that is equal to 100% of sigmoid height (Figure 6 (Top)), the detected inflection point is located at s = 3441.4 which is close to the true inflection point at s = 3550 with 1.67% estimation error. Even at very high noise conditions with $\sigma_\varepsilon^2 = 2500$ so that $3\sigma_\varepsilon = 150$ is equal to 300% of sigmoid height (Figure 6 (Bottom)), the detected inflection point is located at s = 3395 which is only 3% away from the true inflection point location.

Performance of the proposed method

The performance of the proposed method is assessed by comparing the location of the identified inflection point vs. the true inflection point location (D = 5; $r(D)$ = 0.5) through several hundred simulations where the sigmoid $r(s) = \frac{1}{1+e^{-\alpha(s-5)}}$ was corrupted with high level of independent Gaussian noise. The simulations were repeated for six different sigmoid slopes (α). Two hundred simulations were performed for each slope, starting with more challenging case of fairly shallow slope α = 0.5 to sharp slope α = 3.

The detected inflection point, standard error, true mean squared error, and the coverage of estimated confidence interval for the detected inflection points were estimated for different slopes (Figure 7). The mean of the detected inflection points closely follows the true inflection point location (D = 5) for different slopes (Figure 7 (Top)). As it can be observed, while the slope increases from low on the left to high on the right, estimation precision increases (Figure 7 (Middle)). In addition, the estimated standard errors (box plots) closely match the true mean square error computed from the 200 simulations. The 95% coverage depicted by the black disks (Figure 7 (Bottom)) was computed for each slope by the percentage of the number of simulation instances for which the true inflection point (D = 5) satisfies

$$D \in \{\hat{D} \pm 1.96 * Se(D)\} \quad (19)$$

The point-wise confidence interval of the coverage was then estimated (red disks) for each slope using the Binomial distribution applying the number of simulation instances (Figure 7 (Bottom)).

Insights

The proposed method selects the optimal bandwidth to ensure a single inflection point. This avoids the uncertainty due to detection of zero or multiple inflection points when applying standard bandwidth selection methods such as cross validation. Therefore the proposed method increases the accuracy of the estimated inflection point location for such applications where the true function has only one inflection point. This method permits a flexible fit to estimate the underlying function by taking advantage of non-parametric regression while avoiding over-smoothing.

Figure 8 illustrates the basic mechanism under a low noise with

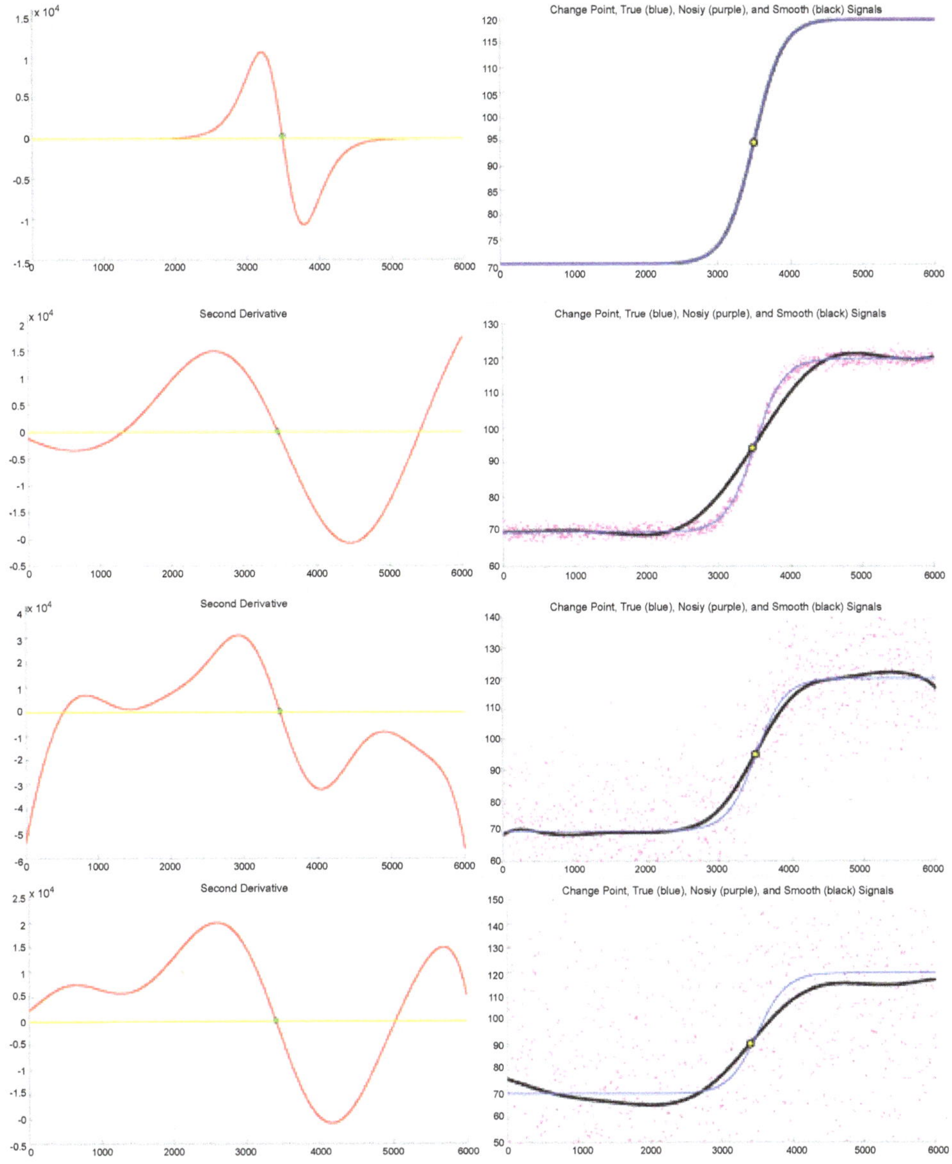

Figure 5: Located inflection point by the proposed method for different levels of independent Gaussian noise: first row (no noise), second row (low noise), third row (high noise), and fourth row (very high noise). (Left column) zero down-crossings of the second derivative of the estimated sigmoid (red) vs. the zero line (yellow). (Right column) True sigmoid (blue), noisy observations (purple dots), and estimated sigmoid (black) using the optimal constrained bandwidth where the located inflection points are superimposed as yellow disks.

$\sigma_\varepsilon^2 = 25$ or equivalently $3\sigma_\varepsilon = 15$ that is equal to 30% of sigmoid height. Starting from a small bandwidth, the proposed method gradually increases the bandwidth to smooth out false inflection points. The bandwidth increases from left to right: left = 100, middle = 300, and right = 600. Shallow inflection points disappear by applying larger bandwidths. We should point out that there is a wide range of large bandwidths for which there is only one inflection point. We propose selecting the smallest bandwidth within the range that guarantees only one inflection point in order to reduce over-smoothing caused by applying the larger bandwidths.

Note from Figure 6 that the confidence bands do not necessarily cover the true function. In other words, the estimation of the function is biased. However, this is not a concern in our case because our interest is in estimating the inflection point location, not the underlying function. The gain in accuracy in estimating the inflection point location is worth the extra bias in the estimation of the function.

The estimation of the inflection point location and its standard error, however, are nearly unbiased. This can be observed in Figure 7 (Top) and Figure 7 (Middle), which show that the estimated inflection point locations are close in average to the true value, and that the estimated log squared standard errors are close in average to the true log mean square errors. Figure 7 (Bottom) further shows for all slopes larger than $\alpha = 1.5$ that the coverage of the 95% confidence interval for the inflection point location is indeed at least the nominal (95%). For shallow slopes ($\alpha = 0.5$ & 1), the region in which the true function transitions from low to high is larger, making the detection of the inflection point harder. In this case, the estimation of the inflection point is still unbiased but the coverage of the confidence intervals drops below 90% for such challenging cases.

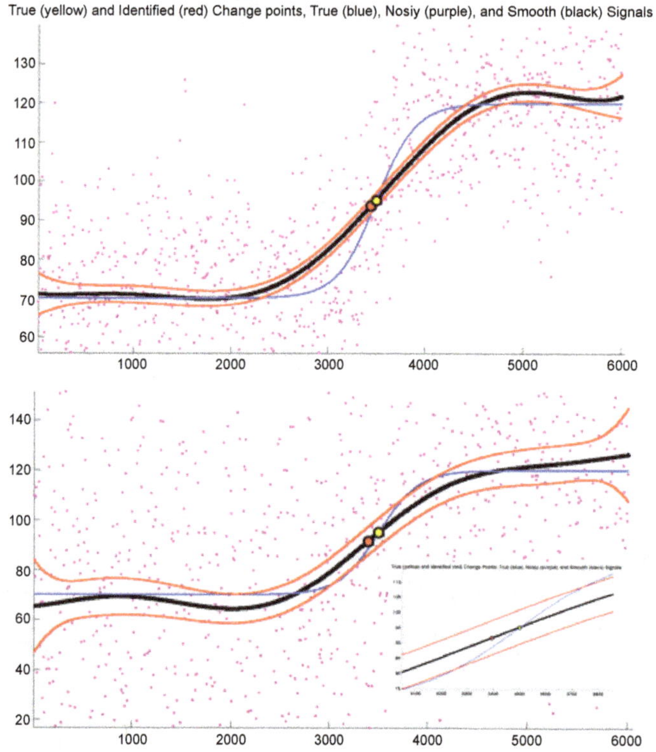

Figure 6: True sigmoid (blue), noisy observations (purple), estimated sigmoid applying the proposed method (black), confidence interval (red), and the true (yellow disk) and the detected inflection point (red disk) where the true sigmoid function is corrupted with Gaussian noise. (Top) High noise with $3\sigma_{\varepsilon} = 50 = 100\%$ of sigmoid height. (Bottom) Very high noise with $3\sigma = 150 = 300\%$ of sigmoid height.

References

1. Basseville M, Benveniste A (1986) Detection of abrupt changes in signals and dynamical systems. Lecture notes in control and information sciences. Springer-Verlag, Berlin, Germany, New York, USA.
2. Desobry F, Davy M, Doncarli C (2005) An Online Kernel Change Detection Algorithm. IEEE T Signal Proc 53: 2961-2974.
3. Kawahara Y, Sugiyama M (2009) Change-Point Detection in Time-Series Data by Direct Density-Ratio Estimation. Proceedings of society of industrial and applied mathematics (SIAM).
4. Ke Y, Sukthankar R, Hebert M (2007) Event Detection in Crowded Videos. Proceedings of the 11th IEEE International Conference on Computer Vision.
5. Lai TL (1995) Sequential change-point detection in quality control and dynamical systems (with discussion). J Roy Stat Soc B Met 57: 613-658.
6. Mei Y (2006) Sequential change-point detection when unknown parameters are present in the pre-change distribution. Ann Stat 34: 92-122.
7. Basseville M, Nikiforov (1993) Detection of abrupt changes: theory and application. Prentice-Hall information and system sciences series. Prentice Hall, Englewood Cliffs, New Jersey, USA.
8. Gustafsson F (1996) The Marginalized Likelihood Ratio Test for Detecting Abrupt Changes. IEEE T Automat Contr 41: 66-78.
9. James B, James KL, Siegmund D (1987) Tests for a change-point. Biometrika 74: 71-83.
10. Olshen AB, Venkatraman ES, Lucito R, Wigler M (2004) Circular binary segmentation for the analysis of array-based DNA copy number data. Biostatistics 5: 557-572.
11. Smith AFM (1975) A Bayesian approach to inference about a change-point in a sequence of random variables. Biometrika 62: 407-416.
12. Takeuchi J, Yamanishi KA (2006) Unifying Framework for Detecting Outliers and Change Points from Non- Stationary Time Series Data. IEEE T Knowl Data En 18: 482-489.
13. Yakir B, Krieger AM, Pollak M (1999) Detecting a change in regression: First order optimality. Ann Stat 27: 1896-1913.
14. Yao YC (1988) Estimating the number of change-points via Schwarz' criterion. Stat Probabil Lett 6: 181-189.
15. Joo J, Qiu P (2009) Jump detection in a regression curve and its derivative. Technometrics 51: 289-305.
16. Gijbels I, Goderniaux A (2004) Bandwidth selection for change point estimation in nonparametric regression. Technometrics 46: 76-86.

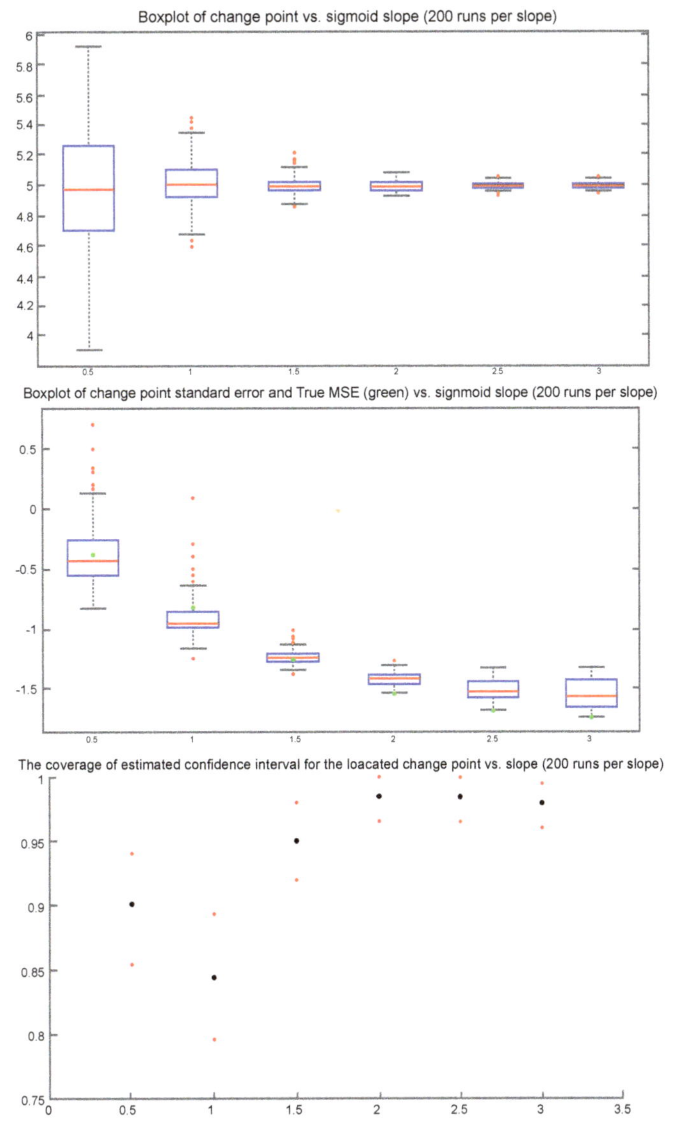

Figure 7: (Top) Estimated location of the inflection points for six different slopes with 200 simulations instances per slope. (Middle) Estimated log squared standard error associated with the detected inflection point locations and log true mean squared error (green disks) for the same simulation instances. (Bottom) The coverage of confidence interval of estimated inflection point locations for the same simulation instances.

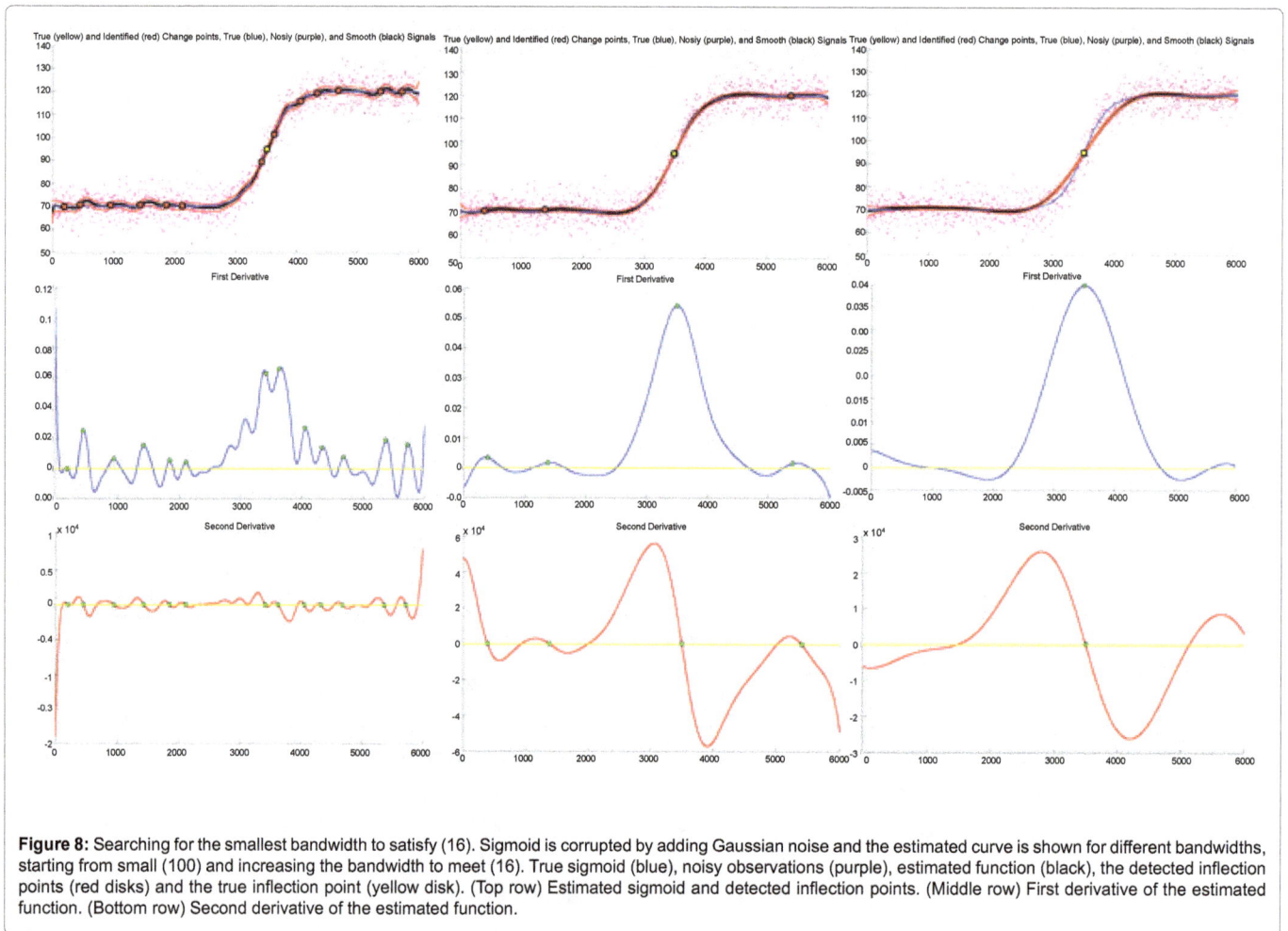

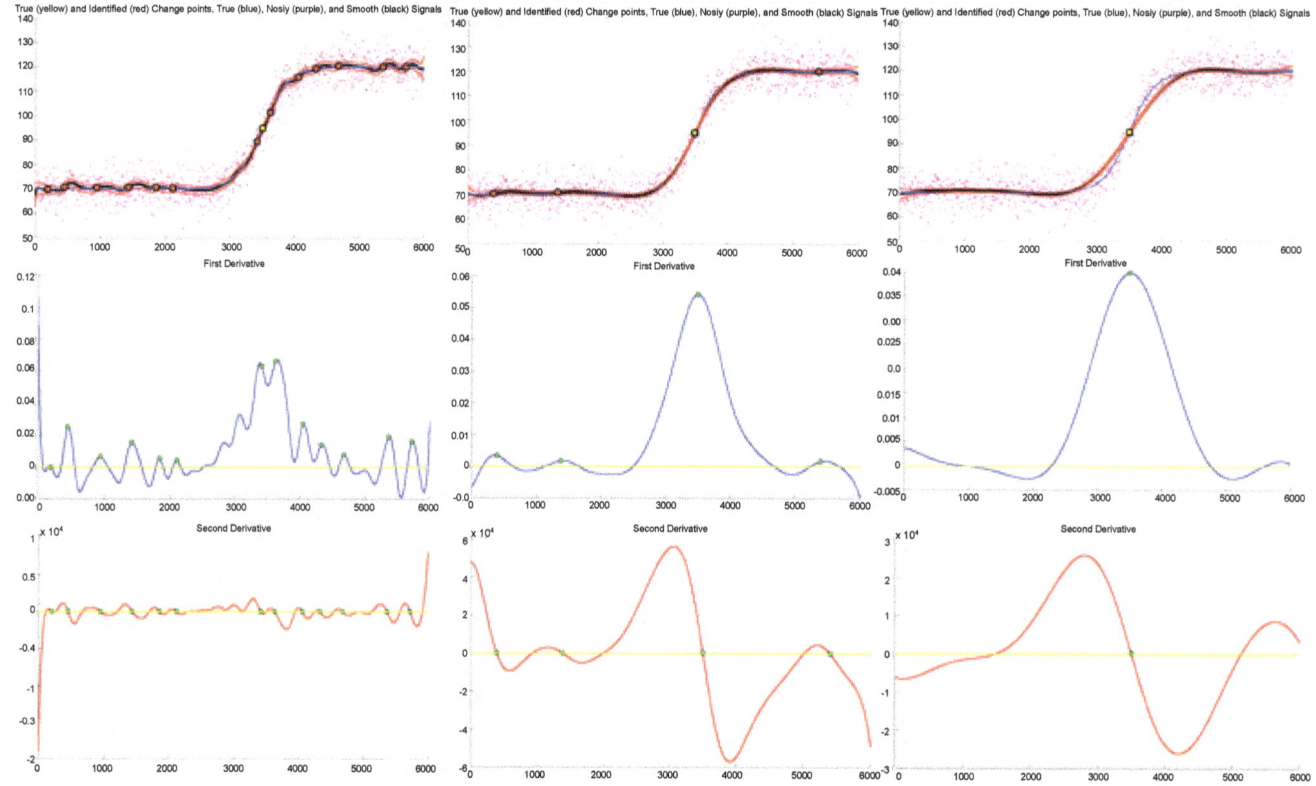

Figure 8: Searching for the smallest bandwidth to satisfy (16). Sigmoid is corrupted by adding Gaussian noise and the estimated curve is shown for different bandwidths, starting from small (100) and increasing the bandwidth to meet (16). True sigmoid (blue), noisy observations (purple), estimated function (black), the detected inflection points (red disks) and the true inflection point (yellow disk). (Top row) Estimated sigmoid and detected inflection points. (Middle row) First derivative of the estimated function. (Bottom row) Second derivative of the estimated function.

15

Stability Studies of the Nigerian 330 KV Integrated Power System

Omorogiuwa Eseosa[1]* and Samuel Ike[2]

[1]*Electrical/Electronic Engineering, Faculty of Engineering, University of Port Harcourt, Nigeria*
[2]*Electrical/Electronic Engineering, Faculty of Engineering, University of Benin, Nigeria*

Abstract

The Nigeria 330 KV integrated power network consisting of seventeen (17) generating stations, sixty four (64) transmission lines and fifty two (52) buses is studied, to investigate the time limits of stability before, during and after occurrence of a three phase (3-θ) fault on the largest generating station (Egbin) and to determine also the most affected generating stations and buses in the network.Theswing and torque equations expressed in time domain was used for the study and the network was modeled in ETAP transient analyzer environment. Transient stability time limit of the system was set to operate at maximum value of ten (10) seconds. Before the fault (between 0.000secs-0.0006secs), the system dynamics was not affected and the peak values of terminal current, rotor angle, frequency,mechanical and electrical power obtained were observed to be within stability region. However, as the fault occurred between 0.0006 secs-0.042 secs, the system dynamics changes, thus affecting the quadrature axis. This change in quadrature axis affected the individual generator's exciter current, exciter voltage, electrical power, mechanical power, frequency, rotor angle and terminal current, though still remain within stability boundary. However, when the fault is cleared within this time, the system returns to its stability region. When the fault lasted beyond 0.042 secs, there is loss of system synchronism. Generating stations that were majorly affected are Omotosho, Sapele, AES and Delta stations. It was observed that the bus voltages connected to these stations deviated from the statutory limit of 313.45 KV-346.5 KV. Their bus voltage values were: Omotosho (361.42KV), Sapele (358.42 KV), AES (350.43 KV) and Delta (364.32 KV). The other buses connected to the other generating stations were however not affected.

The province's population is expected to grow by about 28 percent – or about 3.7 million people – by 2030 and become increasingly urbanized. The structure of the economy will also change as the high-tech and service sectors grow and demand from large industries is expected to grow moderately.

Keywords: PHCN; NIPP; IPP; ETAP; EGBIN; FCT; 3-θ FAULT

Introduction

Transient stability is concerned with the effect of large disturbances due to fault(s) that occur in power system. The most severe is the three phase fault. One way of improving transient stability in power systems is by increasing bus voltage above its nominal value [1], thus making the machine to decelerate fast. This maintains the transmission voltage at the most optimal point or mid-point after a fault is cleared. Transient stability studies involves the determination of whether or not synchronism is maintained after the machine has been subjected to severe disturbance, and this depends on the location and kind of fault in the network. This may be sudden application of load, loss of generation, loss of large load or a fault on the system [2]. Transient stability studies determines the machine power angles, speed deviations, electrical frequency, power flow of machines, lines, transformers and the bus voltage levels [1]. The swing equation is used for transient stability studies (rotor angle stability) and it is seriously affected by the type and location of fault. The nature of the fault means whether it is a three phase or a single phase short circuit fault and the location means whether the fault is on the largest machine in the network or on any other machine(s). Three-phase short circuit fault at the generator bus is the most severe type, as it causes maximum acceleration of the connected machine [3].

Literature Review

The swing equation is the basic equation used to investigate power system transient stability studies.Various numerical methods based on the relevant mathematical modeling of the network and differential evaluation of the machines are applied to solve this equation. It issolved using various numerical methods expressed mainly in time domain. These include: Euler method, Modified Euler method, Runge-Kutta method using implicit integration, equal area criterion, point by point method and Transient Energy Function (TEF) method [4]. Runge-Kutta method is used to determine the first swing stability limit of power system through checking the existence of peaks of rotor angles of severely disturbed generators in the post-fault period [5]. This method is very fast and accurate in determining the critical clearing time. Transient energy function technique is a valuable tool for use in power systems transient analysis needed for both planning and operating functions [6]. The equal-area criterion predicts power system stability and determines critical clearing angle, however it is only applicable to either a one or two machine system connected to an infinite bus bar [7]. Lyapunov's stability criteria, though applicable to stability studies, but as the network gets more complex and large, it becomes inaccurate [2]. The three-phase short circuit at the generator bus is the most severe type [3]. Considering Nigeria power system consisting of seventeen generating stations, fifty two (52) buses and sixty four (64) transmission lines, is gradually becoming large and more complex, effective analysis of such system could only be carried out effectively, efficiently and accurately by the aid of digital computers and the required software. Matlab/Simulink and Etap Transient analyzerare some of such Soft wares/Programs used to study transient

***Corresponding author:** Omorogiuwa Eseosa, Electrical/Electronic Engineering, Faculty of Engineering, University of Port Harcourt, Rivers State, Nigeria
E-mail: oomorogiuwa@yahoo.com

stability. Matlab/Simulink isused to consider stable, critically stable and unstable state of a multi-machine power system and obtained the individual generator angles [8].

Various control methods and controllers have been developed over time to ensure that, after a very sudden and large disturbance, the system still maintains stability by adjusting its protective schemes and control actions [9]. The transient stability control scheme is very difficult to implement because a disturbance that causes instability can only be controlled if a significant amount of computation (analysis) and communication is accomplished [10]. Time domain simulation method using the swing equation was used by [8] to assess the transient stability by setting the fault clearing time (FCT) randomly to determine whether the system is stable or unstable after a fault is cleared.

Equations relevant to the study

The swing equation, torque equation, equal area criteria, and the determination of the time response equivalent to the rotor angle of the machines.

Swing equation (Rotor angle determination)

This is the fundamental equation that determines rotor dynamics in transient stability studies and it is a non linear differential equation that is solved accurately using digital computer program. The equation is given below:

$$M\frac{d^2\delta}{dt^2}+D\left[\frac{d\delta}{dt}\right]=P_m-P_e=P_a \tag{1}$$

Where:

M = angular momentum (joules-sec/rad)

D = damping coefficient

P_m = mechanical power; $p_m = p_{mo} - \Delta p_{mo}$

P_{mo} = input mechanical power

Δp_{mo} = change in input mechanical power due to governor action

P_e = output mechanical power as modified by voltage regulator

P_a = net accelerating power

δ = power angle

T = time

P_e= electrical power

Torque equation

$$T=\frac{\pi p^2}{8}\phi_{air}f_r\sin\delta \tag{2}$$

Where:

T = mechanical torque

P = number of poles

φair = air-gap flux

f_r = rotor field MMF

δ = power (rotor) angle

This equation gives the relationship between mechanical torque, stator voltage, rotor angle, excitation current and voltage of the generator.

Changes in any of these quantities in the two (2) equations causes the rotor speed and acceleration to fall into the three conditions as shown in the below Table 1.

Equal area criteria

Thiscriteria means all energy gained from the turbine during acceleration period must be returned back to the system by the rotor. It is also used to determine the critical clearing angle. This is given in the equation below

$$E_1=\int_{\delta_0}^{\delta_{cr}}(P_m-P_e)d\delta=A_1 \tag{3}$$

$$E_2=\int_{\delta_{cr}}^{\delta_m}(P_e-P_m)d\delta=A_2 \tag{4}$$

$$\left[\frac{d\delta}{dt}\right]^2=\frac{2}{M}\int_{\delta_0}^{\delta}d\delta \tag{5}$$

The critical clearing time is given as

$$t_{cr}=\sqrt{(\delta_{cr}-\delta_0).\frac{4H}{\omega_0 P_m}} \tag{6}$$

Basic numerical equation used to determine time response equivalent to rotor angle

It gives the corresponding time of the rotor angular displacement. It is obtained based on the mathematical concept and it makes use of numerical integration packages based on mathematical concepts. The equation for the basis of the numerical method used is given below

$$\Delta\delta_n=\Delta\delta_{n-1}+\frac{\Delta P_{n-1}+k_4}{M}(\Delta t)^2 \tag{7}$$

Where δ_n and $\Delta\delta_{n-1}$ are rotor angle changes. Δt is a very small time interval (usually 0.05 sec). It gives a detailed numerical solution of the swing equation and requires little iteration. Hence, it is simple, easy to understand and gives minimal round off errors.

Etap (Transient analyzer)

ETAP is a dynamic stability program that incorporates comprehensive dynamic models of prime movers and other dynamic systems. It has an interactive environment for modeling, analyzing, and simulating a wide variety of dynamic systems. It provides the highest performance for demanding applications, such as large network analysis which requires intensive computation, online monitoring and control applications. It is particularly useful for studying the effects of nonlinearity on the behavior of the system, and as such, it is an ideal research tool [9].

Performing power system transient stability is a very comprehensive task, that requires the knowledge of machine dynamic models, machine control unit models (such as excitation system and

Condition		Net Power (pa)	Rotor Speed	Rotor acceleration
1	$P_e = p_m$	Zero	Constant	Zero
2	$P_m > p_e$	Positive	Increasing	Positive
3	$P_m < p_e$	negative	Decreasing	Negative

Table 1: Description of rotor speed and acceleration under certain conditions.

automatic voltage regulators, governor and turbine/engine systems and power system stabilizers) numerical computations and power system electromechanical equilibrium phenomenon.

Aim

To investigate transient stability limits of the generators in the Nigeria 330 Kv integrated power project (NIPP) before, during and after system disturbances due to fault of the generating station(s) in the network in time domain. Etap 4.0 (power station transient stability analyzer) is used to model and solve the network and machine differential equation interactively.

Methodology

The stability limits of Nigeria 330 KV power network consist of Seventeen (17) Generating Stations, Fifty Two (52) buses and Sixty Four (64) Transmission lines is studied to investigate the limits of stability before, during and after disturbance. A one line diagram as well as the transmission line parameters and generator installed and available capacitiesof the network are shown in appendix A, B and C respectively. In this study, three phase (3-θ) short circuit fault on the largest generator (EGBIN with available output capacity of 1320 MW) is considered and analyzed graphically.The impact it will have on other generators rotor angles, exciter currents, exciter voltages, electrical power, mechanical power, frequencies, terminal currents and the entire stability of the network are also determined. Theswing equationexpressed in time domain is solved using the Runge-kutta (using the predictor-corrector routines) and equal area criteria approach in ETAP environmentis used for the analysis. In cases of fault(s), it randomly set up a fault clearing time (FCT) for the machines in the network and solves all the complex differential equations of all equipment in the network.

Modeling the Nigeria 330 KV Integrated Power Network Using Etap

The Integrated Nigerian 330 KV Integrated Power Project (NIPP) interconnects all the generating stations and load centers. The system consists of synchronous generators, motors, transmission lines, transformers, loads and protective devices. Figure 1 below shows the model of the network used for this study. This is obtained by modeling the parameters (bus voltages, transmission line parameters transformers ratings and their loadings, generators ratings and their power limits) as obtained from Power Holding Company of Nigeria (PHCN), while Figure 2 shows the transient stability simulation results obtained.

The simulation time is set from 0 sec-10.0 secs while subjecting EGBIN to a three-phase short circuit fault, and the behavior of the network is obtained graphically as shown in Figures 2-13 below.

Results and Discussions

The Nigeria 330KV integrated power network consisting of seventeen (17) generating stations, sixty four (64) transmission lines and fifty two (52) buses is studied, to investigate the limits of stability before, during and after three phase (3-θ) fault on Egbin power station. When 3-θ short circuit fault occurred at Egbin generator, the system dynamics changes, thus affecting the quadrature axis. This change in the quadrature axis affects the generator's exciter current, exciter voltage, electrical power, mechanical power, frequency, rotor angle and terminal current. Tables 2a, 2b and 2c shows the results obtained and gives the stability limits of these electrical quantities before, during and after a three phase (3-θ) fault. Figures 2-13 shows a plot of how this fault affects the behavior of the other generators in the network. It was observed that before the 3-θ pre-fault, the peak values obtained between time range of 0.000 Secs-0.0600 Secs operates within the

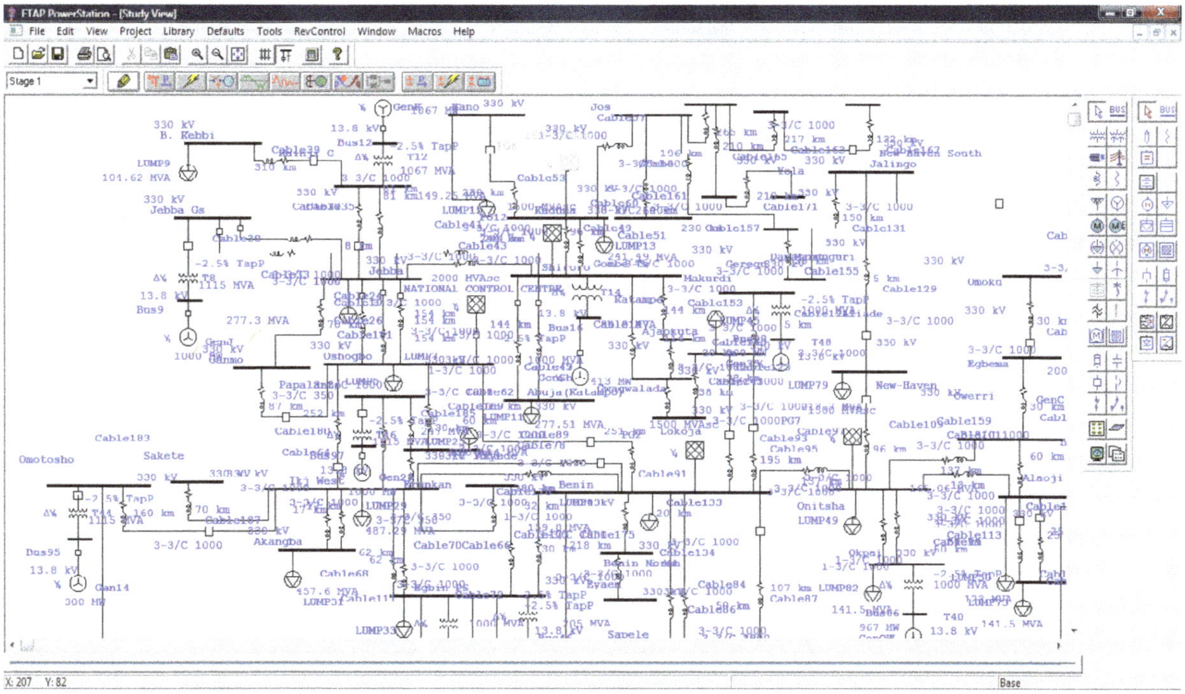

Figure 1: Model of the network used.

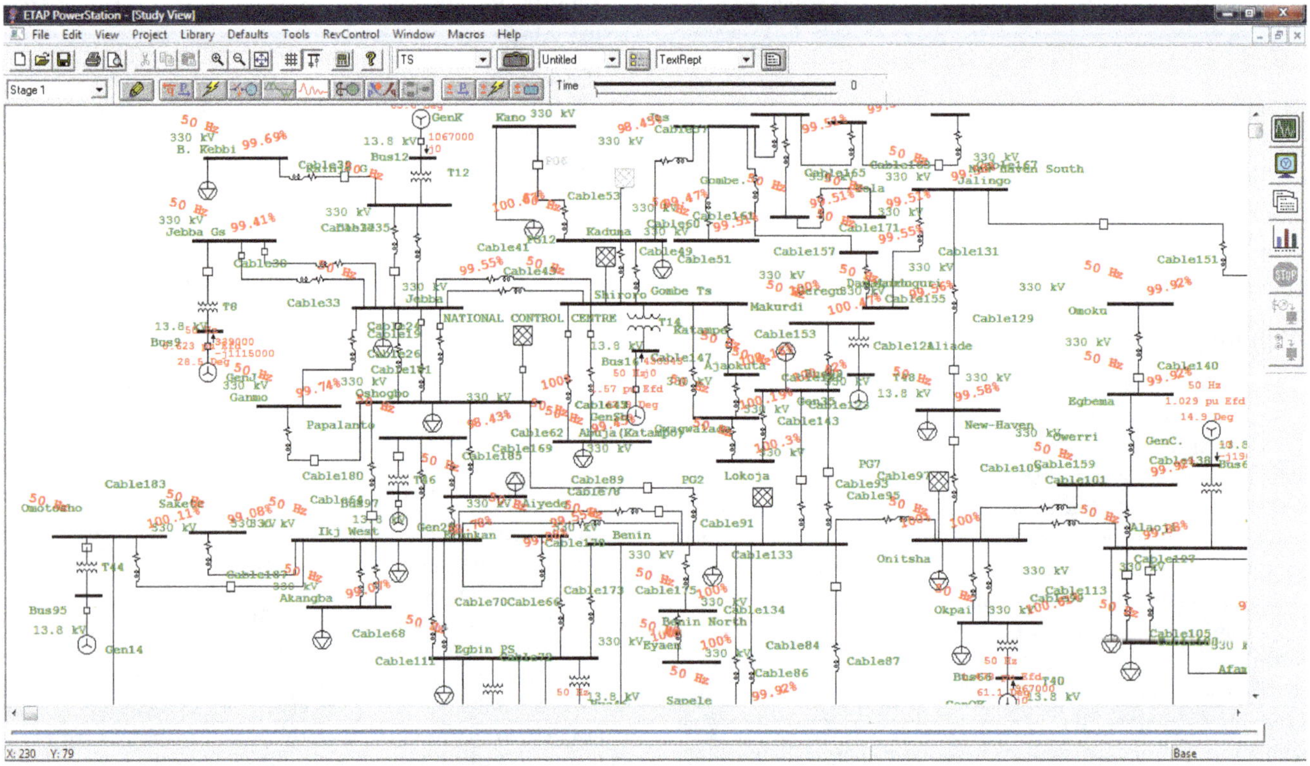

Figure 2: Model of the proposed 330kv network of Nigeria 330kv integrated power system after simulation.

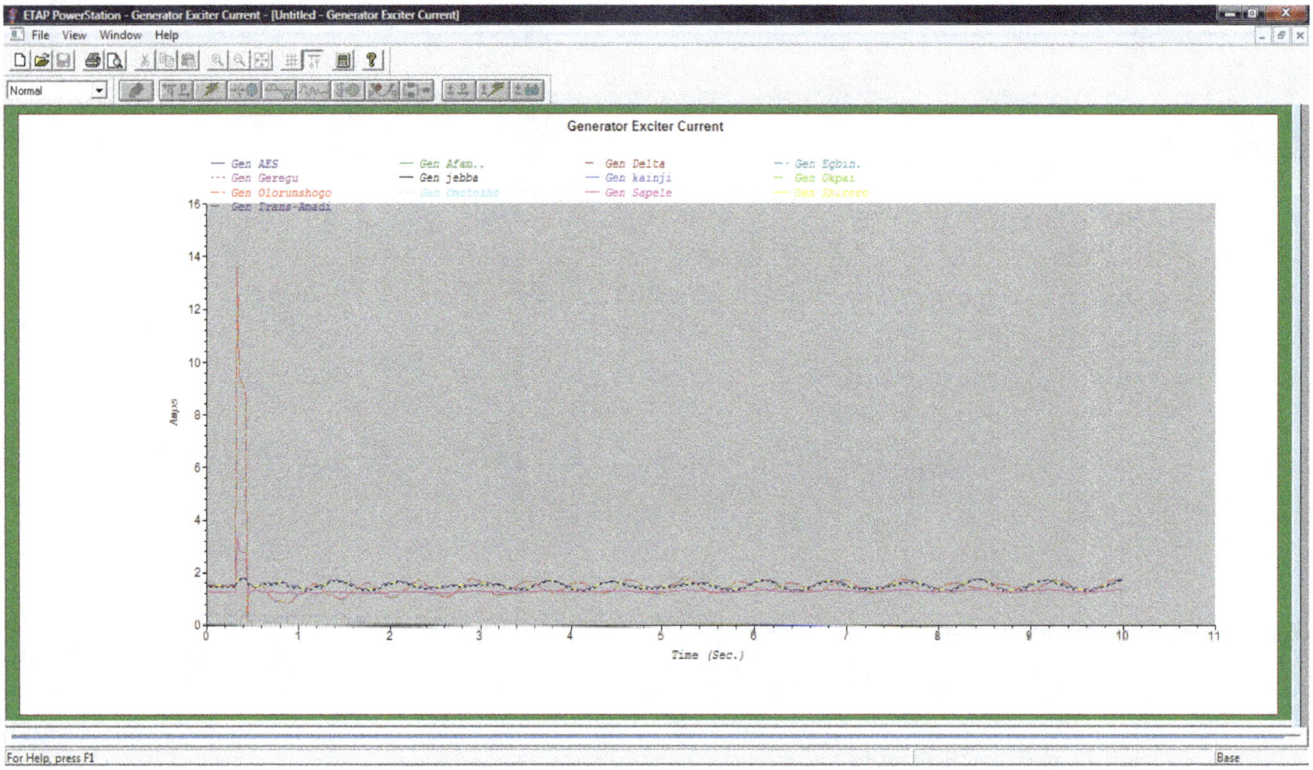

Figure 3: Plots of Generators Exciter Current versus Time.

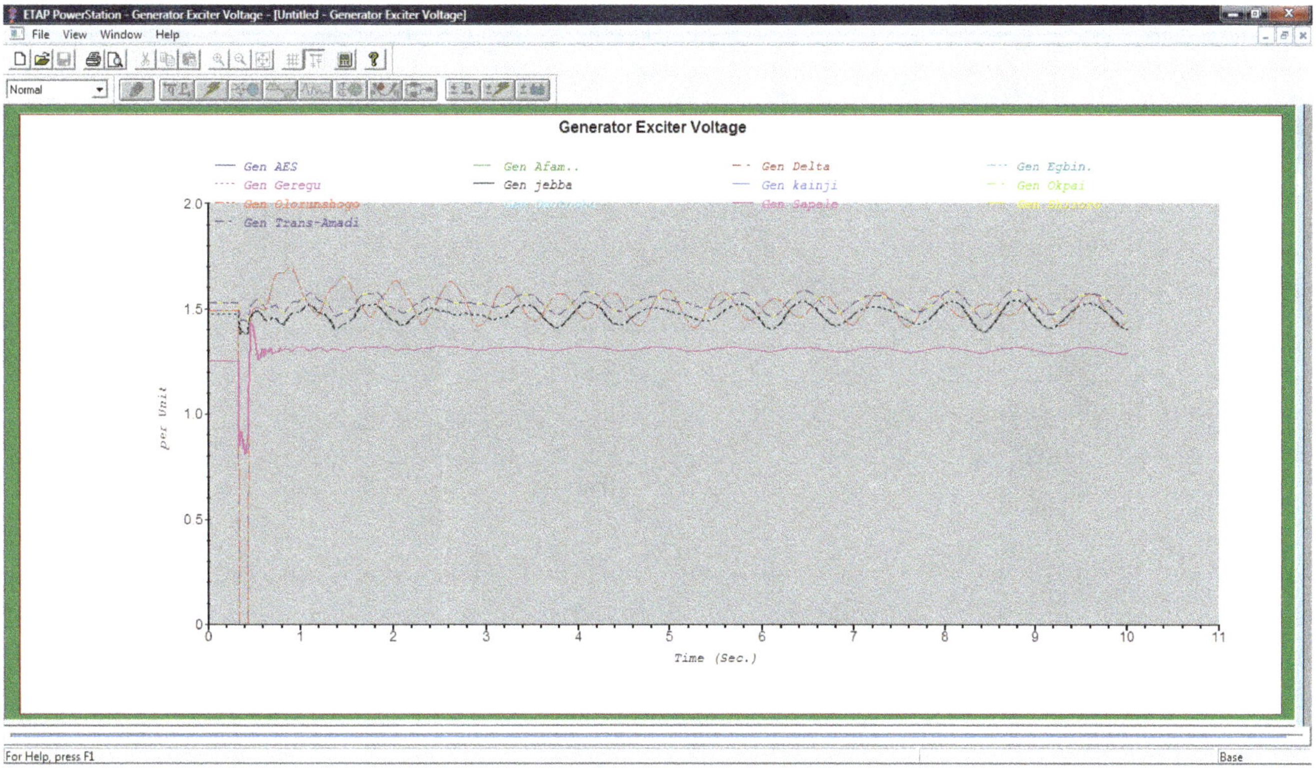

Figure 4: Plots of Generators Exciter voltages versus Time.

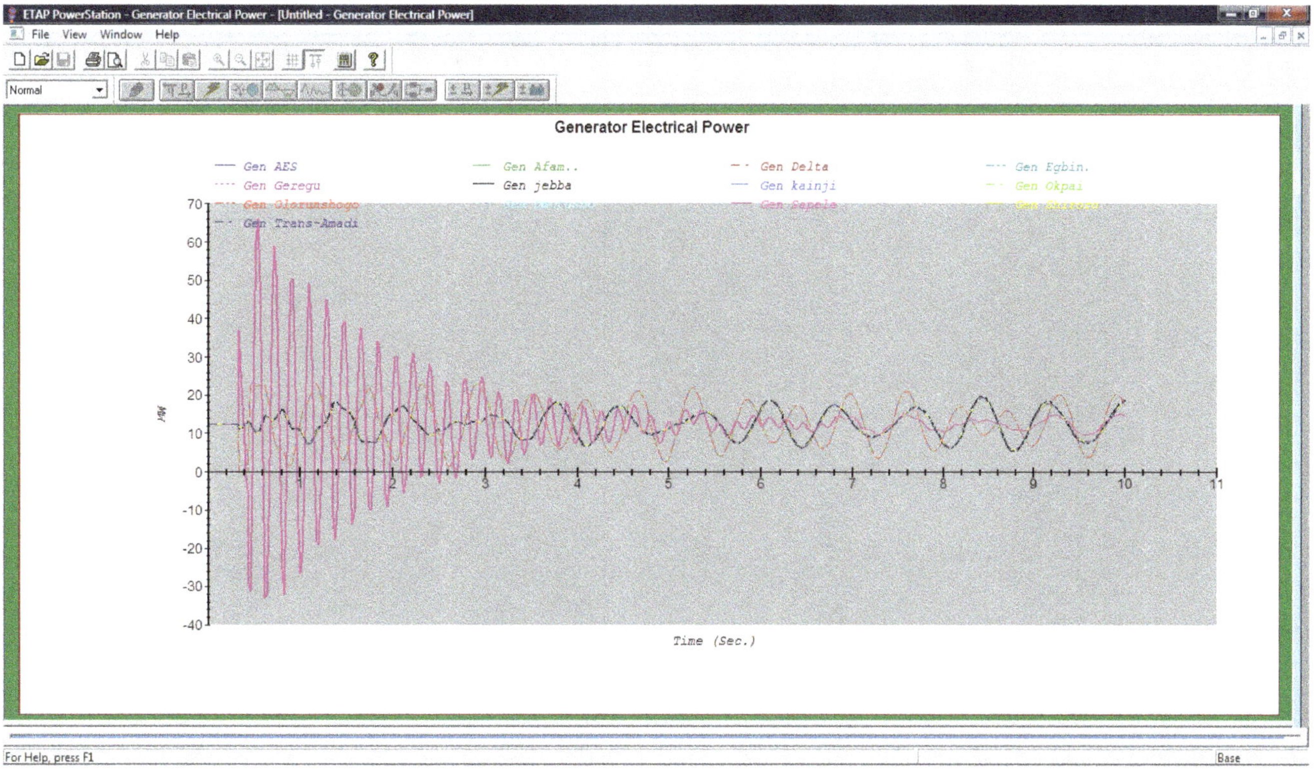

Figure 5: Plots of Generators electric power versus Time.

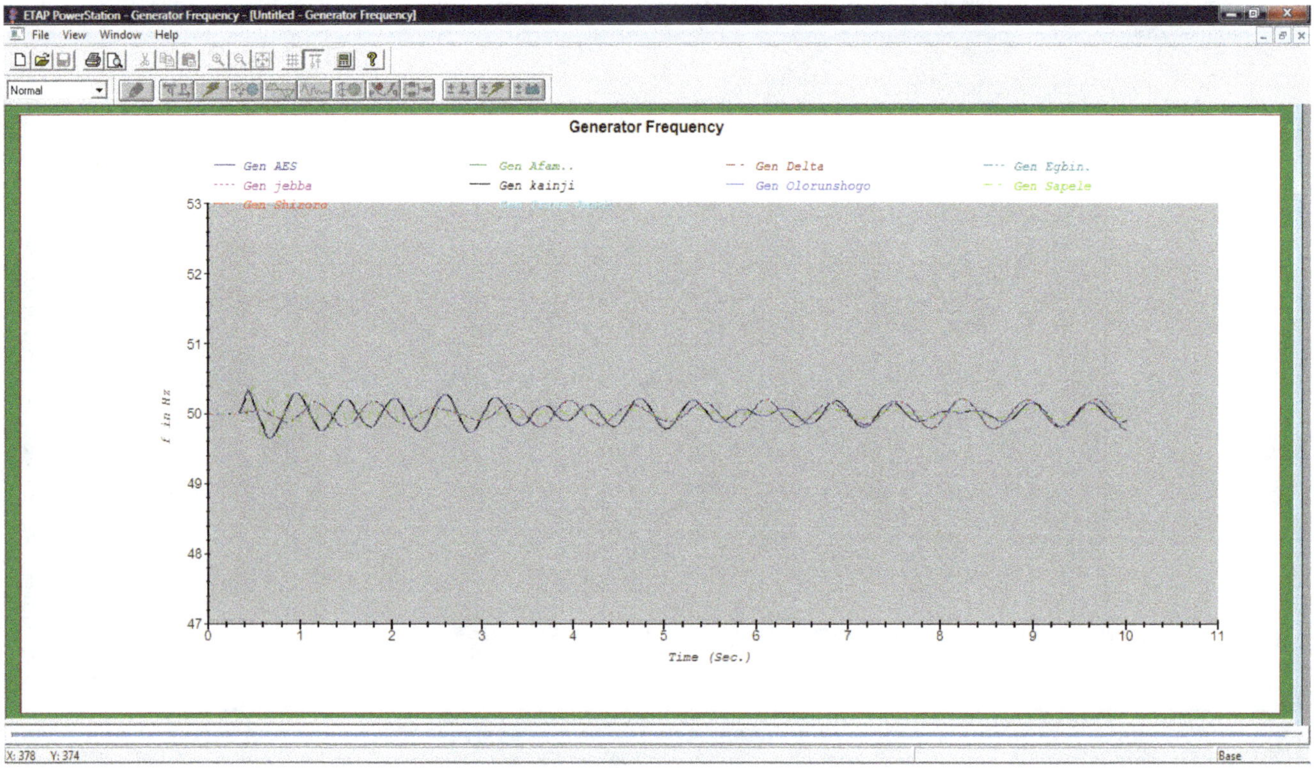

Figure 6: Plots of Generators frequency versus Time.

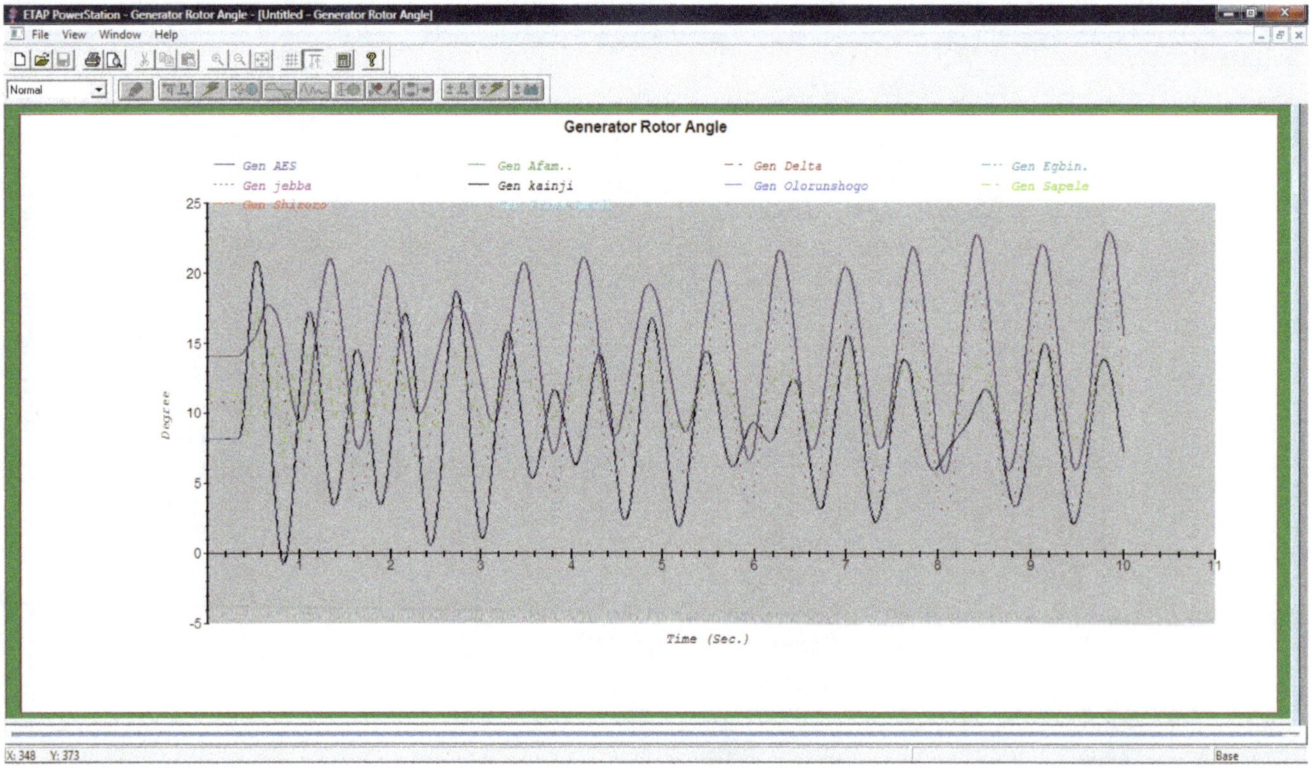

Figure 7: Plots of Generators rotor angle versus Time.

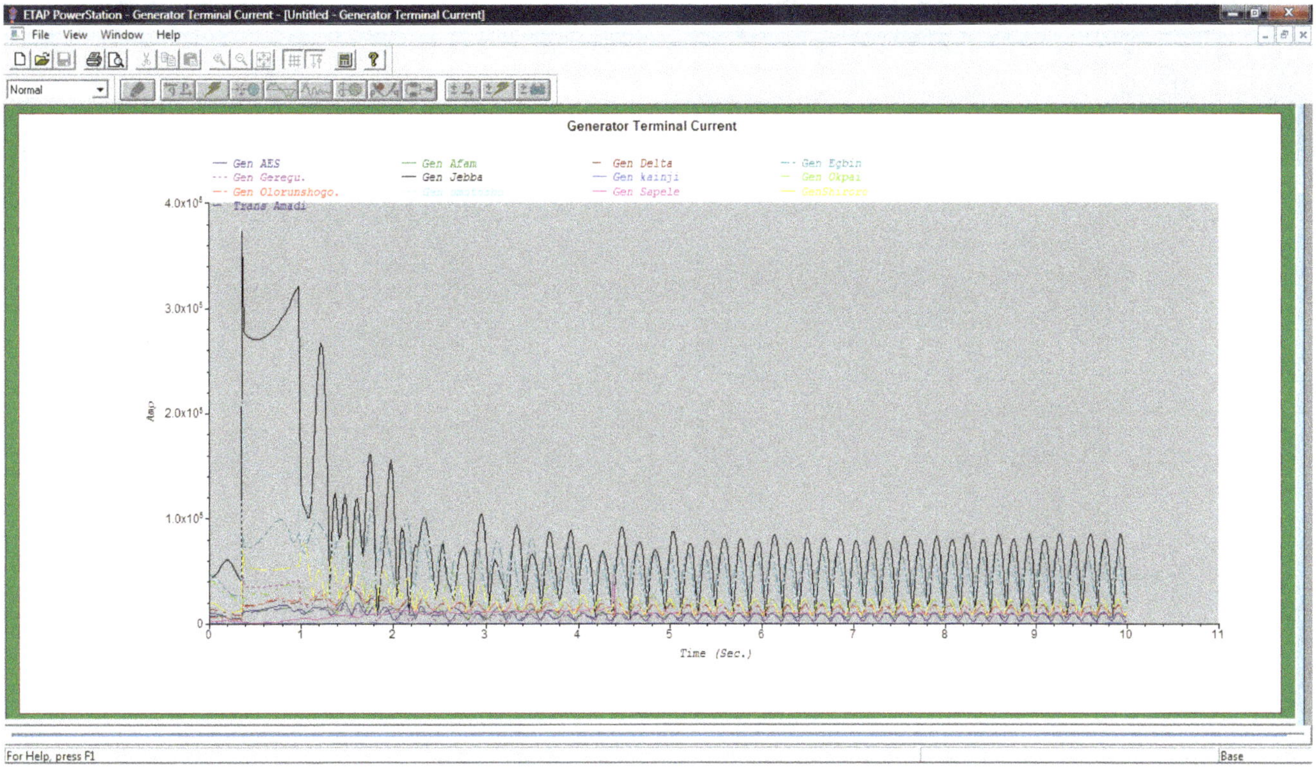

Figure 8: Plots of Generators terminal current versus Time.

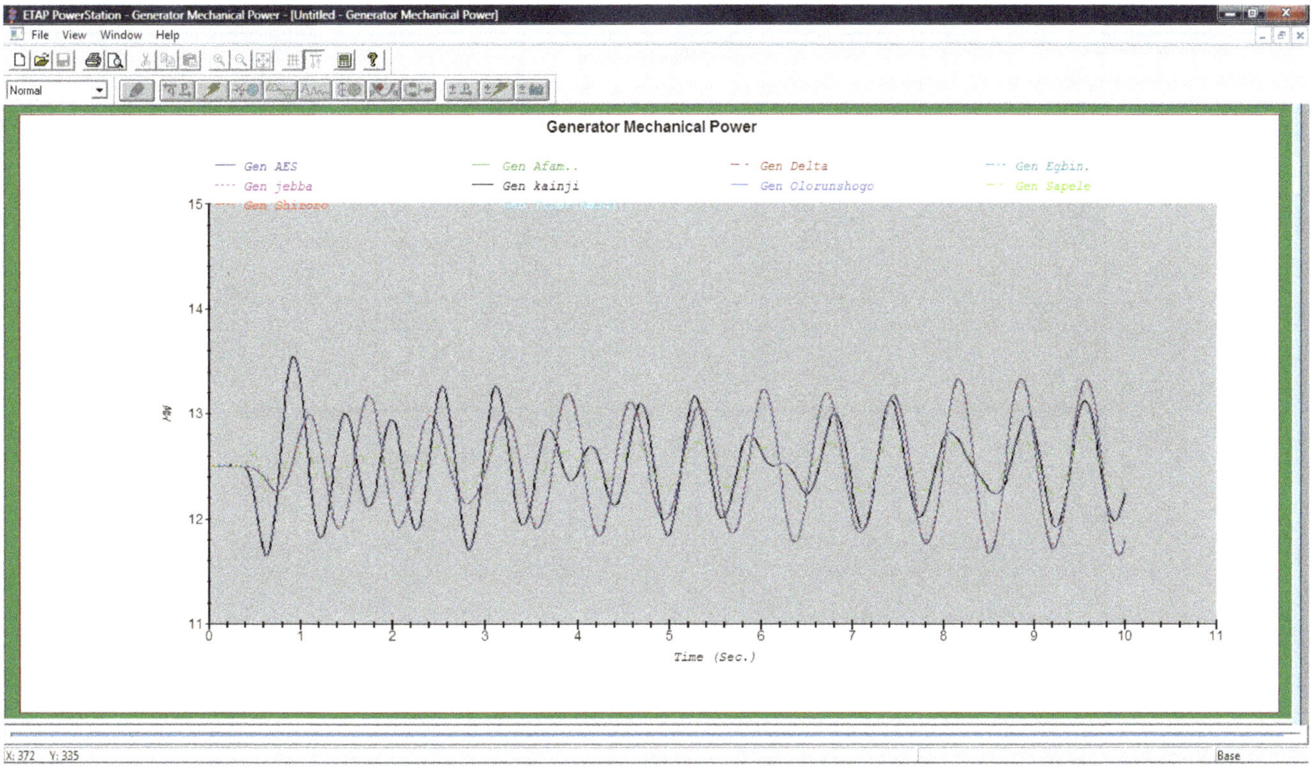

Figure 9: Plot of Generators mechanical powers versus Time.

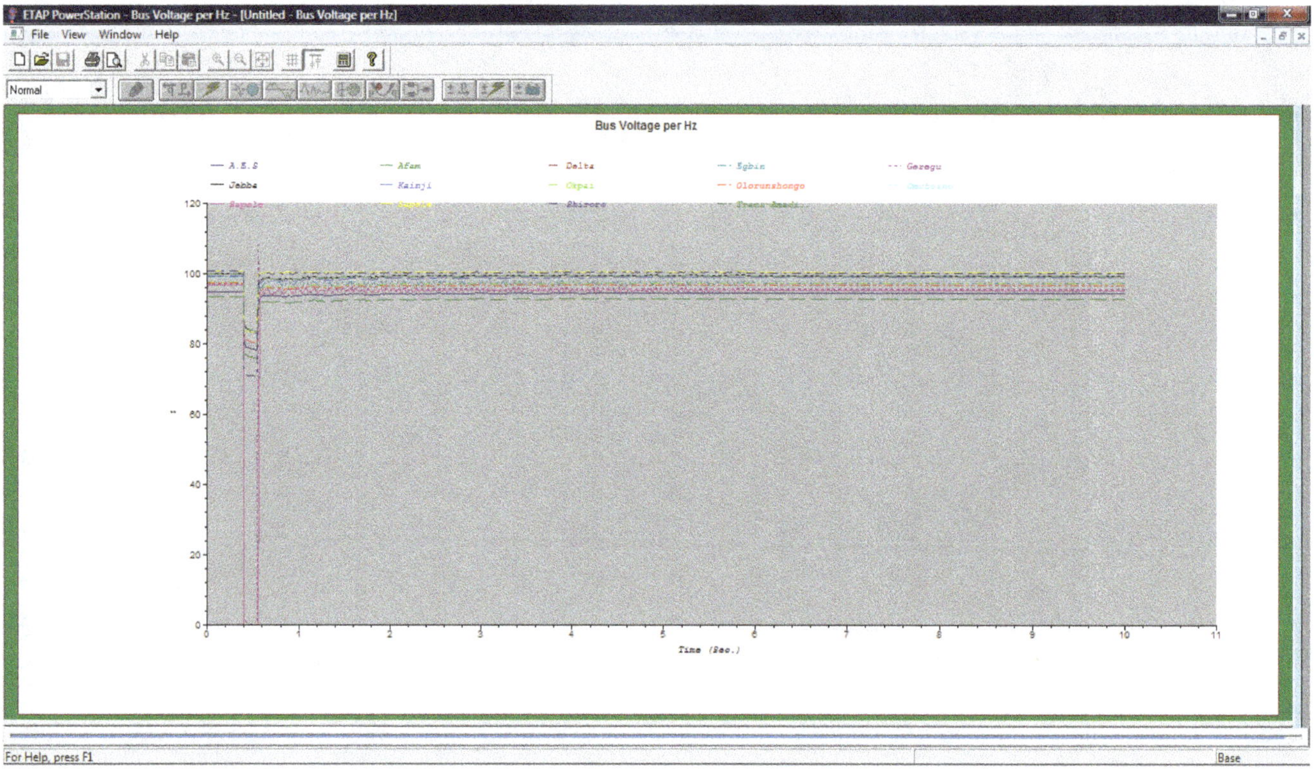

Figure 10: Plot of generators bus voltages per Hz Versus Time.

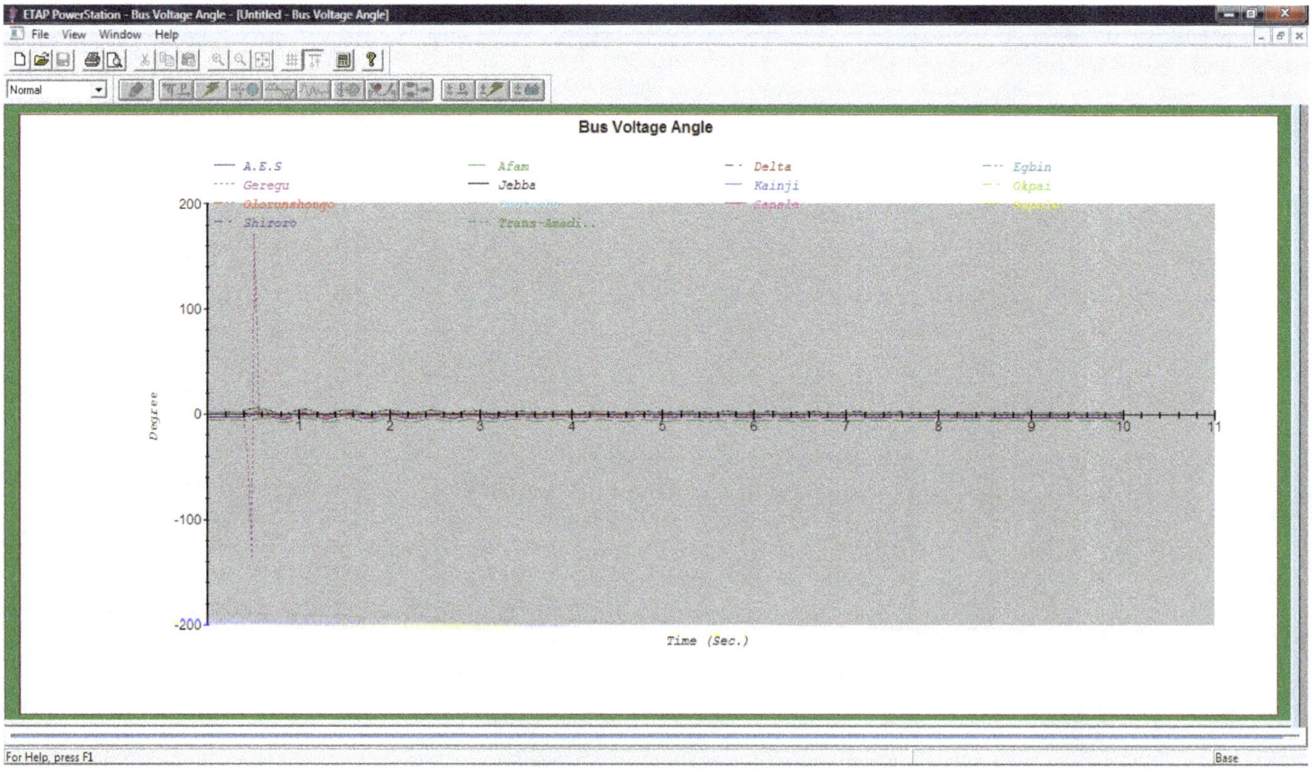

Figure 11: Plots of Generators Bus Voltages Angles Versus Time.

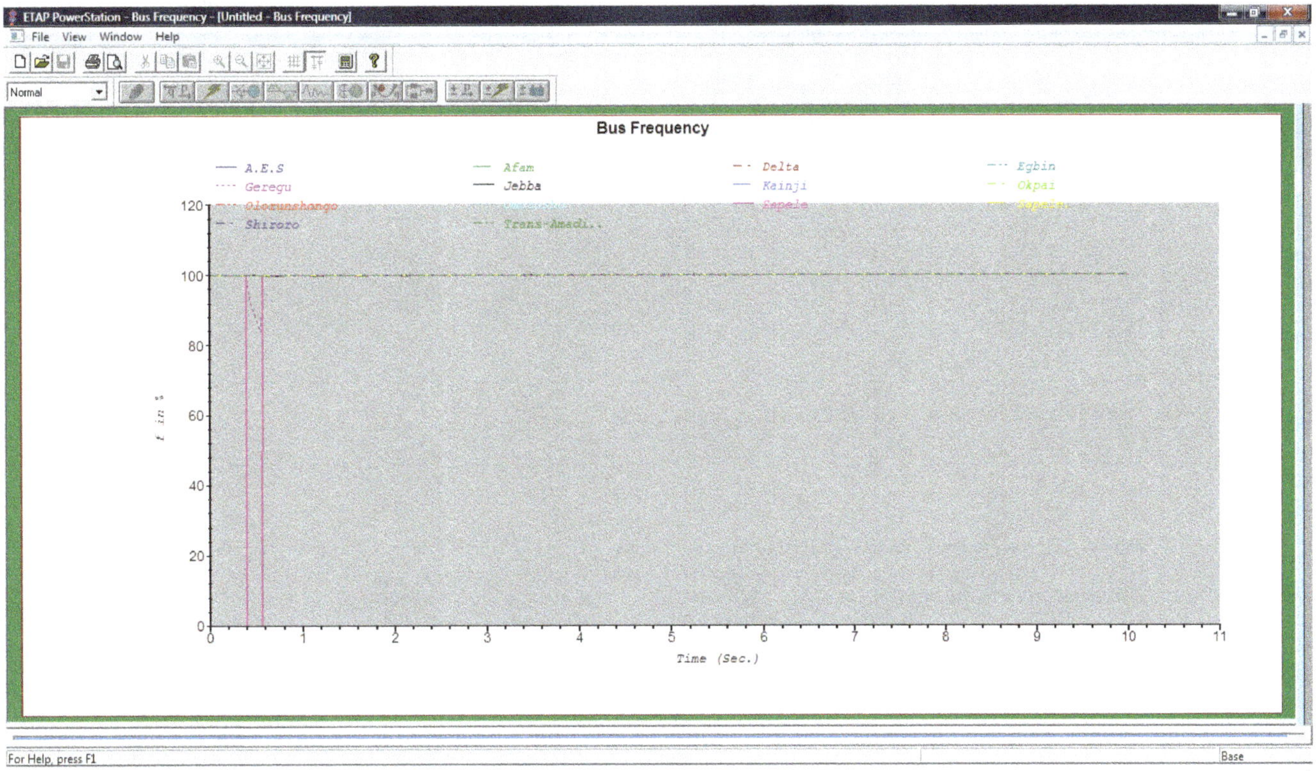

Figure 12: Plots of Generators Bus Frequency versus Time..

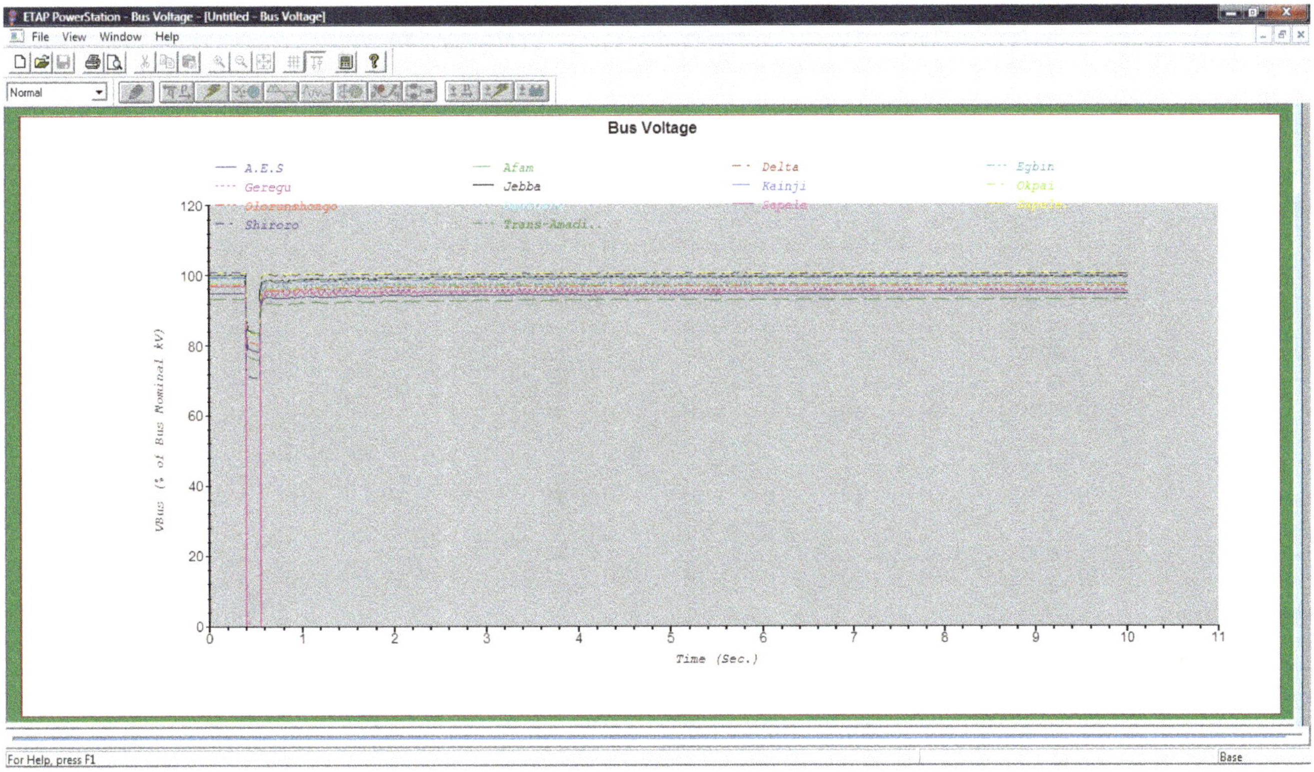

Figure 13: Plots of Generators Bus Voltages Versus Time.

Generators	Mechanical Power (MW)	Electrical Power (MW)	Terminal Current(A)	Rotor angle (Degree)	Frequency (Hz)	Exciter Current (p.u)	Exciter Voltage (p.u)
AES	203.27	203.34	10,200	21	49.55	1.55	3.83
Afam I-V	46.41	46.24	3,080.23	32	50.01	1.49	1.70
Afam VI	371.86	371.72	25,992.49	19	50.00	1.52	3.68
Egbin	788.42	788.50	53,732.21	23	49.90	1.92	5.12
Geregu	62.22	62.16	4,441.23	28	49.98	1.05	3.20
Jebba	585.72	585.92	37,873.82	32	49.96	0.70	5.02
Kainji	214.46	214.60	14,678.85	43	50.00	2.10	3.92
Okpai	150.53	150.03	9,031.28	38	50.00	1.48	3.68
Olorunshogo phase 1	50.28	50.32	3,285.72	32	49.99	1.50	2.57
Olorunshogo phase 2	106.99	107.01	6,505.20	16	49.87	1.41	3.63
Sapele	161.94	161.97	10,540.02	41	49.86	0.40	3.72
Shiroro	377.53	377.06	24,320.26	38	50.00	1.80	4.38
Trans-Amadi	77.78	78.00	4,763.4	26	49.76	1.57	3.62
Omotosho	45.27	45.37	2,986.23	29	49.65	1.60	0.98
Ibom	66.45	66.24	4,273.58	23	50.00	1.30	3.45
Omoku	68.46	68.48	4,449.87	20	50.00	1.80	3.57
Delta	381.01	381.03	24,900.48	24	49.54	2.05	4.70

Table 2a: 3-Phase Pre-Fault Peak Values at 0.000 to 0.060 Sec, During Fault Peak Values at 0.061 Secs -0.4220 Secs.

Generators	Mechanical Power (MW)	Electrical Power (MW)	Terminal Current(A)	Rotor angle (Degree)	Frequency (Hz)	Exciter Current (p.u)	Exciter Voltage (p.u)
AES	239.7	543.56	37,292.02	72	51.15	2.27	1.62
Afam I-V	54.65	92.60	10,675.59	75	50.48	0.99	0.73
Afam VI	440.21	741.27	89,414.17	64	50.60	2.56	1.84
Egbin	938.23	1796.65	186,838.01	88	51.40	2.98	2.16
Geregu	74.04	125.95	15,277.83	72	50.10	1.86	1.34
Jebba	697.01	1166.75	130,285.94	82	49.48	2.92	2.11
Kainji	255.21	426.25	50,495.24	81	50.63	2.32	1.62
Okpai	178.81	302.18	31,067.60	81	49.20	2.17	1.58
Olorunshogo phase 1	59.72	101.14	11,302.88	77	50.32	1.51	1.07
Olorunshogo phase 2	119.72	202.08	22,377.89	62	50.07	2.14	1.52
Sapele	193.05	454.19	38,257.67	84	52.26	2.20	1.59
Shiroro	452.53	765.67	83,904.89	83	50.45	2.57	1.87
Trans-Amadi	93.41	84.05	16,481	72	49.96	2.12	1.52
Omotosho	55.07	126.03	13,302.49	86	51.35	0.57	0.41
Ibom	79.71	93.72	14,743.85	73	50.21	2.01	1.45
Omoku	82.04	137.78	15,396.55	65	50.57	2.11	1.55
Delta	461.01	974.68	87,831.96	83	51.34	2.75	1.98

Table 2b: During Fault Peak Values at 0.061 Secs -0.4220 Secs.

Generators	Mechanical Power (MW)	Electrical Power (MW)	Terminal Current(A)	Rotor angle (Degree)	Frequency (Hz)	Exciter Current (p.u)	Exciter Voltage (p.u)
AES	202.13	201.02	10,200	25	48.51	1.23	3.23
Afam I-V	42.11	41.14	3,080.23	37	49.24	1.29	1.60
Afam VI	359.23	358.72	25,992.49	26	49.08	1.27	3.38
Egbin	772.62	770.51	53,732.21	34	48.45	1.42	5.04
Geregu	58.13	57.13	4,441.23	37	49.56	0.97	2.98
Jebba	576.14	573.22	37,873.82	38	48.66	0.57	4.94
Kainji	208.21	206.45	14,678.85	42	49.32	1.95	3.92
Okpai	146.21	150.41	9,031.28	43	49.21	1.24	3.68
Olorunshogo phase 1	46.16	45.35	3,285.72	37	48.23	1.42	2.97
Olorunshogo phase 2	103.21	101.45	6,505.20	22	48.67	1.32	3.61
Sapele	158.22	161.97	10,540.02	45	49.23	0.31	2.84
Shiroro	371.05	369.06	24,320.26	43	49.21	1.74	3.98
Trans-Amadi	74.28	72.15	4,763.4	34	49.23	1.53	3.08
Omotosho	43.19	42.21	2,986.23	32	49.15	1.48	1.08
Ibom	63.24	61.29	4,273.58	32	48.56	1.26	3.35
Omoku	64.24	63.21	4,449.87	28	50.00	1.72	2.97
Delta	376.21	372.25	24,900.48	30	49.54	1.98	3.82

Table 2c: After fault (Stability with Damped Oscillation).

allowable tolerable frequency limit of 48.45 Hz-51.45 Hz at nominal frequency of 50 Hz. Moreso, the net power (difference between electrical and mechanical power) is small and the rotor angle is within 90 degrees, thus satisfying stability criteria. During the fault (0.061 secs-0.042 secs), it was observed that the net power was large and the frequency was gradually moving out of its allowable tolerable limit. However, the various electrical quantities peak values indicates that if the fault is not cleared after 0.042 secs, the system become unstable and loss synchronism between the generators, that will eventually lead to system collapse as shown in Figures 4-13. Egbin had its electrical and mechanical power before the fault to be 788.42 MW and 788.50 MW, while during the fault to be 938.23 MW and 1596.65 MW respectively. Generators that were majorly affected are Omotosho, Sapele, AES and Delta stations. The bus voltages connected to these stations, were deviating from the statutory limit of 313.45 KV-346.5 KV at 0.042 secs until the oscillation was damped. Their bus voltage values after 0.042 secs are Omotosho (361.42 KV), Sapele (358.42 KV), AES (350.43 KV) and Delta (364.32 KV). These values are as a result of the three-phase fault in the network. The buses connected to the other generating stations were however not affected.

Conclusion/Recommendation

Transient stability study of the Nigeria 330 KV integrated power network consisting of Seventeen (17) generating stations, Fifty Two (52) buses and 64 Transmission lines was carried out using ETAP 4.0 Transient analyzer. The impact of three phase short circuit fault of the largest power station (EGBIN) on the entire system stability was considered. The system was analyzed before, during and after the fault. It was observed that before the fault (0.000-0.0060 secs), the system was still stable until it got to 0.042 sec. At this time, four generating stations (Omotosho, Sapele, AES and Delta) were almost getting out of synchronism. However, when the fault was cleared, the system returned to its stability. The obtained result showed that the fault when allowed to last beyond 0.042 seconds causes buses connected to four (4) generating stations (AES, Sapele, Delta and Omotosho) to swing away from the stability region hence operating outside the allowable tolerable voltage limit of 314.45 KV-346.45 KV.

References

1. Eseosa O (2011) "Efficiency Improvement Of The Nigeria 330KV Network Using Facts Device" University Of Benin, Benin City.
2. Kundur P (1994) 'Power System Stability and Control' Tata McGraw-Hill, New Delhi.
3. Nagrath J, Kothari DP (1994) 'Power System Engineering' Tata McGraw-Hill, New Delhi.
4. Weedy BM, Cory BJ (1998) "Electric Power Systems: Wiley Student Edition, London.
5. Haque MH, Rahim AHMA (2002) 'Determination of first swing stability limit of multi machine power systems through Taylors series expansion'. IEE proceedings 136: 373-380
6. Ruiz-Vega D, AL Bettiol, Ernst D, Wehenkel L, Pavella M (1998) "Transient stability-constrained generation rescheduling," in Bulk Power System Dynamics and Control IV—Restructuring, Santorini, Greece, 105-115.
7. Saadat H (2002) 'Power System Analysis, New Delhi,' Tata McGraw-Hill Publishing Company Limited. PP 189, 486.
8. Noor IAW, Azah M (2008) 'Transient stability assessment of a power system using probabilistic neural network' Am J Appl S.
9. Pavella MD Ernst, Ruiz-Vega D (2000) Transient Stability of Power Systems: a Unified Approach to Assessment and Control: Kluwer Academic Publishers.
10. Anjan B (2003) 'Power System Stability: New Opportunities for Control1'Washington State University Pullman, WA 99164-2714.

Peculiarities of Propagation of Electromagnetic Excitations through Nonideal 1d Photonic Crystal

Vladimir V. Rumyantsev*

A.A. Galkin Donetsk Institute for Physics and Engineering, National Academy of Sciences, Ukraine

Abstract

Nonideal 1D photonic crystal is modeled as a macroscopically homogeneous layered system (which is one-dimensional Si-liquid crystal superlattice with two elements-layers in the cell) with randomly included admixture layers. The virtual crystal approach which is the method to describe quasi-particle excitations in disorder media is used. Peculiarities of the dependence of photonic band gap width on admixture layers concentration have been studied. The results are the evidence of substantial polariton spectrum reconstruction caused by presence of defect layers.

Keywords: 1D photonic crystal; Admixture elements; Virtual crystal approximation; Band gap width

Introduction

At present, propagation of electromagnetic waves in thin films and layered crystal media [1-4], in particular, in photonic magnetic crystals [5-7] and composite crystals based on silicon and liquid crystal [8-12], are being investigated extensively. Interest in investigation of these objects is motivated, on the one hand, by the demand for different layered structures with given properties in solid-state electronics and, on the other hand, by achievements in technologies providing the possibility of growing such films and periodic structures with controlled characteristics by the molecular-beam epitaxial method. A large number of works [13-15] have been devoted to theoretical and experimental investigation of exciton-type excitations in dielectric ideal superlattices. The general theory of optical waves in anisotropic crystals, including those composed of macroscopic layers, is considered in [16]. In [11], the forbidden photonic bands of a crystal made of alternating silicon and liquid crystal layers are calculated. 1D photonic band gap structures attract lot of attention of researches due to their omnifarious using for optical filtering [17], sensing [18] and so on. The logic of further development of the theory of layered structures requires consideration of more complex systems-superlattices with impurity layers, as well as with layers of variable composition and thickness. In [6], the dispersion of polaritons in a superlattice with a single impurity layer has been studied. At the same time, of considerable interest are investigations of nonideal superlattices with an arbitrary number of impurity layers and of the dependence of the polariton spectrum on the concentration of corresponding defects, which make it possible to expand the capabilities of modeling the properties of such systems and to create layered materials with given characteristics.

The method of calculation of polariton excitations has much in common with methods of finding other quasiparticle excitations (electron, phonon, etc.) in solids. In the present chapter authors offer for description of photon modes in macroscopically heterogeneous ambience approach, based on configuration averaging (it specifically is new in this case), which was used before [19] for the microscopic calculation of disordered systems quasiparticles spectra. A relatively simple approximation in framework of this approach for calculation polariton spectra and corresponding optical characteristics of disordered ambiences is the Virtual-Crystal Approximation (VCA) [19-21]. VCA (first it was used by Nordheim and Parmenter [19]) consist of replacing of correct one-electron potential (appropriate to a given configuration of atoms of the alloy) by its average which is taken over all possible random configurations. The approximation is a widely used for study of disordered structures. For example, based on the pseudopotential scheme under the VCA in which the effect of compositional disorder is involved, the dependence of optoelectronic properties of $GaAs_x Sb_{1-x}$ on alloy composition x have been studied in [22]. Within this approximation the configurationally dependent parameters of the Hamiltonian are replaced with their configurationally averaged values. Description of transformation of a polariton spectrum in a sufficiently simple superlattice, using VCA, is the first step towards the study of imperfect systems. However investigation of properties of polariton spectra and the related physical quantities (density of elementary excitation states, characteristics of the normal electromagnetic waves etc.) in less simple systems requires application of more complex methods. Such are the method of the coherent (one- or many-site) potential [21], the averaged T-matrix method [23] and their numerous modifications used for various particular problems.

1D photonic crystal may be modeled as a set of macroscopically homogeneous layers with randomly included extrinsic (with respect to the ideal superlattice) layers. Corresponding configuration-dependent material tensors in our model of an imperfect superlattice are represented in terms of random quantities. After configuration-averaging the translational symmetry of a considered system is "restored" that allows us obtain the system of equations which define normal modes of electromagnetic waves, propagating in one-dimensional (1D) "periodic" medium. Within the VCA the peculiarities of the dependence of the band gap width and refractive index upon concentration of admixture layers for the non-ideal photonic crystalline system (that is layered structure or striped thin film) is studied by authors [24]. The photon modes spectrum of a non-ideal superlattice with an arbitrary number of layers (strips) per elementary cell, obtained within the VCA, is concretized for the Si - liquid crystal system. Dependence of the band gap width upon concentration of admixture elements and refractive index peculiarities was analyzed.

***Corresponding author:** Vladimir V. Rumyantsev, A.A. Galkin Donetsk Institute for Physics and Engineering, National Academy of Sciences, Ukraine
E-mail:rumyants@teor.fti.ac.donetsk.ua

Propagation of Electromagnetic Waves in Nonhomogeneous Structures

Dielectric $\hat{\varepsilon}(\vec{r})$ and magnetic $\hat{\mu}(\vec{r})$ permeability, which determine optical characteristics of a periodic medium, must satisfy the periodic boundary conditions:

$$\hat{\varepsilon}(x,y,z)=\hat{\varepsilon}(x,y,z+d),\quad \hat{\mu}(x,y,z)=\hat{\mu}(x,y,z+d) \tag{1}$$

where $d=\sum_{j=1}^{\sigma}a_j$ is the period of the superlattice, σ is the number of layers per elementary cell, a_j are the thicknesses of the layers which form a one-dimensional chain of elements oriented along the z-axis. The material tensors $\hat{\varepsilon}$ and $\hat{\mu}$ of a crystalline superlattice with an arbitrary number of layers σ have the following form in the coordinate representation:

$$\begin{pmatrix}\hat{\varepsilon}(z)\\ \hat{\mu}(z)\end{pmatrix}=\sum_{n,\alpha}\begin{pmatrix}\hat{\varepsilon}_{n\alpha}\\ \hat{\mu}_{n\alpha}\end{pmatrix}\left\{\theta\left[z-(n-1)d-\left(\sum_{j=1}^{\alpha}a_{n_j}-a_{n\alpha}\right)\right]-\theta\left[z-(n-1)d-\sum_{j=1}^{\alpha}a_{n_j}\right]\right\} \tag{2}$$

In Eq. (2) $\theta(z)$ is the Heaviside function, $n=\pm1,\ \pm2,\ \ldots$ is the number of a one-dimensional crystal cell, index $\alpha=1,2,\ldots,\sigma$ designates the elements of the cell. Below we consider an imperfect system, in which disordering is connected with variation of the composition (rather than of the thickness) of admixture layers, so that $a_{n\alpha}\equiv a_{\alpha}$. Within our model, the configurationally dependent tensors $\hat{\varepsilon}_{n\alpha}$, $\hat{\mu}_{n\alpha}$ are expressed through the random quantities $\eta_{n\alpha}^{\nu}$ ($\eta_{n\alpha}^{\nu}=1$ if the $\nu(\alpha)$-th sort of layer is in the $(n\alpha)$-th site of the crystalline chain, $\eta_{n\alpha}^{\nu}=0$ otherwise):

$$\begin{pmatrix}\hat{\varepsilon}_{n\alpha}\\ \hat{\mu}_{n\alpha}\end{pmatrix}=\sum_{\nu(\alpha)}\begin{pmatrix}\hat{\varepsilon}_{\alpha}^{\nu(\alpha)}\\ \hat{\mu}_{\alpha}^{\nu(\alpha)}\end{pmatrix}\eta_{n\alpha}^{\nu(\alpha)} \tag{3}$$

Calculation of a polariton spectrum for the imperfect superlattice is realized within the VCA (similarly to the solid quasi-particle approach) through the following replacement: $\hat{\varepsilon}\to\langle\hat{\varepsilon}\rangle$, $\hat{\mu}\to\langle\hat{\mu}\rangle$ (angular parentheses designate the procedure of configuration averaging). In addition, from Eq. (3) and [21] we have:

$$\begin{pmatrix}\langle\hat{\varepsilon}_{n\alpha}\rangle\\ \langle\hat{\mu}_{n\alpha}\rangle\end{pmatrix}=\sum_{\alpha,\nu(\alpha)}\begin{pmatrix}\hat{\varepsilon}_{\alpha}^{\nu(\alpha)}\\ \hat{\mu}_{\alpha}^{\nu(\alpha)}\end{pmatrix}C_{\alpha}^{\nu(\alpha)} \tag{4}$$

where $C_{\alpha}^{\nu(\alpha)}$ is the concentration of the $\nu(\alpha)$-th sort of admixture layer in the α-th sublattice. There is a normalization condition $\sum_{\nu(\alpha)}C_{\alpha}^{\nu(\alpha)}=1$. It follows from Eq. (2) that the Fourier-amplitudes $\hat{\varepsilon}_l$, $\hat{\mu}_l$ and the averaged dielectric $\langle\hat{\varepsilon}_{n\alpha}\rangle$ and magnetic $\langle\hat{\mu}_{n\alpha}\rangle$ permeabilities of layers (4) are related as

$$\begin{pmatrix}\hat{\varepsilon}_l\\ \hat{\mu}_l\end{pmatrix}=-\frac{i}{2\pi l}\sum_{\alpha}\begin{pmatrix}\langle\hat{\varepsilon}_{n\alpha}\rangle\\ \langle\hat{\mu}_{n\alpha}\rangle\end{pmatrix}\left\{\exp\left(i\frac{2\pi}{d}l\sum_{j=1}^{\alpha}a_j\right)-\exp\left[i\frac{2\pi}{d}l\left(\sum_{j=1}^{\alpha}a_j-a_{\alpha}\right)\right]\right\} \tag{5}$$

Since the configurational averaging "restores" the translational symmetry of a crystalline system, in the considered case of imperfect superlattice the "acquired" translational invariance of the one-dimensional chain allows us to present Maxwell equations (for harmonic dependency of the electric and magnetic field strengths $\vec{E}(\vec{r},\omega)$, $\vec{H}(\vec{r},\omega)$ on a time) in the form:

$$\nabla\times\vec{E}(\vec{r},\omega)=\frac{i\omega}{c}\langle\hat{\mu}(z)\rangle\cdot\vec{H}(\vec{r},\omega),\quad \nabla\times\vec{H}(\vec{r},\omega)=-\frac{i\omega}{c}\langle\hat{\varepsilon}(z)\rangle\cdot\vec{E}(\vec{r},\omega) \tag{6}$$

Hence, according to the Floquet theorem, Fourier-amplitudes $\vec{f}_{K,p}^{(E,H)}$ of the electric and magnetic field strengths satisfy the following relation:

$$\left[\vec{\beta}+\left(K+p\frac{2\pi}{d}\right)\vec{e}_z\right]\times\begin{pmatrix}\vec{f}_{K,p}^{(H)}\\ \vec{f}_{K,p}^{(E)}\end{pmatrix}=\frac{\omega}{c}\begin{bmatrix}-\sum_l\hat{\varepsilon}_l\cdot\vec{f}_{K,p-l}^{(E)}\\ \sum_l\hat{\mu}_l\cdot\vec{f}_{K,p-l}^{(H)}\end{bmatrix} \tag{7}$$

Here $\vec{\beta}$ is an arbitrary planar (in the XOY plane) wave vector, $\vec{e}_z$ is a unit vector along the z-axis, $\vec{K}=(0,0,K)$ is the Bloch vector. The system (7) defines normal modes of electromagnetic waves, propagating in the considered "periodic" medium. Below, for simplicity, we shall restrict our study to the case of light, propagating along the z-axis ($\vec{\beta}=0$) in a nonmagnetic lattice, $\hat{\mu}=\hat{I}$ is a unit matrix; the liquid-crystal layers we shall treat as uniaxial, $\varepsilon_{ij}=\varepsilon_x\delta_{i\hat{x}}\delta_{j\hat{x}}+\varepsilon_y\delta_{i\hat{y}}\delta_{j\hat{y}}+\varepsilon_z\delta_{i\hat{z}}\delta_{j\hat{z}}$; obviously, that for $\vec{K}\parallel z$, zz-components of the tensor $\hat{\varepsilon}$ do not appear in final formulas, and $\varepsilon_x=\varepsilon_y\equiv\varepsilon$. Furthermore, we shall assume [16], that K is close to the value, defined by the Bragg's condition: $\left|K-\frac{2\pi}{d}\right|\approx K$, $c^2K^2\approx\omega^2\varepsilon_0$. This case corresponds to a resonance of plane waves between the components $\vec{f}_{K,p}^{(E,H)}$ at $p=0,\ -1$ (these terms dominate in the system (7)). After eliminating the $\vec{f}^{(H)}$ variables, Eqs. (7) with respect to $\vec{f}^{(E)}$ take the form:

$$\begin{bmatrix}K^2-\dfrac{\omega^2}{c^2}\varepsilon^{(0)} & -\dfrac{\omega^2\varepsilon^{(1)}}{c^2}\\ -\dfrac{\omega^2\varepsilon^{(-1)}}{c^2} & \left(K-\dfrac{2\pi}{d}\right)^2-\dfrac{\omega^2}{c^2}\varepsilon^{(0)}\end{bmatrix}\begin{pmatrix}\vec{f}_{K,0}^{(E)}\\ \vec{f}_{K,-1}^{(E)}\end{pmatrix}=0 \tag{8}$$

where $\varepsilon_{l=0}\equiv\varepsilon^{(0)}$, $\varepsilon_{l=\pm1}\equiv\varepsilon^{(\pm1)}$. Putting the determinant of the system (8) equal to zero we obtain the dispersion relations $\omega_{\pm}=\omega(K)$. Two roots of this equation $\omega_{\pm}$ define the boundaries of the spectral band: at frequencies $\omega_-(K)<\omega<\omega_+(K)$ (band gap) the roots are complex and electromagnetic waves decay (Bragg's reflection); frequencies $\omega<\omega_-$, $\omega>\omega_+$ correspond to propagating waves.

Results and Discussion

We shall confine ourselves to the case of propagation of electromagnetic radiation in a nonmagnetic superlattice with the two layers-elements (Si-layer and liquid crystal layer) per elementary cell. Concentration and dielectric permeability of the base material in the first and the second sublattice are denoted by $C_1^{(1)},\varepsilon_1^{(1)}$ and $C_2^{(1)},\ \varepsilon_2^{(1)}$ ($\varepsilon_1^{(1)}=11.7$, $\varepsilon_2^{(1)}=5.5$) respectively. For admixture this quantities are denoted by $C_1^{(2)},\varepsilon_1^{(2)}$ and $C_2^{(2)},\varepsilon_2^{(2)}$. Simple transformations (with the account that $\left|\varepsilon^{(-1)}\right|=\left|\varepsilon^{(1)}\right|$) lead to the following relations for the refractive index $n\equiv cK/\omega$ of the studied system:

$$n_{\pm}^2\left(C_1^{(2)},C_2^{(2)}\right)=\varepsilon^{(0)}\left(C_1^{(2)},C_2^{(2)}\right)\pm\left|\varepsilon^{(1)}\left(C_1^{(2)},C_2^{(2)}\right)\right|\cong\varepsilon^{(0)}\left[1\pm\frac{\Delta\omega_1\left(C_1^{(2)},C_2^{(2)}\right)}{\omega}\right] \tag{9}$$

$(n_+^2-n_-^2)/2\varepsilon^{(0)}\cong\Delta\omega_1/\omega$, $\Delta\omega_1=|\omega_+-\omega_-|$ - is the lowest band gap width. It follows from Eq. (9) that the quantity $\Delta\omega_1$ is determined by the corresponding coefficient of the Fourier expansion (5), which in this case is $\left|\varepsilon^{(1)}\right|$. The band gaps of higher orders are as well determined by corresponding Fourier-coefficients of the dielectric permeability.

$$\begin{gathered}\varepsilon^{(0)}=\left(\varepsilon_1^{(1)}f_1a_1+\varepsilon_2^{(1)}f_2a_2\right)/d\\ |\varepsilon^{(1)}|=\frac{1}{\pi}\left|\varepsilon_2^{(1)}f_2-\varepsilon_1^{(1)}f_1\right|\sin\pi a_1/d\end{gathered} \tag{10}$$

The functions $f_1\left(C_1^{(2)},\frac{\varepsilon_1^{(2)}}{\varepsilon_1^{(1)}}\right)=1-C_1^{(2)}\left(1-\frac{\varepsilon_1^{(2)}}{\varepsilon_1^{(1)}}\right)$, $f_2\left(C_2^{(2)},\frac{\varepsilon_2^{(2)}}{\varepsilon_2^{(1)}}\right)=1-C_2^{(2)}\left(1-\frac{\varepsilon_2^{(2)}}{\varepsilon_2^{(1)}}\right)$ depend on the concentration of admixture layers and their relative dielectric permeability.

Figure 1 shows the concentration dependence of the refractive index $n_{\pm}\equiv cK/\omega_{\pm}$ of the studied composite superlattice. It is readily seen, that the form of the corresponding surfaces has a non-monotone character if the dielectric permeability of both admixtures is $\varepsilon_i^{(2)}/\varepsilon_i^{(1)}\ll1\,(i=1,2)$ (case a) or $\varepsilon_i^{(2)}/\varepsilon_i^{(1)}\gg1\,(i=1,2)$. The dependence of n_+ and n_- on $C_1^{(2)}$ and $C_2^{(2)}$ becomes monotonous in an other cases (b). The latter fact determines the behavior of the lowest band gap.

In Figure 2 the lowest energy gap width is plotted vs. the concentrations $C_1^{(2)}$, $C_2^{(2)}$ of admixture layers for a superlattice with alternating silicon and liquid-crystal layers. The energy gap $\Delta\omega_1$

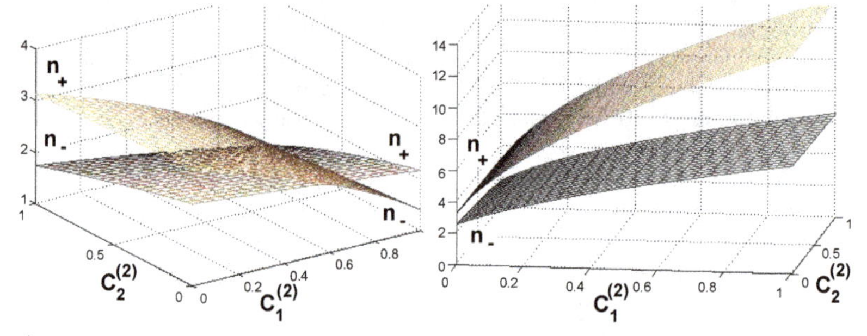

Figure 1: Refractive index $n_{\pm} = cK/\omega_{\pm}$ of the composite superlattice (with alternating silicon and liquid-crystal layers) *vs* the concentrations of admixture layers: a) $\varepsilon_1^{(2)}/\varepsilon_1^{(1)} = 0.1$, $\varepsilon_2^{(2)}/\varepsilon_2^{(1)} = 0.2$; b) $\varepsilon_1^{(2)}/\varepsilon_1^{(1)} = 20$, $\varepsilon_2^{(2)}/\varepsilon_2^{(1)} = 0.2$; $a_1/a_2 = 1$.

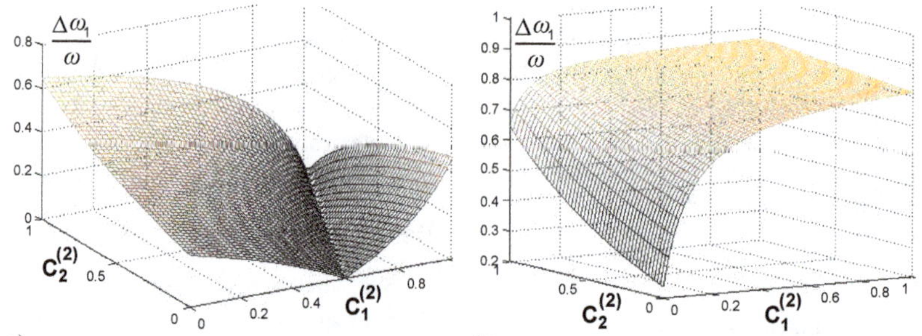

Figure 2: Relative width of the lowest band gap $\Delta\omega_1/\omega$ of the composite superlattice (with alternating silicon and liquid-crystal layers) *vs* the concentrations of admixture layers. Surface a) for the case $\varepsilon_1^{(2)}/\varepsilon_1^{(1)} = 0.1$, $\varepsilon_2^{(2)}/\varepsilon_2^{(1)} = 0.2$; surface b) for the case $\varepsilon_1^{(2)}/\varepsilon_1^{(1)} = 20$, $\varepsilon_2^{(2)}/\varepsilon_2^{(1)} = 0.2$; $a_1/a_2 = 1$.

vanishes at $\varepsilon_1^{(1)} f_1 = \varepsilon_2^{(1)} f_2$ for the case a) in Figure 2.

Conclusion

Our present study shows that optical characteristics of an imperfect superlattice may be significantly altered as a result of transformation of its polariton spectrum due to presence of admixture layers. The developed theory is a basis for phenomenological description of a wide class of optical processes in nonideal multilayered systems. Formulas (2-5, 7) allow a numerical calculation of the concentration dependence of relevant optical characteristics. The essential quantities governing the propagation of electromagnetic waves through the studied media are the refractive indices, the photon gap width and the directly measured quantities, which they define (for example, the light transmission coefficient). Graphic representation of $n_{\pm}, \Delta\omega/\omega(C_1^{(2)}, C_2^{(2)})$ (Figure 1 and 2) shows, that for the considered binary systems the character of the concentration dependence is different in different concentration intervals. The study carried out by authors [24] shows that optical characteristics of the nonideal photonic crystals may vary due to transformation of its photon modes spectrum caused by the presence of admixture elements. The case of nonideal multilayered systems with bigger number sublattices and components of alien layers allows even for a greater variety in behavior of the refractive index and the gap width. This circumstance considerably widens the opportunities for modeling such composite materials with prescribed properties.

References

1. Zhang C, Hirt DE (2007) Layer-by-layer self-assembly of polyelectrolyte multilayers on cross-section surfaces of multilayer polymer films: A step toward nano-patterning flexible substrates. Polymer 48: 6748-6754.
2. Liu H, Zhu SN, Dong ZG, Zhu YY, Chen YF, et al. (2005) Coupling of electromagnetic waves and superlattice vibrations in a piezomagnetic superlattice: Creation of a polariton through the piezomagnetic effect. Phys Rev B 71: 125106.
3. Baraban LA, Lozovski VZ (2004) Reflection and absorption of light by a thin semiconductor film. Opt Spectrosc 97: 810-816.
4. Rumyantsev VV, Shunyakov VT (1992) Exciton-polariton dispersion in ultrathin atomic cryocrystals. Physica B 176: 156-158.
5. Figotin A, Vitebsky I (2001) Nonreciprocal magnetic photonic crystals. Phys Rev E 63: 066609.
6. Lyubchanskii IL, Dadoenkova NN, Lyubchanskii MI, Shapovalov EA, Lakhtakia A, et al. (2004) One-dimensional bigyrotropic magnetic photonic crystals. Appl Phys Lett 85: 5932-5934.
7. Belotelov VI, Kotov VA, Zvezdin AK (2005) Abstracts of the International Conference on Functional Materials. Partenit, Cremea, NASU: Kiev, Ukraine.
8. Ha YK, Yang YC, Kim JE, Park HY, Kee C-S, et al. (2001) Tunable omnidirectional reflection bands and defect modes of a one-dimensional photonic band gap structure with liquid crystals. Appl Phys Lett 79: 15-17.
9. Yi Y, Bermel P, Wada K, Duan X, Joannopoulos JD, et al. (2002) Tunable multichannel optical filter based on silicon photonic band gap materials actuation. Appl Phys Lett 81: 4112-4114.
10. Tolmachev VA, Perova TS, Berwick K (2003) Design Criteria and Optical Characteristics of One-Dimensional Photonic Crystals Based on Periodically Grooved Silicon. Appl Optics 42: 5679-5683.
11. Tolmachev VA (2005) Tuning of the photonic band gaps and the reflection spectra of a one-dimensional photonic crystal based on silicon and a liquid crystal. Opt Spectrosc 99: 765-769.

12. Tolmachev VA, Perova TS, Astrova EV (2008) Thermo-tunable defect mode in one dimensional photonic structure based on grooved silicon and liquid crystal. Phys Status Solidi (RRL) 2: 114-116.

13. Bass FG, Bulgakov AA, Tetervov AP (1989) High-Frequency Properties of Semiconductors with Superlattices; Moscow, Russia.

14. Pokatilov, EP, Fomin, VM, Beril SI (1990) Vibrational Excitations, Polarons, and Excitons in Multilayer Systems and Superlattice. Shtiintsa, Chisinau, Moldova.

15. Tyu NS (1994) Local field effects and tensors of dielectric permeability in organic superlattices. Solid State Commun 90: 667-675.

16. Yariv A, Yeh P (1984) Optical Waves in Crystals: Propagation and Control of Laser Radiation. Wiley, New York, USA.

17. Awasthi SK, Ojha SP (2008) Wide-Angle Broadband Plate Polarizer with 1D Photonic Crystal. PIER 88: 321-335.

18. Banerjee A (2009) Enhanced Refractometric Optical Sensing By Using One-Dimensional Ternary Photonic Crystals. PIER 89: 11-22.

19. Parmenter RH (1955) Energy Levels of a Disordered Alloy. Phys Rev 97: 587-698.

20. Dargan TG, Capaz RB (1997) Koiler Belita Braz J Phys 27/A: 299-304.

21. Ziman JM (1979) Models of disorder. The theoretical physics of homogeneously disordered systems, John Willey & Sons, Inc, New York, USA.

22. Mezrag F, Aouina NY, Bouarissa NJ (2006) Optoelectronic and dielectric properties of $GaAs_xSb_{1-x}$ ternary alloys. J Mater Sci 41: 5323-5328.

23. Los' VF (1987) Projection operator method in the theory of disordered systems. I. Spectra of quasiparticles Theor Math Phys 73: 85-102.

24. Rumyantsev VV, Fedorov SA, Gumennyk KV (2011) Peculiarities of Band Gap Width Dependence Upon Concentration of Admixtures Randomly Included in 1D Photonic Crystal. Photonic Crystals: Optical Properties, Fabrication and Applications. Nova Science Publishers, Inc, New York, USA.

Design of a Mobile Phone Controlled Door: A Microcontroller based Approach

Bamisaye AJ* and Adeoye OS

Department of Electrical and Electronic Engineering, The Federal Polytechnic, Ado-Ekiti, Nigeria

Abstract

This paper presents a microcontroller based approach of a mobile phone and keypad controlled door. The door can be remotely controlled either by receiving set of instructions through the mobile phone or the keypad acting as the transmitter. The design consists of four main functional modules, which include: the mobile communication, controlling, decoding and the switching module. The decoding module and controlling module are made using integrated circuit chips ensuring proper conversion of signal to binary codes, which enables the microcontroller to effectively communicate with the switching device handling opening and closing of the door. The mobile communication module act as the transceiver unit which employs the use of a mobile phone serving as the communication device between the user at one end and the door at the other end. The decoding module and the controlling module are made possible using modern integrated circuit chips ensuring proper conversion of signal to binary codes, enabling the microcontroller to communicate properly with the switching device responsible for opening and closing the door. Only the right code can open the door, in case of sending wrong codes over three consecutive times, the password will have to be reset, because the system has sensed an intruder attempts.

Keywords: Microcontroller; Codes; Binary; Decoder; Keypad; Mobile communication

Introduction

Mobile communication system is an essential entity which provides the ability to disseminate information to a far distance, though depending on the coverage area and capacity of the network [1]. Based on these features that mobile communication provides, it has been of great advantage in business, security, banks, companies, institutions etc., [2]. Nowadays, Security has been a prime concern in the home and office management. Digital door lock security system provides security and safety to house or office owners, belongings, assets from being damaged by external agent or unwanted strangers. The mobile phone controlled door lock security system is an access control system that allows only authorized persons to access restricted area.

Ushie JO constructed a prototype security door that can be remotely controlled by a GSM phone set acting as the transmitter and another GSM phone set with dual tone multi-frequency (DTMF) connected to the door motor through a DTMF decoder interfaced with microcontroller unit and a stepper motor [3]. There was no provision for keypad assess in case the GSM signal fluctuates or fail which is a limitation.

Prince NN developed a security door system that can either receive command through the mobile phone or through the computer system (configured to output data through the parallel port) [4]. The use of keypad interfaced with the system is cheaper, easily assessable and affordable especially in developing areas and also easier to maintain than computer system interfaced.

Mohammad A developed a Microcontroller Based Reprogrammable Digital Door Lock Security System by Using Keypad and GSM [5]. He explained the idea of using a microcontroller and GSM to open a door, he used the GSM to send message to the device thereby opening the door. He explained further that Tones generated from DTMF keypad can identify what unit to be controlled as well as unique function to be performed. The system can be improved on by making it user friendly and easy control and accessible.

The proposed approach for designing this system is to implement a microcontroller-based control module that receives its instructions and command from a Mobile phone over the mobile network. The microcontroller then will carry out the issued commands.

The main purpose of this system is to lock and unlock a door by a mobile phone and matrix keypad, using a unique code entered through the matrix keypad and mobile phone. Opening and closing of doors involves human to be physically involved in the task. The authorized person can message the mobile phone stacked to the system which in turn is connected to the door motor that can open/close the door by entering the correct password. This method is very convenient as one doesn't have to get down of his door post to open the door physically. The cell phone is set to understand the message been sent to it. So, after the message have been send to the cell phone it sense it weather the code is from the desired number and whether the code is correct. The system will control the door through Short Message Service (SMS) effectively, receiving and transmitting data via SMS, this will eliminate the need of being physically present in any location for tasks involving the opening of door within a household and is power reliable because the power source is backed up with battery in case of mains failure. For example, if I have a visitor waiting for me at home and I am still in the office, I can open the door of my house from my office for them by sending the authentic code and the door will be open automatically, I can do the same if I want to lock it.

***Corresponding author:** Bamisaye, Department of Electrical and Electronic Engineering, The Federal Polytechnic, Ado-Ekiti, Nigeria
E-mail: ayobamisaye@gmail.com

This work is based on the concept of DTMF technology. All numeric buttons on the keypad of a mobile phone generates a unique frequency when pressed. These frequencies are decoded by the DTMF Decoder IC at the receiving end which is fed to the microcontroller. If this decoded values i.e., code pressed by user matches with the password stored in the microcontroller, then the microcontroller initiates a mechanism to open the door through a motor driver interface.

The Paper is divided into different sections: section 1.2 described the system Architecture, while section 1.3 gives the system flow chat, sections 1.4 described the system hardware and software, Test and Results were carried out in section 1.5 and the conclusion drawn was presented in section 1.6

System architecture

Figure 1 illustrated the block diagram of the system architecture. PIC18F4550 is an 8bit microcontroller of PIC18 family. PIC18F family is based on 16 bit instruction set architecture. PIC18F4550 consists of 32kb flash memory, 2kb SRAM and 256 Bytes EEPROM. This is a 4 in microcontroller consisting of 5 input/output ports (PORT A, PORT B, PORT C, PORT D and PORT E). POR B and PORT D have 8 pin to receive/transmit 8 bit input/output data. The remaining ports have different number of pin for input/output data communications. PIC18F4550 can work on different internal and external clock source. It can work on a varied range of frequency from 31 kHz to 48 kHz. PIC18F4550 has four in-built timers. There are various inbuilt peripherals like ADC (Analogue to Digital converter), comparators etc. in this controller. PIC18F4550 is an advance microcontroller which is equipped with enhanced communication protocol like EUSART, SPI and USB etc. [6,7].

Line buffer is a data structure that holds a fixed amount of data in serial fashion; the oldest data get discarded as new data is added. In this system, the line buffer helps the mobile phone to discard a command when a new command is given to the phone.

Buffers are used for many purposes, which include: Interconnecting two digital circuits operating at different rates, holding data for later use, allowing timing corrections to be made on a datastream, collecting binary data bits into groups that can then be operated upon as a unit and delaying the transit time of a signal in order to allow other operations to occur.

The relay used is a latching relay. A latching relay is a two-position electrically-actuated switch. It is controlled by two momentary-acting switches or sensors, one that 'sets' the relay and the other 'resets' the relay. The latching relay maintains its position after the actuating switch has been released, so it performs a basic memory function. These relay driver communicate with the IC on how to operate in this circuit.

The D.C 12 V relay operates as a switch in the circuit which open or close, according to the need of the needs and its operation. The relay control the solenoid switch which serve as a controlling device for the door latch, it help the door latch to open and close.

12-button key-pad is control by the IC, the 12 button key-pad is 0-9, it also includes cancel and enter button. The LED used in this circuit has two colours which have green and red. The red light shows the system is active and the correct code as not been enter while the green light shows the system is active and the correct code has been entered.

LCD (Liquid Cristal Display) displays the code that is entered through the 12 button key-pad and also displays when the code is correct or wrong. As the button is pressed, the beeper will beep when any wrong code is entered.

System flowchart

Figure 2 showed the system flowchart which starts with the initializing the entire variable from 0-9 including other button on the keypad, when all the variables have being initialized, the system will refresh. If enter button is pressed and there is no password or code that is been pressed, then the cancel button will be pressed thereby clearing the display and clear keypad buffer; if cancel button is not pressed then digit 0-9 will be pressed thereby storing the digit in keypad buffer.

If no button is pressed before the latch is closed, the correct code or password is entered and the latch will open then if the wrong code or password is entered the LCD will display a wrong password after this, the system will refresh.

The second part is using the phone to control the system. Is there any message in the SMS Buffer, if NO, the system will refresh. If YES and SMS having a right password then the CMD will open thereby opening the latch and clearing SMS Buffer and refresh. Is there any message in SMS Buffer?, if NO the system will refresh and if YES and having the right password then CMD will close thereby closing the latch and clearing SMS Buffer and Refresh.

Design of system hardware and software

The system circuit design is shown in Figure 3 the frequency of the microcontroller is regulated by 20MHz crystal capacitor. The 4550 control the LCD which displays the instruction been sent to the IC. The indicator is a double colour LED, Red and Green, when Red is turn on the power is ON and the system is active; when Green light is turn ON the correct password has been entered. The 74HC14 is a Schmitt trigger. In this circuit, it helps in reducing the power that is being supply to the phone to the require voltage through the FBUS-TX.

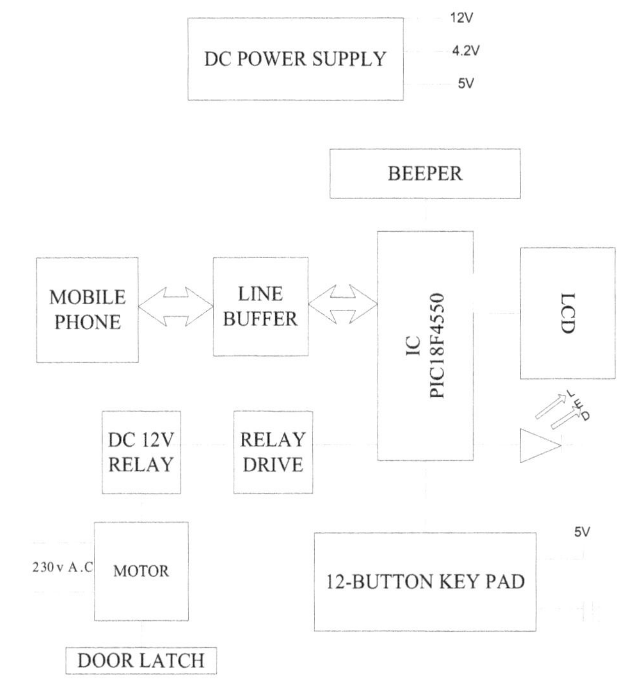

Figure 1. Block diagram of the system Architecture.

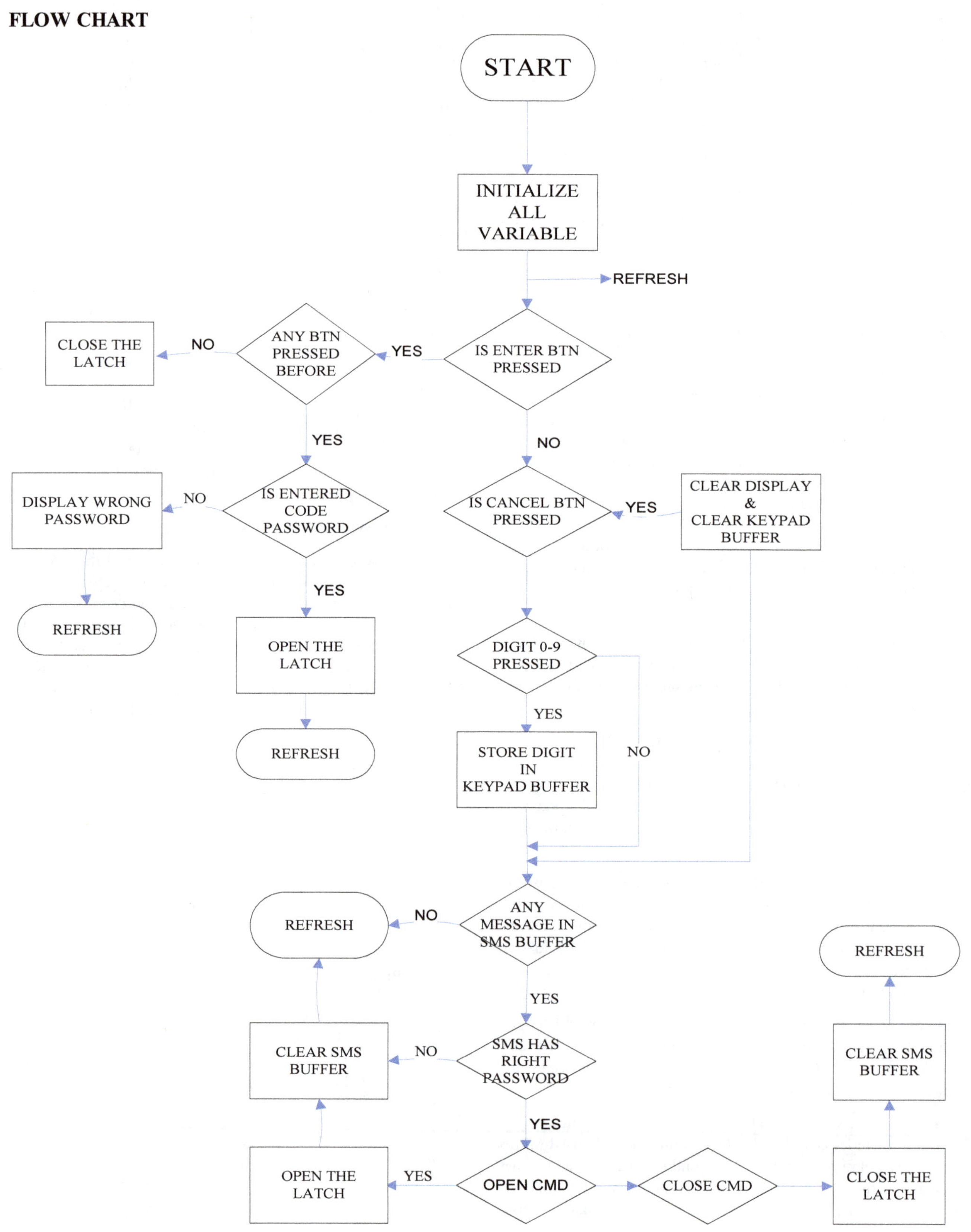

Figure 2. Flow Chart of the System.

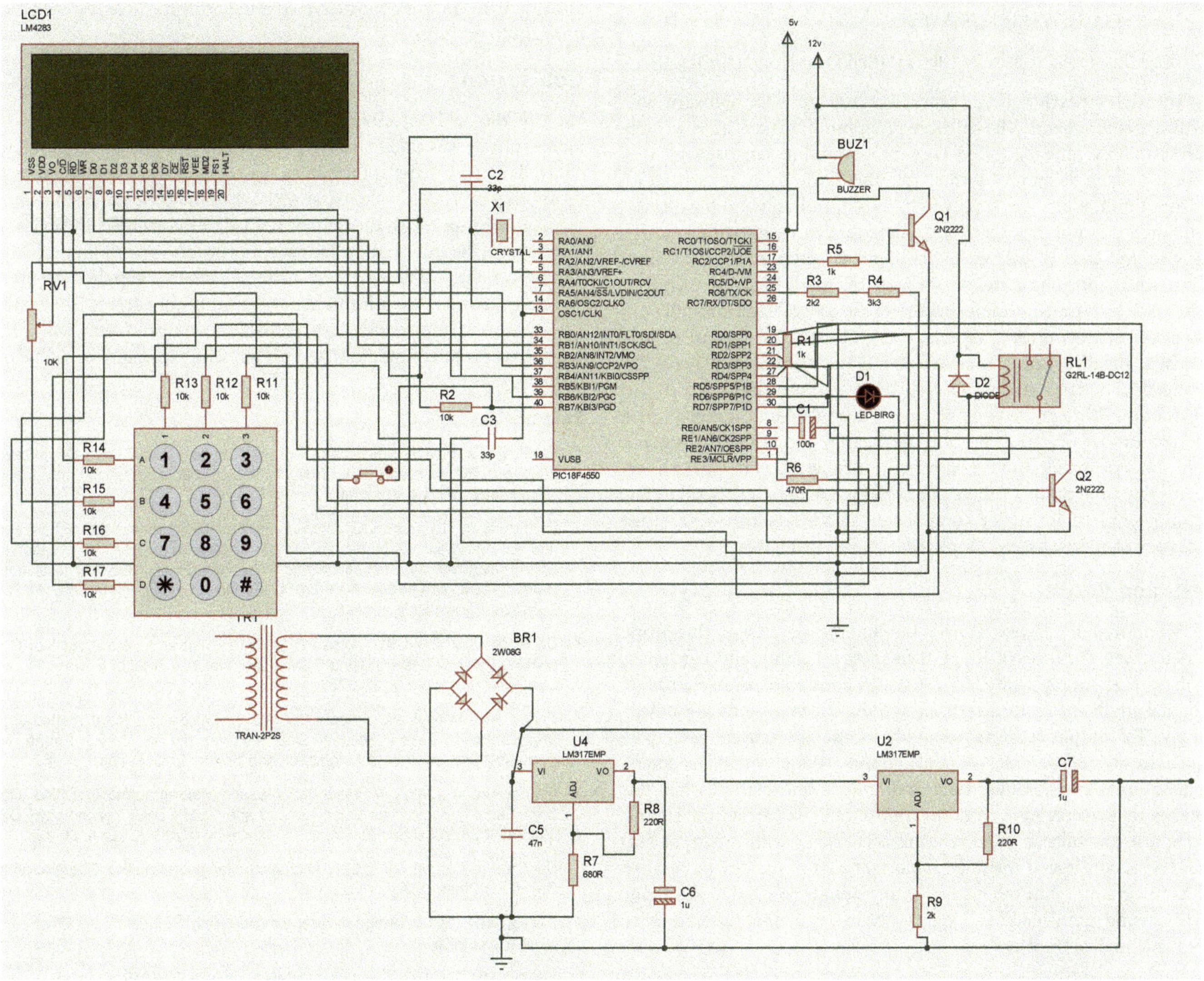

Figure 3. Circuit Diagram of the System.

The beeper is driven by 2N2222 transistor and is connected to pin 17 of the IC. The relay is driven by 2N2222 transistor.

In designing the +5 V and +12 V DC power supply required for the microcontroller and the associated relay control circuits, LM317 adjustable regulators were used. The regulator output voltage is determined by two resistors R1 and R2 which are connected to form a potential divider. This potential dividing network determines the output voltage of the regulator. In operation, the LM317 develops a nominal 1.25 V reference voltage, V_{REF}, between the output and adjustment terminal. The reference voltage is developed across resistor R_1 and since the voltage is constant, a constant current I_1 then flow through the output set resistor R_2, giving an output voltage of

$$V_{out} = V_{REF}\left(1+\frac{R_2}{R_1}\right)+I_{ADJ}R_2 \qquad (1)$$

I_{ADJ} is the current from the adjustment terminal. It represents an error term; the LM317 has been designed to minimize I_{ADJ} and make it very constant with line and load changes; and it has a constant value of about 100 μA. The current set resistor, R_1 connected between the adjustment terminal and the output terminal is usually 220 Ω.

Using the formula in equation in 1 above:

$V_{out} = 1.25(1 + R_2/220) + (100/10^6)\, R_2$

Given that $V_{out} = 5.0$ V

$5.0 = 1.25(1 + R_2/220) + 0.0001R_2$

$5.0 = 1.25 + R_2/176 + 0.0001R_2$

$5.0 = 1.25 + 0.0057818R_2$

$0.0057818R_2 = 5.0 - 1.25 = 3.75$

$R_2 = 3.75/0.0057818 = 648.6$ Ω.2

Since the closest resistor is 680 Ω. 680 Ω is then selected for R_2.

Given that $V_{out} = 12.0$ V

$12.0 = 1.25(1 + R_2/220) + 0.0001R_2$

$12.0 = 1.25 + R_2/176 + 0.0001R_2$

$12.0 = 1.25 + 0.0057818R_2$

$0.0057818R_2 = 12.0 - 1.25 = 10.75$

$R_2=10.75/0.0057818=1859.3\ \Omega$. 3

Since the closest resistor is 1800 Ω, 1800 Ω is selected for R2.

The development of this work is made up with software and hardware components. The software program of this design is written with assembly language. The core programming language of this work is written using C18 format [8]. The hardware is comprised of the input unit, the power supply unit, the control unit and the display unit. The input unit have card slot where cards are inserted for access purpose and it is connected to the control unit through port zero of the microcontroller. The sensor unit (Keypad) is made of micro switches, which transmits information of the control unit through port two of the microcontroller especially when each of them is pressed or punched. The display unit are made with Liquid Crystal Display (LCD) arranged in a serial manner to each other and are connected to the control unit through the line port one. The control unit is made up of microcontroller which can be called the heart of the design as it accepts from the sensor unit and the input unit through the line port one to the display unit. Some of these connections are done using soldering techniques to solder major components to Vero-board while other necessary connections are completed using jumper wires.

Tests and Result

Monitored tests were carried-out on the devices of this work to determine the level of its performance, reliability and efficiency. In the power supply Unit (PSU), test was conducted with the use of a digital multi-meter to determine the output voltage of the power supply. The output terminal result shows the appropriate value and figure. In the sensor unit, we made confirmations so that the micro switches used could deliver right and notable information when the buttons are pressed. Also, from the control unit, we carried-out testing to confirm the output is delivered in accordance with the input fed. This test is considered the most important part of the design because of its integration with other units in the design. We tested the input section using multi-meter and logic probe to ensure that mistakes were not made in data transfer.

After all these tests the results validates the functionality of the system, we were convinced that the work was indeed successful in software development and hardware assembly, and therefore, take the position that the work has delivered and perform efficiently the task it is meant to perform.

Conclusion

The aim and objective of the work has been achieved since the mobile phone was able to transmit information to the control unit, the controller accepts the information transferred and process them by prompting for access code (password) when the code has been entered through the input with the help of keypad, the controller is capable of interpreting the information supplied and process it without problems. The processing will allow access if the information supplied is correct and deny access if the information supplied is not correct. The system operation is user friendly. It is palpable that this system will guide against access to an unauthorized person who will improve safety and security.

References

1. Bamisaye, James A, Kolawole MO (2010) Capacity and Quality Optimization in CDMA 3G Networks. Journal of Telecommunications and Information Technology (JTIT) 4: 101-104.
2. Bamisaye AJ, Ekejiuba CO, Ojo AJ (2011) Telecommunication Engineering and Entrepreneur Opportunities. Paper presented at the 7th Engineering Forum: Engineering Innovations and Entrepreneurship: Gateway to National Development. November 8-11, 2011. School of Engineering, The Federal Polytechnic, Ado-Ekiti, Nigeria.
3. Ushie JO, Donatus EBO, Akaiso E (2013) Design and Construction of Door Locking Security System Using GSM. International Journal of Engineering and Computer Science 2: 2235-2257.
4. Prince NN, Ifeanyi NI, Joseph EC (2013) Design and Implementation of Microcontroller Based Security Door System (Using Mobile Phone and Computer Set). Journal of Automation and Control Engineering 1: 65-69.
5. Mohammad A (2013) Microcontroller Based Reprogrammable Digital Door Lock Security System by Using Keypad and GSM/CDMA Technology. IOSR Journal of Electrical and Electronics Engineering (IOSR-JEEE) 4: 38-42.
6. Sanchez J, Canton MP (2007) Microcontroller programming: The micro-chip PIC, CRC Press USA.
7. Rakesh MR (2012) Microcontroller programming, Robotics and Electronics. RRM Press India.
8. Rakesh MR (2013) C18 compiler installation guide. RRM Press India.

Synergetic MPPT Controller for Photovoltaic System

Nadia Mars[1,2]*, Faten Grouz[1], Najib Essounbouli[2] and Lassaad Sbita[1]

[1]Research Unit of Photovoltaic, Wind and Geothermal Systems (SPEG), the National Engineering School of Gabes (ENIG), University of Gabes, Av. Omar Ibn El Khattab, Zrig Eddakhlania (6072), Tunisia
[2]Center for Research in Information and Communication Sciences and Technology (CReSTIC), IUT, 9 Av Quebec, 10000 Troyes, France

Abstract

A novel non-linear power point tracking method of photovoltaic system (PV) based on synergetic control strategy is proposed. This technique uses a synergetic control strategy to achieve the maximum power point output without a chattering phenomenon. The PV system consists of a PV panel, DC-DC boost converter, MPPT controller and an output load. Synergetic control is easy to implement, has controllable dynamics towards the origin and gives a good maximum power operation under environmental changes (solar radiation and PV cell temperature).To show the robustness and validity of this approach a mathematical model is presented and simulated using Matlab/SIMULINK under different atmospheric conditions. Indeed, the simulation results showed the advantages of this new algorithm, especially the reduction of the chattering problem.

Keywords: PV system; Synergetic control; Maximum power point tracking; Boost converter; PV panel

Introduction

Nowadays, solar energy has great importance because of its sustainability and environmental friendly characteristic. The photovoltaic module represents the fundamental power conversion unit of PV system, the fact that the output characteristic depends on the solar radiation and the cell temperature [1]. In order to get the maximum of power from solar panels and enhance the PV system's efficiency, the selection of a maximum power point tracker (MPPT) algorithm is necessary [2].

A significant number of MPPT control systems have been developed for years to extract the maximum power that the photovoltaic module can provide [3]. Each MPPT has its own advantages and disadvantages. Due to their simplicity, perturbation and observation (P&O) algorithm and incremental conductance (IC) method are the most widely used [4,5]. The P&O algorithm consists of perturbing the PV output voltage and observing the output power to determine the peak power direction. The IC methods compare between the instantaneous and the incremental conductance to track MPP [6]. However, these methods have some disadvantages: P&O control fails to track the MPP during the rapid solar irradiation changes and IC method around the maximum power point [7].

To solve these problems many solutions have been reported in literature. To get better performance to the PV system, an adaptive perturbation step size has been proposed [8]. This method is effective but it is complex in implementation as it needs the operation point location. In addition, the implementation of this control is switched between adaptive duty cycle and fixed duty cycle control.

There are also other techniques such as fractional short circuit current method [9] and fractional open circuit voltage method [10]. However, these two methods have a weaker and less accurate performance.

In addition, a several number of intelligent methods have been adopted to estimate the voltage and the load current values such as fuzzy logic, artificial neural networks and genetic algorithms [11-13]. Such methods are frequently complex and require considerable knowledge in control system design.

Recently, sliding mode control (SMC) is used in photovoltaic systems [14,15]. SMC is a non-linear control strategy which has several advantages such as robustness, good dynamical response and simplicity in its implementation. On the other hand, its major drawback is a chattering phenomenon. Hence, this phenomenon induces many undesirable oscillations in control signal [16]. Synergetic control (SC) as a solution is proposed in this paper to ensure stability of PV system with fast dynamic response. Synergetic control theory is introduced in ref. [17]. SC, like sliding mode control, is a non-linear control strategy. It allowed changing the system structure by switching from one set of continuous functions of state to another at any instant [18]. Synergetic control has the advantage of finite time convergence and tiny steady state error. In addition, it should achieve similar performance as SMC without chattering phenomenon [19]. A MPPT control strategy based on SC has been presented in ref. [20]. However this algorithm has the drawback such as parameter tuning difficulties and complexity. Therefore, the aim of this paper is to develop a novel SC scheme that is simple to implement and possesses an excellent steady state performance.

Inspired from the above work, this paper proposes a novel MPPT using synergetic approach for stand-alone photovoltaic system. The outline of the paper is as flows. The PV panel model is described in section II. Section III presents the mathematical model for a boost DC-DC converter. In the following section, synergetic approach procedure is exposed and the proposed synergetic MPPT controller is given. Section V presents the simulation result and discussion. In the last section, conclusions are given.

Photovoltaic Generator

How a PV cell working

Photovoltaic cell is basically a p-n junction which converts directly

***Corresponding author:** Nadia Mars, SPEG, ENIG, Tunisia, Center for Research in Information and Communication Sciences and Technology (CReSTIC), IUT, 9 Av Quebec, 10000 Troyes, France, E-mail: nedyamars@hotmail.fr

sunlight to electricity. When the cell exposed to the sunlight, the cell photons are absorbed by the semiconductor atoms, freeing electrons from the negative layer. These electrons find their path through an external circuit toward the positive layer resulting an electric current [21,22]. Monocrystalline, polycrystalline and thin film technologies are the major families of PV cells. The absorption depends mainly on the cell surface, semiconductor or band gap, the temperature and the solar radiation [23].

PV panel model

To model the PV Panel, the scientific community offer several models. The single diode model is the most classical one described in literature [24]. The equivalent circuit (Figure 1) consists of current source to model the incident luminous flux, a diode for cell polarisation phenomena, a parallel resistance due to leakage current and a series resistance representing various contacts [6].

The general mathematical PV cell equation is given by the following equation [25]:

$$I = I_{ph} N_p - I_d - I_{sh} \quad (1)$$

Where, I_{ph} is a photo current, I_d is a diode current and I_{sh} a shunt current. The module photo-current evaluated as:

$$I_{ph} = G_k [I_{sc} + k_I (T_{op} - T_{ref})] \quad (2)$$

The shunt current is described by the following equation:

$$I_{sh} = \frac{V_{pv} + I_{pv} R_s}{R_{sh}} \quad (3)$$

The dark current I_d is calculated by the following equation:

$$I_d = I_s \left[\exp\left(q \frac{V_{pv} + R_S I_{pv}}{N_S V_t} \right) - 1 \right] \quad (4)$$

The reserved saturation current varies with temperature is described according to the following equation:

$$I_s = I_{rs} \left(\frac{T_{op}}{T_{ref}} \right)^3 e^{\left[\frac{qEg}{nk} \left(\frac{1}{T_{op}} - \frac{1}{T_{ref}} \right) \right]} \quad (5)$$

Finally the output current can be expressed as:

$$I = N_p I_{ph} - N_p I_s \left[\exp\left(\frac{q V_{pv} + R_S I_{pv}}{N_S V_t} \right) - 1 \right] - N_p \frac{V_{pv} + I_{pv} R_S}{R_{Sh}} \quad (6)$$

An ideal PV cell has very high equivalent shunt resistance and very low equivalent series resistance (Simplification $R_{sh} >>> R_s$)

Equation (1) with R_s=0 and $R_{sh}=\infty$ becomes:

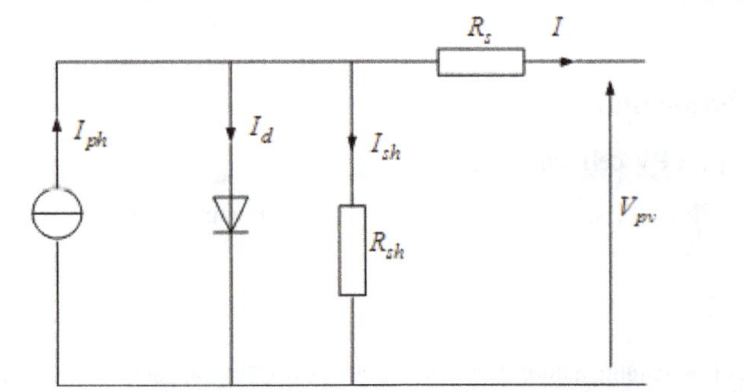

Figure 1: Equivalent circuit of solar cell.

$$I = N_p I_{ph} - N_p I_s \left[\exp\left(\frac{A V_{pv}}{N_S n} \right) - 1 \right] \quad (7)$$

PV characteristics

The PV panel specifications used for simulations are presented in Table 1.

Figure 2 shows the typical current versus voltage and power versus voltage curves for the photovoltaic module. Figures 3 and 4 show the I-V and P-V characteristics of PV module under different temperature levels (fixed irradiance). Figures 5 and 6 show the I-V and P-V characteristics of PV module under different solar radiation levels (fixed temperature) [20].

It is clear that the PV module inherits nonlinear characteristics at its MPP. This MPP varies with temperature and irradiance. Hence, PV panel power increases with irradiance increasing or temperature decreasing. Therefore, in order to ensure that the photovoltaic system work at its MPP, a control algorithm is needed.

Maximum power (Pmax)	60 W
Open-circuit voltage (Voc)	21.1 V
Short-circuit current (Isc)	3.8 A
Optimum operating voltage (Vmpp)	17.1 V
Optimum operating current (Impp)	3.5 A

Table 1: Specifications of PV Panel.

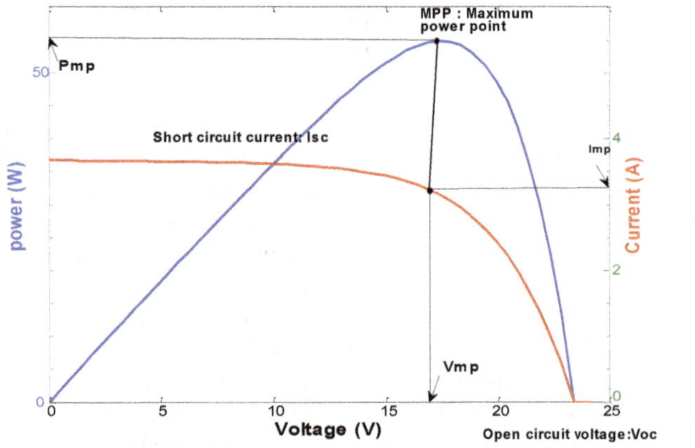

Figure 2: I-V and P-V characteristic curves at a fixed irradiation (G=1000W/m²) and a fixed temperature (T=25°C).

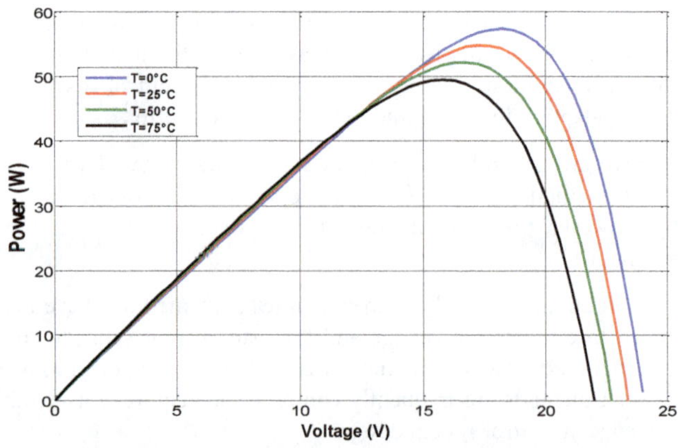

Figure 3: Power-Voltage curve in different temperature (G=1000W/m²).

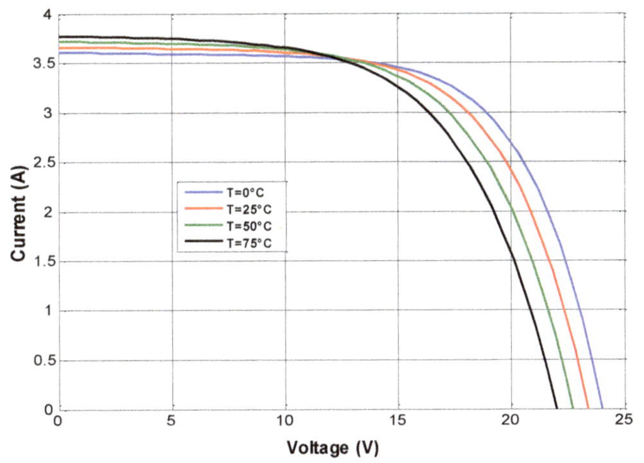

Figure 4: Current-Voltage curve in different temperature (G=1000W/m²).

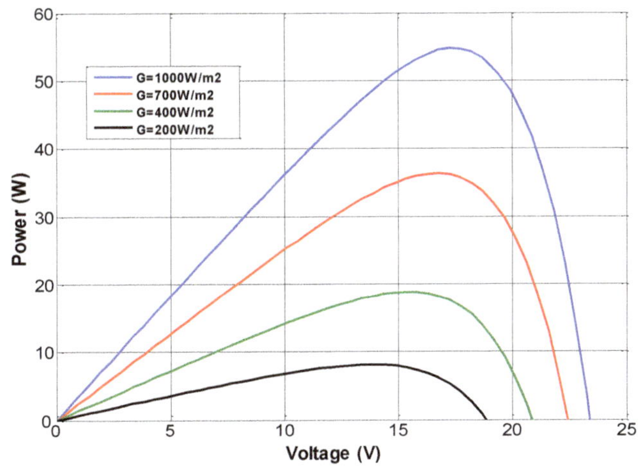

Figure 5: Power-Voltage curve in different irradiance (T=25°C).

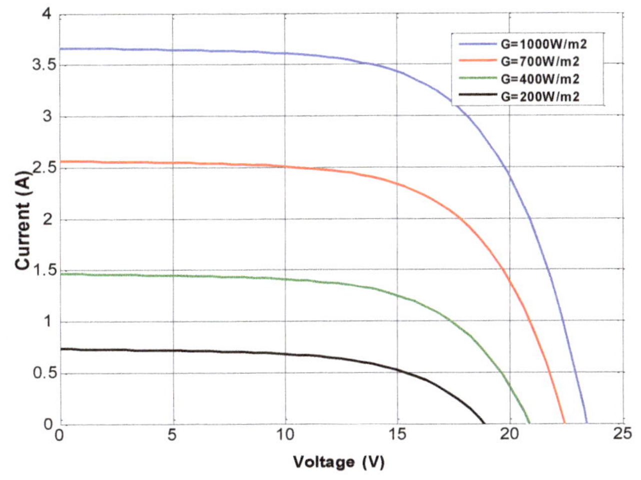

Figure 6: Current-Voltage curve in different irradiance.

DC-DC Converter

In order to extract the maximum power from the PV module, the DC-DC converter allows adaptation between the PV module and the load [3]. Figure 7 shows the circuit of the boost converter, whose output voltage (V_0) is more than or equal to the input voltage V_{pv} (PV generator voltage).

The switch S operates at a high frequency to produce a chopped output voltage [26]. The power flow is controlled by adjusting the ON and off of the switching. Hence, When S=1, the switch is ON. The equations can be written as:

$$\begin{cases} \frac{dV_{pv}}{dt} = \frac{1}{C1}(i_{pv} - i_L) \\ \frac{dI_L}{dt} = \frac{1}{L}V_{pv} \\ \frac{dV_0}{dt} = \frac{1}{C2}(i_L - i_0)\frac{1}{C2}i_L \end{cases} \tag{8}$$

When S=0, the switch is OFF the equation can be expressed in equation:

$$\begin{cases} \frac{dV_{pv}}{dt} = \frac{1}{C1}(i_{pv} - i_L) \\ \frac{dI_L}{dt} = \frac{1}{L}V_{pv} - \frac{1}{L}V_0 \\ \frac{dV_0}{dt} = \frac{1}{C2}(i_L - i_0) \end{cases} \tag{9}$$

The boost converter can be used to drive a high voltage load from a low voltage PV module. The dynamic model of the used boost converter can be derived as:

$$\begin{cases} \frac{dV_{pv}}{dt} = \frac{1}{C1}(i_{pv} - i_L) \\ \frac{dI_L}{dt} = \frac{1}{L}V_{pv} + \frac{1}{L}(S-1)V_0 \\ \frac{dV_0}{dt} = \frac{1}{C2}(i_L - i_0) - \frac{1}{C2}Si_L \end{cases} \tag{10}$$

If we set $x=[x_1\, x_2 x_3]^{Tr}=[V_{pv} i_L\, V_0]^{Tr}$

With Tr matrix transpose, the above expression can be written as:

$$\dot{x} = \frac{dx}{dt} = f(x,t) + g(x,t)S \tag{11}$$

Where,

$$x = \begin{bmatrix} V_{pv} \\ i_L \\ V_0 \end{bmatrix}; f(x) = \begin{bmatrix} \frac{1}{C1}(i_p - i_L) \\ \frac{1}{L}(V_{pv} - V_0) \\ \frac{1}{C2}(i_L - i_0) \end{bmatrix}; g(x) = \begin{bmatrix} 0 \\ \frac{1}{L}V_0 \\ -\frac{1}{C2}i_L \end{bmatrix}$$

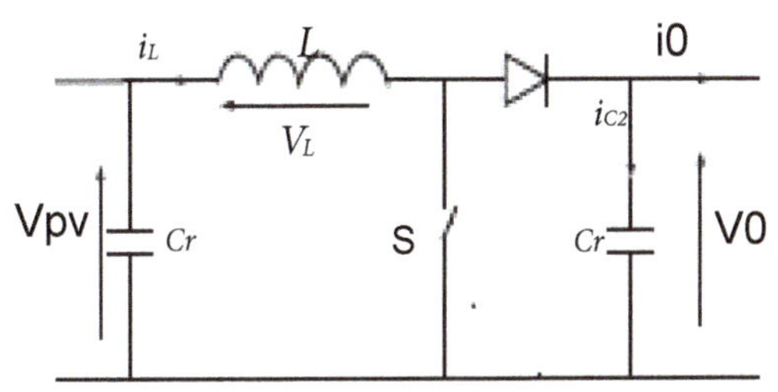

Figure 7: Diagram of boost converter.

Synergetic MPPT Controller

Synergetic control procedure

The general synergetic control procedure is reviewed in this section. Synergetic control theory is a nonlinear approach, it uses a non-linear model of the power system to overcome the above mentioned problems of the linear controls [27]. The nonlinear differential equation of the system can be described by the following form [28,29].

$$\dot{x} = \frac{dx}{dt} = f(x,s,t) \tag{12}$$

Where $x=(x_1,x_2,\ldots,x_n)$ is the state variable vector of size n, $s=(s_1,s_2,\ldots,s_m)$ is the control input of size m=1 and t is time.

The SC approach is based on a particular choice of the macro variable. Therefore, we started by defining a nonlinear macro-variable as follows:

$$\Psi=\Psi(x,t) \tag{13}$$

The controller objective is to force the system to operate the manifold (Ψ=0). However, using the same procedure as in the synergetic approach [30,31]. The macro-variable can be a simple linear combination of the state variables. Hence, the designer can select the characteristics of this macro-variable according to the control specifications such as the settling time and the control objective. The desired dynamic evolution of the macro-variable is:

$$T\dot{\Psi}+\Psi=0, T\succ 0 \tag{14}$$

Where T is a specific designer chosen that determines the rate of convergence speed to the manifold specified by the macro-variable. Taking into account the chain rule of differentiation that is given by:

$$\dot{\Psi}=\frac{d\Psi}{dx}\dot{x} \tag{15}$$

Combining 12, 14 and 15

$$T\frac{d\Psi}{dx}f(x,s,t)+\Psi=0 \tag{16}$$

Finally upon solving equation 16, the control law can be described as flow:

$$S=\delta[x,t,\Psi(x,t),T] \tag{17}$$

As can be seen, the control output S depends not only on the system state variables but also on the time constant T and the macro variable Ψ as well giving the designer latitude to choose the characteristics of the controller by selecting a suitable macro-variable and a constant time T. By suitable selection of macro-variable, the designer can obtain many interesting characteristics such as:

- Stability
- Noise suppression
- Parameters ffinsensffitffivffity.

The procedure summarized here can be performed by hand for simple systems, such as the DC-DC boost converter used for this study, which has a small number of state variables.

Synergetic MPPT controller

Maximum power point tracking is an essential stage of a photovoltaic system for tracking the maximum power. The system studied in this paper is a stand-alone PV system. As shown in Figure 8, it consists of a PV panel, a DC-DC converter, a load and a synergetic controller.

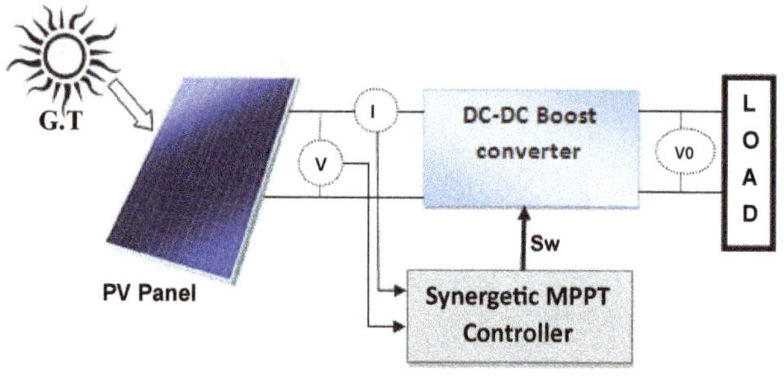

Figure 8: Block diagram of PV system with synergetic controller.

Like all other MPPT, the modelling of the synergetic MPPT controller is based on the output power of the cell which is P=V*I. The optimisation of the output power is achieved as shown in Figure 2, by selecting the manifold as:

$$\frac{\partial P}{\partial V}=0 \tag{18}$$

Hence, the manifold is defined as:

$$\Psi=\frac{\partial P}{\partial V}=V\frac{\partial I}{\partial V}+I \tag{19}$$

In studied DC-DC boost converter, there are two states: the output voltage (x1) and the inductor current (x2). As Ψ is function of x1 only, hence the chain rule of differentiation becomes:

$$\dot{\Psi}=\frac{d\Psi}{dx}\dot{V}=2\left[\frac{\partial I}{\partial V}+\frac{\partial^2 I}{\partial V^2}V\right]\dot{V} \tag{20}$$

Then the desired dynamic evolution of the macro-variable can be expressed:

$$T\dot{\Psi}+\Psi=0; T\geq 0 \tag{21}$$

So,

$$T\left[2\frac{\partial I}{\partial V}+\frac{\partial^2 I}{\partial V^2}V\right]\dot{V}+V\frac{\partial I}{\partial V}+I=0 \tag{22}$$

$$d=1-\frac{V}{V_0}=\frac{V\frac{\partial I}{\partial V}+I}{\frac{V_0}{L}\left[2\frac{\partial I}{\partial V}+\frac{\partial^2 I}{\partial V^2}V\right]} \tag{23}$$

Stability proof

Based on Lyapunov function the macro-variable requires that:

$$\dot{V}=\frac{1}{2}\Psi^2\succ 0;\dot{V}=\Psi=\frac{d\Psi}{dx}\prec 0;\Psi\dot{\Psi}\prec 0 \tag{24}$$

Results and Discussion

The PV model system has been implemented in Matlab/Simulink shown in Figure 9, which includes the PV array, the DC-DC boost converter with an MPPT controller connected to a load. The PV modules specifications' are shown in Table 1.The converter circuit topology is designed to be compatible with given load to achieve the maximum power transfer from the solar panel. The MPPT system specification' used in the simulation is shown in Table 2.

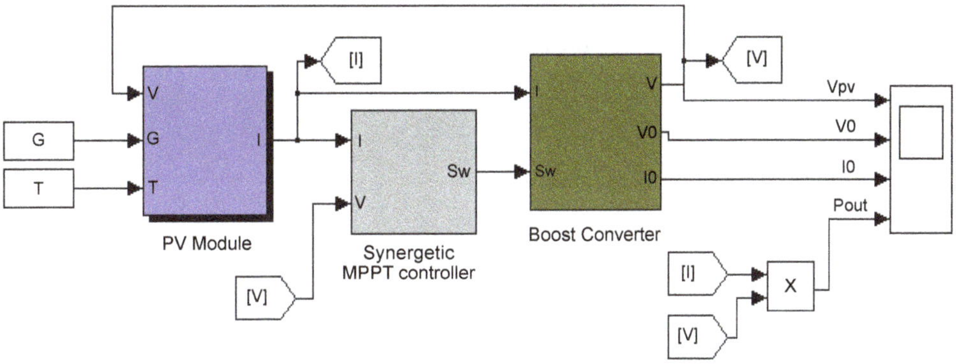

Figure 9: Simulation diagram of the stand-alone PV system.

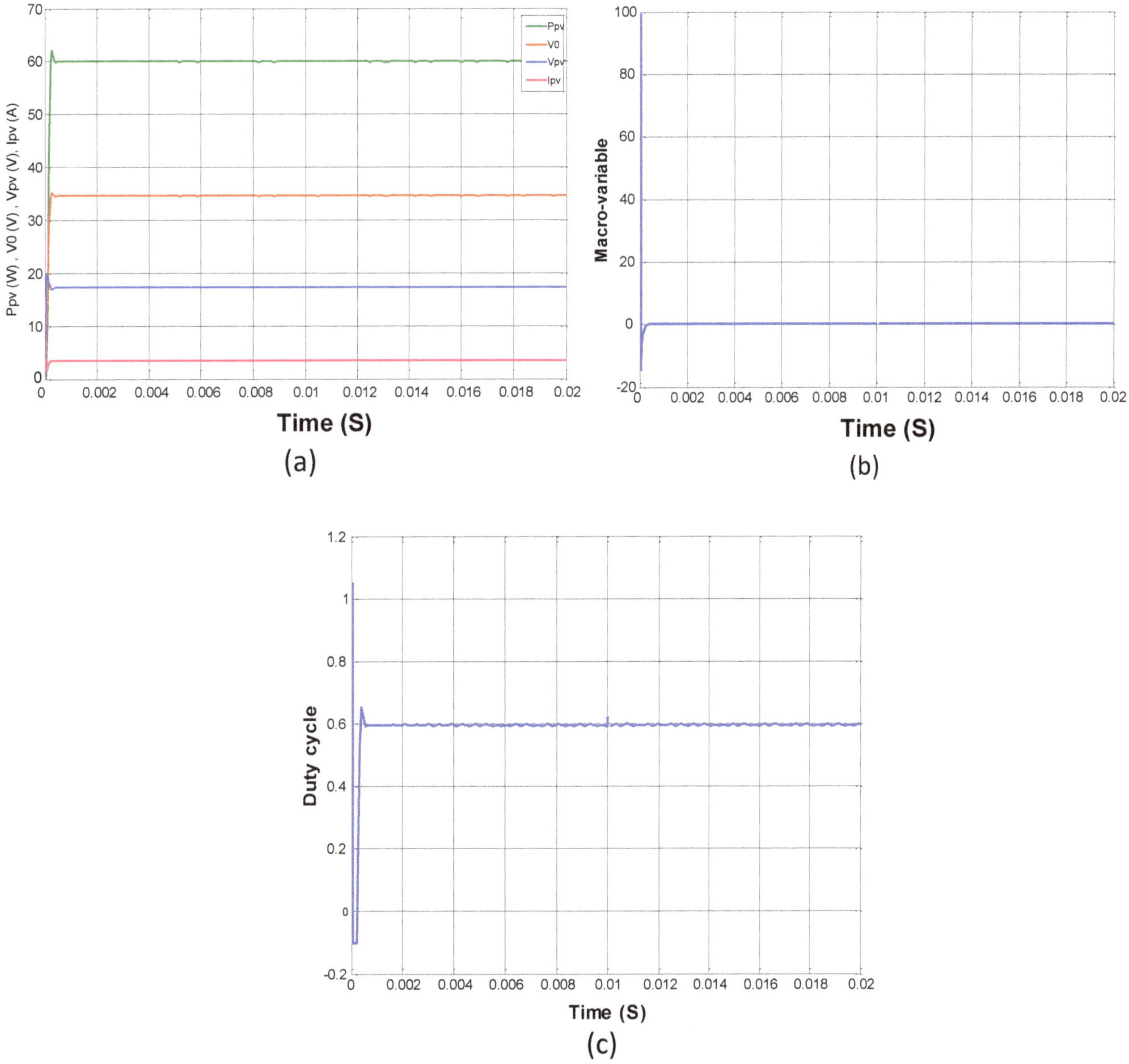

Figure 10: Simulation with standard condition (a) Ppv,V0, Vpv and Ipv, (b) Duty Cycle and (c) macro variable.

To evaluate the effectiveness of proposed MPPT, We consider the PV cell with irradiance is 1000W/m^2 and temperature is 25°C. The simulation results of the output power, the PV panel voltage, the output voltage and the PV panel current are shown in Figures10a-10c show the duty cycle and the macro-variable.

The photovoltaic system is dependent on the temperature and irradiation conditions and the power will be maximum for along the variations of temperature and irradiation in the PV panel which is giving to input to DC-DC converter. Therefore, we will evaluate the robustness of the proposed MPPT according to temperature and irradiance variation. In each figure, two different values of temperature or irradiance are presented in order to show the robustness (Figures 11-14).

L	480 (mH)
R	5 (Ω)
K	1.38e-23 (J/K)
Q	1.6e-19 (C)
Rs	0.18 (Ω)
Rp	360 (Ω)
N	36
T	0.003

Table 2: System specifications.

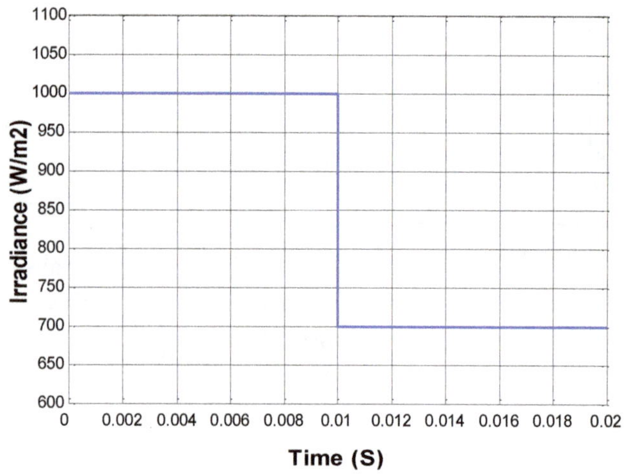

Figure 11: Solar irradiance variation.

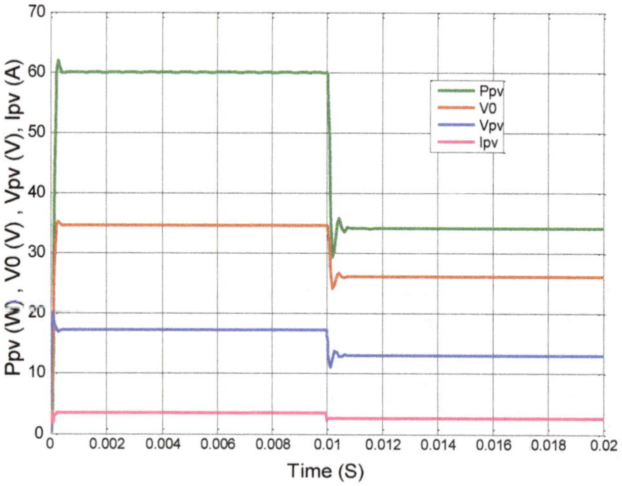

Figure 12: Simulation with step irradiance change (1000W/m^2, T=25°C).

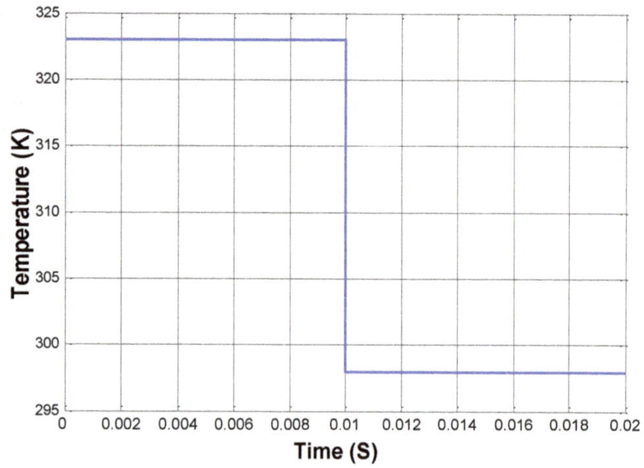

Figure 13: Temperature variation.

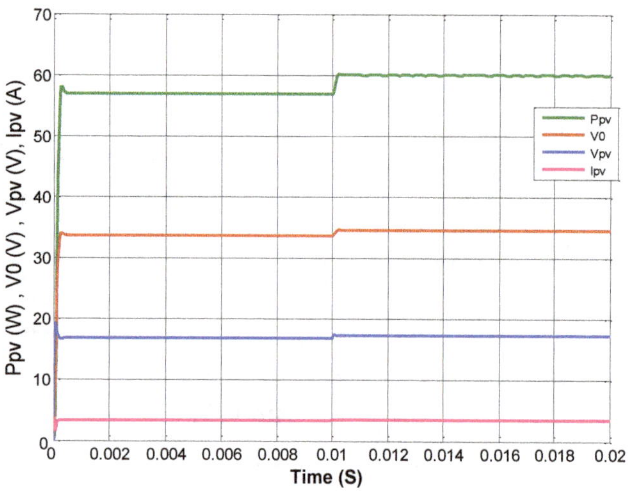

Figure 14: Simulation with step irradiance change (1000W/m^2, T=25°C).

All the obtained results show that the use of synergetic MPPT controller is effective. The SC ensures very fast convergence to the MPP and there are no oscillations around the MPP. Moreover it provides a high robustness with the variation of the external conditions (temperature and solar irradiance).

Conclusion

In this paper, a whole PV system with optimal control strategy has been presented. The Synergetic control is formulated and applied to the PV system. This system consists of a solar panel, DC-DC boost converter, synergetic MPPT controller and an output load. The effectiveness and the robustness of the proposed MPPT are proven by simulation results.

It is proved using Simulation Matlab/Simulink that the designed SC showed good results as it successfully and precisely tracked the MPP with a significantly higher efficiency. Hence, synergetic control eliminates chattering effects and robust to abrupt change of solar radiation and temperature.

In further work, these results will be experimentally validated.

References

1. Kachhiya K, Lokhande M, Patel M (2011) Matlab/simulink model of solar PV module and MPPT algorithm. National Conference on Recent Trends in Engineering and Technology.
2. Rao GJ, Mangal DK, Shrivastava SK, Baig MG (2016) Modeling and Simulation of Incremental Conductance Maximum Power Point Tracking (MPPT) Algorithm for Solar PV Array Using Boost Converter. IJSRSET.
3. Atik L, Petit P, Sawicki JP, Ternifi ZT, Bachir G, et al. (2016) Comparison of four MPPT techniques for PV systems. AIP Conference Proceedings 1758(1): 10.1063/1.4959443.
4. Zegaoui A, Aillerie M, Petit P, Sawick JP, Charles JP, et al. (2011) Dynamic behaviour of PV generator trackers under irradiation and temperature changes. Solar Energy 85: 2953-2964.
5. Zegaoui A, Aillerie M, Petit P, Sawick JP, Jaafar A, et al. (2011) Comparison of Two Common Maximum Power Point Trackers by Simulating of PV Generators. Energy Procedia 6: 678-687.
6. Belkaid A, Gaubert JP, Gherbi A (2016) An Improved Sliding Mode Control for Maximum Power Point Tracking in Photovoltaic Systems. CEAI 18: 86-94.
7. Gomes de Brito MA, Galotto L, Sampaio LP, Melo GA, Canesin CA, et al. (2013) Evaluation of the Main MPPT Techniques for Photovoltaic Applications. IEEE Transactions on Industrial Electronics 60: 1156-1167.
8. Jiang Y, Abu Qahouq JA, Haskew TA (2013) Adaptive Step Size with Adaptive-Perturbation-Frequency Digital MPPT Controller for a Single-Sensor Photovoltaic Solar System. IEEE Transactions on Power Electronics 28: 3195-3205.
9. Kollimalla SK, Mishra MK (2013) A new adaptive P&O MPPT algorithm based on FSCC method for photovoltaic system. International Conference on Circuits, Power and Computing Technologies (ICCPCT), pp: 20-21.
10. Murtaza AF, Sher HA, Chiaberge M, Boero D, Giuseppe MD, et al. (2013) A novel hybrid MPPT technique for solar PV applications using perturb & observe and Fractional Open Circuit Voltage techniques. 15th International Conference Mechatronika.
11. Esram T, Chapman PL (2007) Comparison of Photovoltaic Array Maximum Power Point Tracking Techniques. IEEE Transactions on Energy Conversion 22: 439-449.
12. Reisia AR, Moradib MH, Jamasb S (2013) Classification and comparison of maximum power point tracking techniques for photovoltaic system. Renewable and Sustainable Energy Reviews 19: 433-443.
13. Subudhi B, Pradhan R (2013) A Comparative Study on Maximum Power Point Tracking Techniques for Photovoltaic Power Systems. IEEE Transactions on Sustainable Energy 4: 89-98.
14. Fan L, YU Y (2011) Adaptive Non-singular Terminal Sliding Mode Control for DC-DC Converters. Advances in Electrical and Computer Engineering 11: 119-122.
15. Romdhane NMB, Damak T (2011) Adaptive Terminal Sliding Mode Control for Rigid Robotic Manipulators. International Journal of Automation and Computing 8: 215-220.
16. Rezkallah M, Sharma SK, Chandra A, Singh B, Rousse DR (2017) Lyapunov Function and Sliding Mode Control Approach for the Solar-PV Grid Interface System. IEEE Transactions on Industrial Electronics 64: 785-795.
17. Kolesnikov AA (2000) Modern applied control theory: Synergetic Approach in Control Theory. TRTU, Moscow, Taganrog, pp: 4477-4479.
18. Utkin VI (1977) Variable structure systems with sliding mode. IEEE Transactions on Automatic Control 22: 212-222.
19. Abderrezek H, Harmas MN (2016) Comparison study between the terminal sliding mode control and the terminal synergetic control using PSO for DC-DC converter. 4th International Conference on Electrical Engineering (ICEE).
20. Attoui H, Khaber F, Melhaoui M, Kassmi K, Essounbouli N (2016) Development and experimentation of a new MPPT synergetic control for photovoltaic systems. Journal of Optoelectronics and Advanced Materials 18: 165-173.
21. Salmi T, Bouzguenda M, Gastli A, Masmoudi A (2012) MATLAB/Simulink Based Modelling of Solar Photovoltaic Cell. International Journal of Renewable Energy Research- IJRER.
22. Pandiarajan N, Muthu R (2011) Mathematical modelling of photovoltaic module with Simulink. 1st International Conference on Electrical Energy Systems.
23. Kumari JS, Babu CS (2012) Mathematical Modeling and Simulation of Photovoltaic Cell using Matlab-Simulink Environment. International Journal of Electrical and Computer Engineering (IJECE) 2: 26-34.
24. Mehimmedetsi B, Chenni R (2016) Modelling of DC PV system with MPPT. 3rd International Renewable and Sustainable Energy Conference (IRSEC).
25. Ramos-Hernanz JA, Campayo JJ, Larranaga J, Zulueta E, Barambones O, et al. (2012) Two photovoltaic cell simulation models in matlab/Simulink. International Journal on Technical and Physical Problems of Engineering (IJTPE) 4: 45-51.
26. Bendiba B, Krim F, Belmilia H, Almia MF, Bouloumaa S (2014) Advanced Fuzzy MPPT Controller for a Stand-alone PV System. Energy Procedia, pp: 383-392.
27. Jiang Z (2007) Design of power system stabilizers using synergetic control theory. IEEE Power Engineering Society General Meeting.
28. Laribi M, Aït Cheikh MS, Larbès C, Barazane L (2010) Application de la commande synergetique au contrôle de vitesse d'une machine asynchrone. Revue des Energies Renouvelables 13: 485-496.
29. Santi E, Monti A, Li D, Proddutur K, Dougal RA (2004) Synergetic Control for Power Electronics Applications: A Comparison with the Sliding Mode Approach. Journal of Circuits, Systems and Computers.
30. Santi E, Monti A, Li D, Proddutur K, Dougal R (2003) Synergetic control for dc-dc boost converter implementation option. IEEE Transactions on industry applications 39: 1803-1813.
31. Bezuglov A, Kolesnikov A, Kondratiev I, Vargas J (2005) Synergetic Control Theory Approach For Solving Systems Of Nonlinear Equations. World Multi-Conference, Systemics Cybernetics and Informatics.

19

Implementation of LMS-ALE Filter Using Vedic Algorithm

Joseph Jintu K[1]* and Purushotham U[2]

[1]VLSI & Embedded systems PESIT Bangalore, India
[2]Department of Electronics & Communications PESIT Bangalore, India

Abstract

ALE or adaptive line Enhancers are special kinds of adaptive filters widely used in noise cancellation circuits. In circuits where we don't have any prior knowledge of signal and noise, fixed filters unit never works good. Among adaptive filter ring algorithms LMS algorithm is very common, in our work also we use LMS algorithm. LMS-ALE filters removes the sinusoidal noise signals present in the channel by calculating the filter coefficients in every iteration. LMS-ALE filter has large number of multiplier units. FFT or Fast Fourier Transform blocks present in LMS algorithm again consist of large array of multiplier units. Optimization of LMS-ALE filter lies must start from optimization of multiplier blocks. Here we use Vedic "Vertical and crosswise algorithm" for multiplier design. When compared to conventional booth multiplier based LMS-ALE filter units, Vedic multipliers gives more performance in areas like resource utilization, power requirement, delay etc. The work includes designing Vedic multipliers, complex Vedic multipliers, redesigning Radix-8 FFT using Vedic multipliers, redesigning LMS block using Vedic FFT, redesigning LMS ALE filter using Vedic multipliers and Vedic LMS blocks. Major part of design is done in verilog using Xilinx ISE design suite. ADC block present in LMS-ALE filter is done in Matlab version 2013.

Keywords: ALE filter; LMS algorithm; Vedic algorithm; Fast multipliers; Booth multipliers; Radix 8 FFT; Verilog; Xilinx ISE Design suite; Matlab

Introduction

In physical environment noise is automatic signal, in all kinds of signal generated there is some kind of unwanted noise signals, when such a signal is amplified in a communication channel both noise and desired signal gets amplified equally, it reduces the clarity of communication system, hence noise cancellation is inevitable. ALE Filter is a most common noise cancellation system. It uses some kind of adaptive algorithm [LMS algorithm]. LMS adaptive block consist of FFT and inverse FFT blocks as it handles the signal in frequency domain. Multipliers are the most repeated block in LMS-ALE filter, to optimize the performance we need to do the optimization from basic multiplier block. In this work we use Vedic algorithm for doing the multiplication. Compared to conventional booth algorithm Vedic multipliers requires less partial product adders, hence it improves the performance in terms of delay, resource utilization and power requirement [1-5].

Methodology

ALE filter

Adaptive Line Enhancer [ALE] filter optimized to remove sinusoidal noise signals present in the channel. In Figure 1, s(n) represents the desired signal, n(n) denotes the noise signal, z^{-} represents de-correlation function, and the block is followed by an adaptive predictor unit block. De-correlation eliminates any kind of correlation that may exist between the noise samples. Predictor can make prediction on the sinusoidal component of the noise signals and system will adaptively minimize the instantaneous squared error output. Inputs to the adaptive filter or predictor unit is units behind the original input signal. Therefore, in order to time align the enhanced signal, $\hat{s}(n)$ with the input signal, x(n) the adaptive filter must be able to 'predict ahead' in time by optimizing its filter coefficients in a least squares sense, hence the instantaneous squared error is minimized [5].

Vedic maths

Vedic mathematics or else called Indian mathematic is originated in Indian sub-continent at 1200BC. Ancient time significant growth to this field is done by great scholars like Aryabhata, rahmagupta, Mahavira, Bhaskara II, Nilakantha Somayaji etc. the decimal number system that we are using today is first record in Indian mathematic books. Vedic maths are the list of mental mathematical calculation techniques described in Vedas. Those mental techniques are combined together and described in special text called "vedic mathematics" [1].

Vedic multiplication algorithm

Vertical and crosswise algorithm is one among the 16 Vedic sutras mentioned in the Vedic mathematics. Vertical and crosswise algorithm is also called as Urdhva Tiryagbhyam. I use this sutra to optimize the multiplier performance. Multiplication example using vertical and crosswise algorithm is given below. We want to multiply 33 by 44:

3 3

4 4 ×

1 4 $_{2}$**5** $_{1}$**2** answer

Multiplying vertically on the right we get 3 × 4=12, so we put down 2 and carry 1 (written $_{1}$2 above). Then we multiply crosswise and add the two results: 3 × 4 + 3 × 4=24. Adding the carried 1 gives 25 so we put 5 and carry 2 ($_{2}$5). Finally we multiply vertically on the left, get 3 × 4=12 and add the carried 2 to get 14 which we put down [2].

Complex vedic multiplication

The flow chart in Figure 2 shows the procedure to compute the

***Corresponding author:** Joseph Jintu K, VLSI & Embedded systems PESIT Bangalore, India, E-mail: Jintuk.joseph@gmail.com

addition of any two signed numbers, Where x1 denotes both the input say a, b have same sign, if both have same sign then x1 take 1, Else x1=0. When x1 is high it looks like simple unsigned number addition. The result takes the sign of the inputs. If x1=0, we need to compute y1 value. Y1 is equal to 1 if first input says 1 > second input b, else it takes 0. If y1=1, then I perform a-b operation as a is larger than b, else I perform b-a operation, which implies b is larger than a. The output takes the same sign as a larger input operand.

The flow chart in Figure 3 shows the procedure to compute the subtraction of any two signed numbers, Where x1 denotes both the input say a, b have same sign, if both have same sign then x1 take 1, Else x1=0. When x1 is high, I need to compute y1 value. Y1 is equal to 1 if first input says 1 > second input b, else it takes 0. If y1=1, then I perform a-b operation as a is larger than b, else I perform b-a operation, which implies b is larger than a. The output takes the same sign as a larger input operand. If x1=0, it looks like simple unsigned number addition. The result takes the sign of the inputs [6].

Vedic 4 × 4 complex multiplier

Figure 4 shows the 4 × 4 bit multiplication using Vedic algorithm [3]. It consists of four 2 × 2 Vedic multiplier units, two 4-bit [N bit adder] adders and one 2-bit [N/2 bit adder] units. The orange colored circles indicate the selected operand for multiplication. q [1:0] is equal to Q [1:0], the final result [7].

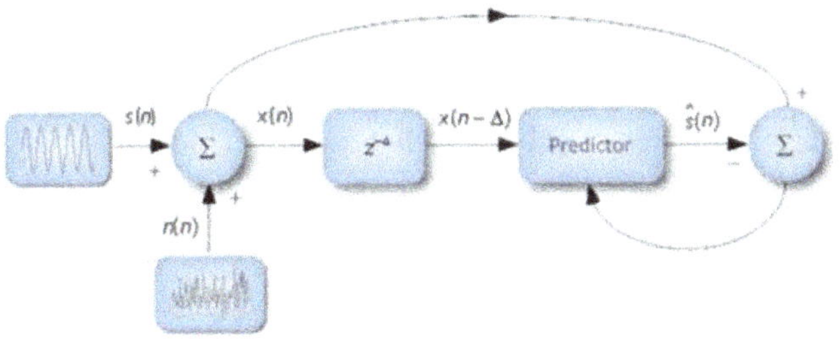

Figure 1: LMS-ALE filter block diagram.

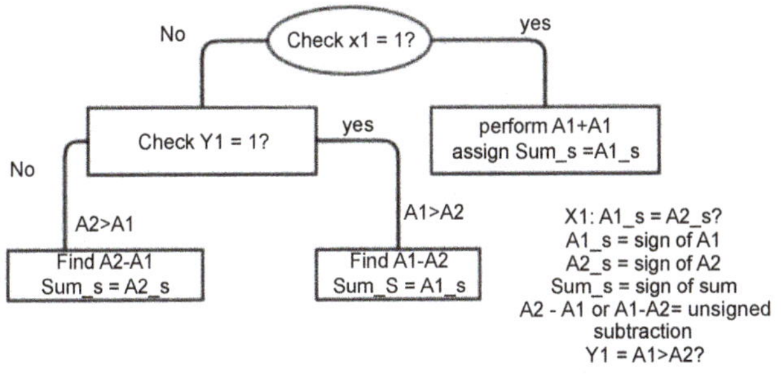

Figure 2: Signed multiplication algorithm [signed addition].

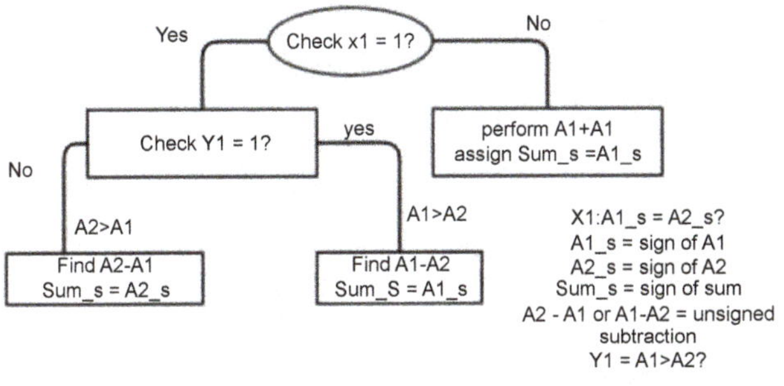

Figure 3: Signed multiplication algorithm [signed subtraction].

LMS algorithm

LMS algorithm is the basic adaptive algorithm available, this algorithm helps any system to mimic a desired filter by generating weights adaptively according to the value to error signal. Figure 5

The LMS algorithm is very useful and easy to compute [8,9]. The LMS algorithm will perform Ill, if the adaptive system is an adaptive linear combiner, as Ill as, if both the n-dimensional input vector X(k) and the desire output

d(k) are available in each iteration, where X(k) is

$$X(k)=\begin{bmatrix} x_1(k) \\ x_2(k) \\ \cdot \\ \cdot \\ \cdot \\ x_n(k) \end{bmatrix}$$

And the n-dimensional corresponding set of adjustable weights W(k) is

$$W(k)=\begin{bmatrix} w_1(k) \\ w_2(k) \\ \cdot \\ \cdot \\ \cdot \\ w_n(k) \end{bmatrix}$$

By having the input vector X(k), the estimated output y(k), can be computed as a linear combination of the input vector X(k) with the weight vector W(k) as

$$y(k) = X^T(k)W(k)$$

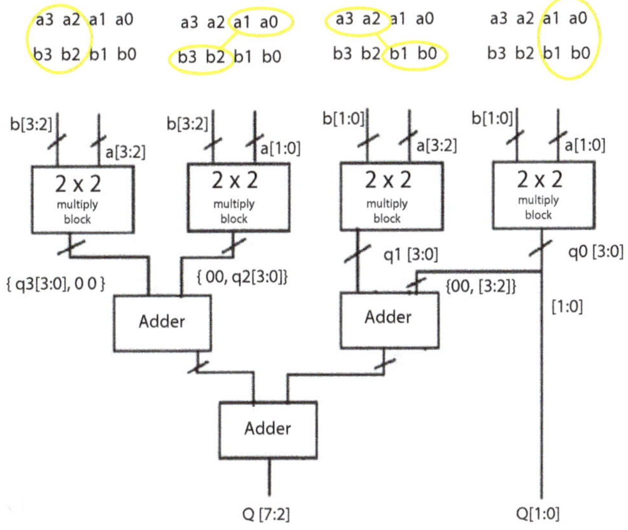

Figure 4: 4 x 4 Vedic multiplier block diagram.

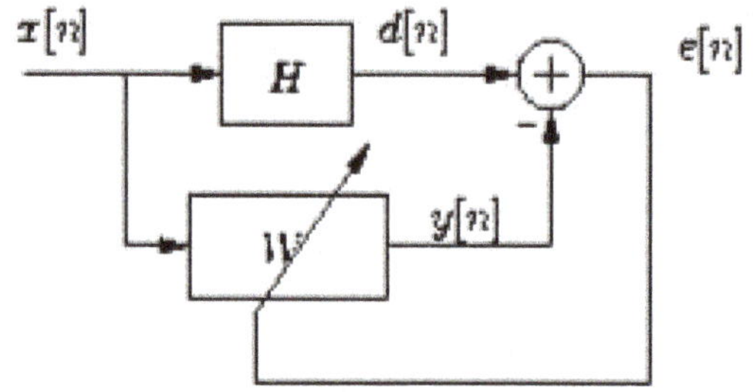

Figure 5: LMS algorithm block diagram.

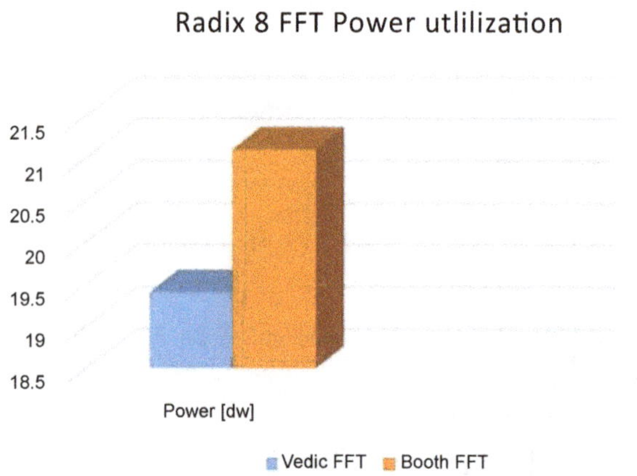

Figure 6: Power utilization comparison Graph of Radix 8 FFT implemented through Vedic algorithm and Booth algorithm.

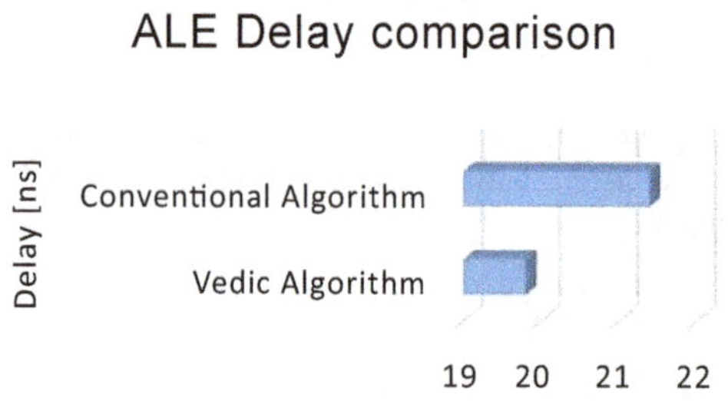

Figure 7: Delay comparison table for ALE implemented using conventional FFT and Vedic FFT.

Thus, the estimated error e(k), the difference between the estimated output y(k), and the desired signal d(k), can be computed as

$$e(k) = d(k) - y(k) = d(k) - X^{T}(k)W(k)$$

Results

The above figure is the simulation result obtained when 2 complex numbers each of 6 bit wide are multiplied, a and b are the input, each has two components a_r, a_i and b_i, b_r. B_r, a_r represents the real part and a_i, b_i represents the imaginary part. A_r_s, a_i_s and b_i_s, b_r_s represents the sign of real part and imaginary part of inputs respectively. The output obtained is c, it also has two components c_r [real part] and c_i [imaginary part].

For smaller modules booth algorithm based multipliers consumes larges resources, for 4 bit multiplier it uses 73 and 19, 4 input LUTs when implemented Booth algorithm and Vedic algorithm. 19 and 42 numbers of occupied slices when implemented using Vedic algorithm and Booth algorithm. When the input gets wider or module becomes larger Vedic algorithm based design consumes larges resources and booth algorithm based design consumes lesser resources.

Figure 6 shows the power utilization report for Radix 8 FFT [10] implemented using Vedic algorithm and Booth algorithm. These results are obtained from the X-power analyzer tool of Xilinx ISE software. For radix 8 FFT designed using Vedic algorithm the total power utilization is 194mw, it consist of three 16 bit Vedic multiplier and two 8 bit multiplier. Hence the total power utilization is sum of power utilization of each multiplier units. For Radix 8 FFT implemented using Booth algorithm the total power utilization is around 211mw (Figures 7 and 8).

Conclusion

The Fast multiplier design using Vedic algorithm has outstanding performance features in resources utilization, power requirement, delay taken, and area requirement. In this work a generic N × N bit Vedic multiplier which can perform both signed and unsigned multiplication are designed in Xilinx using verilog. An FFT module which can perform N × N bit Fourier transformation is designed in Xilinx using the Vedic multipliers designed earlier. An ADC module which takes audio input from system, converts the floating point value to 18 bit binary value are modeled in Matlab. This binary value becomes the input for the ALE. ALE block is designed in Xilinx using verilog language. The multiplier units in Ale are redesigned using Vedic multipliers. The

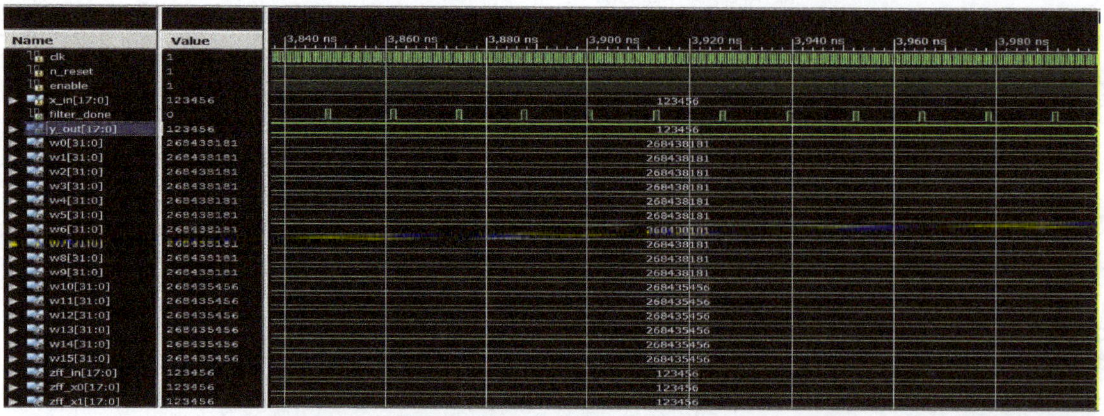

Figure 8: Simulation results for adaptive line enhancing filter using Vedic FFT.

comparison results between Vedic implementations and conventional implementation are also generated for each stage.

References

1. Tirtha SBK (1965) Vedic Mathematics. Motilal Banarsidass, Delhi, India.
2. Mehta P, Gawali D (2009) Conventional Versus Vedic Mathematical Method for Hardware Implementation of Multiplier. International Conference on Advances in Computing, Control and Telecommunication Technologies, IEEE computer society, Trivandrum, Kerala, pp: 640-642.
3. Sudeep MC, Sharath BM, Vucha M (2014) Design and FPGA Implementation of High Speed Vedic Multiplier. International Journal of Computer Applications 90: 6-9.
4. Agrawal J, Matta V, Arya D (2013) Design And Implementation Of FFT Processor Using Vedic Multiplier With High Throughput. International Journal of Emerging Technology And Advanced Engineering 3: 207-211.
5. He Y, He H, Li L, Wu Y, Pan H (2008) The Applications And Simulation Of Adaptive Filter In Noise Canceling. International Conference on Computer Science and Software Engineering, IEEE, Wuhan, Hubei 4: 1-4.
6. Premananda BS, Samarth SP, Shashank B, Bhat SS (2013) Design And Implementation Of 8 Bit Vedic Multiplier. International journal of advanced research in electrical and electronics and instrumentation engineering 2: 5877-5882.
7. Saha P, Banarjee A, Bhattacharya P, Dandapat A (2011) High Speed ASIC Design of Complex Multiplier Using Vedic Mathematics. IEEE students technical symposium, IIT Kharagpur.
8. http://eewiki.net
9. http://www.xilinx.com
10. Mittal N, Kumar A (2011) Hardware Implementation of FFT Using Vertically and Crosswise Algorithm. International Journal of Computer Applications 35: 17-20.

An Electric Fence Energizer Based on Marx Generator

Maryam Minhas*, Tanveer Abbas, Reeja Iqbal and Fatima Munir

Department of Electrical Engineering, Pakistan Institute of Engineering and Applied Sciences, PIEAS, Islamabad, Pakistan

Abstract

Non-lethal electric fence technologies gained considerable recognition in various application areas ranging from security to live-stock management and farm automation. Fence energizer, which is essentially a High Voltage Pulsed Power Supply (HVPPS), is a pivotal part of a non-lethal fence system with several design options. Marx generator, which is a well-recognized HVPPS, has never been tried as a fence energizer. This paper investigates potential of Marx generator as a fence energizer and identifies specific requirements and challenges that need to be addressed. Finally, a design of Marx generator is presented for a non-lethal electric fence having voltage rating of 40 kV, output pulse duration of 15 µs, pulse rate of up to 100 Hz and maximum output energy to be lower than 20 mJ per pulse to ensure its non-lethal nature. To meet the design requirements, a novel gate driver design is presented which is compact, cost effective and meets specific requirements of non-lethal nature of the fence.

Keywords: Non-lethal; Fencing; Security; Marx generator; Gate driver

Introduction

An electric fencing system applies electric shock to a person or an animal that touches the fence wire. Electric fences were used for the first time by Germans during World War I to stop illegal border crossing [1]. These fences were lethal due to the flow of high amperage AC current, thus, resulting in many human causalities as well as livestock damage. The drawbacks of lethal electric fences attracted researchers to work on non-lethal electric fencing for its constructive use. In this regard, initial work was carried out in late 1930s. Non-lethal electric fencing systems were initially developed for livestock and wild life management. Modern electric fencing systems use repetitive pulses of high voltage (in the range of several kV) and have various applications such as livestock management and depredation [2-5], crop protection [6-8], along highways [9,10] and security [11,12]. It is still an attractive field of research.

Non-lethal electric fencing finds a wide range of applications that require control over unauthorized/unwanted movement of humans or animals across a certain boundary. Conventionally, livestock managers use lethal methods for control of predation such as shooting, trapping, or use of devices such as M44's. These methods do not only demand more effort but their excessive use can also cause extermination of predator's specie [2,3]. Non-lethal electric fencing is an effective and more humane alternative of conventional depredation methods. It is used around aquaculture facilities to save fish from predator birds [4]. It is successfully used to protect crop against buffalo or elephant predation [6-8]. It is installed along highways to reduce human fatalities and wildlife destruction due to road accidents caused by animal vehicle collision [9-10]. Electric fencing along with an intrusion detection unit is employed for security of sensitive sites such as military bases, strategic instalments, prisons and airports [11,12]. Moreover, non-lethal fencing is used to protect private properties such as land and houses, and around private swimming pools to avoid accidents such as drowning of children [13]. It can also be used by law enforcement agencies for crowd management.

An electric fencing system has various parts such as energizer, fence wire, fence posts, insulators and ground rods. Energizer is the central part that generates high voltage pulses for the system. A simplified block diagram showing working of an electric fencing system is shown in Figure 1. One end of the energizer is connected to the fence wire while its other end is connected to ground. Fence wire is supported by fence posts through insulators. Insulators prevent current leakages to ground through fence posts. As shown in Figure 2, when a subject makes contact between fence wire and ground, it completes the current path and suffers from a high voltage shock that causes it to deter instantly.

The lethal or non-lethal behaviour of the system depends on the nature of electric pulse. For non-lethal system, pulse duration is small (in micro-seconds) and pulse current and energy transferred to the subject are kept below a safe limit. In this regard, researches have been carried out by medical specialists and standards have been developed that define non-lethal/harmless limits of current and energy when applied to a subject [14-17]. As lethal or non-lethal behaviour of

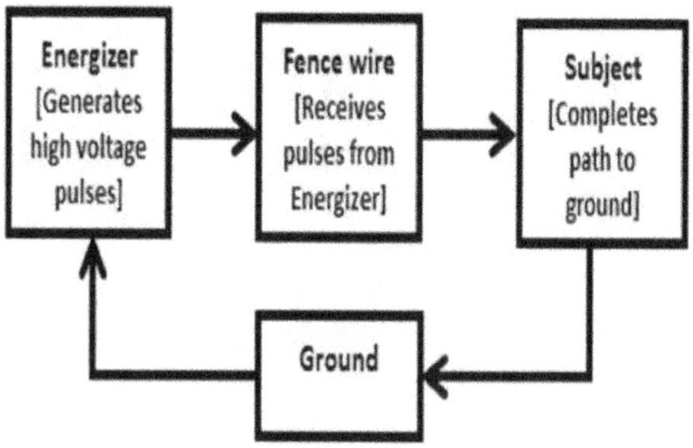

Figure 1: Working of electric fencing system.

***Corresponding author:** Maryam Minhas, Department of Electrical Engineering, Pakistan Institute of Engineering and Applied Sciences, PIEAS, Islamabad, Pakistan, E-mail: maryamkd@gmail.com

fencing system is a function of pulse nature, so energizer design is the most important part in development of an electric fencing system.

An energizer is a high voltage pulsed power supply that consists of two functional subunits which are energy storage unit and switching unit [18]. The energy storage unit stores electrical energy which is then released by switching unit during the output pulse. Design of an energizer for a non-lethal electric fencing system involves requirements of high output voltage, low output current, controllable pulse duration over various pulse rates and control over release of electrical energy during output pulse. In addition to that, the system must be power efficient and capable enough to store huge amount of energy so that it can work without an input source for several hours if employed in a remote area. Other requirements for energizer design are compactness and reliability.

Design options of electric fence energizer are listed below [19,20]:

i. Direct discharge type pulse generator: In this scheme, a capacitor is charged to a very high voltage by an input DC power supply. The capacitor is connected to the fence through a semiconductor switch. The requirement of a semiconductor switch with high voltage rating demands cascading of multiple semiconductor devices (such as IGBTs). A stack of semiconductor devices need highly synchronized gate drivers for reliable switching operation. As the stack consists of series connected switches, damage or failure of one switch can lead to failure of the whole system.

ii. Pulse transformer type pulse generator: This scheme employs a capacitor which is charged to a lower voltage as compared to the capacitor in direct discharge type. A switch connects the capacitor to the pulse transformer. This transformer steps up the low voltage pulse (from switch) to a higher voltage. The disadvantage of this scheme is loss of power efficiency due to use of pulse transformer which is undesirable in a fencing system.

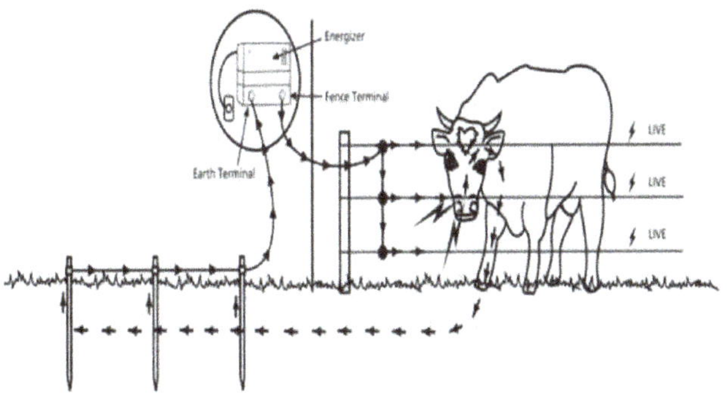

Figure 2: Electric shock applied by a fencing system.

iii. Vector inversion type pulse generator: this scheme employs a number of capacitors and transformers. During their charging, the series capacitors are connected alternatively in opposite polarities through pulse transformers and during discharging these capacitors get connected with the same polarity. This technique involves problems of complexity and inefficiency due to use of pulse transformers.

iv. Marx generator type pulse generator: This scheme, as shown in Figure 3, employs capacitors that are charged in parallel and discharged in series giving a high voltage output pulse. Its circuit basically consists of a number of stages with each stage comprising of a switch and a capacitor, and connected to the next stage through upper and lower diodes.

Comparative study of abovementioned HVPPS schemes showed that Marx generator is a superior technology due to a number of desirable features such as low voltage stress and low current stress on semiconductor switches, diodes and capacitors, high efficiency and controllability of pulse width and pulse rate [19,20]. Marx generator has applications in various areas such as UWB radar systems [21], Microwave and X-ray generation [22], plasma sources [23,24], particle acceleration [25], water treatment, food processing and air pollution control [26] but it has never been used as a fence energizer, perhaps, due to its circuit complexity. As mentioned above, it employs diodes, capacitors and switches. If the switches are implemented using semiconductor devices, they require gate driver circuits making the overall system quite complex for a fence energizer. However, Marx generator has many superior features to make fence energizers more effective.

Our work investigates potential of Marx generator as a non-lethal electric fence energizer expanding the applications of Marx generator and broadening the design options for fence energizers. This paper presents design of a Marx generator for a 40 kV non-lethal electric fence. The novelty of this work lies in the use of Marx generator as a fence energizer with reduced circuit complexity and design of a passive gate driver that not only solved problems of complexity, high voltage isolation and synchronized switching but also provided control over release of electrical energy during output pulse. Moreover, the total energy stored in energizer is much greater than the energy released per pulse so the system can work for some time if input power source is removed.

Requirement Analysis

The non-lethal and non-injurious electric shock applied by a

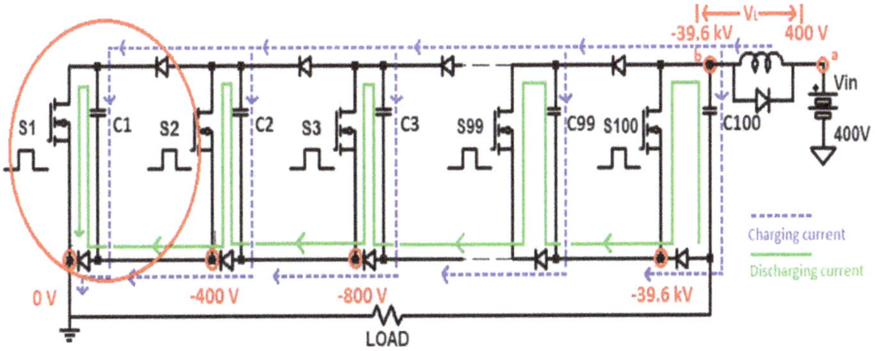

Figure 3: Schematic diagram of Marx generator.

fencing system must also be severe enough to deter the intruder instantly. A fencing system with more severe shock would be a stronger psychological and physical barrier for the intruder. Thus, severity of the shock is a measure of effectiveness of the system. In addition to that, the intruder must not be able to disable the system by cutting or damaging the fence wire by some normally insulated equipment. This has been achieved by keeping the output voltage as high as 40 kV (conventional systems have output up to 10 kV). High voltage will not only enhance system's effectiveness but will also render it failsafe by making the use of normally insulated equipment difficult. Moreover, to ensure an insurmountable barrier, the system must not allow any intrusion between two successive pulses. This requirement has been met by keeping the maximum pulse rate as high as 100 Hz. To save the subject of the shock from fatality or getting harmed, the pulse duration should be very short which is taken to have a maximum value of 15 µs. Moreover, energy delivered to the subject during output pulse is also an important factor in determining the harmless nature of shock. The safe limit is 5 J/s [14]. At a pulse rate of 100 Hz, the energy limit becomes 5/100=50 mJ per pulse. The system should be cost effective and compact so the use of step up/ down transformer is avoided. The input to the energizer is obtained from 220 V AC.

Design of Marx Generator

This section presents working of Marx generator and design of its circuit as a fence energizer. Design of gate driver for semiconductor switches is explained in the next section. A circuit diagram of Marx generator is shown in Figure 3. The circuit consists of cascaded similar stages. The first stage is highlighted by a circle in Figure 3. Each stage employs an energy storage unit and a switching unit. Energy storage unit consists of a capacitor that gets charged by input through a pair of diodes while the switching unit employs a semiconductor switch. The circuit multiplies the input DC voltage to the number of stages implemented and generates pulsed voltage at output. During charging, the switches are turned off and the diodes are forward biased, so the capacitors get connected in parallel. These parallel capacitors get charged to the input voltage. The charging current is shown in Figure 3 by dotted blue lines. During discharging, the switches are on and the diodes (that connect capacitors and switches in parallel) are off (discussed later) to connect the capacitors in series. The discharging current is shown in Figure 3 by solid green lines. Thus, the series connected capacitors get discharged by switches to produce a high voltage output pulse.

As shown in Figure 3, when the first switch S1 is turned on, the positive terminal of its corresponding capacitor C1 (which was previously at 400 V) drops to 0 V. As a result, the source of switch S2 (which is connected with negative terminal of C1) drops to -400 V and the diodes of first stage get reversed biased (discussed in detail later). Similarly, when the second switch S2 is turned on, the positive terminal of C2 drops to – 400 V and -800 V appear at source of S3.

In this way, the negative voltage keeps on increasing such that when all the switches are closed, a potential difference of 40 kV appears across load. The rest of the section deals with design of Marx generator circuit and selection of its components.

Input Supply

The input power supply consists of a bridge rectifier and a capacitive filter. The AC mains voltage is rectified by the bridge circuit. Its output is then fed to capacitive filter to obtain DC voltage for charging of Marx generator circuit. The DC voltage comes out to be $220 \times \sqrt{2} = 310V$.

Selection of diodes

The capacitors are charged by an input supply of 310 V. The peak charging voltage can vary due to changes in supply voltage. Moreover, the charging voltage can exceed the input voltage as charging takes place through an inductor (effect of inductor is discussed later). These variations have been accommodated by considering the charging voltage as 400 V instead of 310 V. The forward current rating of diodes is not a problem because capacitances are small and capacitors do not get completely discharged during switching. The diodes get off when the switches are turned on. So, peak inverse voltage is a major consideration in selection of diodes. As shown in Figure 4, node 'a' drops to 0 V and node 'b' drops to -400 V when switch S1 is turned on causing turning off of diode D1'. When switch S2 is turned on, node 'c' also drops to -400 V resulting in turning off of diode D1.

As shown in Figure 4, the maximum reverse voltage appearing across any diode is 400 V. Thereby, the peak inverse voltage of diodes must be greater than 400 V. Moreover, they must be fast recovery diodes to support the operation of system at faster pulse rates. To satisfy the abovementioned conditions, diodes FR305 have been selected.

Selection of switches

The switches, during their off time, are connected in parallel with capacitors. The capacitors are considered to be charged at 400 V so the maximum voltage appearing across any switch during its off time is 400 V. Thus, the selected switch must have a voltage rating greater than 400 V. The current rating of switch is not a problem because capacitances are small and little discharging takes place during switching. So, MOSFET IRF840 is selected to meet the requirements of design.

Design of capacitors

Design of a capacitor is based on evaluation of capacitance value, determination of its voltage rating and selection of its type. The charging of capacitors takes place by a DC input power supply of 310 V. The output of DC supply can fluctuate from 310 V causing variation in charging voltage of capacitors. Moreover, the inductor employed at input side can also cause overcharging of capacitors. So, to compensate these variations, the capacitors are considered to be charged at 400 V and their voltage rating is selected to be higher than 400 V. Moreover, polar electrolytic capacitors are chosen to minimize cost.

Evaluation of capacitance is a critical part of design because it dictates the energy released per pulse by each capacitor. The total energy E released during output pulse, number of stages n to be implemented and the drop in capacitor voltage (due to discharging during switching) are related by following expression

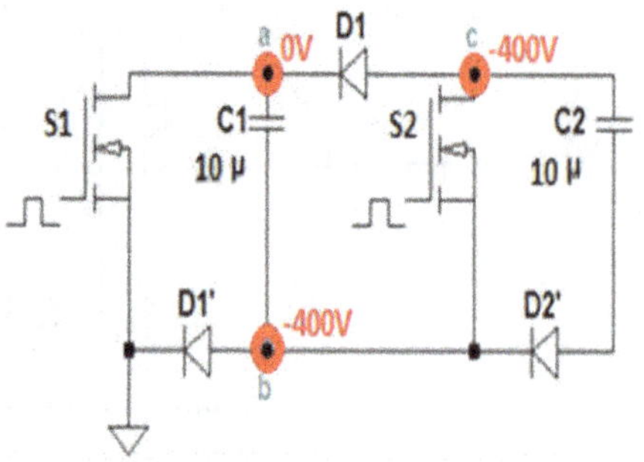

Figure 4: Turing on and off of diodes.

$$E = \frac{1}{2} nC\Delta V_c^2 \quad (1)$$

The number of stages n is ratio of output voltage to input voltage. If an output voltage V_{out}=40 kV is to be obtained from a DC input voltage V_{in}=400 kV the number of stages n is given as $n = \frac{v_{out}}{v_{in}} = \frac{40kV}{400V} = 100$

The voltage across each capacitor can decrease up to a maximum of 6 V ($\Delta V_{c=}$6 V) due to intrinsic property of gate driver design. To ensure non-lethal behaviour of fence, the energy released during one pulse for a pulse rate of 100 Hz must be less than 50 mJ and is selected as 18 mJ. For n=100 (100 stages) capacitance comes out be

$$C = \frac{2E}{nC\Delta V_c^2} = 10\mu\text{F}$$

Design of input inductor

As shown in Figure 3, an inductor is employed between input source and the rest of the circuit. This inductor provides power efficiency and protection to the input power supply. When all the switches are turned on, a negative voltage of -39.6 kV appears at node 'b'. If the power supply is directly connected with the circuit without an inductor, its positive terminal will connect to the highly negative node (node 'b' at -39.6 kV). This will result in huge input current surge which will not only reduce the power efficiency but can also damage the supply. On the other hand, a properly designed input inductor would block changes in input current due to high potential difference created across its terminals during output pulse.

The value of inductance *L* is related to the voltage across terminals of the inductor V_L, pulse duration *d* and increase in current ΔI as follows:

$$L = V_L \frac{d}{\Delta I} \quad (2)$$

The current profile of inductor is shown in Figure 5. There is a linear increase (with slope $\frac{V_L}{L}$) in current during output pulse of width and period. This current will charge the capacitors to a voltage higher than the input voltage. Freewheeling diodes tend to decay the inductor current when the charging voltage of capacitors exceeds the input voltage (discussed later). Thus the rising ramp in Figure 5 shows the current flowing from the supply into the circuit through inductor. The average current drawn by the circuit from the supply I_{avg} can be calculated from peak current Δ I flowing through inductor, duration of output pulse d and time period T as

$$I_{avg} = \frac{\Delta I \times d}{2T} \quad (3)$$

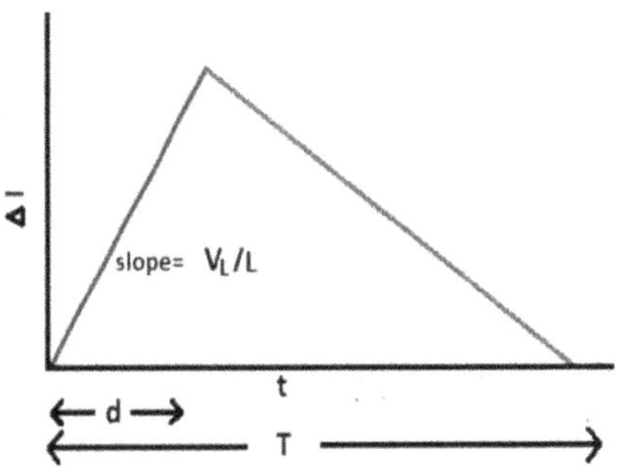

Figure 5: Current through inductor.

The power drawn by the circuit from the input P_{in} is related to average current from input I_{avg} and input voltage V_{in} as

$$P_{in=} V_{in} I_{avg} \quad (4)$$

It can be seen from Equation 2 that peak current through inductor is inversely related to value of inductance. Thus, a smaller inductance will result in high peak current which would in turn result in increased average current from the input. This excessive current will be dissipated in freewheeling diodes resulting in loss of power efficiency. Moreover, higher peak current through inductor will result in an increase in size of inductor. The inductor size (taken as core area product $A_C A_w$), value of inductance L, peak current through inductor Î and current (rms) I_{rms} are related as

$$L\hat{I}\, I_{rms} = K_{cu} J_{rms} B_{max} A_C A_W \quad (5)$$

Where K_{cu} is copper fill factor, J_{rms} is current density and B_{max} is saturation limited flux density. As shown in Figure 5, Î=ΔI and $I_{rms} = \frac{\Delta I}{\sqrt{3}}$ so by putting values of I_{rms} and Î we get

$$\frac{1}{\sqrt{3}} L\Delta I^2 = K_{cu} J_{rms} B_{max} A_C A_W \quad (6)$$

To obtain a direct relation between rise in current Δ I and core size, we substitute L from Equation 2 into Equation 6. So the above expression becomes

$$\frac{1}{\sqrt{3}} \times \left(V_L \frac{d}{\Delta I} \right) \Delta I^2 = K_{cu} J_{rms} B_{max} A_C A_W \quad (7)$$

On simplification Equation 7 takes the following form

$$\frac{1}{\sqrt{3}} \times V_L \Delta Id = K_{cu} J_{rms} B_{max} A_C A_W \quad (8)$$

The above equation clearly shows that if we halve the value of inductance the size of inductor will become two times its original value.

Thus, the input inductor should be designed such that it fulfils the conditions listed below:

i. Inductance should be high enough to minimize the peak current ΔI which would in turn increase power efficiency of system and reduce size of the inductor.

ii. The power drawn by the circuit from input supply must be higher than the output power to compensate for losses in the diodes. Otherwise, the system will be unable to generate an output pulse with required parameters.

The output power can be calculated from the energy delivered to the load per pulse. For the presented design, the energy released per pulse is 18 mJ. When the system is operated at a pulse rate of 100 Hz this energy corresponds to an output power of 1.8 W.

As discussed before, the input power must be greater than or equal to 1.8 W. Taking V_{in}=400 V, Equation 4 becomes

$$1.8\text{ W} = 400\text{ V} \times I_{avg}$$

$$I_{avg=} 4.5\text{ mA}$$

Using Equation 3, peak current Δ I comes out to be

$$\Delta I = \frac{2 \times T \times I_{avg}}{d}$$

$$\Delta I = 6A$$

The value of inductance can now be determined from Equation 2 as

$$L = \frac{40000 \times 15 \times 10^{-6}}{6} = 100mH$$

Any value of inductance lower than as calculated above will render the system power inefficient and bulky.

Freewheeling diode

As shown in Figure 3, a freewheeling diode is connected in parallel to the inductor with its anode connected with the positive terminal of last capacitor and cathode connected towards the supply. The diode protects the capacitors from getting damaged due to overcharging and prevents the circuit from producing undesirable resonance effects.

In absence of freewheeling diode, the inductor will continue to transfer excessive electrical energy into the capacitors due to high potential difference across its terminals during output pulse. As in case of a fencing system, no load is connected with the output (except when the subject touches the fence) the capacitors will get overcharged. This can damage the capacitors if their charging voltage exceeds their rated voltage. Moreover, the excess energy transferred to the capacitors during on time of switches will be returned to the supply during off time of switches. So, the circuit will act as a series LC resonator producing undesirable effects at output as shown in Figure 6.

The freewheeling diodes must be employed so as to meet the following essential conditions.

i. As shown in Figure 5, the inductor current must fall to zero during off time of pulse (which equals 10 ms for pulse rate of 100 Hz and pulse duration of 15 μs).

ii. During on time of pulse, the reverse voltage across diode must not exceed its rated peak inverse voltage.

A stack of series connected diodes is used to fulfil the abovementioned conditions. The first condition requires the diode to provide a forward voltage drop of $V = L\frac{\Delta I}{\Delta t}$. If forward voltage drop offered by one diode is assumed to be 0.5 V then 12 diodes can meet the requirement. The second condition needs the number of diodes to be equal to or greater than 40 kV/PIV (PIV of one diode). For FR305, PIV equals 600 V. So, the diode stack requires 40 kV/600=67 FR305 diodes.

Gate Driver Design

When employed as an electric fence energizer, Marx generator uses a large number of semiconductor switches to produce an output pulse of several kilovolts in accordance with its design requirement of a severe shock. Consequently, the design of gate drivers for Marx generator switches becomes a major challenge that needs to be addressed ingeniously.

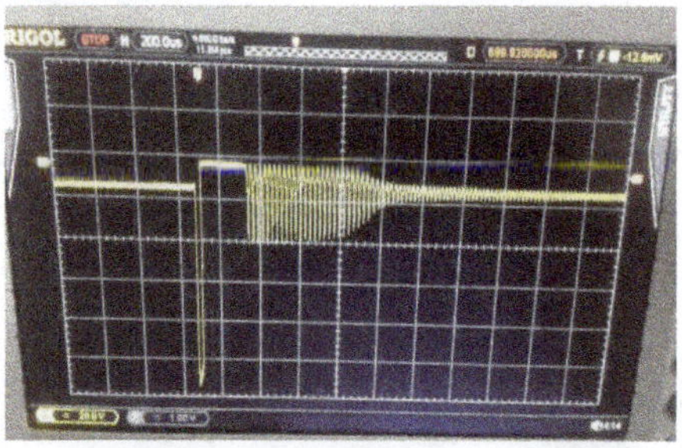

Figure 6: Resonance effect due to inductor.

Requirements

As shown in Figure 3 the source voltage increases with every stage such that when S1 is turned on, the source voltage of switch S2 drops to -400 V. Similarly, turning switch S2 on pulls the source voltage of switch S3 down to -800 V. In this way, when all the switches are turned on the source of last switch drops to -39.6 kV. This switch now requires a gate voltage with respect to -39.6 kV which demands a gate driver capable of providing voltage isolation of 39.6 kV.

A case of unsynchronized switching is shown in Figure 7. When switch S1 is on and switch S2 is off, it causes the capacitor C2 to discharge through forward biased diode D1 and switch S1, thereby reducing power efficiency. Moreover, the voltage of C2 does not add up in the output voltage. Therefore, synchronized switching is an important gate driver design requirement that ensures maximum power efficiency and a complete output pulse.

Keeping in view the electric fencing requirements, the gate driver should allow a non-lethal amount of electrical energy to be transferred to the fence wire. Moreover, it should render the system efficient for continuous operation even in the absence of an input source.

Therefore, it can be concluded that the gate driver for semiconductor switches based Marx generator (as a fence energizer) must fulfil the following design requirements:

- High voltage isolation
- Synchronized switching
- Controlled release of output energy
- Compactness of design and high efficiency

Related work

Generally used gate driver schemes for semi-conductor switches based Marx generators employ either optocouplers or pulse transformers. Two of the related designs are discussed below.

Self-supplied gate driver is an optocoupler based design that is referred from [27] and shown in Figure 8. In case of this type of gate driver, Marx generator circuit itself provides power to the gate driver through capacitor C_p which is charged to zener voltage when the switch is off. A resistor R controls the charging current of capacitor while a diode D prevents it from discharging when the switch is on.

Another gate driver scheme is referred from [28] and presented in Figure 9. It uses a pulse transformer to drive Marx generator switches. A full bridge inverter generates pulses and feeds them to the pulse transformer. These positive and negative pulses turn the main switch

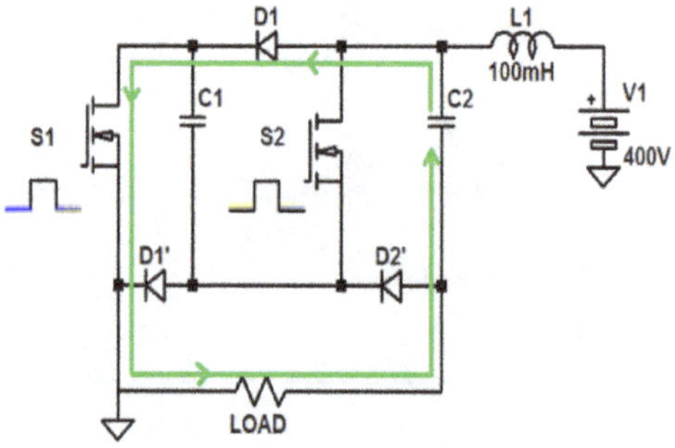

Figure 7: Discharging of capacitor due to unsynchronized switching.

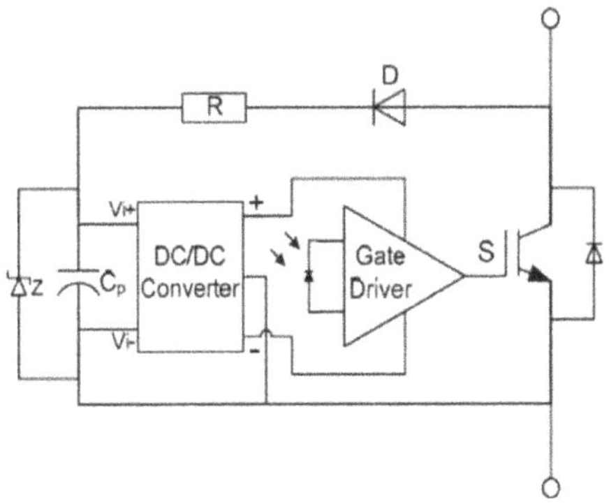

Figure 8: Self supplied gate driver [27].

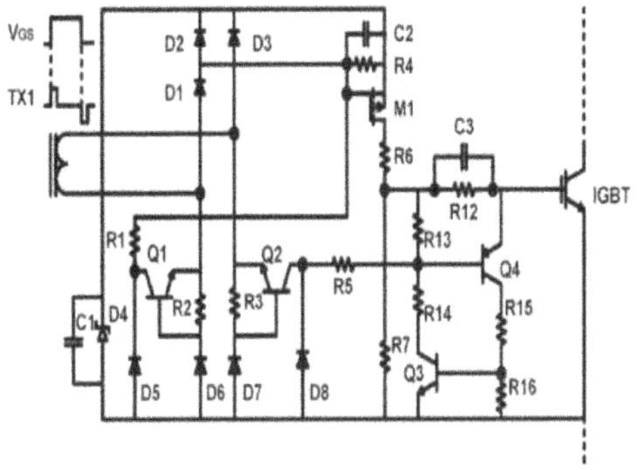

Figure 9: Pulse transformer based gate driver.

on and off respectively by the combined action of different elements in the gate driver circuitry.

The abovementioned designs are suitable gate drivers for pulsed generators (that employ semiconductor switches) that are required to produce high power. But when used for low power applications that use low power rating MOSFETs and diodes, and small capacitors (10 μF), such gate drivers increase circuit cost and complexity.

Therefore, a novel, compact and cost-efficient gate driver based on passive components such as diodes, capacitors and resistors has been proposed that ensures high voltage isolation as well as synchronized switching and, unlike the abovementioned schemes, makes Marx generator an attractive design option for electric fence energizer.

Proposed gate driver design

The proposed scheme employs one active driver (optocoupler) to drive the first switch and passive drivers to drive the rest of the switches. Each passive driver comprises a gate capacitor Cgn, a zener diode Zn connected in series with a diode (with opposite polarity) and another diode Dgn connected between gate and source of the corresponding switch. The gate to source voltage Vgs that is required to turn the switch on is provided through the combined action of zener diode and gate capacitor by pulling down the source to a negative voltage with respect to the gate voltage. Diode, on the other hand, contributes in turning off of the switch. This scheme is shown in Figure 10. During discharging mode of Marx generator, with its capacitors charged to input voltage Vc, the source of nth stage MOSFET is pulled down to -(nVC) volts due to which zener of voltage rating V_{gs} gets reverse biased and discharges the gate capacitor (initially charged to 0 V) to $-(nV_C - V_{gs})$ volts, thus developing the required gate to source voltage Vgs to turn the MOSFET on. This is shown in Figure 10, when opto-coupler turns the first switch S1 on, the source of second switch S2 drops to -400 V, reverse biasing the zener Z1 and discharging the gate capacitor Cg1 to -(400-Vgs) volts, thus developing the required gate to source voltage Vgs.

During charging mode of Marx generator, the source terminals of all MOSFET switches get connected to the ground. The negative voltage of $-(nV_C-V_{gs})$ volts at gate of any switch, due to the discharged gate capacitor, forward biases diode Dg. This enables the charging of gate capacitor to 0 V by utilizing the leakage currents flowing towards the forward biased diode Dg, thus turning off the switch. As discussed earlier, when the first switch S1 is turned off, the diode D1' gets forward biased, thus connecting the source of S2 to ground. As a result, the anode of diode Dg1 also gets connected with ground while its cathode is at - (400-Vgs) volts thus forward biasing the diode and charging the gate capacitor Cg1 to 0 V, thus turning off the switch S2. The diode connected in series with zener gets reverse biased during charging of gate capacitor to block forward current surge through zener. As shown in Figure 11, the negative voltage across gate capacitor keeps increasing with each stage. This introduces problems of increase in voltage rating of gate capacitors and high forward surge currents (during charging of gate capacitor) through diodes Dgn. The first problem is solved by connecting the terminal of gate capacitor, previously connected to ground, with source of previous stage instead. This modification in gate driver circuitry, shown in Figure 11, limits the maximum potential difference across any gate capacitor to approximately 400 V.

The second problem arises due to the absence of some controlling element, which causes the current through gate capacitors to become very high during their discharging period and keep increasing for higher stages as the voltage at source of the switch with which gate capacitor is connected to becomes more negative. Therefore, a resistor is connected in series with gate capacitor to limit this surge current. Waveform of such currents (in absence of resistors) in different colours simulated for few initial stages is shown in Figure 12.

Since, zener limits gate to source voltage Vgs across the switches, the given design uses a zener diode of 10 V to obtain Vgs of 10 V as shown in Figure 13.

The selection of gate capacitor depends on its capacitance, voltage rating and current rating. A greater capacitance would draw surge current from the circuit while smaller would discharge earlier due to leakage currents resulting in turning off the switch before time. So, the value of capacitance is optimized through simulations to be 1000 pF. Its voltage rating is selected to be 450 V for DC input supply of 400 V. Accordingly, the current rating of capacitor becomes 0.45 A for gate resistance of 1 kΩ. The diode Dg1 and the diode connected in series with zener are fast recovery diodes FR305 having the voltage rating of 600 V.

The passive gate driver circuit has been simulated in LT-spice. When a control pulse is applied through opto-coupler to turn the first switch on, a negative voltage of -400 V appears at the source of second switch and the reverse biased zener discharges the gate capacitor to develop gate to source voltage of 10 V to turn the second switch on. The control pulse along with the gate to source voltage obtained through simulations is shown in Figure 13. On the other hand, in the absence of control pulse the source of switch gets connected to ground and the gate capacitor charges to 0 V due to leakage current, thus dropping the gate to source voltage to 0 V. Figure 14 shows the charging and

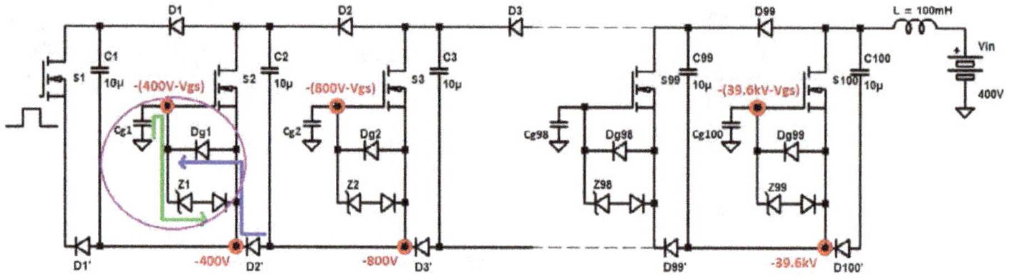

Figure 10: Proposed gate driver scheme.

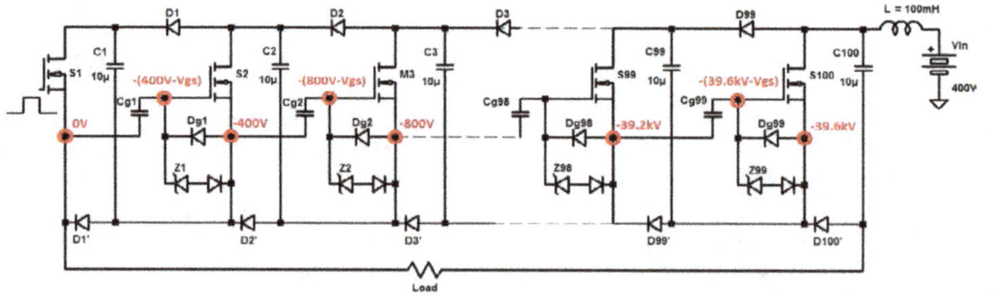

Figure 11: Marx generator circuit with modified passive gate driver.

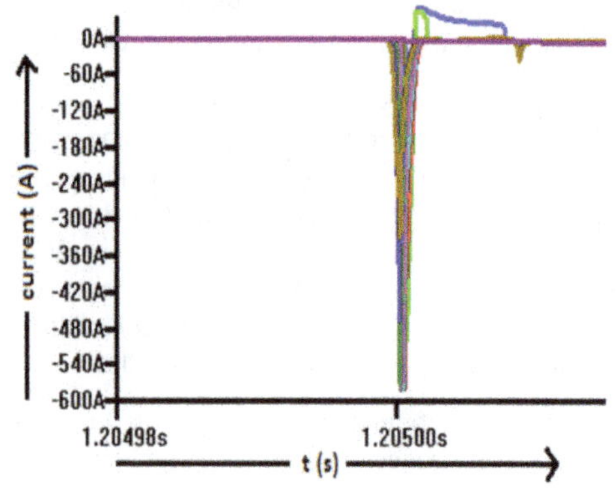

Figure 12: Surge current through gate capacitors.

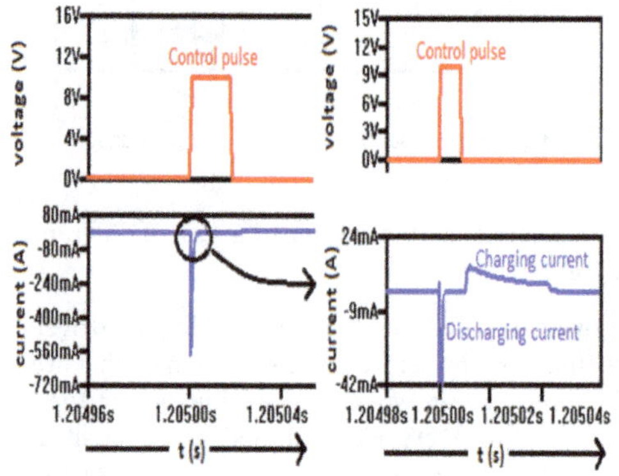

Figure14: Charging and discharging current through gate capacitor.

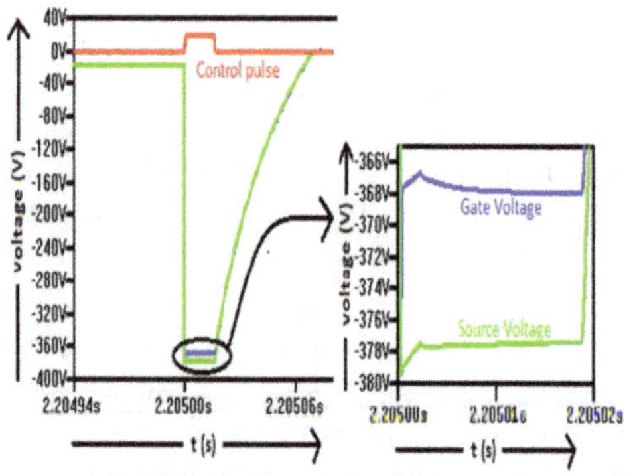

Figure 13: Gate to Source voltage.

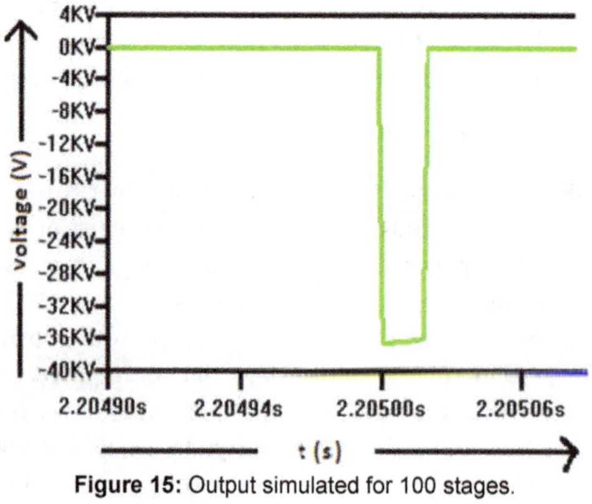

Figure 15: Output simulated for 100 stages.

discharging current waveforms of gate capacitor with reference to control pulse. The final output pulse of -40 kV produced by 100 stages with each stage at 400 V is shown in Figure 15.

The proposed gate driver design fulfills the requirement of controlled transfer of electrical energy to the output. It is the inherent feature of this design that the Marx generator capacitors can discharge

Figure 16: Final form of Marx generator circuit.

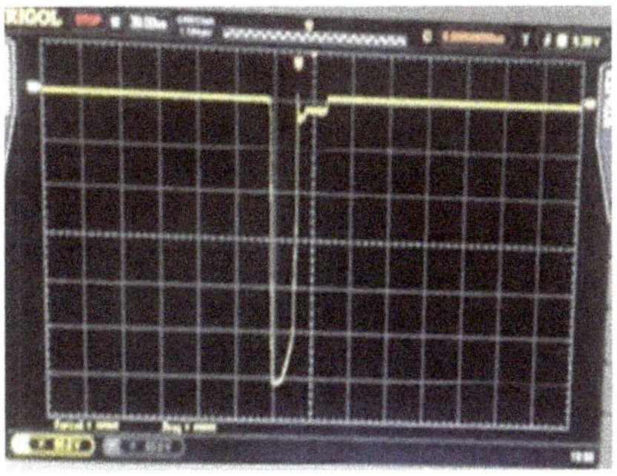

Figure 17: Output voltage measured on oscilloscope across a capacitive divider of 1/100.

a maximum voltage of 6 V (zener diode rating) only. When they tend to discharge more than 6 V, the gate to source voltage drops below 4 V and the switch gets turned off. It results in controlled release of energy per pulse, hence the maximum deliverable value of energy corresponding to a maximum drop in capacitor's voltage is 18 mJ which is well below the lethal limit of 50 mJ [14]. This unique feature of gate driver renders the overall system suitable for low power high voltage applications such as non-lethal electric fencing.

Implementation and Testing

Marx generator with proposed gate driver design is implemented and its final form is shown in Figure 16. Its implementation includes development of PCBs, winding of input inductor, and employment of a capacitive bank (as a voltage divider) for measurement of high voltage output pulse on oscilloscope.

Design of PCBs

Major consideration during PCB design of proposed circuit is to keep high potential nodes as far as possible while also maintaining compactness of the overall system.

Winding of input inductor

From Section II, it was found that Δ I=6 A, I_{rms}=3.46 A, K_{cu}=0.1, J_{rms}=3 A/mm^2 and B_{max}=0.2 T (for ferrite core). Putting these values in Equation 5 we get area product $A_C A_w$=34600000 mm^4. As I_{rms}=3.46 A, copper wire with SWG-18 is used for winding of inductor.

The potential difference between two ends of inductor is 40 kV so it needs protection against insulation breakdown of winding wire. This is done by dividing the total number of turns in multiple layers such that the potential difference between two consecutive turns is 250 V and between two adjacent layers is 10 kV. Moreover, additional insulation is provided by placing fish paper (voltage breakdown rating of 15 kV) between adjacent layers.

Capacitive bank

To measure high voltage (up to 40 kV) on oscilloscope a capacitive voltage divider (with 100 series connected capacitors) is used. The reason behind employing capacitive divider instead of resistive divider (of value 50 MΩ) is that high resistances have inductive behaviour. So a resistive divider cannot allow flow of current during a short pulse (of duration 15 μs). The output waveform of 40 kV is shown in Figure 17.

Testing

The working of the system is tested by measuring the output voltage using capacitive voltage divider and its waveform is shown in Figure 17. Moreover, air breakdown test has been performed in a spark gap at different frequencies between 1 Hz and 100 Hz for pulse durations up to 15 μs. The reliability of system is verified by its continuous operation for more than 10 hours in the lab. Air breakdown in a spark-gap of 10 mm is performed.

Conclusions

Marx generator is one of the design options for non-lethal electric fence energizers with several attractive features such as being energy efficient and reliable. Marx generator is superior to other techniques because electronic components such as capacitor and switches are subjected to lower voltage stress and lower current stress. Its other features that make Marx generator (with proposed gate driver) a better option for a non-lethal electric fence energizer are compactness, higher efficiency, better pulse width control and most importantly control over release of output energy. It was never tried before because of its hardware complexity. In a semiconductor switches based Marx generator implementation, gate drivers are the major contributors of hardware complexity. This issue is addressed by a novel gate driver design which is simple, compact and cost effective. Moreover, it controls output energy to ensure non-lethal nature of the fence.

References

1. Maartje A (2006) The art of staying neutral: the Netherlands in the First World War. Amsterdam University Press 1914-1918.
2. Roger DN, Theade J (1988) Electric fences for reducing sheep losses to predators. Journal of Range Management Archives 41: 251-252.
3. Shivik JA (2004) Non-lethal alternatives for predation management. Sheep & Goat Research Journal.
4. Kimberly SC, Pitt WC, Conover MR (1996) Overview of techniques for reducing bird predation at aquaculture facilities. Archived USU Extension Publications.
5. Turner LW, Absher CW, Evans JK (1986) Planning fencing systems for intensive grazing management. University of Kentucky, College of Agriculture, Cooperative Extension Service.
6. Ludwig S, Baldus RD (1998) Assessment of Crop Damage and Application of Non-Lethal Deterrents for Crop Protection East of Selous Game Reserve. No 24 Tanzania Wildlife Discussion Paper.
7. Scott EH, Craven SR (1988) Electric fences and commercial repellents for reducing deer damage in cornfields. Wildlife Society Bulletin (1973-2006) 16: 291-296.
8. Masayuki S, Momose H, Mihira T (2011) Both environmental factors and countermeasures affect wild boar damage to rice paddies in Boso Peninsula, Japan. Crop Protection 30: 1048-1054.
9. Scott CW, Ramakrishnan U, Ward JS (2006) Deer damage management options. Connecticut Agricultural Experiment Station.

10. Lauren LM, Conover MR, Frey SN (2008) Deer–vehicle collision prevention techniques. University of Nebraska - Lincoln.
11. Haim P (1998) Intrusion detection system. US Patent No. 5,852,402.
12. Asaf G (2009) Vibration sensor for boundary fences. US Patent No. 7,532,118.
13. John L (1983) Fencing of private swimming pools in New Zealand. Community health studies 7: 285-289.
14. Martino MGBD, Reis FSD, Dias GAD (2006) An electric fence energizer design method. International Symposium on Industrial Electronics, IEEE.
15. Mark WK, Perkins PE, Panescu D (2015) Electric fence standards comport with human data and AC limits. 37th Annual International Conference of the IEEE Engineering in Medicine and Biology Society (EMBC), IEEE.
16. Charles FD, Lagen JB (1941) Effects of electric current on man. Electrical Engineering 60: 63-66.
17. Barnos BW (1989) The electrified fence as a component of a physical protection system. Proceedings of International Carnahan Conference on Security Technology, IEEE.
18. Tao T (2007) Design and Implementation of Full Solid State High Voltage Nanosecond Pulse Generators. ProQuest.
19. Duleepa JT (2008) A novel electric fence energizer: design and analysis. Diss Research Space, Auckland.
20. Thrimawithana DJ, Madawala UK, Woodhead RCB (2006) Pulsed power generation techniques. 32nd Annual Conference on IEEE Industrial Electronics, IEEE.
21. Carey WJ, Mayes JR (2002) Marx generator design and performance. Power Modulator Symposium.
22. Archana S, Senthil K, Sabiyasachi M, Vishnu S, Ankur P, et al. (2011) Development and characterization of repetitive 1-kj marx-generator-driven reflex triode system for high-power microwave generation. IEEE Transactions on Plasma Science 39: 1262-1267.
23. Redondo LM, Silva JF, Tavares P, Margato E (2005) All Silicon Marx-bank topology for high-voltage, high-frequency rectangular pulses. 36th Power Electronics Specialists Conference, IEEE.
24. Kim JH, Ryu MH, Shenderey S, Kim JS, Rim GH (2004) Semiconductor switches based pulse power generator for plasma source ion implantation. Power Modulator Symposium.
25. Kawamura Y, Toyoda K, Kawai M (1984) Generation of relativistic photoelectrons induced by excimer laser irradiation. Applied Physics Letters 45: 307-309.
26. Erwin HWMS, Heesch BEJMV, Paasen SSVBV (1998) Pulsed power corona discharges for air pollution control. IEEE Transactions on Plasma Science 26: 1476-1484.
27. Yifan W, Kefu L, Jian Q, Xiaoxu L, Houxiu X (2007) Repetitive and high voltage Marx generator using solid-state devices. IEEE Transactions on Dielectrics and Electrical Insulation 14: 937-940.
28. Ryoo HJ, Kim JS, Rim GH, Goussev G (2007) Current loop gate driver circuit for pulsed power supply based on semiconductor switches. 34th International Conference on Plasma Science, IEEE.

Effects of Magnetic Field on the Blood Flow in an Aneurismal Aorta

Nezar M[1]*, Aggoune N[2], Nezar DJ[3], Abdessemed R[4], Nezar KS[5] and Nezar A[5]

[1]*Laboratoire d'Innovation en éco-conception, construction et génie sismique (LICEC_GS), Algeria*
[2]*Laboratoirede Mécanique des structures et Matériaux (LaMsM), Algeria*
[3]*Laboratoire de Physique Energétique appliquée de Batna (LPEA), Algeria*
[4]*Laboratoire d'Electrotechnique de Batna (LEB), Algeria*
[5]*Centre hospitalier universitaire de Batna CHU, Algeria*

Abstract

The goal of this paper is to analyze the effect of the magnetic field, applied to a blood tissue whose walls present an aneurism arterial. The interest of this study is to evaluate the rate of influence, if there existed, of the applied magnetic field and to check the preconceived validity of the magneto therapy. The FLUENT computational fluid dynamic code is used to carry out all computation in this work, which is selected to resolve the MHD equations.

Keywords: Aneurism; Blood flow; Magnetic field; Magneto therapy; Harmful effects; CFD

Introduction

The biological fabrics are strongly heterogeneous mediums with very particular properties, and the attack of diseases as aortic aneurism became, unfortunately, very widespread, as caused by arterial hypertension; an essential characteristic of the actual environment. The most unfavorable consequences of this disease are the formation of thrombus (stone of blood); often induced by the slowness of the blood flow inside this aneurism. To mitigate this problem; a library search was undertaken; and public works by CAT and Huang, revealed that an exposed blood flow with a field of intensity 1,3 tesla, will be fluxes; in less than one minute. To check this prejudice, we made a numerical study; on the state of a blood flow of the aorta; by considering the two following principal cases:

- The study of the healthy state of the aorta, which corresponds to the application of a magnetic field in the absence of aneurism;
- The study of the aorta beginning aneurismal state, which makes it possible to ensure a good visibility of the least effects emerging following the interaction of the two phenomena [1].

What an aneurism?

An aneurism is a dilation of an artery wall, in general, of the aorta. This dilation occurs; in general; in a weak area of the arterial wall. The blood pressure involves swelling towards the outside of the weak area. If it is not treated, aneurism can break and involve an internal bleeding. Aneurisms can develop everywhere along the aorta. The most frequent cause of aortic aneurisms is the atherosclerosis (arterial hypertension, diabetes, tobacco... etc), which weaken the wall of the aorta [2,3].

Test bench

To include the side effects of this problem, we considered it useful to reproduce, with real dimensions, the body (Figure 1) with a prolongation of the aorta on the side of the aneurismal zone and thus, to allow as well as possible to include the side effects of this problem, we considered it useful to reproduce, with real dimensions, the body (Figure 1) with a prolongation of the aorta on the side of the aneurismal zone and thus, to allow as well as possible to appreciate the evolution of the blood flow under the action of the two physical phenomena. The interest will be carried, in this case, with the principal channel which is "the aorta" and not with the junctions [4].

Mathematical Formulation

The theory biophysics presents blood, like, a rheofluidifiant fluid; in other words, its dynamic viscosity is described, according to the force of friction F, which a blood layer exerts on its adjacent:

$$F = \tau S \frac{d\vartheta}{dz} \tag{1}$$

Where,

τ: Dynamic viscosity coefficient

S: Surface common to both layers (m^2)

ϑ: Rate of the blood flow

And it's kinematic viscosity is expressed by the relation:

$$v = \frac{\tau}{\rho} \tag{2}$$

Where: ρ : is the density of blood (kg/m^3)

Consequently, its variation laws are described by the Navier Stocks equations:

- Equation of continuity:

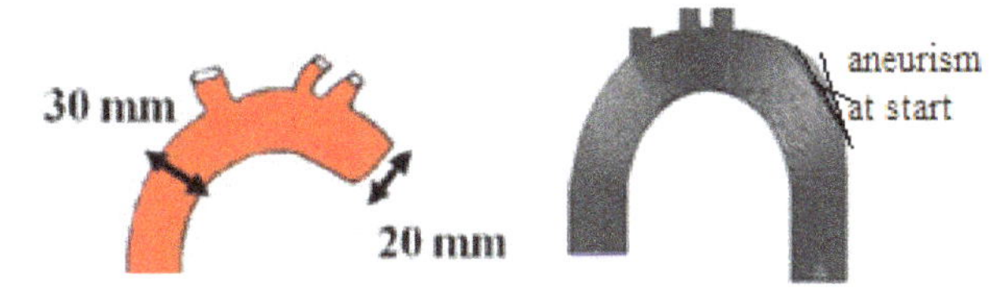

Figure 1: The dimensions of the test bench.

***Corresponding author:** Nezar M, Laboratoire d'Innovation en éco-conception, construction et génie sismique (LICEC_GS), Algeria
E-mail: m.nezar@univ-batna2.dz

$$\frac{\partial \rho}{\partial t}+\sum_{j=1}^{3}\frac{\partial(\rho \bar{u}_j)}{\partial x_j}=0 \qquad (3)$$

- Equations of momentum:

$$\frac{\partial(\rho \bar{u})}{\partial t}+\sum_{j=1}^{3}\frac{\partial(\rho \bar{u}_i \bar{u}_j)}{\partial x_j}=-\frac{\partial p}{\partial x_i}+\sum_{j=1}^{3}\frac{\partial}{\partial x_j}\left[\mu\left[\frac{\partial \bar{u}_i}{\partial x_j}+\frac{\partial \bar{u}_j}{\partial x_i}-\frac{2}{3}\delta_{ij}\sum_{j=1}^{3}\frac{\partial \bar{u}_i}{\partial x_i}\right]\right]+\sum_{j=1}^{3}\frac{\partial}{\partial x_j}\left(-\rho \overline{u_i u_j}\right)+F_v \qquad (4)$$

Where: F_v is the electromagnetic force resulting from the interaction of the induced currents density and the magnetic field which gave its birth. It is written in the form:

$$q=\frac{-}{\sigma}j.j$$

$$q=\frac{1}{\sigma}\vec{j}.\vec{j} \text{ with } \vec{j}=\sigma\vec{E} \text{ (W/m}^3\text{)}$$

σ : is the fluid electric conductivity,

q : induction of the magnetic field

$\vec{E}$: is the local electric field

$\vec{B}$ is the vector of the magnetic field

- Equation of energy:

$$\frac{\partial(\rho E)}{\partial t}+\sum_{j=1}^{3}\frac{\partial(\rho E \bar{u}_j)}{\partial x_j}=\sum_{i=1}^{3}\sum_{j=1}^{3}\left(\frac{\partial}{\partial x_j}\left(\tau_{ij}-\rho \overline{u_i u_j}\right)\bar{u}_i\right)-\sum_{j=1}^{3}\frac{\partial}{\partial x_j}q_j+S_E \qquad (5)$$

Simulation

To try to check the harmful effect, of the magnetic fields on the healthy living organisms, would be the task of this first test. In the second, it would be necessary to check the rate of this harmfulness, in the presence of aneurism. As for the interest granted to aneurism, in particular, and the choice carried to return it the determining reference frame of these effects; it returns, with the increased probability of attack by this disease attends so much favored by the conditions of surrounding current: stress associated permanently with the magnetic field reigning.

On the basis of these reports, the digital simulation will be carried out, by respecting the data of an incompressible medium and an animated blood flow of a speed of 0.035 m/s, and this, to make it possible to approach the maximum of a steady flow. Finally, Table 1, gathers the physical properties used in simulation and the found results are given hereafter [4-8].

Results and Analysis

Healthy state: Figure 2 represents the cartography of the lines speed, which announces a state of driven back flow, with a maximum speed which reduce gradually with 0.015 m/s, while moving away from its source. As for the spectral analysis (Figure 3), it reflects a signal

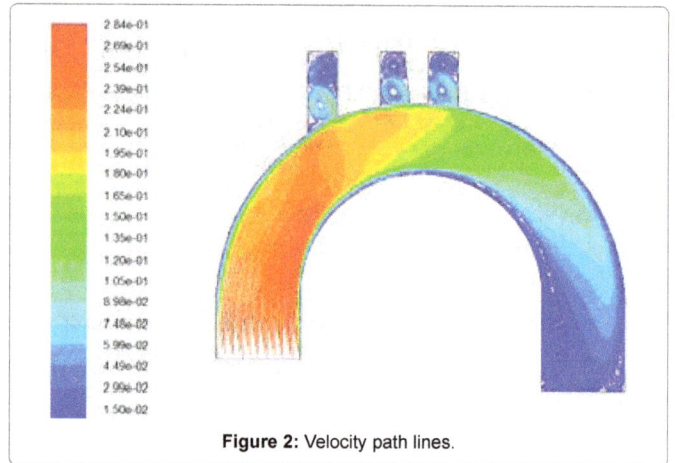

Property of the blood flow	Value
Density ρ	ρ=1010 kg/m³
Kinematic viscosity ν	ν=1.0×10⁻⁵ m²s⁻¹
Dynamic viscosity μ	μ=0.0036 Pa.s
Thermal conductivity k	k=0.50 Wm⁻¹K⁻¹
Heat capacity Cp	Cp=3600 J.kg⁻¹K⁻¹
Electric conductivity σ	σ=0.60 1/Ωm
Magnetic permeability μ_B	μ_B=0.0012 h/m

Table 1: Physical properties of the blood flow.

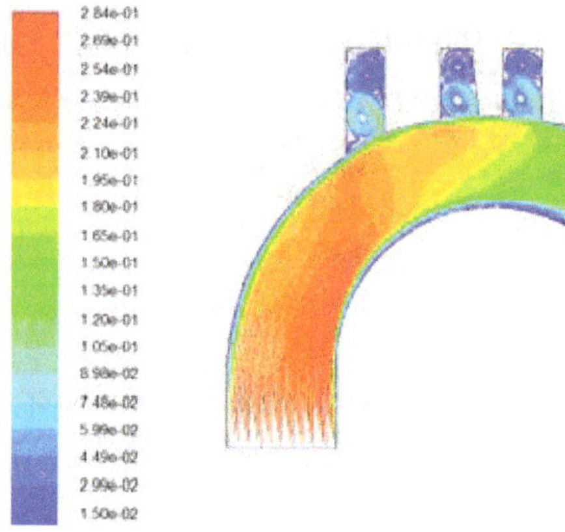

Figure 2: Velocity path lines.

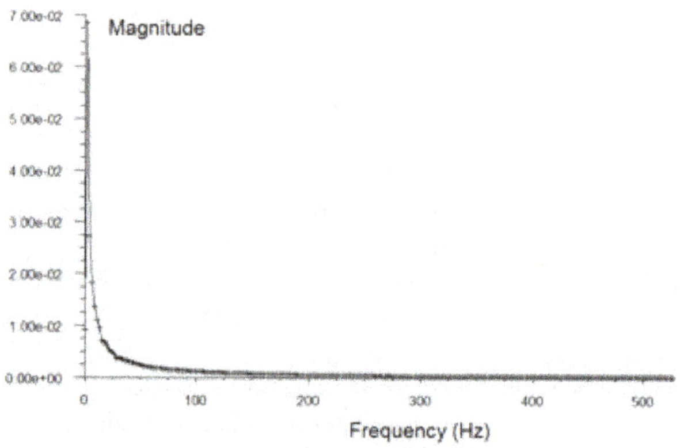

Figure 3: Velocity spectral.

evolving quickly to a mode of stable laminar flow after a peak of 0.07 of amplitude of the fundamental one.

Under these test conditions, the application of a magnetic field and the evaluation of its effects compared to the case of aneurism, are the single means of confirming the harmful effect of the latter. Also, reproduce the "Tao et Huang" test, by applying an inclined magnetic field of 1,3 Tesla, is the other approach which makes it possible to draw up a powerful comparative study.

Figures 4 and 5 are the results ensuing from this test. Its cartography shows clearly that the flow, always, takes the steady form, with almost the same data. But the Fourier spectral analysis, while illustrating, a steady behavior speed, shows a striking attenuation of the fundamental amplitude follow-up of a weak peak 0.003 disturber which can be at the origin of an incipient disorder. This result (Figure 5), can be returned to the gravitational effect and more precisely to the resultant of the two forces: like the force due to the component (B_y=1.3sin (π/4)), is in direct opposition of the force of gravity, a real physical report.

State of aneurismal disorder: In this test, only the effect of the beginning of aneurism was considered.

In the absence of magnetic field: The results (Figures 6 and 7) show clearly the appearance of a turbulent effect in the vicinity of the zone weakened by aneurism. Figure 6, announces an unsteady speed values from the beginning to the end of the sample, presenting turbulent rollers in the aneurismal zone, this report is marked by the succession of spectra (Figure 7) of almost identical size. As for the fundamental

Figure 4: Velocity pathlines.

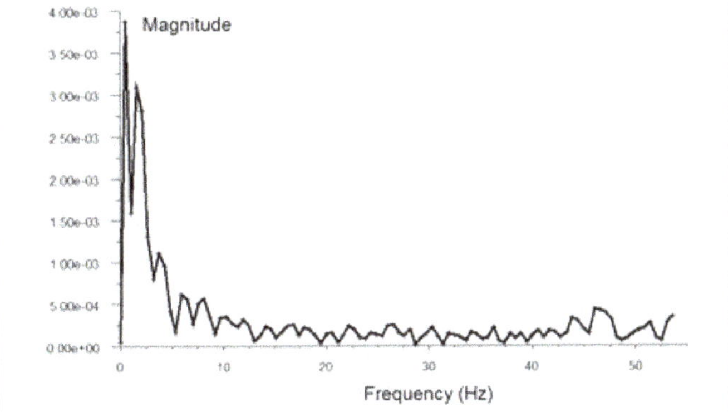

Figure 5: Velocity spectral analysis.

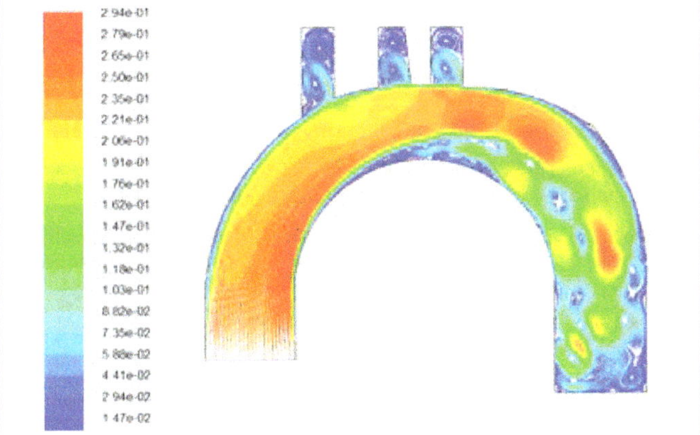

Figure 6: Velocity pathlines.

value, it is almost half than in the first case, which can be translated by the increase in the agitation of the molecules blood supporting, thus, its deceleration. The conclusion which we can draw here is that the rollers formation is the principal cause of the continuous increasing dilation of the aorta walls as the spectra values of the speed magnitude are about initial speed.

Application of a magnetic field: In this second case; we considered it useful to carry out several tests according to the possible orientations of the magnetic field and its intensity.

1er case Bx=0.02T: The application of a horizontal field Bx=0.02T, gave the following results:

Figure 8, shows that speed preserves the same unstable behavior, except that here the size of the roller formed in the aneurismal zone is larger than the preceding one.

As for Figures 8 and 9, we see that the fundamental one and the associated harmonics have negligible magnitudes compared to the preceding cases. By projecting this case on the disorders which can take place, we can say that the consequences of this last, according to a speed review of data, are the accentuated weakening of the aorta wall or the formation of deposits (thrombus) [9].

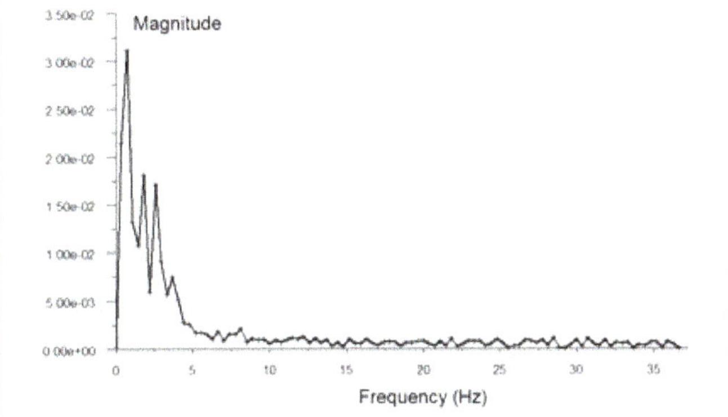

Figure 7: Velocity spectral analysis.

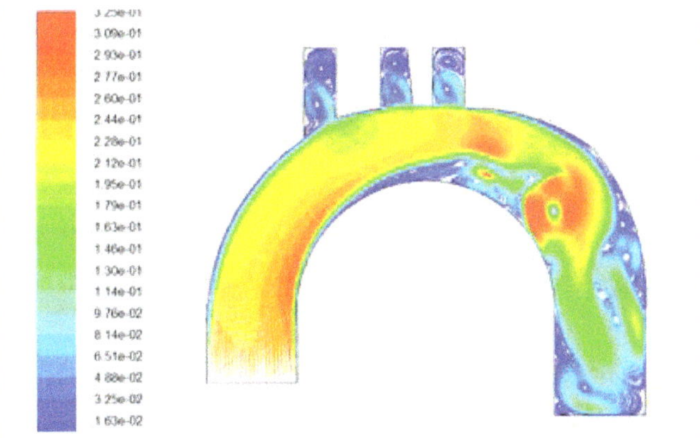

Figure 8: Velocity pathlines.

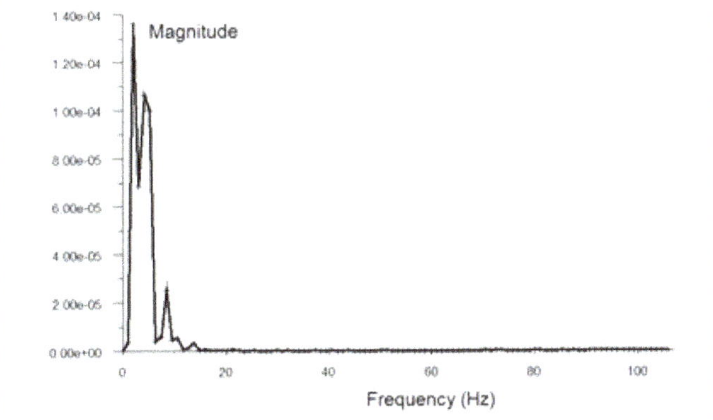

Figure 9: Velocity spectral analysis.

2nd Case: By=0.02T: The vertical orientation of the magnetic field may be considered as a particular case of this study, because, it is interested in the effect resulting from the gravitational force and the Lorentz one; on the blood flows displacement. The found results are given on Figures 10 and 11.

The interpretation of Figure 10, shows the same remarks of unsteady speed and of rollers formation in the aneurismal zone with small variations of form and at exist flow speed slightly higher than that of the top.

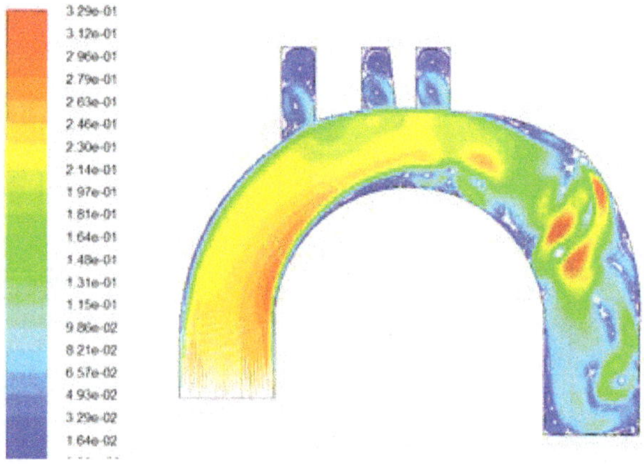

Figure 10: Velocity pathlines.

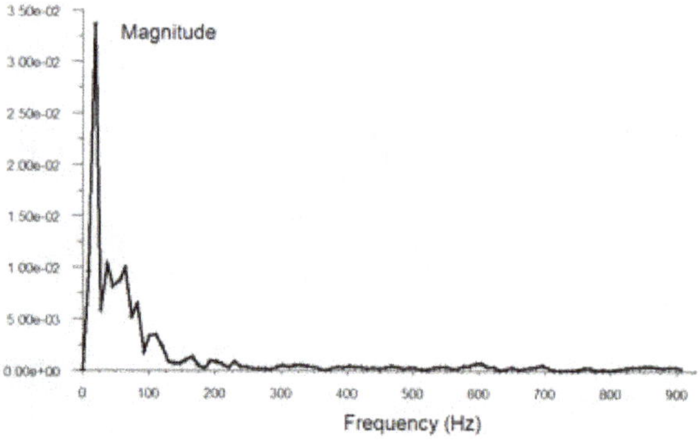

Figure 11: Velocity spectral analysis.

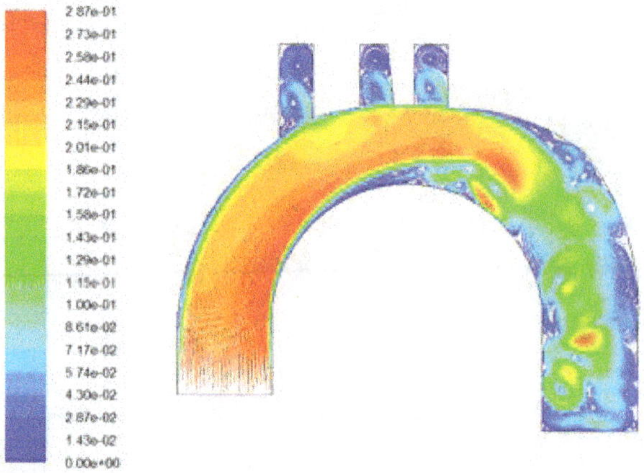

Figure 12: Velocity pathlines.

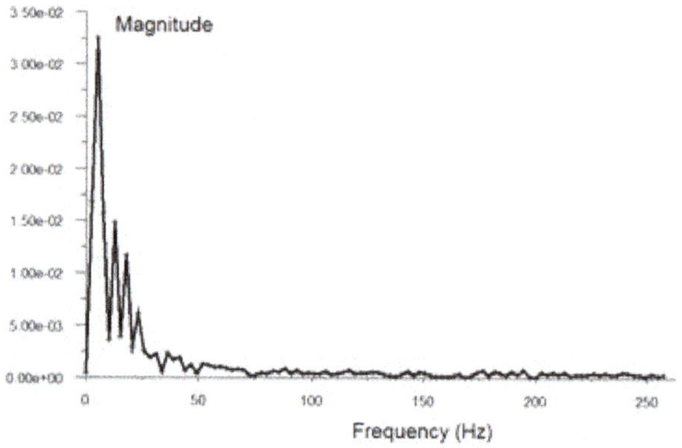

Figure 13: Velocity spectral analysis.

As for the spectral analysis, a return of fundamental towards the value of 0.035 is followed of weak harmonics which attenuate gradually [10].

3rd case: Bx=By=1.3T: The results consequential from this test show clearly that the application of the tilted field supports the progressive return towards steady speed (Figure 12) initially observed (Figure 5). But the presence of harmonic on the layout of Fourier (Figure 13), always, shows the adverse effect of the fluctuations even of low amplitudes [11,12].

Conclusion

At the end of this short presentation, we can say: even if the magneto therapy has some advantages to relieve certain diseases, its application must be limited and re-examined in order to avoid all kinds of complications for the patients suffering from serious diseases like the case from the aortic aneurism, which generally carries out towards unquestionable death, while passing by pains associated with the fluctuations generated by the new channels and the formed thrombi.

References

1. HAS (2006) Evaluation des endoprothèses dans le traitement des anévrismes et des dissections de l'aorte thoracique. Service évaluation en santé publique, Service évaluation économique février, France.
2. Item 131:Anévrismes (2010) Collège des Enseignants de Médecine vasculaire et Chirurgie vasculaire. Université Médicale Virtuelle Francophone.
3. Hoang LH (2007) Contribution à la modélisation tridimensionnelle des interactions champ électromagnétique- corps humain en basses fréquences. CCSD.
4. Connes P () La viscosité sanguine, Fiche outil, Rubrique: VII / Physiologie. Section: 3 / Biologie cellulaire, Numéro : A / Techniques.
5. MIA Cardiologie (2006) Item 134: L'anévrysme de l'aorte abdominale- Montpellier. Année Universitaire.
6. Lehéricy PS (2011) IRM: effets biologiques, implications de la Directive 2004/40 sur les champs électromagnétiques. SFRP.
7. Quemada D (1976) Hydrodynamique Sanguine: Hemorheologie et ecoulement du sang dans les petits vaisseaux. Hal Jpa-00210431.
8. Attar A, Nataf P (2010) Anevrismes de l'aorte thoracique ascendante. Cardiologie, Presse Med 39: 26-33.
9. Scorretti R (2003) Caractérisation numérique et expérimentale du champ magnétique B.F. généré pardes systèmes électrotechniques en vue de la modélisation des courants induits dans le corps humain. Autre. Ecole Centrale de Lyon, Français.
10. Bernard L (2007) Caractérisation électrique des tissus biologiques et calcul des phénomènes induits dans le corps humain par des champs électromagnétiques

de fréquence inférieur au GHz. Ecole Centrale de Lyon; Universidade federal de Minas Gerais, Français.

11. Scorretti R, Burais N (2010) Numerical dosimetry of currents induced in body by ELF magnetic fields. The International Journal for Computation and Mathematics in Electrical and Electronic Engineering.

12. Parker CB (1994) McGraw Hill Encyclopaedia of Physics.

Effect of Wind Energy Participation in AGC of Interconnected Multi-source Power Systems

Ibraheem Nasiruddin and Saab B Altamimi*

Department of Electrical Engineering, Engineering College, Qassim University, Kingdom of Saudi Arabia

Abstract

This paper presents the investigations on the effect of wind energy system's participation on dynamic stability margins available on AGC of interconnected power system. A two area power system model interconnected via EHV-AC tie-line is considered for the study. Each of the areas is consisting of hybrid sources of power generation like; hydro, thermal, gas and wind power plants. Various participation factors for electrical energy received from wind power plants, along with the energy from thermal, gas and hydro plants, are considered for the investigations. Moreover, any reduction of generation from thermal power plant is supposed to be supplied by wind power plant for fuel saving and to reduce emissions to environment. The optimal AGC regulators are designed using full state vector feedback control theory. Following the achievement of optimal gains of AGC regulators, the system closed-loop system eigenvalues are obtained for various case studies. The investigations of the closed loop eigenvalues carried out reveal that all the closed-loop eigenvalues are lying in the negative half of s-plan for all case studies and thus ensure the closed-loop system stability. Also, closed-loop eigenvalues are found to be sensitive to reduction in thermal generation and subsequent increase in electrical energy from wind power plants. It is also observed that the computed complex eigenvalues have shown a considerable decrease in the magnitude of its imaginary part when reduction of thermal generation is met by wind power generation. The reduced magnitudes of imaginary parts of closed-loop eigenvalues result in cost effective controller realization and improvement in system stability. On the other hand, the replacing the deficit caused in the supply with wind energy has no undesirable emissions to environment.

Keywords: Electrical energy; Automatic generation control; Wind energy; Solar energy; Eigenvalues

Introduction

The structure of today's power systems is huge and complex. A typical power system consists of large number of generators interconnected via networks of transmission lines, which provide power to consumers at nominal voltage and frequency. The maintenance of these parameters at the nominal values is necessary to achieve satisfactory operation of connected equipment with high efficiency and minimum wear and tear of the consumer equipment during their operation. Therefore, main parameters to be maintained properly are the system frequency and voltage profile. These parameters also responsible to dictate and determine the system stability and quality of the power supply. In a power system, frequency deviations are mainly due mismatch between real power generation and its demand, whereas voltage variations are function of reactive power imbalance in the system. The active and reactive power balance in the power system can be achieved by tracking the real and reactive power generation with continuously varying load demands. This can be done by designing and implementing effective schemes called automatic generation control (AGC) schemes. The control loops of these two parameters, assumed to be decoupled in nature and can be handled separately [1-3].

The modern power systems, from the operational and control point of view, are generally divided into control areas to form a coherent group of generators for sharing their technical, economic and operational benefits. Further to mitigate mismatch between generation and demand effectively and easily, these control areas are interconnected through tie lines for providing contractual exchange of power under normal operating conditions and even a quick assistance in emergency situations. Therefore, the control problem in power system is to maintain frequency and power exchanges between the areas at their rated values. The frequency deviation (Δ F) and tie-line flow deviation (ΔP_{tie}) can occur due to sudden area load changes. To minimize these deviations as soon as possible; a signal is generated by a linear combination of ΔF and ΔP_{tie}; known as area control error (ACE). Through the implementation of properly designed AGC schemes, the necessary change in generation is carried out by manipulation of speed changer of various generating units based on ACE minimization principles.

The fossil fuels such as coal, oil and natural gas, nuclear energy, hydro energy are commonly used energy sources at power plants for power generations. Since fossil fuels are depleting day by day, therefore, it is the need of the hour to go for non-conventional fuels like; solar energy, wind energy and many others for electricity generation. The power engineers have been engaging themselves for technology development in this area to harness electrical power from these fuels. The trend of adding a considerable amount of power from non-conventional energy sources is encouraging. With these developments, the control areas of power systems may supposed to have conventional and non-conventional sources of energy. The electricity generation scenario all over the world has different set of fuels exploited for electrical energy generation. In most of the countries, generally electricity is generated by hydro, thermal, gas and wind power plants. Among the nonconventional sources wind energy is considered to have a lion's share. Wind power is extracted from air flow using wind turbines for generating electricity. One of the major objectives

***Corresponding author:** Saab B Altamimi, Department of Electrical Engineering, Engineering College, Qassim University, POB 6677-51452, Kingdom of Saudi Arabia, E-mail: saab_tam@hotmail.com

achieved by harnessing electrical energy from wind is reduce eliminate the emissions associated with thermal power plants. However, the operation of wind power plants with conventional power plants has a system stability problems due to uncertainty in the flow of natural wind energy. In addition to this, the task of AGC is carry out by conventional sources power plants but disturbance can also be influenced by deviation in wind energy [4-6]. Therefore, system stability analysis of power systems with hybrid energy sources is of prime importance before proposing any AGC scheme for such systems. In this work, a comprehensive system stability analysis is carried out considering different combination of wind and thermal power plants. The power contribution from other power plants is considered as constant due to economic and environmental considerations.

Brief Literature Survey

In literature, there are large number of publications appeared by considering various aspects of design and implementation of AGC schemes [1,2,7-28]. Most of the works on AGC of power systems are reviewed comprehensively [4-6]. The first attempt in this area was aimed to control the frequency of a power system via the flywheel and speed governor of the prime-mover of the generator set. Immediately the scheme was noticed insufficient to meet the objectives of AGC and therefore, the technique was augmented with a supplementary control and adjoined to the speed governor of the prime mover with the combination of PI control strategy based on frequency deviation (Δf) signal [10,11,29]. Later, the conventional AGC schemes are described by Cohn [30]. The modern control concept based on optimal control theory was presented by Elgerd and Fosha [1,2] to design AGC regulators power systems. They suggested a PI structured full state feedback form of controller for developing optimal AGC regulators. The use of sate feedback control has a serious drawback that all states must be measured, therefore, an idea of sub-optimal control designs mooted to circumvent these drawbacks of optimal AGC regulators [12,13]. Usually, sub-optimal AGC regulators failed to provide the desired system dynamic performance. The problem was handled by researchers to design optimal AGC regulators by reconstructing the unavailable states from the available outputs and controls by using an observer [14,15]. Over the last two decades have seen the application of artificial intelligent techniques such as; Fuzzy Logic, Artificial Neural Networks, Genetic Algorithm, Particle Swarm Optimization, Bacteria Foraging and Hybrid Intelligent Techniques as powerful tools for designing of AGC regulators in power systems. Many studies exploiting artificial intelligent techniques for the design of AGC regulators in power systems considering various system aspects are appeared in references [16-31].

Most of these AGC studies have been carried out for interconnected power systems by considering single source of power generation in a control area [1,2,7-28]. However, in practical situations, a control area may comprise of a mix of hydro, thermal, gas and non-conventional energy sources based power plants. Only few studies on AGC of power systems considering multi sources power plants in a control area are appeared [4,5,31-33]. However, in these power system models, the dynamics of the wind power plants is missing [31-33]. Since, wind energy has been regarded as one of the most popular renewable energy sources for electricity generation, a due attention must be paid to consider its effect on dynamics of a control area having multi-sources for energy generation [6].

The optimal AGC regulator designs based on optimal control concept are simple to design, less costly and offer robust performance, therefore, in this paper, vector feedback control theory is adapted for designing and implementation of AGC schemes in power system model under consideration. The paper presents the design of full state feedback PI structured optimal AGC regulators for a 2-area interconnected power system consists of hydro, thermal, gas and wind power plants. Since wind power plants faced stability problems while operating with conventional power plants, therefore a comprehensive stability analysis is carried out by achieving various patterns of closed-loop system eigenvalues with the implementation of designed optimal AGC regulators. Various participation factors for wind energy are considered in overall power generation to meet the load demand on the system.

Power system model and case studies

A 2-area power system model consisting of hybrid source power plants with hydro, thermal, gas and wind turbines interconnected via EHV-AC tie line is selected for the study. Figure 1 represents the transfer function model of the system under consideration. The nomenclature and numerical data are given in reference [4].

In the power system model given above, all the thermal power plants are considered lumped together and represented by a single thermal plant dynamics. Similarly, hydro, gas and wind power plants are represented by respective single plant dynamics. The case studies identified for the study are given in Table 1. These case studies are identified based on different combination of sharing factor of power plants participating in AGC schemes. In this work, the reduction of generation from thermal power plants is considered to be supplied by wind power plants for saving of fuel and to reduce air pollution from thermal power plant.

Dynamic Modeling of Power System Under Investigation

The power system model under investigation is a linear continuous time-invariant system which can be represented by the following standard state space equations;

$$\frac{d}{dt}\underline{X} = A\underline{X} + B\underline{U} + \Gamma\underline{P_d} \tag{1}$$

$$\underline{Y} = C\underline{X} \tag{2}$$

Where, $\underline{X}$, $\underline{U}$, $\underline{P_d}$ and $\underline{Y}$ are state, control, disturbance and output vectors respectively. A, B, Γ and C are system, control, disturbance and output matrices of compatible dimensions. The matrices are developed based on system parameters and the operating point. The various state variables of power system model under investigation are described in Figure 1. For the power system model, defining the state variables as shown in transfer function model of the system as;

$x_1 = \Delta F_1$, $x_2 = \Delta P_{tie12}$, $x_3 = \Delta F_2$, $x_4 = \Delta P_{Gt1}$, $x_5 = \Delta P_{Rt1}$, $x_6 = \Delta X_{t1}$,

$x_7 = \Delta P_{Gh1}$, $x_8 = \Delta X_{h1}$, $x_9 = \Delta X_{RH1}$, $x_{10} = \Delta P_{Gg1}$, $x_{11} = \Delta P_{FC1}$, $x_{12} = \Delta P_{VP1}$,

$x_{13} = \Delta X_{g1}$, $x_{14} = \Delta P_{Gt2}$, $x_{15} = \Delta P_{Rt2}$, $x_{16} = \Delta X_{t2}$, $x_{17} = \Delta P_{Gh2}$, $x_{18} = \Delta X_{h2}$,

$x_{19} = \Delta X_{RH2}$, $x_{20} = \Delta P_{Gg2}$, $x_{21} = \Delta P_{FC2}$, $x_{22} = \Delta P_{VP2}$, $x_{23} = \Delta X_{g2}$, $x_{24} = \int ACE_1 dt$, $x_{18} = \Delta X_{h2}$

, $x_{19} = \Delta X_{RH2}$, $x_{20} = \Delta P_{Gg2}$, $x_{21} = \Delta P_{FC2}$, $x_{22} = \Delta P_{VP2}$, $x_{23} = \Delta X_{g2}$, $x_{24} = \int ACE_1 dt$,

$x_{25} = \int ACE_2 dt$, $x_{26} = \Delta P_{GW1}$, $x_{27} = \Delta P_{GW2}$

The system state, control and disturbance vectors for power system model under investigation are as;

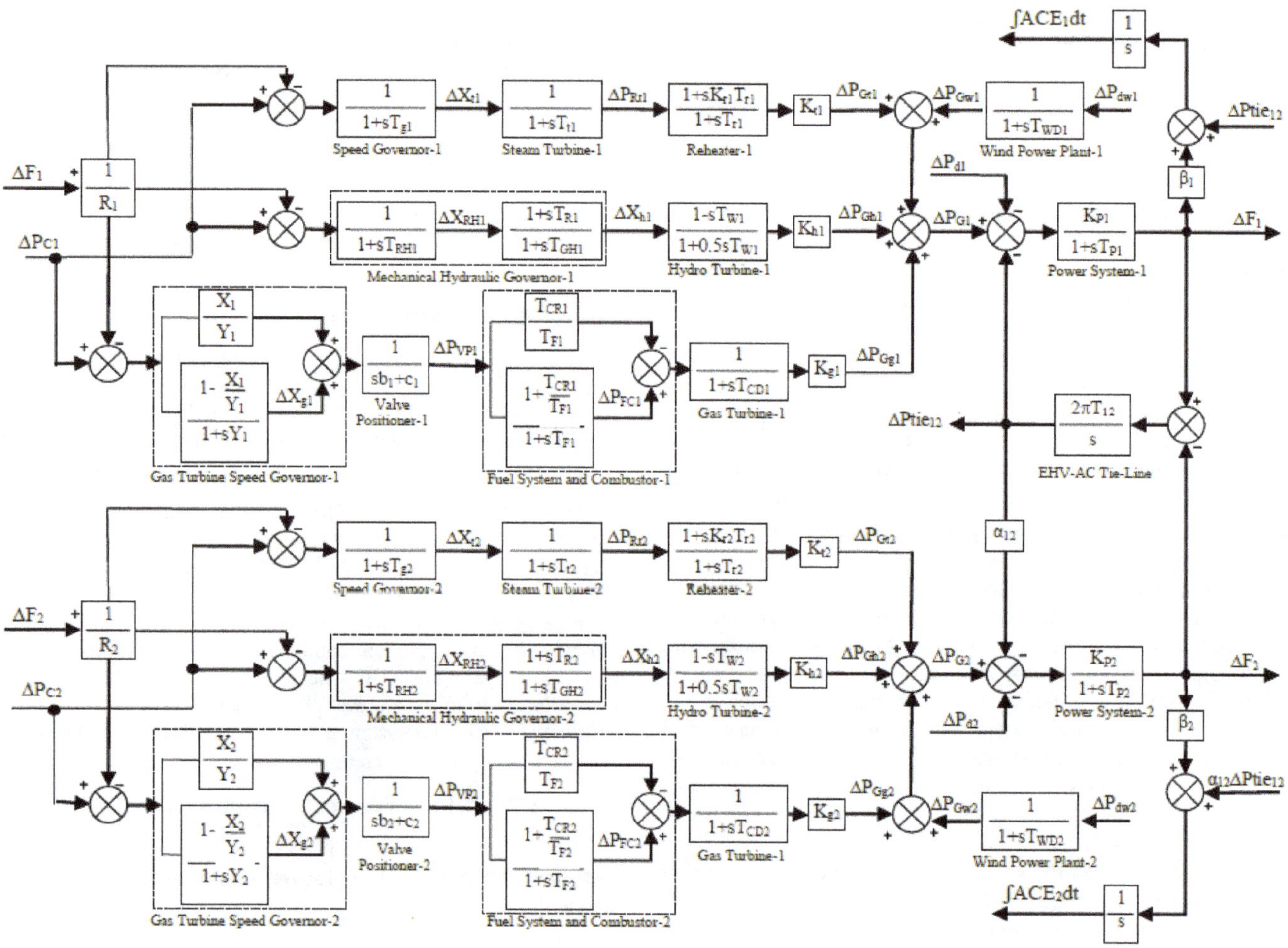

Figure 1: Transfer function model of power system under investigation.

➢ **State Vector**

$$[X]^T = \begin{bmatrix} x_1 & x_2 & x_3 & x_4 & x_5 & x_6 & x_7 & x_8 & x_9 & x_{10} \\ x_{11} & x_{12} & x_{13} & x_{14} & x_{15} & x_{16} & x_{17} & x_{18} & x_{19} & x_{20} \\ x_{21} & x_{22} & x_{23} & x_{24} & x_{25} & x_{26} & x_{27} \end{bmatrix}$$

Or

$$[X]^T = \begin{bmatrix} \Delta F_1 & \Delta P_{tie12} & \Delta F_2 & \Delta P_{Gt1} & \Delta P_{Rt1} & \Delta X_{t1} & \Delta P_{Gh1} & \Delta X_{h1} & \Delta X_{RH1} & \Delta P_{Gg1} \\ \Delta P_{FC1} & \Delta P_{VP1} & \Delta X_{g1} & \Delta P_{Gt2} & \Delta P_{Rt2} & \Delta X_{t2} & \Delta P_{Gh2} & \Delta X_{h2} & \Delta X_{RH2} & \Delta P_{Gg2} \\ \Delta P_{FC2} & \Delta P_{VP2} & \Delta X_{g2} & \int ACE_1 dt & \int ACE_2 dt & \Delta P_{Gw1} & \Delta P_{Gw2} \end{bmatrix} \quad (3)$$

➢ **Control Vector**

$$[\underline{U}]^T = [\Delta P_{C1}\ \Delta P_{C2}] \quad (4)$$

➢ **Disturbance Vector**

$$[\underline{P}_d]^T = [\Delta P_{d1}\ \Delta P_{d2}\ \Delta P_{dw1}\ \Delta P_{dw1}] \quad (5)$$

Dynamic equations

The following differential equations can be derived from transfer function model shown in Figure 1.

$$\frac{d}{dt}(x_1) = -\frac{1}{T_{P1}}x_1 - \frac{K_{P1}}{T_{P1}}x_2 + \frac{K_{P1}}{T_{P1}}x_4 + \frac{K_{P1}}{T_{P1}}x_7 + \frac{K_{P1}}{T_{P1}}x_{10} + \frac{K_{P1}}{T_{P1}}x_{26} - \frac{K_{P1}}{T_{P1}}\Delta P_{d1} \quad (6)$$

$$\frac{d}{dt}(x_2) = 2\pi T_{12}x_1 - 2\pi T_{12}x_3 \quad (7)$$

$$\frac{d}{dt}(x_3) = -\frac{\alpha_{12}K_{P2}}{T_{P2}}x_2 - \frac{1}{T_{P2}}x_3 + \frac{K_{P2}}{T_{P2}}x_{14} + \frac{K_{P2}}{T_{P2}}x_{17} + \frac{K_{P2}}{T_{P2}}x_{20} + \frac{K_{P2}}{T_{P2}}x_{27} - \frac{K_{P2}}{T_{P2}}\Delta P_{d2} \quad (8)$$

$$\frac{d}{dt}(x_4) = -\frac{1}{T_{r1}}x_4 + K_{t1}\left(\frac{1}{T_{r1}} - \frac{K_{r1}}{T_{t1}}\right)x_5 + \frac{K_{t1}K_{r1}}{T_{t1}}x_6 \quad (9)$$

$$\frac{d}{dt}(x_5) = -\frac{1}{T_{t1}}x_5 + \frac{1}{T_{t1}}x_6 \quad (10)$$

$$\frac{d}{dt}(x_6) = -\frac{1}{R_1 T_{g1}}x_1 - \frac{1}{T_{g1}}x_6 + \frac{1}{T_{g1}}\Delta P_{C1} \quad (11)$$

$$\frac{d}{dt}(x_7)=\frac{2K_{h1}T_{R1}}{T_{GH1}R_1T_{RH1}}x_1-\frac{2}{T_{w1}}x_7+(\frac{2K_{h1}}{T_{w1}}+\frac{2K_{h1}}{T_{GH1}})x_8+(\frac{2K_{h1}T_{R1}}{T_{GH1}T_{RH1}}-\frac{2K_{h1}}{T_{GH1}})x_9-\frac{2K_{h1}T_{R1}}{T_{GH1}T_{RH1}}\Delta P_{C1} \quad (12)$$

$$\frac{d}{dt}(x_8)=-\frac{T_{R1}}{T_{GH1}R_1T_{RH1}}x_1-\frac{1}{T_{GH1}}x_8+(\frac{1}{T_{GH1}}-\frac{T_{R1}}{T_{GH1}T_{RH1}})x_9+\frac{T_{R1}}{T_{GH1}T_{RH1}}\Delta P_{C1} \quad (13)$$

$$\frac{d}{dt}(x_9)=-\frac{1}{R_1T_{RH1}}x_1-\frac{1}{T_{RH1}}x_9+\frac{1}{T_{RH1}}\Delta P_{C1} \quad (14)$$

$$\frac{d}{dt}(x_{10})=-\frac{1}{T_{CD1}}x_{10}+\frac{K_{g1}}{T_{CD1}}x_{11}-\frac{K_{g1}T_{CR1}}{T_{F1}T_{CD1}}x_{12} \quad (15)$$

$$\frac{d}{dt}(x_{11})=-\frac{1}{T_{F1}}x_{11}+(\frac{1}{T_{F1}}+\frac{T_{CR1}}{T_{F1}^2})x_{12} \quad (16)$$

$$\frac{d}{dt}(x_{12})=-\frac{X_1}{b_1R_1Y_1}x_1-\frac{c_1}{b_1}x_{12}+\frac{1}{b_1}x_{13}+\frac{X_1}{b_1Y_1}\Delta P_{C1} \quad (17)$$

$$\frac{d}{dt}(x_{13})=(\frac{X_1}{R_1Y_1^2}-\frac{1}{R_1Y_1})x_1-\frac{1}{Y_1}x_{13}+(\frac{1}{Y_1}-\frac{X_1}{Y_1^2})\Delta P_{C1} \quad (18)$$

$$\frac{d}{dt}(x_{14})=-\frac{1}{T_{r2}}x_{14}+K_{t2}(\frac{1}{T_{r2}}-\frac{K_{r2}}{T_{t2}})x_{15}+\frac{K_{t2}K_{r2}}{T_{t2}}x_{16} \quad (19)$$

$$\frac{d}{dt}(x_{15})=-\frac{1}{T_{t2}}x_{15}+\frac{1}{T_{t2}}x_{16} \quad (20)$$

$$\frac{d}{dt}(x_{16})=-\frac{1}{R_2T_{g2}}x_3-\frac{1}{T_{g2}}x_{16}+\frac{1}{T_{g2}}\Delta P_{C2} \quad (21)$$

$$\frac{d}{dt}(x_{17})=-\frac{2K_{h2}T_{R2}}{T_{GH2}R_2T_{RH2}}x_3-\frac{2}{T_{w2}}x_{17}+(\frac{2K_{h2}}{T_{w2}}+\frac{2K_{h2}}{T_{GH2}})x_{18}+(\frac{2K_{h2}T_{R2}}{T_{GH2}T_{RH2}}-\frac{2K_{h2}}{T_{GH2}})x_{19}-2\frac{K_{h2}T_{R2}}{T_{GH2}T_{RH2}}\Delta P_{C2} \quad (22)$$

$$\frac{d}{dt}(x_{18})=-\frac{T_{R2}}{T_{GH2}R_2T_{RH2}}x_3-\frac{1}{T_{GH2}}x_{18}+(\frac{1}{T_{GH2}}-\frac{T_{R2}}{T_{GH2}T_{RH2}})x_{19}+\frac{T_{R2}}{T_{GH2}T_{RH2}}\Delta P_{C2} \quad (23)$$

$$\frac{d}{dt}(x_{19})=-\frac{1}{R_2T_{RH2}}x_3-\frac{1}{T_{RH2}}x_{19}+\frac{1}{T_{RH2}}\Delta P_{C2} \quad (24)$$

$$\frac{d}{dt}(x_{20})=-\frac{1}{T_{CD2}}x_{20}+\frac{K_{g2}}{T_{CD2}}x_{21}-\frac{K_{g2}T_{CR2}}{T_{F2}T_{CD2}}x_{22} \quad (25)$$

$$\frac{d}{dt}(x_{21})=-\frac{1}{T_{F2}}x_{21}+(\frac{1}{T_{F2}}+\frac{T_{CR2}}{T_{F2}^2})x_{22} \quad (26)$$

$$\frac{d}{dt}(x_{22})=-\frac{X_2}{b_2R_2Y_2}x_3-\frac{c_2}{b_2}x_{22}+\frac{1}{b_1}x_{23}+\frac{X_2}{b_2Y_2}\Delta P_{C2} \quad (27)$$

$$\frac{d}{dt}(x_{23})=(\frac{X_2}{R_2Y_2^2}-\frac{1}{R_2Y_2})x_3-\frac{1}{Y_2}x_{23}+(\frac{1}{Y_2}-\frac{X_2}{Y_2^2})\Delta P_{C2} \quad (28)$$

$$\frac{d}{dt}(x_{24})=\beta_1x_1+x_2 \quad (29)$$

$$\frac{d}{dt}(x_{25})=\alpha_{12}x_2+\beta_2x_3 \quad (30)$$

$$\frac{d}{dt}(x_{26})=-\frac{1}{T_{W1}}x_{26}+\frac{1}{T_{W1}}\Delta P_{dw1} \quad (31)$$

$$\frac{d}{dt}(x_{27})=-\frac{1}{T_{W2}}x_{27}+\frac{1}{T_{W2}}\Delta P_{dw2} \quad (32)$$

Using these differential equations and the numerical data given in [32], all the system matrices; [A], [B] and [Γ] can be obtained. With the help of these matrices, the optimal AGC regulators of the power system model under consideration are designed using full state vector feedback control strategy. The derivation of PI structured optimal AGC regulators is described in [33].

Simulation of results

The state space model developed in previous section is simulated on MATLAB platform for all case studies as given in Table 1. The optimal feedback gains and associated performance index (J) values are obtained for PI structured optimal AGC regulators for all case studies. These are given in Table 2. Using these optimal gains, case-wise closed-loop system eigenvalues are computed to investigate the closed-loop system stability. These closed-loop system eigenvalues are shown in Table 3.

Results and Discussion

The performance index values shown by Table 2 are indicative of the cost aspects of physical realization of AGC regulators of the power system. The lower value of performance index will result in the cheaper optimal AGC regulator design. From the inspection of Table 2, it can be revealed performance index value is not increased with a reduction in share of thermal generation which is supplied by wind power plant under all case studies. However, optimal gains of AGC regulators are sensitive to variation in sharing factors of wind power plants for all case studies.

The observation of optimal closed-loop eigenvalues shown in Table 3 shows that system is stable under all case studies. However, all the eigenvalues corresponding to state variables of both areas show no significant change in the magnitude of their real parts. The magnitudes of eigenvalues of state variables corresponding to Sr. Nos. (1-13) are increased considerably but opposite trend is seen in the magnitudes of eigenvalues of state variables corresponding to Sr. No. (14-25). Moreover, the magnitudes of eigenvalues corresponds to Sr. no. (26-27) have no considerable change. The complex eigenvalues have shown a considerable reduction in the magnitude of its imaginary part for case studies 1-4. The reduced magnitudes of imaginary parts of closed-loop eigenvalues result in fast and smooth decay of dynamic system response. This is an additional contribution provided to system dynamics when reduction in thermal generation is supplied by wind power plants.

Conclusions

This paper presents a comprehensive stability analysis based on closed-loop system eigenvalues of a 2-area interconnected power system consisting of hybrid sources of power generation in each area. To carry out stability analysis of power system model under investigation, the reduction of generation from thermal power plant is supplied by wind power plant for fuel saving and to reduce emissions from thermal plants. The optimal AGC regulators are designed using full state vector feedback control theory. Following the achievement of optimal gains of AGC regulators, the system closed-loop eigenvalues

Case Study No.	Sharing factor of various energy sources for a scheduled generation in area-1				Sharing factor of various energy sources for a scheduled generation in area-2			
	Kt1	Kh1	Kg1	Kw1	Kt2	Kh2	Kg2	Kw2
1	0.6	0.30	0.10	0.00	0.6	0.30	0.10	0.00
2	0.5	0.30	0.10	0.10	0.5	0.30	0.10	0.10
3	0.4	0.30	0.10	0.20	0.4	0.30	0.10	0.20
4	0.3	0.30	0.10	0.30	0.3	0.30	0.10	0.30

Table 1: Case Studies.

x	Optimal Gains	Performance Index (J)
Case Study-1	0.6571 -1.8591 0.1430 6.7836 -0.6908 0.4749 2.2330 1.4607 0.2166 0.1430 1.8591 0.6571 1.2217 -0.1847 0.0053 0.3074 0.3645 -0.1487	513.7227
	1.0207 0.2644 0.2882 0.3052 1.2217 -0.1847 0.0053 0.3074 0.3645 0.1110 0.0068 0.0001 -0.0286 6.7836 -0.6908 0.4749 2.2330 1.4607	
	-0.1487 0.1110 0.0068 0.0001 -0.0286 1.0000 -0.0000 5.9267 1.3380 0.2166 1.0207 0.2644 0.2882 0.3052 -0.0000 1.0000 1.3380 5.9267	
Case Study-2	0.6358 -1.9126 0.1659 7.2603 -0.6090 0.4645 2.2418 1.5203 0.3395 0.1659 1.9126 0.6358 1.4444 -0.1820 0.0050 0.3533 0.3988 -0.1430	608.0909
	1.0049 0.2673 0.2901 0.3409 1.4444 -0.1820 0.0050 0.3533 0.3988 0.1341 0.0087 0.0004 -0.0206 7.2603 -0.6090 0.4645 2.2418 1.5203	
	-0.1430 0.1341 0.0087 0.0004 -0.0206 1.0000 -0.0000 6.2827 1.5551 0.3395 1.0049 0.2673 0.2901 0.3409 0.0000 1.0000 1.5551 6.2827	
Case Study-3	0.6094 -1.9633 0.1937 7.8374 -0.5115 0.4538 2.2435 1.5838 0.5045 0.1937 1.9633 0.6094 1.7306 -0.1747 0.0045 0.4098 0.4385 -0.1308	741.8015
	0.9835 0.2699 0.2918 0.3807 1.7306 -0.1747 0.0045 0.4098 0.4385 0.1629 0.0111 0.0008 -0.0106 7.8374 -0.5115 0.4538 2.2435 1.5838	
	-0.1308 0.1629 0.0111 0.0008 -0.0106 1.0000 -0.0000 6.7036 1.8282 0.5045 0.9835 0.2699 0.2918 0.3807 -0.0000 1.0000 1.8282 6.7036	
Case Study-4	0.5761 -2.0071 0.2283 8.5530 -0.3927 0.4427 2.2344 1.6509 0.7325 0.2283 2.0071 0.5761 2.1096 -0.1605 0.0036 0.4809 0.4854 -0.1082	941.2288
	0.9544 0.2720 0.2935 0.4255 2.1096 -0.1605 0.0036 0.4809 0.4854 0.1996 0.0144 0.0014 0.0019 8.5530 -0.3927 0.4427 2.2344 1.6509	
	-0.1082 0.1996 0.0144 0.0014 0.0019 1.0000 0.0000 7.2107 2.1801 0.7325 0.9544 0.2720 0.2935 0.4255 0.0000 1.0000 2.1801 7.2107	

Table 2: Optimal Feedback Gains.

Sr. No.	Case Study-1	Case Study-2	Case Study-3	Case Study-4
1	-24.6963	-24.6947	-24.6931	-24.6914
2	-24.6962	-24.6947	-24.6931	-24.6914
3	-15.4985	-15.4933	-15.4879	-15.4823
4	-15.4978	-15.4927	-15.4874	-15.4819
5	-0.4175 + 1.9743i	-0.3709 + 1.9674i	-0.3258 + 1.9601i	-0.2827 + 1.9526i
6	-0.4175 - 1.9743i	-0.3709 - 1.9674i	-0.3258 - 1.9601i	-0.2827 - 1.9526i
7	-2.3003	-2.2823	-2.2649	-2.2479
8	-2.2129	-2.2041	-2.1953	-2.1866
9	-1.1605 + 0.7841i	-1.1189 + 0.7322i	-1.0795 + 0.6742i	-1.0428 + 0.6088i
10	-1.1605 - 0.7841i	-1.1189 - 0.7322i	-1.0795 - 0.6742i	-1.0428 - 0.6088i
11	-1.4949	-1.5114	-1.527	-1.5421
12	-0.1547 + 0.0937i	-0.1445 + 0.0875i	-0.1337 + 0.0801i	-0.1221 + 0.0711i
13	-0.1547 - 0.0937i	-0.1445 - 0.0875i	-0.1337 - 0.0801i	-0.1221 - 0.0711i
14	-5.4948	-5.4966	-5.4984	-5.5002
15	-5.4688	-5.4703	-5.4719	-5.4734
16	-4.1996	-4.2074	-4.2163	-4.2263
17	-4.0908	-4.0912	-4.093	-4.0973
18	-3.8327	-3.8757	-3.9147	-3.9488
19	-3.8131	-3.8361	-3.8572	-3.8763
20	-1.2014	-1.1972	-1.1937	-1.1907
21	-1.0945	-1.0988	-1.1035	-1.1086
22	-0.3357	-0.3287	-0.32	-0.3091

23	-0.2537	-0.2453	-0.2344	-0.2198
24	-0.0312	-0.0323	-0.0339	-0.0361
25	-0.0312	-0.0324	-0.0339	-0.0363
26	-0.2	-0.2	-0.2	-0.2
27	-0.2	-0.2	-0.2	-0.2

Table 3: Closed-loop System Eigenvalues.

are obtained for various case studies under investigation. It has been found that the closed-loop system stability is ensured under all operating conditions as identified in various case studies. The system closed-loop eigenvalues are found to be sensitive to reduction in thermal generation when this reduced generation is supplied by wind power plants. It is also observed that the complex eigenvalues have a considerable reduction in the magnitude of its imaginary part when there is a reduction of power generation from thermal plants and this power is supplied by wind power plants. The reduced magnitudes of imaginary parts of closed-loop eigenvalues result in fast and smooth decay of system dynamic response. The high participation factor of wind power plants is an additional merit to reduce the emissions from thermal plants to the environment.

References

1. Elgerd, Olle I, Fosha CE (1970) Optimum megawatt-frequency control of multi-area electric energy systems. IEEE Transactions on Power Apparatus and Systems PAS-89 4: 556-563.
2. Fosha CE, Olle EI (1970) The Megawatt Frequency Control Problem: A New Approach Via Optimal Control Theory. IEEE Transactions on Power Apparatus and Systems PAS-89 4: 563-577.
3. Kundur P (1994) Power System Stability and Control, McGraw-Hill: New York, USA.
4. Ibraheem, Nasiruddin, Bhatti TS, Hakimuddin N (2015) Automatic generation control in an interconnected power system incorporating diverse source power plants using bacteria foraging optimization technique. Electric Power Components and Systems 43: 189-199.
5. Nizamuddin (2013) Automatic Generation Control of Multi Area Power Systems with Hybrid Power Generation Sources, Ph. D Thesis, Jamia Millia Islamia, New Delhi, India.
6. Völler S, Al-Awaad AR, Verstege JF (2009) Wind farms with energy storages integrated at the control power market. CIGRE/IEEE PES Joint Symposium on Integration of wide-scale Renewable Resources into the Power Delivery System, Calgary, Paris pp: 1-13.
7. Ibraheem, Kumar P, Kothari DP (2005) Recent philosophies of automatic generation control strategies in power systems. IEEE Transactions on Power Systems 20: 346-357.
8. Kumar P, Ibraheem (1996) AGC strategies: a comprehensive review. Int Journal of Power and Energy Systems 16: 371-376.
9. Ibraheem, Kumar P, Khatoon S (2006) Overview of power system operational and control philosophies. Int Journal of Power and Energy Systems 26: 1-11.
10. Concordia C, Kirchmayer LK (1953) Tie line power and frequency control of electric power systems. AIEE Transactions on Power Apparatus and Systems 72: 562-572.
11. Concordia C, Kirchmayer LK (1954) Tie-line power and frequency control of electric power systems-Part III. AIEE Transactions on Power Apparatus and Systems 73: 133-146.
12. Nasiruddin I, Kumar P, Hasan N, Nizamuddin (2012) Sub-optimal AGC of interconnected power system using output vector feedback control strategy. Electric Power Components and Systems 40: 977-994.
13. Hasan N, Nasiruddin I, Kumar P, Nizamuddin (2012) Sub-optimal automatic generation control of interconnected power system using constrained feedback control strategy. Int Journal of Electrical Power & Energy Systems 43: 295-303.
14. Rubaai A, Udo V (1992) An Adaptive Control Scheme for Load Frequency Control of Multi Area Power Systems Part-I: Identification and functional design. Electric power Systems Research 24: 183-188.
15. Rubaai A, Udo V (1992) An adaptive control scheme for load frequency control of multiarea power systems Part-II: Implementation and test results by simulation. Electric power Systems Research 24: 189-197.
16. Wu QH, Hogg BW, Irwin GW (1992) A Neural Network Regulator for Turbogenerators. IEEE Transactions on Neural Networks 3: 95-100.
17. Douglas LD, Green TA, Kramer RA (1994) New approaches to the AGC nonconforming load problem. IEEE Transactions on Power Systems 9: 619-628.
18. Chaturvedi DK, Satsangi PS, Kalra PK (1999) Load frequency control: A generalized neural network approach. Int Journal of Electrical Power & Energy Systems 21: 405-415.
19. Zeynelgil HL, Demiroren A, Sengor NS (2002) The application of ANN technique to automatic generation control for multi-area power system. Int Journal of Electrical Power & Energy Systems 24: 345-354.
20. Saikia LC, Mishra S, Sinha N, Nanda J (2011) Automatic generation control of a multi area hydrothermal system using reinforced learning neural network controller. Int Journal of Electrical Power & Energy Systems 33: 1101-1108.
21. Oysal Y (2005) A comparative study of adaptive load frequency controller designs in a power systems with dynamic neural network models. Energy Conversion and Management 46: 2656-2668.
22. Talaq J, Al-Basri F (1999) Adaptive fuzzy gain scheduling for load-frequency control. IEEE Transactions on Power Systems 14: 145-150.
23. Kocaarslan I, Çam E (2005) Fuzzy Logic Controller in Interconnected Electrical Power Systems for Load-Frequency Control. Int Journal of Electrical Power & Energy Systems 27: 542-549.
24. Çam E (2007) Application of fuzzy logic for load frequency control of hydro electrical power plants. Energy Conversion and Management 48: 1281-1288.
25. Yeşil E, Güzelkaya M, Eksin I (2004) Self Tuning Fuzzy PID Type Load and Frequency Controller. Energy Conversion and Management 45: 377-390.
26. El-Sherbiny MK, El-Saady G, Yousef AM (2002) Efficient fuzzy logic load-frequency controller. Energy Conversion and Management 43: 1853-1863.
27. Abdel-Magid YL, Dawoud MM (1996) Optimal AGC Tuning with Genetic Algorithms. Electric Power Systems Research 38: 231-238.
28. Abdennour A (2002) Adaptive optimal gain scheduling for the load frequency control problem. Electric Power Components and Systems 30: 45-56.
29. Kirchmayer LK (1959) Economic Control of Interconnected Systems, New York: John Wiley & Sons.
30. Cohn N (1961) Control of Generation and Power Flow on Interconnected Systems, New York: Wiley.
31. Ramakrishna KSS, Sharma P, Bhatti TS (2010) Automatic generation control of interconnected power system with diverse sources of power generation. Int Journal of Engineering, Science and Technology 2: 51-65.
32. Ramakrishna KSS, Bhatti TS (2007) Sampled-data automatic load frequency control of a single area power system with multi-source power generation. Electric Power Components and Systems 35: 955-980.
33. Ibraheem, Nizamuddin, Bhatti TS (2014) AGC of Two Area Power System Interconnected by AC/DC Links with Diverse Sources in each Area. Int Journal of Electrical Power & Energy Systems 55: 297-304.

A Real time Alternative to the Hilbert Huang Transform Based on Internal Model Principle

Edris Mohsen*, Lyndon J Brown and Jie Chen

Department of Electrical and Computer Engineering Western University, London, Ontario, Canada

Abstract

This article presents a new tuning approach for an adaptive internal-model-principle based signal identification algorithm whose computational costs are low enough to allow a realtime implementation. The algorithm allows an instantaneous Fourier decomposition of non-stationary signals that have a strongly predictable component. The algorithm is implemented as a feedback loop resulting in a closed loop system with a frequency response of a bandpass filter with notches at the frequencies of the Fourier decomposition. This is achieved through real time selection of the coefficients of the transfer functions in the feedback loop. Previously these coefficients were selected by solving a large set of coupled linear equations. Rules for explicitly solving for these parameters are given that only involve evaluating frequency responses at the frequencies of the instantaneous Fourier decomposition. This allows realtime implementation on a low cost lap top with sampling rates up to 10 kHz.

Keywords: Internal model principle; Frequency identification; Adaptive multiple notch filters; Periodic disturbance; State variables; Bandpass filter; Instantaneous Fourier decomposition

Introduction

In this article, we are interested in the problem of identifying signals of the following form

$$d(t) = \sum_{i=1}^{n}\sum_{j=1}^{m_i} \bar{A}_{ij}(t)\sin\phi_{ij}(t) + n(t) \tag{1}$$

$$\phi_{ij}(t) = \int_0^t jw_i(t)dt + \phi_{ij}(0) \tag{2}$$

and n(t) is measurement noise. These are signals that are the sums of n periodic components with each component composed of m_i harmonics. The periods, the harmonic amplitudes and relative phases can vary slowly in time. By identification, we mean determining the values ω_i, $\bar{A}_{ij}$ and φ_{ij} - φ_{11}.

Several techniques have been developed in the literature to solve this problem. The most traditional technique is the fast Fourier transform. Newer techniques include wavelet analysis. These approaches suffer from not allowing continuous estimations of the frequencies and have difficult trade-offs between time and frequency resolutions. Other approaches are based on the use of adaptive notch filters [1] and output regulation [2]. A new approach that has been widely applied is the Hilbert Huang Transform (HHT) [3]. Control engineers treat similar problems where exact tracking of reference signals or rejection of disturbances is required. Approaches that accomplish this include repetitive controllers [4] and adaptive feed-forward cancellation (AFC) [5]. The repetitive controller is based on a fundamental control theory principle called the internal model principle (IMP). This principle was presented by Francis and Wonham and states that the output error can be driven asymptotically to zero by placing a model of exogenous signals in a stable feedback loop [6]. Unfortunately small errors in this model can lead to significant degradation in the performance of internal model principle controllers. This problem of uncertainty in the signal model can be overcome with adaptive controllers [7]. In achieving asymptotically perfect rejection of disturbances it is inherent that the disturbance is completely identified. Thus, these types of controllers can be turned into signal processing algorithms by replacing the process to be controlled with tuning functions [8].

Unfortunately, to successfully implement this algorithm requires being able to tune a stable feedback control loop for the entire range of possible frequencies in the model given by equation (1). Fortunately, it has been shown that in the signal processing framework, the simplest tuning solution, i.e. selecting all of the gains to be one, is guaranteed to be stable. This algorithm has been successfully applied to the problem of the repeatable disturbances seen in disk drive head control [9]. Unfortunately, by resorting to this simple tuning approach, there is no control over the dynamics and noise rejection characteristics of the algorithm.

When the frequencies are known a priori, the report [10] shows how the dynamics of the algorithm can be completely specified. Unfortunately this article requires solving a set more than $2n_t = 2\sum m_i$ coupled linear equations which are a function of the signal's frequencies. Unless the sample rate is less than 1Hz this will not be feasible to do each sample. This article shows how these parameters can be explicitly solved by simply evaluating some frequency response functions at certain frequencies.

In Section II, an instantaneous Fourier decomposition (IFD) algorithm [11] that is similar in approach to the HHT is presented. In Section III an updated formula for calculating the instantaneous frequencies are given. In Section IV, the new realtime tuned algorithm is presented. In Section V, the ability of the proposed algorithm to identify the periodic signal with uncertain frequencies is demonstrated. Conclusions are drawn in Section VI.

A preliminary version of this article was presented at the 30th annual IEEE Canadian Conference on Electrical and Computer Engineering (IEEE 2017 CCECE) in Windsor [12].

***Corresponding author:** Edris Mohsen, Department of Electrical and Computer Engineering Western University, London, Ontario, Canada
E-mail: emohsen2@uwo.ca

Adaptive Algorithm and Comparison to HHT

The HHT proceeds from the realization that the Hilbert transform gives a mathematically precise definition of instantaneous frequency that agrees with our intuitive understanding when applied to narrowband signals. In this narrowband case, the instantaneous frequency can be approximated as the derivative of the angle of the narrowband signal and $\sqrt{-1}$ times the quadrature of that signal where the quadrature can be approximated by either a scaled version of the derivative or integral of the signal. The HHT uses an empirical method to break down signals into narrowband signals. This empirical method is numerically intensive and not compatible with a realtime implementation.

Our algorithm uses the same approximations to estimate the instantaneous frequencies as the HHT but uses an alternative, notch filter based approach that simultaneous calculates the quadrature signals and decomposes the signal into narrow band signals. The structure of the adaptive instantaneous frequency decomposition is shown in Figure 1, where *G(s)* is a tuning function.

Each of the transfer functions $IM_{i,j}$ are an internal model for a sinusoid of frequency $j*\hat{\omega}_i$. When the model frequencies and the signal frequencies match, i.e., $\hat{\omega}_i = \omega_i$ and the closed loop system is stable, each u_{ij} will be a single sinusoidal and meet the HHT definition of an intrinsic function. The basic algorithm is the state space based implementation of the internal models given by

$$\dot{X}_{ij} = A_{ij}X_{ij} + \begin{bmatrix} K_{1ij} \\ K_{2ij} \end{bmatrix} e \tag{3}$$

$$u_{ij} = [01]X_{ij}$$

where $X_{ij}=[x_{1ij}\ x_{2ij}]^T$, i=1,2,……, n and j=1,2,……, m_i. A_{ij} is expressed as follows

$$A_{ij} = \begin{bmatrix} 0 & -j\hat{\omega}_i \\ j\hat{\omega}_i & 0 \end{bmatrix} \tag{4}$$

This is taken from [11] with minor modifications to fit the signal model that was given in equations (1) and (2). The gains K_{1ij}, K_{2ij} have been moved to the input vector from the output vector so that adjustments in their value do not directly change u, i.e., a bumpless transfer. Consequently, the responses at $x_{1ij}(t)$ *and* $x_{2ij}(t)$ in steady state are:

$$x_{1ij}(t) = \bar{A}_{ij}\cos(j\omega_i t + \phi_i) \tag{5}$$

$$x_{2ij}(t) = \bar{A}_{ij}\sin(j\omega_i t + \varphi_i) \tag{6}$$

i.e., the second state is the sinusoidal component of the original signal and the first state is its quadrature. While the states are time varying, when the signal parameters are time invariant $\bar{A}_{ij} = \sqrt{x_{1ij}^2(t) + x_{2ij}^2(t)}$ is time invariant as is $\varphi_i = \tan^{-1}(x_{2ij}(t)/x_{1ij}(t)) - j\omega_i t$

Frequency Estimation ($\hat{\omega}_i$)

Since the state variables x_{1i1} and x_{2i1} are orthogonal to each other then, as with the HHT, the derivative of the angle of $x_{1i1} + \sqrt{-1}x_{2i1}$ is ω_i. It can be shown that when $j\hat{\omega}_i \neq \omega_i$ then in steady state, without noise

$$\omega_i = \frac{d}{dt}\angle\left(x_{2i1} + \sqrt{-1}x_{1i1}\right)$$

$$\omega_i - \hat{\omega}_i = \frac{(K_{2i1}x_{1i1} - K_{1i1}x_{2i1})e}{x_{1i1}^2 + x_{2i1}^2} \tag{7}$$

Thus, using an integral controller

$$\dot{\hat{\omega}}_i = K_{ai}\frac{(K_{2i1}x_{1i1} - K_{1i1}x_{2i1})e}{x_{1i1}^2 + x_{2i1}^2}$$

can be used to update the frequency estimates.

Thus a quasi-periodic signal can be decomposed into a sum of narrow band signals, $\{u_{ij}\}=\{x_{2ij}\}$, and a real time Fourier representation of the reference can be obtained. The signal *u(t)* is the estimate of the signal of interest and can be represented by

$$u(t) = \sum_{i=1}^{n}\sum_{j=1}^{m_i} u_{ij}(t)$$

$$u_{ij}(t) = x_{2ij}(t) \tag{8}$$

$$\hat{A}_{ij} = \sqrt{\hat{x}_{1ij}^2(t) + \hat{x}_{2ij}^2(t)} \tag{9}$$

In ref. [11], it is establish for sufficiently small K_{ai} the algorithm is locally exponentially stable when *G(s)* and the K_{1ij}, K_{2ij} are chosen so that the feedback loop in Figure 1 is stable at each point in time. Designing these controller parameters is a challenging problem as it is assumed that there is limited knowledge about the $\{\omega_i\}$ and during transients there can be a significant difference between $\{\omega_i\}$ and $\{\hat{\omega}_i\}$.

Control Parameter Selection

Off-line tuning

As with ref. [10], we satisfy the above stability assumption by designing the closed loop system to incorporate a bandpass filter with notch filter. Let a 2nd order desirable bandpass filter be given by

$$T_{bp}(s) = \frac{d_1 s^2}{s^4 + c_1 s^3 + c_2 s^2 + c_3 s + c_4} \tag{10}$$

We choose the controller parameters to be such that the transfer function from d to e is

$$T_{de} = \frac{d_1 s^2}{s^4 + c_1 s^3 + c_2 s^2 + c_3 s + c_4} \times$$

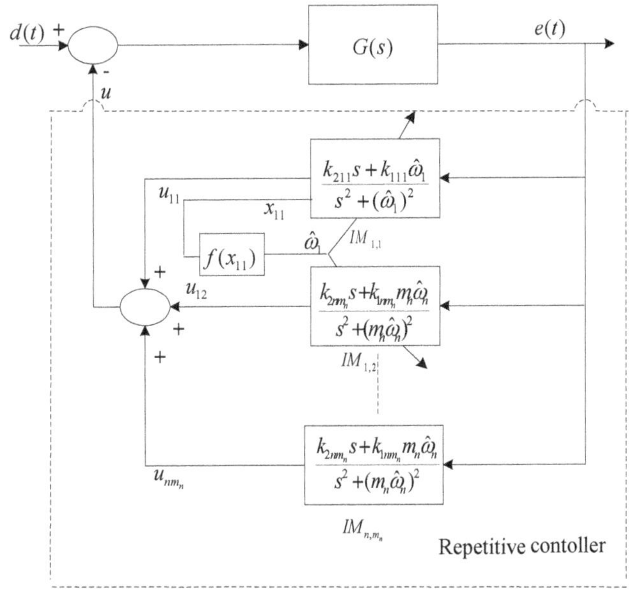

Figure 1: Structure of adaptive instantaneous frequency decomposition.

$$\prod \frac{s^2+(j\hat{\omega}_i)^2}{s^2+2\varepsilon_{ij}j\hat{\omega}_i s+(j\hat{\omega}_i)^2} \tag{11}$$

where ε_{ij} are small real numbers, and $j\hat{\omega}_i$ are the notches frequency. The presence of the numerator of the second term is a fundamental consequence of the internal model principle. Therefore, the ability of the algorithm to improve noise rejection is achieved.

An analysis of Figure 1 gives

$$T_{de} = \frac{G(s)}{1+G(s)\sum_{i=1}^{n}\sum_{j=1}^{m_i}\left(\frac{K_{2ij}s+K_{1ij}(j\hat{\omega}_i)}{s^2+(j\hat{\omega}_i)^2}\right)} = \frac{b_1 s^2 \Pi\left(s^2+(j\hat{\omega}_i)^2\right)}{a(s)\Pi\left(s^2+(j\hat{\omega}_i)^2\right)+b_1 s^2\sum\left(K_{2kl}s+K_{1kl}l\hat{\omega}_k\right)Y_{kl}(s)} \tag{12}$$

Where,

$$Y_{kl} = \prod_{i=1}^{n}\prod_{\substack{j\neq l\\ i=k}}\left(s^2+(j\hat{\omega}_i)^2\right)$$

And

$$G(s) = \frac{b_1 s^2}{s^4+a_1 s^3+a_2 s^2+a_3 s+a_4} = \frac{b_1 s^2}{a(s)} \tag{13}$$

Note in equations (11,12) Π represents $\prod_{i=1}^{n}\prod_{j=1}^{m_i}$ and Σ represents $\sum_{k=1}^{n}\sum_{l=1}^{m_k}$. The terms Y_{kl} are the product of all the terms $s^2+(j\hat{\omega}_i)^2$ except the $i=k$, $j=l$ term. Now, all the controller parameters can be calculated by matching the coefficients of numerators and denominators in equations (11) and (12). Note, the only controller parameters in the numerator is b_1 hence we get $b_1=d_1$. A unique solution for a_i, where $i=(1,2,\ldots,4)$ for the tuning function $G(s)$ and the feedback gains $\{K_{111}K_{211},\ldots,K_{1nm_n}K_{2nm_n}\}$ for each internal model can be derived from the denominator. Unfortunately we get a set of $2n_t+4$ coupled equations with $2n_t+4$ unknowns where $n_t=\sum_{i=1}^{n} m_i$, which is possible to solve off-line and/or theoretical but not practicable to solve in real time. The contribution of this article is to develop a less computationally intensive algorithm for calculating the controller parameter to meet the realtime requirement.

On-line frequency identification

Now the crucial question is how to choose $G(s)$ and K_{1jk}, K_{2jk} and implement the algorithm without needing to solve a set or $2n_t+4$ linear equations. It can be seen that all of the terms in the denominator except the term containing Y_{kl} will be zero if $s=\pm\sqrt{-1}l\hat{\omega}_k$. Thus when

$$s=\pm\sqrt{-1}l\hat{\omega}_k$$

$$b_1 s^2\left(K_{2kl}s+K_{1kl}l\hat{\omega}_k\right)Y_{kl}(s) = (s^4+c_1 s^3+c_2 s^2+c_3 s+c_4)\prod\prod\left(s^2+2\varepsilon_{ij}j\hat{\omega}_i s+(j\hat{\omega}_i)^2\right) \tag{14}$$

This generates 2 complex and complementary conjugate equations with 2 unknowns, i.e., the real part of either equation gives K_{1jk} and the imaginary gives K_{2jk}. The 4 a_i parameters can be explicitly solved by equating the coefficients of the degree 0, 1, $2n_t+2$, $2n_t+3$ terms of the denominator. Note the second term of the denominator of equation (12) contribute nothing to these four terms. These coefficients can be calculated by utilizing the relationships between the coefficients of a polynomial and the roots of a polynomial. We have that

$$\prod_{i=0}^{n_t}(s+r_i) = s^{n_t}\sum_i r_i s^{n_t-1}+\sum_i\sum_{j>i} r_i r_j s^{n_t-2}+\ldots+s\sum_i\prod_{j\neq i} r_j+\prod r_i$$

Extending this to the following product

$$\prod_{i=1}^{n_t}(s^2+2\varepsilon_i w_i+w_i^2) = s^{2n_t}+\sum 2\varepsilon_i w_i s^{2n_t-1}$$
$$+\left(\sum w_i^2+\sum_i\sum_{j>i}4\varepsilon_j w_j\varepsilon_i w_i\right)s^{2n_t-2}$$
$$+\cdots+s\sum_i 2\varepsilon_i w_i\prod_{j\neq i} w_j^2+\prod w_i^2$$

and equating the following sets $\{w_i\}=\{j\omega_i\}$ and $\{\varepsilon_i\}=\{\varepsilon_{ij}\}$, we get

$$a_1 = c_1\sum_i 2\varepsilon_i w_i$$

$$a_2 = \sum_{i=1}^{n_t}\sum_{j>i}4\varepsilon_j w_j\varepsilon_i w_i + c_1\sum_{i=1}^{n_t}2\varepsilon_i w_i + c_2$$

$$a_3 = c_3 + c_4\sum_{i=1}^{n_t}2\varepsilon_i / w_i$$

$$a_4 = c_4$$

Linear dependency of equations: When $j\hat{\omega}_i = l\hat{\omega}_k$ when $i\neq k$ then the equations to be solved become linearly dependent. With our solution technique this is reflected in the fact that the denominator of equation (12) will be zero when we substitute in $s=\pm\sqrt{-1}l\hat{\omega}_k$ and it will not be possible to calculate two pairs of internal model gains. Further, while it is theoretical possible to solve when the frequencies are extremely close, we get solutions that lead to unstable results because of numerical stability issues. To solve this problem, while calculating the controller gains, we drop the approximately redundant internal model when the frequencies become close, i.e., within 0.1%. After calculating the controller gains, the two redundant models are each assigned half of the gain. That is when $j\omega_i = l\omega_k$, we drop Internal model $IM_{l,k}$ from the design stage. Let $\bar{K}_{1ij}$ and $\bar{K}_{2ij}$ be the calculated controllers gains. Then $K_{1ij}=K_{1kl}=0.5\bar{K}_{1ij}$ and $K_{2ij}=K_{2kl}=0.5\bar{K}_{2ij}$. It should be noted that the threshold for HHT to distinguish between close frequencies is 10%.

Simulation Results

In this particular section, the effectiveness of our real time implementation of our proposed adaptive algorithm is verified via simulation. The model configuration parameters that are used with the matlab/simulink (R2016) environment are as follows: Solver ode5 (Dormand-prince) selection with fundamental sample time is 0.0025 s. Therefore, the sampling rate in our case is selected to be 400Hz, then the Nyquist frequency is 200 Hz. The code generation with C language

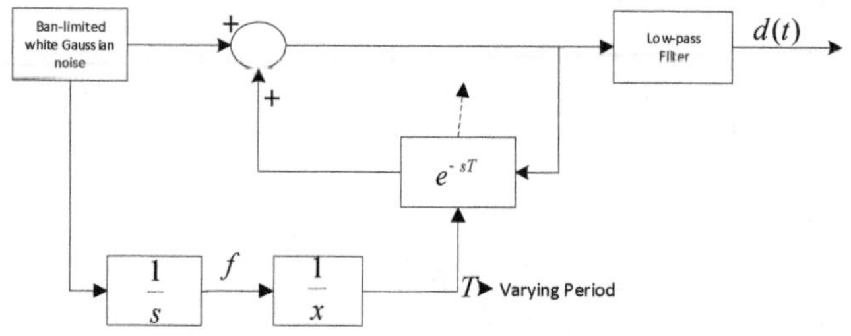

Figure 2: Structure of periodic signals generator.

and tool chain (Microsoft visual C++ 2012 V11.1 n-make 164-bit windows). All random numbers were zero mean.

Our signal to be identified was produced by summing the outputs of two copies of the model shown in Figure 2. The feedback loop containing the pure delay is called a repetitive controller and is capable of producing any periodic disturbance with period T. The value T was an integrated band limited white disturbance. The frequency cutoff of this noise was 20 rad/s and the variance was 0.5. The initial conditions for both fundamental frequencies are 4.2 and 5 Hz. The disturbance input to the repetitive controller causes the amplitudes and relative phases to vary slowly with time as well. This random signal was band limited to 50 Hz and had variance 0.1. Additional measurement noise was added to the sums of these two signals. This noise was band limited to 50 Hz and had a variance of 0.1. The low pass filter had a cutoff frequency of 100 rad/s concentrating the energy in the harmonics to below the 4th and third harmonic, respectively though signal was present in all harmonics up to the Nyquist frequency.

The frequency adaption gains were chosen as K_a=1.95 or with frequency 7.5% to 10% of the fundamental frequencies (Table 1). The closed loop transfer function was chosen to be a second order Chebyshev band-pass filter with 1 dB band-pass ripple, and low and high band-pass frequencies are 1 and 50 Hz, respectively. So the bandpass filter transfer function is given by

$$T_{bp}(s) = \frac{3.262.10^4 s^2}{s^4 + 200s^3 + 3.897.10^4 s^2 + 2.369.10^5 s + 1.403.10^6} \tag{15}$$

$d1 = b1 \times 10^{04}$	$c1$	$c2 \times 10^{04}$	$c3 \times 10^{05}$	$c4 \times 10^{06}$
3.2624	200	3.897	2.369	1.403

Table 1: Bandpass filter parameters.

$b1 \times 10^{04}$	$a1$	$a2 \times 10^{04}$	$a3 \times 10^{05}$	$a4 \times 10^{06}$
3.2624	373.4367	8.6845	2.8156	1.4027

Table 2: Values of simple tuning function *G(S)* (b_1; a_1; a_2; a_3 and a_4).

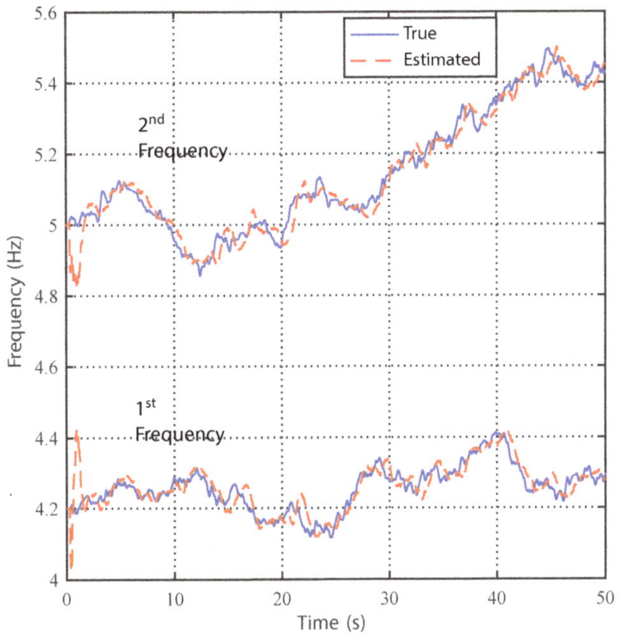

Figure 3: Frequency identification for both fundamental components.

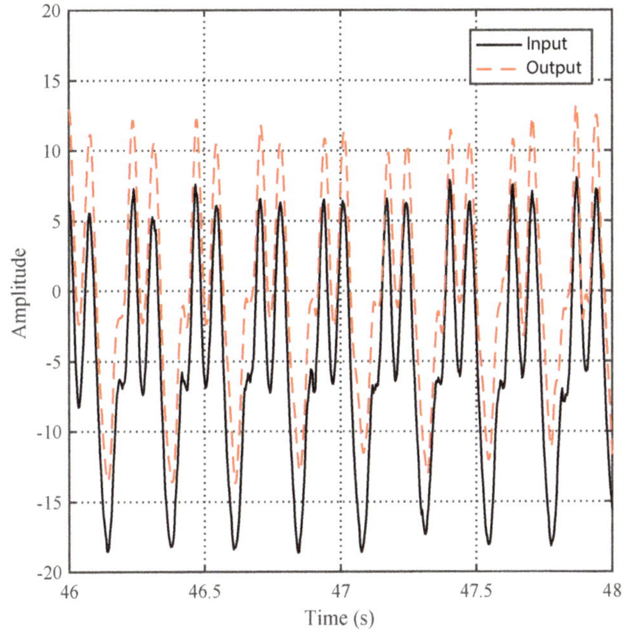

Figure 4: Identified versus true signal for the first set of the internal model.

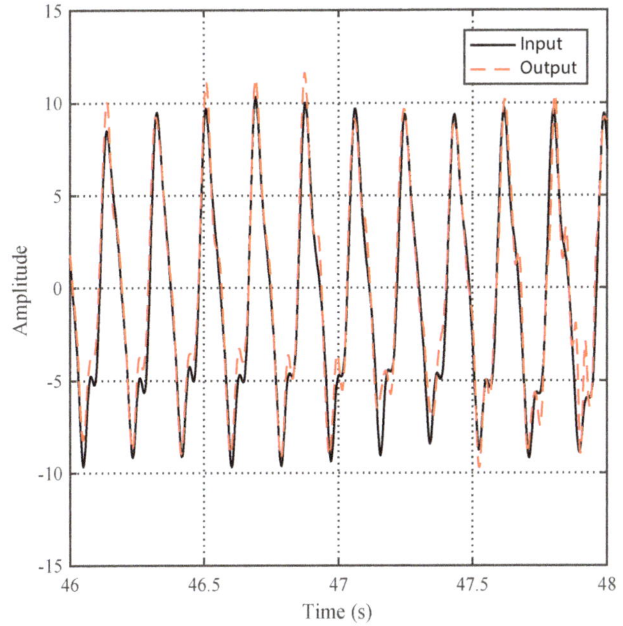

Figure 5: Identified versus true signal for the second set of the internal model.

For $\hat{\omega}_1 = 4.2 * 2 * \pi$ and $\hat{\omega}_2 = 5 * 2 * \pi$ and ε_{ij} are small real numbers $\varepsilon_{ij} = 0.1$ the coefficients of the simple tuning function are given in the Table 2.

Under these conditions, a 50 s Matlab simulation could be performed in under 5 s. The identified frequencies are shown in Figure 3. We can see good identification and tracking of the fundamental frequencies. Figures 4 and 5 show a close up of the actual outputs of the signal generators and the identified signals (very good tracking of amplitude and relative phases). The first component has significant DC which we have not attempted to identify. In particular, there is no

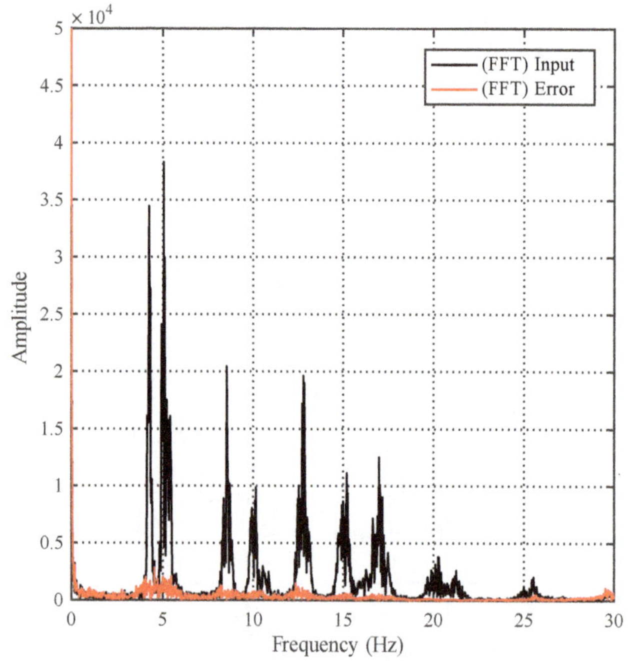

Figure 6: Fast Fourier transform of the input signal and error.

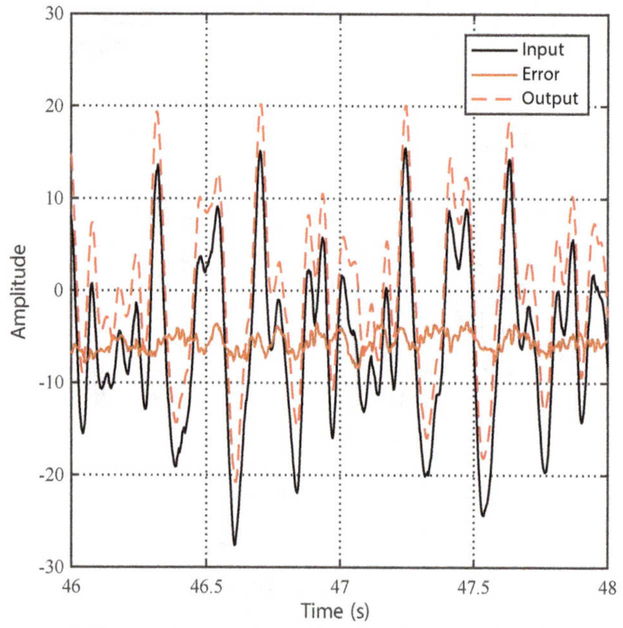

Figure 7: Comparison of identified signal and real signal.

way to distinguish and hence identify the DC content of the two true signals. Again we show good matches and thus we are able to identify these periodic signals in real time.

To get an overview of the signal frequency content and the accuracy of the identified models, the FFT transforms of the signal to be identified and the error signal are shown in Figure 6. It can be seen that most of harmonics have been identified although there is a huge DC component in both the signal and (*e*) in the proposed algorithm, which is as anticipated as we did not attempt to identify it.

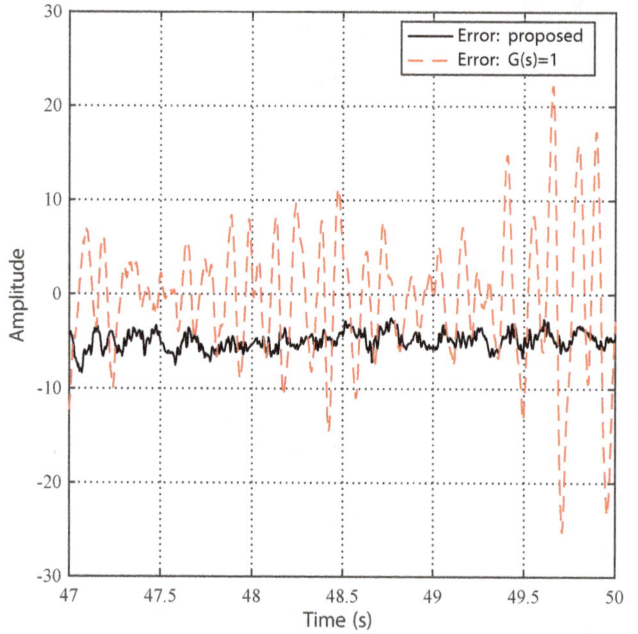

Figure 8: Error signals for proposed tuning versus simple tuning.

Figure 7 displays the quasi-periodic signal *d(t)*, the identified signal *y(t)* and their difference *e(t)*. After a brief transient we see that e becomes quite small. Note at 42 s the 5th harmonic of the 1st signal and the 4th harmonic of the second signal both had frequencies of 21.61 Hz. When a threshold of 0.01% was chosen for eliminating the redundant internal model, the algorithm went unstable. At the threshold of 0.1% there was a brief loss (<0.1 s) of performance in the signal estimation. At a threshold of 0.5% there was no noticeable loss in quality of the signal estimation.

It can be shown that the choice for controller parameters of *G(s)*=1, K_{1ij}=0 and K_{2ij}=1 always results in a stable feedback loop for any possible values of $\hat{\omega}_i$. Unfortunately, this leaves the dynamics of the closed loop system uncontrollable and uncertain which may require more conservative selections of the adaption gain and increase amplification of measurement noise. Figure 8 demonstrates the low performance of the simply tuned algorithm compared with the proposed algorithm. This simple approach resulted in much longer initial transient response (not shown). The steady state error was about 10 times larger in magnitude.

Acknowledgment

The authors would like to thank Libyan government and Libyan Ministry of higher education and scientific research for their fund and support for his research despite the hard environment that they work at due to unstable situation in Libya. In addition we would like to acknowledge the research funding provided by Western University.

Conclusion and the Future Work

This article has shown the instantaneous Fourier decomposition algorithm that is based on the orthogonal state variables of an internal model principle controller. First we examined how we implemented this algorithm off-line by matching the coefficients of numerators and denominators in both equations (11,12). Second, and the main contribution in this article is to develop a means of calculating the controller parameters that has a lower computational burden such that it can be successfully implemented in realtime. As a result of our work in section IV-B, the schema has been successfully implemented online after solving for the issue of overlapping harmonics from different signal components. One of the models is removed from the design process eliminating the dependent equations. The associated controller gains are distributed equally in the implemented controller.

Thus, the instantaneous Fourier decomposition is implemented in real time, the frequency is identified with high speed of convergence and the predictable disturbance is identified as well as the system stability is guaranteed.

Our future work will be conducted to identify the uncertain frequencies of periodic signals and eliminate periodic disturbances in discrete state space form.

References

1. Regalia PA (1991) An improved lattice-based adaptive IIR notch filter. IEEE Transactions on Signal Processing 39: 2124-2128.
2. Kim H, Shim H, Jo NH (2014) Adaptive add-on output regulator for rejection of sinusoidal disturbances and application to optical disc drives. IEEE Transactions on Industrial Electronics 61: 5490-5499.
3. Huang NE, Shen Z, Long SR, Wu MC, Shih HH (1998) The empirical mode decomposition and the hilbert spectrum for nonlinear and non-stationary time series analysis. Proceedings of the Royal Society A 454: 903-995.
4. Hara S, Yamamoto Y, Omata T, Nakano M (1988) Repetitive control system: a new type servo system for periodic exogenous signals. IEEE Transactions on Automatic Control 33: 659-668.
5. Bodson M, Sacks A, Khosla P (1992) Harmonic generation in adaptive feed forward cancellation schemes. Proceedings of the 31st IEEE Conference on Decision and Control 2: 1261-1266.
6. Francist BA, Wonham WM (1976) The internal model principle of control theory. Elsevier 12: 457-465.
7. Brown LJ, Zhang Q (2001) Control for canceling periodic disturbances with uncertain frequency. Proceedings of the 40th IEEE Conference on Decision and Control 5: 4909-4914.
8. Kim W, Kim H, Chung CC, Tomizuka M (2011) Adaptive output regulation for the rejection of a periodic disturbance with an unknown frequency. IEEE Transactions on Control Systems Technology 19: 1296-1304.
9. Nagashima M, Usui K, Kobayashi M (2007) Rejection of unknown periodic disturbances in magnetic hard disk drives. IEEE Transactions on Magnetics 43: 3774-3778.
10. Zhang Q, Brown LJ (2003) Designing of adaptive bandpass filter with adjustable notch for frequency demodulation. Proceedings of the American Control Conference 4: 2931-2936.
11. Brown LJ, Zhang Q (2003) Identification of periodic signals with uncertain frequency. IEEE Transactions on Signal Processing 51: 1538-1545.
12. Mohsen E, Brown LJ (2017) Realtime implementation of an internal-model-principle signal identifier. IEEE 30th Canadian Conference on Electrical and Computer Engineering (CCECE).

24

Construction of Information Ecosystem on Enterprise Information Portal (EIP)

Allam Maalla*

School of Information Technology and Engineering, Guangzhou College of Commerce, Guangzhou, China

Abstract

Enterprise Information Portal as a unified display platform of Enterprise Information has the capability of integration and display with all kinds of business system. Through bottom layer supporting lots of agreement can realize interconnection of all the business system. The paper takes Life ray EE portal as the Enterprise Information Portal platform, SOA framework and Web Service as business mode. The research on target, framework and key technology of construction of information ecosystem on Enterprise Information Portal realize closed loop of information ecosystem promoting the development of all kinds of enterprise system.

Keywords: Life ray portal; Single sign-on; Virtual portal; Enterprise information ecosystem

Introduction

As the high development of computer and network technology, enterprise Information control and integrated manage the information in manufacturing management activities through network and database technology on depth and breadth to realize the effective use of enterprise internal information, which promote the whole management of enterprise information and sustainable development. However, with the excellent of all kinds of information system on enterprise, new problem come out:

a) The disunion of all the information system increased the cost on use and maintain;

b) The disconnect and non-shared of information resource increased the cost on data consistency;

c) The user can only look up the content, and not customize the content according to the demand; meantime lacking of the single sign-on function made the user sign on the repeated system many times, which limited the work efficiency and influenced the safety of the system.

d) How to integrate the different information system built by different stages, by united interface to log in and out became the main problem of information system development.

As a result, the above problem made how to build Enterprise Information Ecosystem becoming the main problem of information system development. The aim is to realize connection, shared and innovation of enterprise information. Building ecological and balanced information environment and maximum use of Enterprise Information promoted evolution and development of enterprise [1].

The Summary of the Enterprise Information Portal

The paper takes Life ray EE portal as the Enterprise Information Portal platform, SOA framework and Web Service as business mode. Enterprise Portal is the display, publish and management of enterprise basic platform. To solve above problem, Enterprise Portal is taken as the interface of business information acquired and handled. Enterprise Information Portal based on the polymerization of component-based development and deployment of enterprise application system, and connection and shared of information to realize the closed loop of Information Ecosystem. Enterprise Portal as the interface of business information acquired and handled, according to the different of enterprise deployment, the main portal connect information by cascaded and multi-application of portal was built by virtual portal.

According to framework of total enterprise portal, total technology framework of enterprise portal had database, Middleware, business processing, page display and data exchange. Multi-portal application management access business application system and function between different applications in the enterprise portal by configuration, function of business application system was accessed on the framework platform, which can be optional customized. The authority function was visited through enterprise portal, and it was no use to deal with something switching frequently in all kinds of application system [2].

Technical Framework of the EIP

The technology of Enterprise information portal almost includes all the Web technology. For framework of the EIP, total technology framework of enterprise portal had database, Middleware, business processing, page display and data exchange. From logic layer, the whole portal platform can be divided as: client, Presentation Layer, Business Logic Processing Layer, Data storage, Middleware, OS. Every logic layer has their different duty.

a) **Client:** Provide access to portal system for user and technology support for Presentation Layer. Support the normal browser, including IE, Firefox, and Chrome. Portal tray provides entrance to Enterprise information portal for user, new message display and affairs be deal with. It reminds user to look up latest message by the way of pupping message box. Meanwhile, it has the function of integrated address book, Enterprise Search, RSS Subscribers. Channel technology support for Enterprise information portal is provided by Mobile client.

***Corresponding author:** Allam Maalla, School of Information Technology and Engineering, Guangzhou College of Commerce, Guangzhou, China
E-mail: allammaalla@yahoo.com

b) **Presentation layer:** The system follows J2EE and portal technology specification, and considers the distributed technology, grid computing technology, information integration technology. The system follows W3C, CSS, JSR268, WSRP2.0 standard; related portal application in different portal framework is seamless transplant. It follows the integrated standard of enterprise information mode, and avoids the difference brought by different information system.

c) **Business logic processing layer:** It includes the center business application of enterprise information portal, and center function like content management, ESNS Collaboration Application, enterprise application, integrated application, personal work station, enterprise search, grid management, security authentication.

d) **Data storage:** It service for enterprise information portal by data storage. According to the data type, storage is divided as structured and non-structured data storage.

e) **Middleware:** It provides environment for enterprise information portal application.

IT infrastructure service layer: It provides a stable security environment for the server operating system (Figure 1).

Physical model

The function of application execution is realized by services from all kinds of software, including container type software (Web service, application servicer software), platform type software (directory service software, development platform), and Runtime software (JVM, CLR). So, define the software environment when application executing. Suggest define the software environment from 4 levels like application display, application services, integration services and general services [3]. As shown in Figures 2 and 3 below:

Main Function of System

Unified display portal is personalized content display platform which is achieved by internal users through user authentication. Content and processing services can be obtained in the portal through the access control, single point technology, user authentication. Daily office work function module and data for users can be integrated by enterprise portal, like business system to do work, function operation of business system, data permissions, email and SMS platform data, schedule management.

Users according to their own habits, including news subscription and index subscription, personalize the content, style show and custom function of portal. Personalized applications show more characteristics of enterprise information portal meantime improve the user's working efficiency, reduce the operation cost of enterprises.

Application function of system

Portal application is built as a multi node by grid technology and formed a grid enterprise information portal, which can realize unified monitoring and management of the application and resource. The system is based on the concept of hierarchical information. Through the integration of C/S, B/S architecture advantages to establish "information tray (C/S) + personal work station (B/S) + enterprise portal (B/S)" application system (Figure 3), a hierarchical integration platform can provide a single point login entrance, unite abeyance, enterprise search, SMS, KPI index center system the function of entrance and other functions. General layered application is as follows:

a) **Unified display:** Display function is mainly divided into

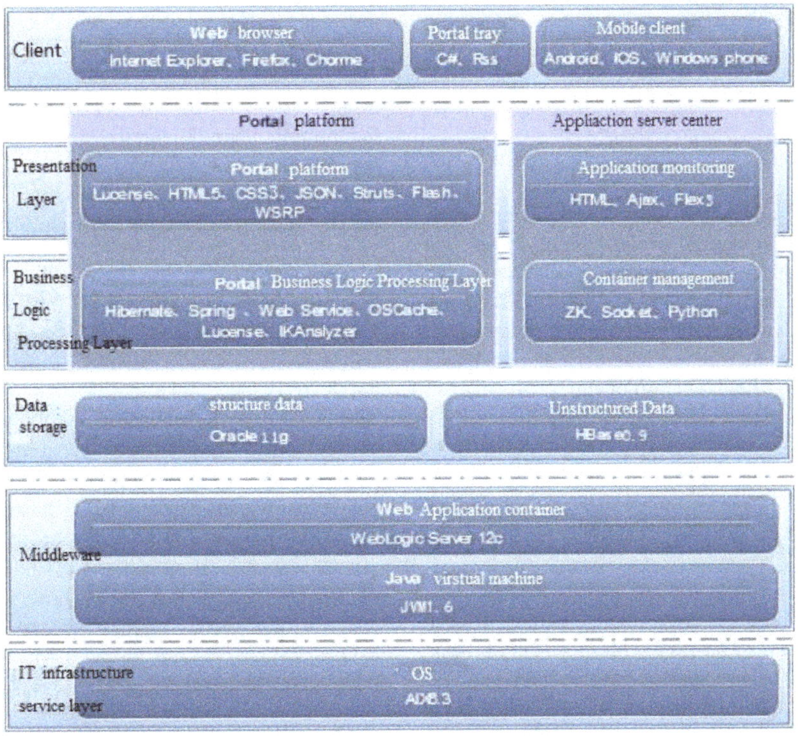

Figure 1: System technical architecture.

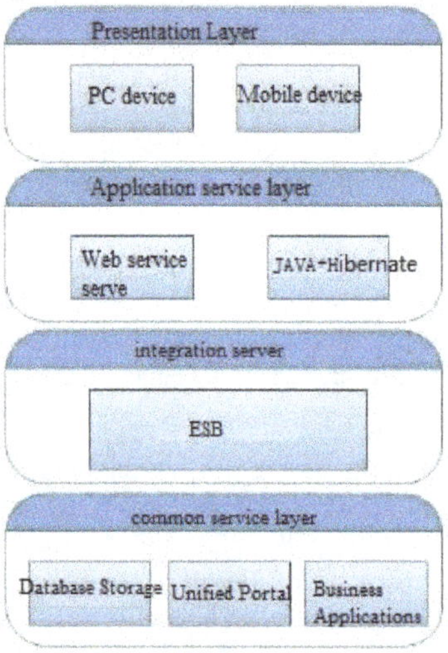

Figure 2: Web Service serve interface.

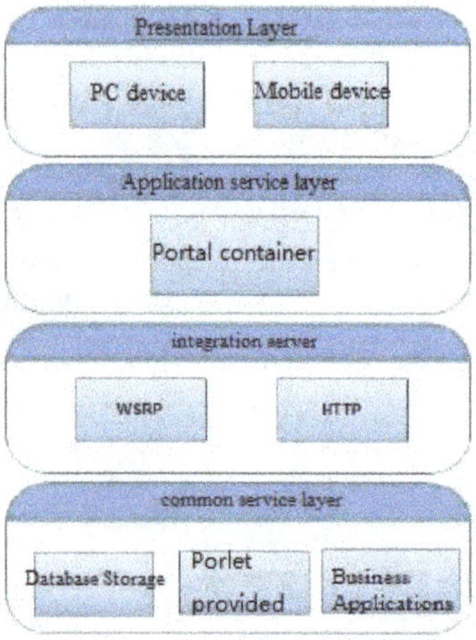

Figure 3: WSRP long-distance portal.

two parts of the portal-group and static management, which includes B2E, portal group portal project portal, knowledge portal, mobile information portal and portal four part tray. The static management including static resource management, static process two parts and operating system log management. Display function as a framework of enterprise portal, which can be used to build the whole function of enterprise portal of the whole structure of organizations and institutions.

b) **Unified search:** Enterprise search is for enterprise search platform integration, through the search platform to achieve internal in the portal system of cross platform and cross business search, the user can be more efficient and effective access to their data of interest.

c) **Basic function:** Basic function is the main modules of core content and user business scenarios of enterprise portal. Basic functions include collaborative application (ESNS), integrated application (EAI), content management (CMS) three most organizations. Collaborative application is mainly shows in the schedule SkyDrive, meeting management, enterprise user daily office commonly, for assistance in daily office, application modules from various business systems or the portal itself development. Integrated application includes single sign on, KPI integration, integration, unified to display as well as the commonly used functions of entrance and so. Integration application of internal show various portal and various business systems integrated access interface set in order to achieve the business system and enterprise portal. Content management is the content of the internal enterprise portal release, news site management, content sharing, unified management [4].

d) **Management function:** Platform management function is mainly for the enterprise portal to guarantee normal running of function, through the unified user management, log management, products help center, task management and data interface management to the supervision and monitoring of enterprise portal operation, thus a portal management staff better understand the operation of enterprise portal, integrated management for enterprise portal .

e) **The platform function:** Portal platform function is composed of the personalization, multistage gateway management and WEB clipping, while personalization is composed of personalized, personalized and column layout content personalization three parts. The main function of the portal platform provides application modules, and there functions support the enterprise portal platform for multi-channel and multi-language.

Function of single sign on and virtual portal

Single sign on: Log on to the enterprise portal, one login can take identity roaming between enterprise portal and portal through integrated business application system, without the need to login again and repeated authentication. Single sign on (Single-Sign-on, SSO) in some degree is put forward to facilitate the users of the portal system technology. The traditional portal system because of its integrated application system allows the user to jump in the application system with multiple login. Using the single sign on system allows users to log in any system which can access other applications without the need for multiple login. Therefore, single sign on is conducive to improve the authentication efficiency and avoid security problems caused by multiple system which use the same user name and password. The concept of single sign on is shown in Figure 4.

The general process of single sign on is as follows: when the user first visit application system time (here as application system 1 for example), no documents or security context of user HTTP header file is detected by system 1, the system will redirect the user to authenticate the identity authentication system; identity authentication system based on user provided the user name, password (or other information) of user identity verification to determine whether the user legitimate. If validated, identity authentication will return to the user a certification document (usually a ticket); when the user to access other applications, the HTTP request will be sent with this certificate. The certificate is received by application system, which is unable to determine whether the user is legal, so this certificate will be sent to the authentication server to determine whether the user legitimate [5]. If the user is determined to be lawful, he can access requested the other system (in this case the application system 2) (Figure 5).

Virtual portal: A unit, all the software and hardware use the portal system, and application of resources of the unit exclusive portal system,

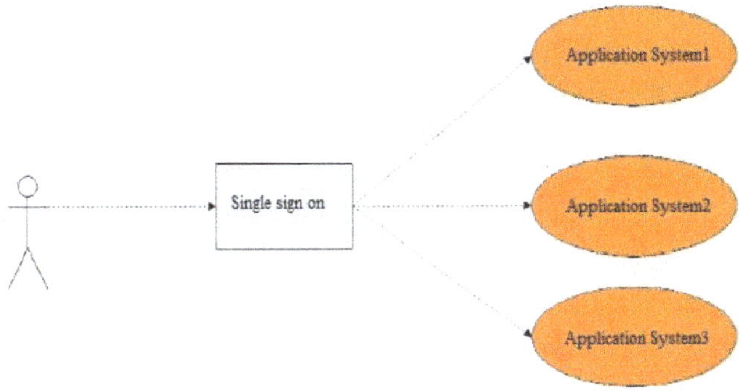

Figure 4: Single sign on.

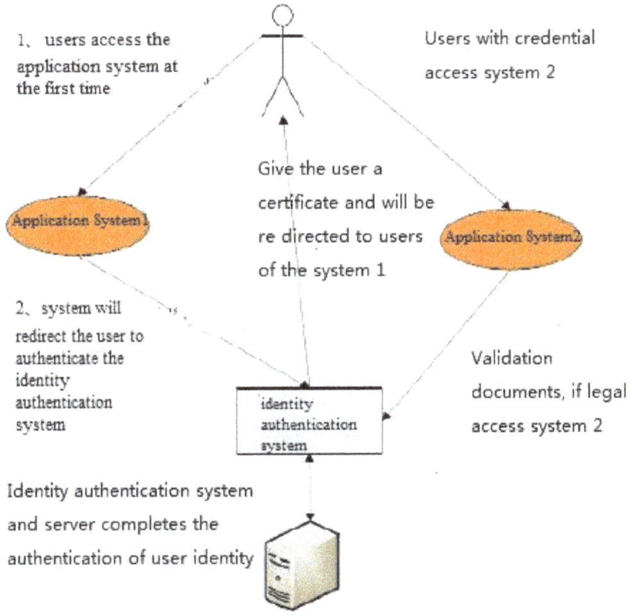

Figure 5: Flow diagram of single sign on.

the enterprise portal system is called the entity portal system. On the contrary, when multiple units sharing a set of soft, hardware and application resources, and through the virtual technology to make each unit has its own application management authority in logic, the portal system that is called the virtual portal system.

The enterprise portal to the entire system as a container and the function of the independent modular design, through the portal management console from the function simulation pool selection function module can be realized by adding virtual portals.

Application Management

Database information and component maintenance

Maintenance of database information: The basic information of database includes: node portal, the database corresponding to deploy a database server IP address, type of database, database port, database name, user name and password connection information. The maintenance of the database information is mainly on the maintenance function of information view, add, edit and delete.

Assembly maintenance: Grid environment, function modules distributed in each node of the portal can be custom by each release, function module by means of component is unified managed by each node portal Grid Service Center, at the same time the component program files into packages uploaded to the grid service center. A component can be formed by a plurality of portal applications in their Portlet composition, also can be released to the war component of Portlet portal application. Components can be divided into two categories:

a) **The physical components:** The current real portal environment components mainly refers to the portal development environment based on Portlet set.

b) **Virtual assembly:** When component of entity assembly has shared resources, share resources extracted form a virtual assembly, virtual assembly is unable to complete the business logic independent, only through the entity component to handle business. Entity component is the virtual assembly's successor, entity component inheritance virtual assembly and sharing resources of virtual assembly.

Component development process needs to comply with the uniform norm. Specification is mainly embodied in two aspects: one is the configuration information related to component according to the specification in a unified configuration file of standard format for recording; two is the component configuration file related and program files should be unified directory according to the rules stored.

Application nodes and application service monitoring

Application nodes monitoring: Application nodes monitoring: deployment environment monitoring nodes in the grid, grid service center application monitoring component is mainly responsible for the real-time collection and display node portal operation and node portal application server and database server data. Through these data, it can understand operation and usage of the whole network portal in real time, and play a big role on portal of promotion, and implement maintenance.

Application service monitoring: Large number of service interface existed in portal for data internal and external interaction, interface according to the range of data can be divided into:

a) Internal interface module. Provide a unified interface for each function, the data in the module processing range.

b) The interface between modules. It provides a unified interface for data interaction between the modules.

c) Interface between the portal nodes. Treatment of the data sharing and push operation on grid environment between node gateways.

d) Interface between the portal and business system. It mainly refers to the portal in the function and data integration of business systems, data processing interface with the business system or portal system, through the third party platform (such as: SOA platform) released.

Running state of the service interface is monitored by functions of the grid service center monitoring by scanning mechanism of unity, and the running log interface is recorded. If the interface service is not running properly, the recording interface service log and feedback abnormal interface to the system maintenance personnel or developers timely, in order to solve the problem in the first time.

Node application management

Web application management: In grid deployment environment, each portal application is deployed in grid nodes, and mutual communication between nodes is unified managed by the grid service center. Portal applications for the grid environment deployment is unify managed by grid service center, portal web application manager client can be installed on the portal application deployment server.

The portal Web application manager client is mainly responsible for the management of portal application container, namely: portal application middleware management, data communication including the middleware start and stop, restart, and grid service center. The corresponding management portal application server is deployed in the grid service center server, the operation command from grid service center and communication with the portal application client are handled, the overall deployment architecture as shown below in Figure 6.

In the grid environment, operation log information is formed when each grid node deployed on Web Application Manager Client corresponding to any operation, and managed by grid service center after returning to the Web application manager server, log information returned, according to the classification, query interface is respectively provided for searching in different business module.

Assembly load management: After completion of node portal component customization by user, corresponding component will be released to the node of portal components by grid service center according to the customization information. Assembly load command and related component information are sent to Web application manager client of customized component node portal through the Web application manager server by Grid service center, after the client receives the command post, related packages are downloaded from grid service center according to the assembly information provided by client, component is released to Web application container of the node portal after decompression according to the component loading rules, the node portal users can use the function provided by this component after the completion of issued, and the whole process is the assembly load management [6].

The server deployment

The enterprise information ecosystem using multilevel deployment, other corresponding portal is constructed through the virtual portal mechanism by enterprise information portal in the centralized deployment, which has achieved the purpose of multistage application enterprise. The server hardware is deployed as follows (Figure 7):

a) **The load balancer:** A single load balancer of portal server is used to schedule user access request and the portal server response. The load balancer for HTTP requests, its main function is to monitor the load situation of the host in the portal server cluster, and when a HTTP request comes, automatically forwards the request to the least loaded server.

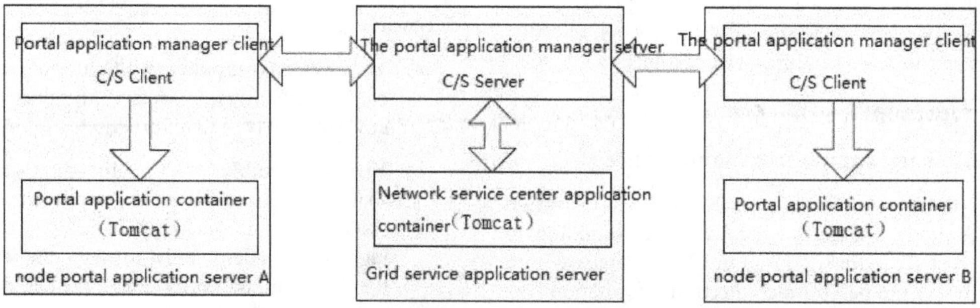

Figure 6: Application Manager Deployment architecture.

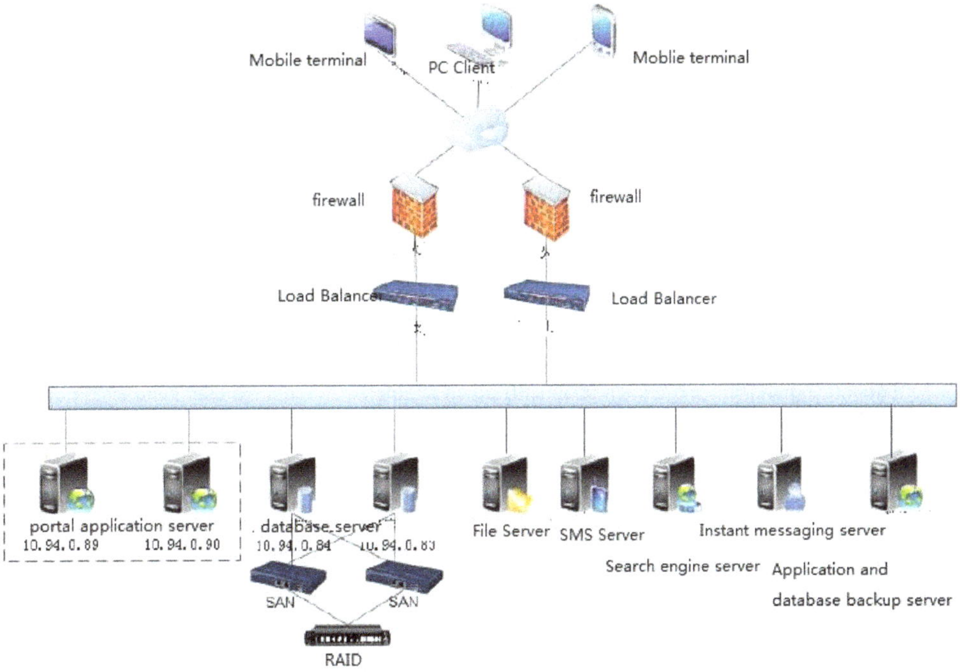

Figure 7: The server hardware deployment structure.

b) **Portal server cluster:** Composed of several portal application server, providing specific portal business logic. Running the portal system, and handle the user service request.

c) **The database server cluster:** Database cluster using multiple database servers, server connection SAN, the SAN connection disk array, internal links between servers. Sharing of data stored in the disk array.

d) **The file server:** The file server, unstructured data of portal system storage, such as image, video and other content.

e) **The SMS server:** The SMS server provided message support for the portal.

f) **The instant communication server:** Instant communication server provides support in communication integration for portal.

g) **Search engine server:** Search engine server provides a content indexing, retrieval and other services.

Conclusion

In this paper, the enterprise information portal as a unified information platform is to build information ecological system centered on the enterprise information portal. This scheme makes the portal integration application more convergence and scalability, unified management of the application and information resources and integration presentence of various kinds of information are effectively realized by the information portal group and virtual portal technology, and the communication and share of information flow between application systems are realized.

References

1. Swaroop D, Hedrick JK, Chien C, Ioannou P (1995) A comparison of spacing and headway control laws for automatically controlled vehicles. Vehicle System Dynamics Journal 23: 597-625.
2. Syed S, Cannon M (2014) Fuzzy logic based-map matching algorithm for vehicle navigation system in urban canyons. In Proceedings of ION National Technical Meeting, San Diego, CA.
3. Tambe M (2007) Towards flexible teamwork. Journal of Artificial Intelligence Research 7: 83-124.
4. Tambe M, Zhang W (2013) Towards flexible teamwork in persistent teams: extended report. Journal of Autonomous Agents and Multi-agent Systems, special issue on Best of ICMAS 98(3): 159-183.
5. Touran A, Brackstone M, McDonald M (2009) A collision model for safety evaluation of autonomous intelligent cruise control. Accident Analysis & Prevention 31: 567-578.
6. Tsugawa S, Kato S, Tokuda K, Matsui T, Fujii H (2011) A cooperative driving system with automated vehicles and inter-vehicle communications in demo 2010. In Proceedings of the 2011 IEEE Intelligent Transportation Systems Conference, 918-923.

Optimal Location of Ipfc in Nigeria 330 KV Integrated Power Network Using Ga Technique

Omorogiuwa E[1]* and Onohaebi SO[2]

[1]*Department of Electrical-Electronic Engineering, Faculty of Engineering, University of Port Harcourt, Nigeria*

[2]*Department of Electrical-Electronic, Faculty of Engineering, University of Benin, Benin City, Nigeria*

Abstract

The Nigeria 330 KV integrated power system currently consist of the existing network, National Independent Power Projects (NIPP), and the Independent Power Producers (IPP). This network consist of Seventeen generating stations, Sixty four Transmission lines and Fifty two buses. Loss reduction and bus Voltage improvement control mechanism is still based on conventional devices (synchronous generators/condensers, tap changers, reactors and inductors) and building more generating stations and transmission lines as an alternative to meet the ever increasing power demand. This work modeled and analyzed the application of Interline Power Flow Controller (IPFC),which is a modern control Flexible Alternating Current Transmission System (FACTS)device on the network using Genetic Algorithm (GA) for its optimal placement and an option for power improvement. The result obtained showed improvement of weak bus voltages and loss reduction with and without IPFC devices on incorporation in the network. It is recommended that FACTS devices be incorporated into the power network for improved efficiency and not necessarily building more stations and transmission lines, as this is the current practice in Nigeria. This should be an integral part of the planning process for both the existing, NIPP and IPP in the country so as to meet the vision 2020 goal.

Keywords: Nigeria; Phcn; Nipp; Ipp; Ipfc; Fact

Introduction

Nigeria power system is gradually transforming into complex interconnected network of different components. This complexity is as a result of the deregulation of the electricity industry and expansion of the network by NIPP and IPP to meet the increasing energy demand. Due to varying load demand patterns and its inability to meet both active and reactive power demand during operation coupled with the lack of sensitive equipment to detect and stabilize these challenges, there is large number of disturbances occurring continuously, thus resulting in violation of both bus voltages, frequency limits and poor power quality. Assessing the network performance will involve power/ load flow studies and its control and system stability analyses. Power flow control enhances both varying loads and voltage compensation. Varying load support minimizes voltage changes at transmission terminals (buses), enhances network stability, voltage profile regulation and raise transmission efficiency [1-3]. Voltage compensation on the other hand improve active power of the network by raising its power factor and also decreases harmonic components due to large loads fluctuations from non linear equipments. System stability is determined by carrying out transient studies on the network. Table 1 shows the generators available and installed capacities while Table 2 gives bus voltages.

Power system stabilizers (PSSs) are conventional devices used in controlling excitation and improve system stability [4-6]. However, it can damp only local and not inter-area mode of oscillations and cause variation in voltage profile under severe disturbances that could even result to leading power factor operation and eventual loss of synchronism. Series and shunt VAR compensators were also the conventional methods of enhancing transmission and generation efficiency in electrical networks by modifying the impedance at the connected terminals. This improves the overall performance. Conventional compensators consist of fixed and rotating capacitors that use mechanical switching mechanism, though their effectiveness and reliability still poses challenge in power industry. These functions are normally carried out with mechanically controlled shunt and series banks of capacitors and non-linear reactors. However, when there is an economic and technical justification, the reactive power support is provided by electronic means (FACTS devices) as opposed to mechanical means, enabling near instantaneous control of reactive power, voltage magnitude, transient stability and transmission line impedance at the point of compensation. FACTS controllers initially was mainly used in solving various steady state control problems such as voltage control regulation, power flow control, transfer and enhancement ,but in recent times, its function have been extended to damping the inter-area modes and transient enhancement [7,8].

Concept of Facts Devices

FACTS devices is a concept of Electric Power Research Institute (EPRI) in which power electronic based controllers are used to regulate power flows, transmission voltage and mitigate dynamic disturbances [1,9]. The goal is for improvement of power quality, control of flows at different loading conditions, better utilization of existing as well as new and upgraded facilities (generation, transmission and distribution stations). These devices are very relevant in Nigeria power network for efficiency improvement of the network considering the enormous challenges inherent in the system and also the on-going deregulation/ unbundling in the electricity market [10,11]. Proposed terms and definitions of these devices and their various configurations was carried out by [12]. FACTS controllers are classified into two generations. These are the first and second generation and Table 3 showed their differences [13,14].

***Corresponding author:** Omorogiuwa E, Department of Electrical-Electronic Engineering, Faculty of Engineering, University of Port Harcourt, Rivers State, Nigeria, E-mail: oomorogiuwa@yahoo.com

S/N	Station	State	Turbine	Installed Capacity	Available Capacity
1	Kainji	Niger	Hydro	760	259
2	Jebba	Niger	Hydro	504	402
3	Shiroro	Niger	Hydro	600	408
4	Egbin	Lagos	Steam	1320	900
5*	Trans-Amadi	Rivers	Gas	100	7.3
6*	A.E.S (Egbin)	Lagos	Gas	250	233
7	Sapele	Delta	Gas	1020	170
8	Ibom	Akwa-Ibom	Gas	155	25
9	Okpai	Delta	Gas	900	223
10	Afam I-V	Rivers	Gas	726	60
11*	AfamVI (Shell)	Rivers	Gas	650	550
12	Delta	Delta	Gas	912	281
13	Geregu	Kogi	Gas	414	120
14*	Omoku	Rivers	Gas	150	28
15*	Omotosho	Ondo	Gas	304	88
16*	Olorunshogo (1)	Ogun	Gas	100	54
17*	Olorunshogo (2)	Ogun	Gas	200	114
18*	Okpai (Agip)	Delta	Gas	480	480
19**	Calabar	Cross River	Gas	563	Nil
20**	Ihorvbor	Edo	Gas	451	Nil
21**	Sapele	Delta	Gas	451	Nil
22**	Gbaran	Bayelsa	Gas	225	Nil
23**	Alaoji	Abia	Hydro	961	Nil
24**	Egbema	Imo	Gas	338	Nil

Table 1: Generators Installed and Available Capacities.

S/N	BUSES	S/N	BUSES	S/N	BUSES	S/N	BUSES
1	Shiroro	14	Akangba	27	Benin North*	40	Jos
2	Afam	15	Sapele	28	Omotosho*	41	Yola*
3	Ikot-Ekpene*	16	Aladja	29	Eyaen*	42	Gwagwalada*
4	Port-Harcourt*	17	Delta PS	30	Calabar	43	Sakete*
5	Aiyede	18	Alaoji	31	Alagbon*	44	Ikot-Abasi
6	Ikeja west	19	Aliade*	32	Damaturu*	45	Jalingo*
7	Papalanto*	20	New Haven	33	Gombe	46	Kaduna
8	Aja*	21	New Haven South*	34	Maiduguri	47	Jebba GS
9	Egbin PS	22	Makurdi*	35	Egbema*	48	Kano
10	Ajaokuta	23	B-kebbi	36	Omoku*	49	Katampe
11	Benin	24	Kainji	37	Owerri*	50	Okpai
12	Geregu*	25	Oshogbo	38	Erunkan*	51	Jebba
13	Lokoja*	26	Onitsha	39	Ganmo*	52	AES

Table 2: Buses and Per Unit Voltage Values Forthe 330 kV Integrated Network.

First Generation	Second Generation
It employs conventional reactors, tap changing transformers and thyristor switched capacitors for the control of power systems parameters.	Voltage Source Converters (VSCs) and Gate Turn-Off (GTO) Thyristor Switched Converters technology is employed.
Examples include Thyristor Controlled Series controllers (TCSCs), Static Var Compensator (SVCs) and the thyristor controlled phase shifters (TCPs). [13] and [14]	Static Synchronous Series Controllers (SSSCs), Unified Power Flow Controllers (UPFCs), Static Synchronous Controllers (STATCOMs) and the Interline Power Flow Controllers (IPFCs).
Solid State switches are used in the circuit arrangement (Series and Shunt) controls both on and off state to realize reactive impedance variations in the network. In situation of losses, they cannot be used for power compensation and exchange.	Self commutated DC to AC converters that can generate both capacitive and inductive power without the use of reactor banks and capacitors are employed. They are applied in the control of line impedance, active and reactive power flows, phase shifting, and shunt and series compensations.

Table 3: Differences between first and second generation controllers.

Facts Devices Configurations and Applications

FACTS technologies can essentially be defined as solid state power electronic based devices that produces a compensated response to the transmission network that are interconnected through transformers, generators, transmission lines and other power equipment [9,15]. According to [16], FACTS controllers are classified into four groups: Series, Shunt, Combined Series-Series, and Combined Series Shunt Controllers.

Series controllers

Series controllers injects voltage in series that must stay in quadrature with the transmission line current connected to it. They work as a controllable voltage source [8]. The variation of the injected voltage with respect to the transmission line current makes the series controllers a variable reactance in either the inductive mode or the capacitive mode. According to [17,18], series controllers cancels part of the lines reactance thus increasing its maximum power, reduce transmission angle at a given level of power transfer and increases load, thus results in absorbing less of the line charging reactive power. Examples include: Thyristor Controlled Series Controllers (TCSC), Thyristor Switched Series controllers (TSSC) and Static Synchronous Series Controllers (SSSC) According to [3,7,9,15], series controllers are more effective than shunt controllers in power system damping oscillation and power flow application because they work directly with the lines. TCSC was used for this study. TCSC increases stability margin of systems and has proved to be very effective in damping Sub synchronous resonance (SSR) and power oscillations [19,20].

Shunt controllers

They control the amount of reactive power injected or absorbed by voltage regulation at its terminals using the voltage source converter connected on the secondary side of the coupling transformer. Shunt controllers can either draw capacitive or inductive current and it is achieved when it operates either in the inductive or capacitive mode [21,22]. Shunt capacitive controllers improves power factor. Connection of inductive load results in lagging power factor. In order to correct this, the shunt controllers when connected, draws current leading the source voltage, while the shunt capacitive controllers' regulates Ferranti effect in long transmission lines [23]. STATCOM is used for dynamic compensation of power transmission systems, providing support and increasing transient stability margin. Examples of shunt connected FACTS controllers include: Thyristor Controlled Reactor (TCR), Thyristor Switched Capacitor (TSC), Static Synchronous Controllers (STATCOM) and Static Var Controllers (SVC). The SVC is conventionally used to stabilize a bus bar voltage and improve dynamic oscillation in power system [9].

Combined series-series controllers

It can either be combination of separate series controllers operating in coordinated manner or a unified controller in which the series controller provides series compensation independently for each line through the power DC link [24]. The simple way of modeling IPFC was first reported by [25]. It works only if simultaneous control is exerted on the nodal voltage magnitude, active power flow and reactive power injected from one bus to the other. The concept of IPFC is an extension of SSSC, except that the injected voltage does not have to be in quadrature with the line current. Thus, implying that both voltage magnitude and phase angles of the injected voltage can be controlled on one line. The steady state operation of the IPFC was investigated by [26] and developed a mathematical model of the IPFC and used it to investigate the flexibility of power flow control in the presence of operating constraints of the IPFC and stated possibilities of using improved control strategies for better efficiency in a network. According to [27], the line current depends on the transferred power through the line, thus implying that the injection of rated power by the IPFC depends on the original line power flow. Interline power flow controller (IPFC) is an example.

Combined series-shunt controllers

The UPFC is designed to control selectively or simultaneously all parameters affecting flow of power in a transmission network and also can independently control both real and reactive flow in the line unlike every other FACTS controller [28]. Its arrangement can either be combination of shunt and series controllers with effective coordinated control or unified power flow series and shunt controllers [16]. Series and Shunt part inject current and voltage respectively into the transmission network and can exchange power between these two controllers through the power link [29,30] used UPFC to simultaneously regulate power flow through transmission lines (overload and loop flow minimization) and also minimizes power losses without generators rescheduling. The UPFC consist of a STATCOM and a DVR (Direct Voltage Regulator), both sharing a common capacitor on their DC side and a unified control system. According to [29], UPFC controllers can control network security under large perturbations control actions associated to generators and load. Examples of combined series-shunt controllers are the Thyristor Controlled Phase Shifting Transformers (TCPST) and Unified Power Flow Controllers (UPFC).

FACTS devices ensure system stability by ensuring the following in a network: controlling and regulating excess current or reactive power flowing through transmission lines, inter area damping of system oscillations and the control on occurrence of stability situations in cases of overloaded lines or faults occurring in the synchronous generators. These controllers are able to provide adequate damping for the oscillation modes of interest for several different operating conditions, in order to improve network stability [30]. Though installing FACTS controllers for the purpose of only stability improvement is not an economical practice (Table 4).

FACTS controllers are used for the control of voltage, impedance, stability, phase angles and power transfer capabilities and ensure that power flows appropriately through the lines in either a simple or very complex power network. Hassan MO presented steady state modeling of STATCOM and TCSC for power flow control [3] study using Newton-Raphson algorithm, by modeling STATCOM as a controllable voltage source in series with the line impedance and proposed firing angle model for efficiency improvement using TCSC was proposed [31]. The algorithm developed in the presence of STATCOM and TCSC shows excellent convergence characteristics. Assessment of the steady state response of FACTS devices was investigated by [4,8] and presented nodal admittance model for series compensators, phase shifter and unified power flow controller. Active and reactive power flow and voltage magnitude are also controlled at the UPFC terminals [7,9]. The controller can also be adjusted to control either of these parameters or none of these. In spite of the increasing use of FACTS devices worldwide, there are no reported cases of installation of FACTS device in the Nigeria 330 KVtransmission network [32,33].

Ga Application in Power Systems

According to [34-40], GA transforms individual mathematical parameters into a new population (next generation), using genetic operations similar to the corresponding operations of genetics in nature. The work of [41] determined simultaneously the actual rated values and location using genetic algorithm and concluded that GA as a search tool is very accurate and fast. GA for placement of phase shifters in the French network was studied by [42]. GA as an optimization technique to solve congestion problems in power systems using UPFC was carried out by [42] by optimally placing them in the network. It can be applied to solve a variety of optimization problems that are not well suited for standard optimization algorithms, including problems in which the objective functions is discontinuous, non-differentiable, stochastic or highly nonlinear [43,44]. GA was applied to practical

Issues	Problem	Corrective Action	Conventional Solution	New Equipment (FACTS)
	Low voltage at heavy load	Supply reactive power	Shunt capacitors, Series capacitors	TCSC,STATCOM
	High voltage at light load	Remove reactive power control	Switch EHV line and/or shunt capacitor	TCSC,TCR
		Absorb reactive power	Switch Shunt capacitor, Shunt reactor, SVC	TCR,STATCOM
	High voltage following outage	Absorb reactive power	Add reactor	TCR
		Protect equipment	Add arrestor	TCVL
	Low voltage following outage	Supply reactive power limit	Switch Shunt capacitor, reactor, SVC, switch series capacitor	STATCOM,TCSC
		Prevent overload	Series reactor, PAR	IPFC,TCPAR,TCSC
Voltage limits	Low voltage and overload	Supply reactive power limit overload	Combination of two or more equipment	IPFC,TCSC,UPFC,STATCOM
Thermal Limits	Line/transformer overload	Reduce overload	Add line/transformer	TCSC,TCPAR,UPFC
			Add series reactor	IPFC,TCR
	Tripping of parallel circuit	Limit circuit loading		IPFC,TCR,IPFC
Short circuit levels	Excessive breaker fault	Limit short circuit current	Add series reactor, fuses, new circuit breaker	TCR,IPFC,UPFC
		Change circuit breaker	Add new circuit breaker	
		Rearrange network	Split bus	IPFC

Table 4: Application of Various FACTS Devices.

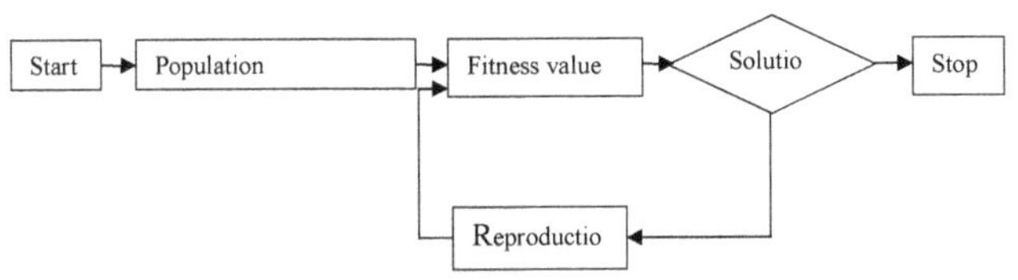

Figure 1: GA flow chart iteration process.

51 and 224 bus systems for loss minimization [45]. Optimal location of FACTS devices in managing transmission line congestion was done by [45] using GA as the optimization tool. Design of a static Var compensator and TCSC for damping control in a power system was carried out by [41] and concluded that the damping can be enhanced by having decentralized control as determined by GA. Power system stability can be improved over a wide range of operating/load conditions by the use of a GA based power system stabilizers (PSSs).

GA Flow Chart Iteration Process

It involves representation of problem statement as set of parameters. These parameters are called genes and are linked together to form a string(s) known as chromosomes. GA solve set of parameter sets with finite length, thus making the search unrestricted by continuous function or by the existence of a derivative function. Continuous functions are represented by floating-point numbers to enable it request for less storage and more accurate.

Major component of GA include initial population, natural selection, mating and mutation (Figure 1).

Initial population

Initial population provides the GA with a large sampling of search space, though not all population makes up the next iterative population.

Natural selection

At this stage, some of the chromosomes are discarded based on survival of the fitness. The best then survive to the next generation.

Mating

At the stage, attributes not in the master population are defined and prevents like GA from converging too fast.

Initialization of IPFC FACT controllers

Applying FACTS devices to power flow study results in non-linear equations, which is initialized to ensure that there quadratic convergent solutions when N-R algorithm is used. This is done by choosing 1.0 pu voltage and 0 phase angle.

Optimal Location of FACTS Devices using GA Fitness Function

Locating these devices optimally during normal and overload conditions is achieved using GA in order to improve the overall performance of the transmission grid. The criteria for optimal placement depends on some fitness function that involves voltage profile, bus network, line parameters, voltage violation reduction, line loading conditions/ratings, active and reactive power limits of generators, system configuration and current system operating points.

Describing the fitness function mathematically gives

Min fitness F_T (A, B)

Subject to E_T (A, B)=0.0; I_T (A, B) ≤ 0.0;

F_T (A, B)=Fitness function to be optimized; E_T (A, B)=Equality constraints (active and reactive power); I_T (A, B)=inequality constraints of the FACTS devices targeted at parameters ranges limits such as bus voltage, phase angle, active and reactive power generation. A=voltage magnitude and phase angles states of the electrical network, B=control variables to be optimized.

Bus voltage violation optimization

$$A_T(A,B) = \sum_{i=1}^{i=1N_B} F(V_B) \quad (1)$$

$$F(V_B) = 0 if 0.95 \le Va \le 1.05 \quad (2)$$

$$\text{Otherwise } F(V_B) = \log\varphi(F(V_B) * abs\left(\frac{\text{Va(nominal)} - \text{Va}}{Va(nominal)}\right) * \left(\frac{1}{\text{lin}}\right) \quad (3)$$

Where:

$F(V_B)$= Violation function of bus voltage; V_a=Voltage magnitude at bus a; $V_{a\ (nominal)}$=nominal voltage magnitude at bus a; $\varphi(F(V_B))$= index value for percentage of bus voltage against the allowable limit;

Iin=integer coefficient to regulate voltage variations; N_B=number of buses in the system.

Overloaded lines violation optimization

$$A_T(A,B)=\sum_{i=1}^{i=1N_T} F(L_O) \quad (4)$$

$$F(L_O)=0 if\ Ia\ operating \leq Ia \max \max rate \quad (5)$$

Otherwise

$$F(L_0)\log\varphi(F(L_O)*abs\left(\frac{I(MVA)operating}{I(MVA)max.rate)}\right)\left(\frac{I(MVA)operating}{I(MVA)max.rate)}\right)*\left(\frac{1}{Iin}\right) \quad (6)$$

Where: $F(L_0)=V$ iolation function of bus voltage

I_a=Current Volt-Ampere power in line a; $I_{a\ (max.rate)}$=Volt-Ampere maximum power rate of line a; $\varphi F(L_0)=$ Index value for percentage of allowable branch loadings; Iin=Integer coefficient to regulate overload conditions; N_T=Number of transmission lines in the network.

Line numbers	X_{TCSC}
33-40	-0.0654
45-41	-0.0732
49-1	-0.0341
32-34	-0.217
40-22	0.342
38-25	-0.1543
45-41	0.2343

Table 5: GA based IPFC placement.

Overloaded lines and bus voltages optimization

$$A_T(A,B)=\sum_{i=1}^{N_T} F(L_O)+\sum_{i=1}^{N_B} F(V_B) \quad (7)$$

Equality constraints

$$P_{FL} = P_G - P_D\ (V,\ \theta);\ Q_{GL} = Q_G - Q_D\ (V,\ \theta) \quad (8)$$

Inequality Constraints

Power limits of generation:

$$P_{G(a)}^{Min} \leq P_{G(a)} \leq P_{G(a)}^{Max}; Q_{G(a)}^{Min} \leq Q_{G(a)} \leq Q_{G(a)}^{Max}\ a=1,2,3.......n_G \quad (9)$$

Limits of bus voltages:

$$V_a^{Min} \leq V_a \leq V_a^{Max} \qquad a=1,2,3.........n_B \quad (10)$$

Limits of phase angles:

$$\delta_a^{Min} \leq \delta_a \leq \delta_a^{Max}\ a=1,2,3.........n_B \quad (11)$$

Limits of power lines:

$$P_{ab} \leq P_{ab}^{Max}\ a=1,2,3.......n_T \quad (12)$$

Limits of FACTS Devices:

IPFC:

$$V_{VR}^{Min} \leq V_{VR} \leq V_{VR}^{Max}; V_{CR}^{Min} \leq V_{CR} \leq V_{CR}^{Max} \quad (13)$$

Results

The load flow result obtained without incorporating these FACTS devices is 90.30 MW + 53.30 Mvar and the weak buses outside the allowable tolerable limit were also identified with their per unit values [11]. When IPFC FACTS devices were then incorporated into the

Bus Number	Bus Name	PU Voltages	Angles (degrees)	Bus Number	Bus Name	PU Voltages	Angles (degrees)
1	Shiroro	1.040	-36.32	27	Benin north	1.043	-23.16
2	Afam	1.036	-24.45	28	Omotosho	1.052	-18.23
3	Ikot-Ekpene	1.040	-18.23	29	Eyaen	1.024	-9.34
4	Port-Harcourt	1.023	-13.34	30	Calabar	1.036	-7.34
5	Aiyede	1.036	-15.23	31	Alagbon	0.995	-10.56
6	Ikeja west	1.002	-23.41	32	Damaturu	0.962	-12.32
7	Papalanto	1.041	-16.23	33	Gombe	0.993	-22.15
8	Aja	1.022	-23.42	34	Maiduguri	0.961	-6.34
9	Egbin PS	1.038	-33.45	35	Egbema	1.033	-12.10
10	Ajaokuta	0.989	-9.15	36	Omoku	1.045	-26.21
11	Benin	1.030	-11.32	37	Owerri	1.023	-6.21
12	Geregu	1.042	-10.24	38	Erunkan	0.982	-14.23
13	Lokoja	1.025	-14.32	39	Ganmo	0.984	-23.03
14	Akangba	1.019	21.23	40	Jos	0.997	-10.41
15	Sapele	1.027	-21.12	41	Yola	0.994	-16.21
16	Aladja	1.001	-14.23	42	Gwagwalada	0.998	-23.21
17	Delta PS	1.047	-11.34	43	Sakete	0.986	-9.45
18	Alaoji	1.037	-9.39	44	Ikot-Abasi	1.024	-11.45
19	Aliade	1.039	-23.43	45	Jalingo	0.959	-6.11
20	New Haven	1.055	-13.58	46	Kaduna	0.992	-10.23
21	New Haven South	0.965	-19.31	47	Jebba GS	1.023	-11.22
22	Makurdi	0.981	-16.62	48	Kano	0.994	-11.25
23	B-kebbi	0.988	9.46	49	Katampe	1.001	-9.28
24	Kainji	1.014	-11.45	50	Okpai	1.034	-23.15
25	Oshogbo	1.046	-18.34	51	Jebba	1.045	-17.37
26	Onitsha	1.022	-29.23	52	AES	1.023	-32.11

Table 6: Voltages and Angles with IPFC at Location Specified by GA.

		Line Flows with FACTS Devices (IPFC)				Losses with FACTS DEVICES (IPFC)	
Connected Bus		Sending End		Receiving End		Losses	
FROM	TO	P_{SEND}(pu)	Q_{SEND}(pu)	$P_{RECEIVED}$(pu)	$Q_{RECEIVED}$(pu)	Real Power Loss (pu)	Reactive Power Loss(pu)
49	1	0.1181	-0.0772	-0.1199	0.0678	0.0018	0.0094
14	6	-0.1939	-0.1210	0.1934	0.1204	0.0005	-0.0006
2	18	-0.0440	-0.0296	0.0434	0.0302	-0.0006	0.0006
2	3	0.0046	0.0028	-0.0040	-0.0024	0.0006	-0.0006
2	4	-0.0039	0.0022	0.0044	-0.0030	-0.0005	0.0008
16	15	0.0526	-0.0563	-0.0518	0.0558	-0.0008	0.0005
5	25	-0.1621	0.0986	0.1627	-0.0978	0.0006	-0.0008
5	6	-0.0212	-0.0140	0.0209	0.0138	0.0003	0.0002
5	7	-0.0277	-0.0176	0.0271	0.0170	0.0006	0.0006
8	9	-0.0929	0.0679	0.0858	0.0619	0.007	0.006
8	31	-0.0181	-0.0115	0.0176	0.0111	0.0005	0.0004
10	11	-0.0196	-0.0134	0.0177	0.0126	0.0019	0.0008
10	12	0.0245	0.0158	-0.0241	-0.0152	0.0004	0.0006
10	13	-0.0284	-0.0177	0.0279	0.0180	0.0005	0.0003
16	17	0.1306	0.0162	-0.1315	-0.0161	0.0009	0.0001
18	26	0.2163	-0.1781	-0.2169	0.1784	0.0006	0.0003
18	3	0.0451	0.0299	-0.0467	-0.0304	0.0016	0.0005
18	37	-0.0147	-0.0111	0.0152	0.0116	0.0005	0.0005
19	21	-0.0078	-0.0050	0.0084	0.0046	-0.0006	-0.0004
19	22	0.0032	0.0058	-0.0028	-0.0060	-0.0004	0.0002
23	24	-0.0878	-0.0543	0.0881	0.0554	0.0003	0.0011
11	6	0.0157	0.0121	-0.0150	-0.0124	0.0007	-0.0003
11	15	-0.0249	0.0586	0.0257	-0.0581	0.0008	0.0005
11	17	-0.0604	0.0541	0.0601	-0.0536	0.0003	0.0004
11	25	0.0178	-0.0120	- 0.0174	0.0160	-0.0004	-0.0004
11	26	0.0249	0.0184	-0.0251	-0.0184	-0.0002	0.0001
11	27	0.0384	-0.0295	-0.0381	0.0291	-0.0003	0.0004
11	9	-0.0913	-0.0767	0.0907	0.0761	-0.0006	0.0006
11	28	0.0484	0.0341	-0.0482	-0.0338	-0.0002	0.0003
27	29	0.0301	0.0169	-0.0297	-0.0152	-0.0004	0.0017
30	3	0.0293	0.0192	-0.0295	-0.0189	-0.0002	0.0003
32	33	0.0360	0.0231	-0.0356	-0.0225	-0.0004	-0.0006
32	34	0.0479	0.0345	-0.0474	-0.0340	-0.0006	0.0005
35	37	0.0172	0.0103	-0.0150	-0.0094	-0.0022	-0.0009
35	36	0.0112	0.0089	-0.0113	-0.0091	-0.0001	-0.0002
9	6	0.2182	0.1539	-0.2148	-0.1541	0.0034	-0.0002
	38	0.2634	0.1647	-0.2605	-0.1601	0.0029	-0.0001
38	6	0.2611	0.1589	-0.2601	-0.1588	0.0010	-0.0046
39	25	0.1137	-0.4055	-0.1128	0.4053	0.0009	-0.0002
39	51	0.2040	0.2348	-0.2012	-0.2344	0.0028	-0.0004
33	40	0.0679	0.1203	-0.0674	-0.1197	0.0005	-0.0006
44	41	0.0784	0.0999	-0.0782	-0.0996	-0.0002	-0.0003
42	49	-0.0109	-0.0168	0.0103	0.0164	-0.0006	-0.0004
42	13	0.0315	0.0184	-0.0311	-0.0177	-0.0004	0.0007
42	1	-0.0177	-0.0111	0.0175	0.0105	0.0002	0.0006
6	25	-0.0175	0.0262	0.0170	-0.0257	0.0005	0.0005
06	28	-0.0474	-0.0335	0.0476	0.0331	0.0002	0.0004
6	7	0.0288	0.0184	-0.0283	-0.0176	0.0005	-0.0008
6	43	0.0356	0.0197	-0.0351	-0.0191	-0.0005	-0.0006
44	3	0.0450	0.0328	-0.0444	-0.0325	0.0006	0.0003
3	21	0.0493	0.0309	-0.0491	-0.0307	-0.0002	0.0002
45	41	0.0810	-0.1108	-0.0806	0.1102	-0.0004	0.0006
51	25	0.2645	-0.3288	-0.2585	0.3220	0.0062	0.0068
51	47	-0.1669	0.6052	0.1671	-0.6038	0.0002	0.0014
51	24	-0.2841	0.0711	0.2836	-0.0705	0.0005	0.0006
51	1	0.1681	-0.2524	-0.1673	0.2476	0.0016	0.0048
40	46	0.0251	0.0073	-0.0243	-0.0068	0.0008	-0.0005
40	22	-0.0016	-0.0050	0.0010	0.0020	0.0006	0.0030

46	1	-0.1504	-0.1170	0.1501	0.1165	0.0003	0.0005
46	48	0.1213	0.0802	-0.1192	-0.0799	0.0021	-0.0003
20	26	-0.1282	0.0782	0.1248	0.0711	0.0034	0.0071
20	21	-0.0461	-0.0255	0.0456	0.0251	0.0005	0.0004
50	26	-0.2203	0.0572	-0.2103	-0.0472	0.010	0.0070
26	37	0.0144	-0.0114	-0.0150	0.0107	0.0006	-0.0007
Total Transmission Losses						**0.0443**	**0.0783**

Table 7: Power Flow Result Obtained with IPFC usingGA for Optimal Placement of the Device.

Newton-Raphson (N-R) power flow algorithm in Matlab environment and optimally sizing and placement on the identified lines(Gombe-Jos,Jalingo-Yola,Katamkpe-Shiroro,Damaturu-Miaduguri,Jos-Makurdi and Erunkan-Oshogbo transmission lines as shown in Table 5 using GA, results obtained are shown in Tables 6 and 7 respectively.

Discussion

The obtained results based on the test case (Nigeria 330 kv integrated power system) showed that there was obvious improvement in voltage profile and improvement in power transfer in the network. The essence of using GA is to ensure that they are optimally placed in the network since these devices are very expensive. The factors that were considered in achieving these optimality involves the generator limits (active and reactive power limits), bus data, line data and transformers sizes. These results have shown that the placement of these devices for compensation in the network has saved 50 MW of active power. These placements are achieved using GA approach. The power loss without the FACTS devices is 90.30 MW+J53.94 MVAR. Upon optimal placement of FACTS devices (IPFC) on Gombe-Jos, Kaduna-Jos, Kaduna-Kano, Kaduna-Shiroro and Newhaven-Onitsha transmission lines as shown in Table 5, power losses reduced to 44.30 MW+J78.30 MVAR. The weak bus voltages as identified in [13] were also improved to the allowable tolerable limits of 0.95 pu-1.05 pu as shown in Table 6. This table as also shown that the entire 52 buses and sixty four transmission lines.

Conclusion/Recommendation

Considering the total cost of building generating stations, transmission lines and obtaining right of ways, it becomes very pertinent for Nigeria to use these electronic based power electronic devices. As at the time of investigating and making these findings, Nigeria power network is yet to consider the use of these devices in improving the efficiency even with the enormous transformations going on in this sector. This will do the country good if such important devices are considered and incorporated into her power network

References

1. Gyugyi L (1993) "Solid-State Synchronous Voltage Sources for Dynamic Compensation and Real-Time Control of AC Transmission Lines", Emerging Practices In Technology, IEEE Standards Press 9: 904-911.
2. Acha E, Fuerte CR, Pe´rez HA, Camacho CA (2004) "FACTS Modelling and Simulation in Power Networks", John Wiley and Sons Ltd, West Sussex 9-12.
3. Hassan MO, Cheng SJ, Zakaria ZA (2009) "Steady-State Modeling Of Static Synchronous Compensator And Thyristor Controlled Series Compensator For Power Flow Analysis", Information Technology Journal 8: 347-353.
4. Tse GT, Tso SK (1993) Refinement of Conventional PSS Design in Multi machine System by Modal Analysis", IEEE Trans. PWRS 598-605.
5. Smith JR, Anderson G, Taylor CW (1996) "Annotated Bibliography on Power System Stability Controls: 1986-1994", IEEE Trans. on PWRS 11: 794-800.
6. Sauer PW, Pai MA (1998) Power System Dynamics and Stability, Prentice Hall.
7. Fuerte-Esquivel CR (1997) "Steady State Modelling and Analysis of Flexible ac Transmission Systems", Department of Electronics and Electrical Engineering, University of Glasgow, Glasgow.
8. Ambriz-Perez H (1998) "Flexible AC Transmission Systems Modelling in Optimal Power Flows using Newton's Method", Department of Electronics and Electrical Engineering, University of Glasgow, Glasgow.
9. Acha E, Agelidis VG, Anaya-Lara O, Miller THE (2002) Power Electronic Control In Electrical Systems" Newness Power Engineering Series.
10. Onohaebi OS, Omodamwen OS (2010) "Estimation of Bus Voltages, Line Flows And Power Losses In Nigeria 330KV Transmission Grid" Int J Academic Res 2.
11. Omorogiuwa E, Ogujor EA (2012) "Determination Of Bus Voltages, Power Losses And Flows In The Nigeria 330kv Integrated Power System"International Journal Of Advances In Engineering and Technology, © IJAET 4: 94-106.
12. Edris (1997) "Proposed Terms and Definitions for Flexible AC Transmission System (FACTS)", IEEE Trans. Power Delivery 12: 1848-1852.
13. IEEE Symposium on Eigenn analysis and Frequency Domain Methods for System Dynamic Performance. IEEE Publication No. 90TH0292-3-PWR.
14. IEEE Symposium on Inter-Area Oscillations in Power Systems, IEEE Publication No. 95TP101, 1994.
15. Li N, Xu Y, Chen H (2000) "FACTS-Based Power Flow Control in Interconnected Power Systems", IEEE Transactions on Power System 15: 257-262.
16. Bhanu C, Krishma K, Kotamarti S, Sankar B, Haranath V (2003) "Power System Operation and Control using FACTS Devices.Barcelona.17th International Conference on Electricity Distribution: CIRED, session 5: 12-15.
17. Glanzmann G, Anderson G (2001) "Coordinated control of FACTS devices based on optimal power flow. Power Engineering Journal 12: 28-34.
18. Song YH, Johns AT (1999) "Flexible AC Transmission Systems(FACTS),IEE Press, London 296-771-3.
19. Larsen E, Bowler V, Damsky C, Nilsson S (1992) Benefits of Thyristor Controlled Series Compensation (TCSC), International Conference on Large High Voltage Electric Systems (CIGRE)
20. Gerbex S, Cherkaoui R, Germund AJ (2001) "optimal location of multi-type FACTS devices in a power system by means of genetic algorithms," IEEE Trans power systems 16: 537-544.
21. Wang HF (2000)"A Unified Model For The Analysis Of FACTS Devices In Damping Power System Oscillations-Part Iii:Unified Power Flow Controller",IEEETrans.On Power Delivery 15: 978-983.
22. Hingorani NG, Gyugyi I (2000) "Understanding FACTS: Concepts And Technology Of Flexible AC Transmission Systems". New York: IEEE Press.
23. Venayagamoorthy GK, Del Valle, Mohagheghi Y (2005) "Effects of a STATCOM, SCRC and UPFC on The Dynamic Behavior Of A 45 Bus Section Of The Brazilian Power System,"Inaugural IEEE PES 2005 Conference And Exposition In Africa, Durban, South Africa 11: 312.
24. Mahdad I, Srairi K (2007) "Dynamic Methodology for Control of Multiple-UPFC to Relieve Overloads and Voltage Violations", The International Conference on (Computer as a Tool) EUROCON 1: 1579-1585.
25. Nabavi-Niaki A, Iravani MR (1996) Steady-State and Dynamic Models Of Unified Power Flow Controller (UPFC) For Power System Studies, IEEE Transactions On Power Systems 11: 1937-1943.
26. Chang CT, Hsu YY (2002) "Design Of UPFC Controllers And Supplementary Damping Controller For Power Transmission Control And Stability Enhancement Of A Longitudinal Power System", IEEE Proceedings 149.

27. Uzunovic E, Faardanesh B, Macdonald Z, Schauder SJ (2000) "Interline Power Flow Controller (IPFC): A Part of Convertible Static Compensators(CSC)" North American Power Symposium.

28. Rajarama R, Alvardo F, Camfield R, Jalali S (1998) "Determination of location and amount of series compensation to increase power transfer capability", IEEE Transactions on Power systems 13: 294-299.

29. Canizares CA (2000) "Power Flow and Transient Stability Models of FACTS Controllers for Voltage and Angle Stability Studies", IEEE/PES Panel on Modeling, Simulation and Application of FACTS Controller in Angle and Voltage Stability Studies, Singapore 1-8.

30. Kuiava R, Rodrigo A, Newton GB (2009) "Robust Control Methodology For The Design of Supplementary Damping Controllers for Facts devices. RevistaControle and Automation, 20.

31. Gotham D, Heydt GT (1998) "Power Flow control and Power Flow Studies for Systems with FACTS devices", IEEE Transaction on Power Systems 13: 60-65.

32. Holland JH (1992) "Adaptation in Natural and Artificial Systems, MIT Press, Cambridge.

33. Koza JR (1992) "Genetic Programming: On Programming Computers by Means of Natural Selection and Genetics", The MIT Press, Cambridge, MA.

34. Michalewicz Z (1992) "Genetic Algorithms + Data Structures=Evolution Programs, Artificial Intelligence", Springer-Verlag, New York.

35. Reeves CR (1993) Modern Heuristic Techniques for Combinatorial Problems, Blackwell Scientific Publications, Oxford.

36. Grefenstette JJ (1994) Genetic Algorithms for Machine Learning, Kluwer Academic Publishers.

37. Back T (1995) Evolutionary Algorithms in Theory and Practice, Oxford University Press, New York.

38. Mitchell M (1996) An Introduction to Genetic Algorithms, MIT Press, Cambridge, MA.

39. Dasgupta D, Michalewicz Z (1997) Evolutionary Algorithms in Engineering Applications, Springer-Verlag, Berlin.

40. Gen M, Cheng R (1997) Genetic Algorithms and Engineering Design, Engineering Design and Automation, John Wiley and Sons, New York.

41. Haupt RL, Haupt SE (1998) "Practical Genetic Algorithms," John Wiley and Sons, Inc., New York.

42. Bhattacharyya A, Goswami B (2011) "Optimal Placement of FACTS Devices by Genetic Algorithm for the Increased Load Ability of a Power System" World Academy of Science, Engineering and Technology 75.

43. Leung HC, Chung TS (2000) "Optimal Power Flow with a Versatile FACTS controller by Genetic Algorithm Approach", Power Engineering Society IEEE Winter Meeting 4: 2806-2811.

44. Karr CL, Freeman LM (1998) Industrial Applications of Genetic Algorithms, CRC Press, Boca Raton, FL.

45. Bagchi TP (1999) Multi objective Scheduling by Genetic Algorithms, Kluwer Academic Publishers, Dordrecht (The Netherlands).

Assess the Modeling Effects of PSS and Governor on Voltage Stability of Power System

Amir Sharifian*

Department of Electrical Engineering, Ahrar Institute of Technology and Higher Education, P.O.Box 41931-63584, Rasht, Iran

Abstract

In this paper, we assess the modeling effects of PSS and governor on voltage stability of power system by applied accurate dynamic model of power system. This model consists of the detailed models for the synchronous machines, automatic voltage regulators (AVRs), Prime mover and speed governor, Power system stabilizer (PSS) is used for voltage stability assessment.

Voltage stability assessment is done by calculation of voltage stability margin (VSM). The VSM is an index that describes distance between the current operation state point and the maximum voltage stability limit point. The IEEE 14-bus standard test system is used for simulation. The obtained results from preformed simulation show effects of PSS and governor on voltage stability.

Keywords: Voltage stability assessment; Voltage stability margin; Power system modeling

Introduction

Due to open market of electricity, economic reasons and environmental constraints the modern power systems are utilized closer to their stability limits and the voltage stability problem has become a major challenge in the utilization of power systems. Hence, the on-line voltage instability assessment has a pivotal contribution in protecting and management of power systems [1,2]. The voltage stability is described capability of a power system to preserve both transferred power and voltage controllable during different contingencies. Voltage instability leads to uncontrollable voltage decrease in a considerable part of the power system. Voltage collapse occurs when the power system is heavily loaded and cannot preserve its generation and transmission scheme.

Various methods have been presented for evaluation the voltage stability problem. They are mainly organized into two discrete categories encompassing static method (such as L-index and modal analysis) and dynamic method (such as time domain simulation) [3]. In the static methods, the steady state power flow model of the power systems is used to evaluate voltage stability. This method is very fast and Practical in online application, but its obtained results is not reliable and accurate as shown in [4]. In the dynamic method [5-7], dynamic model of power system is employed to the voltage stability assessment. Therefore, we need to solve a set of differential and algebraic equations (DAEs) of power system in time domain. Although this approach is very reliable and accurate, it need large calculations time and is not suitable for on-line applications [8-10].

A separate MLP NN has been utilized to monitor online voltage stability for different contingencies [11]. In addition, in order to choose features for training the MLP NN, a regression method has been employed. In order to estimate a long-term voltage stability margin, an artificial neural network (ANN) has been used to the accompaniment of sequential forward selection approach for feature selection [12]. A separate MLP NN is applied to estimate voltage stability margin (VSM) [13]. The load buses voltage magnitudes and active and reactive powers are used the ANN inputs. An artificial feed forward neural network (FFNN) is applied to estimate L-index as an index for voltage stability monitoring in power systems [14]. In this method voltage, real power and reactive power of load and generator buses are utilized to train FFNN. In order to classify the contingencies that result in voltage instability in steady state RBF NN is used [15]. In this method, a correlation coefficient and class reparability index are applied to reduce the number of RBF NN inputs. A single enhanced radial basis function network is utilized to evaluate voltage stability by estimating MW margins for various contingencies [16]. In order to estimate VSM, a RBF NN is employed to the accompaniment of wavelet feature extraction approach [17]. In this method, both Multi-Resolution Wavelet Transform (MRWT) and principle component analysis (PCA) are used to compact the voltage profiles features. Maximum L-index of the load buses is estimated by RBF NN for voltage stability analysis [18]. In this method, pre-contingency state power flows compressed by mutual information method for feature selection are employed as inputs of RBF NN. In ref [7], in order to assess voltage stability, a MLP NN is utilized to approximate voltage stability margin (VSM). In this method, two procedures inclusive the Gram–Schmidt orthogonalization and ANN-based sensitivity are employed to minimize the number of input variables [19].

In this paper, we assess the modeling effects of PSS and governor on voltage stability of power system by applied accurate dynamic model of power system. This model consists of the detailed models for the synchronous machines, automatic voltage regulators (AVRs), Prime mover and speed governor, Power system stabilizer (PSS) is used for voltage stability assessment.

Voltage stability assessment is done by calculation of voltage stability margin (VSM). The VSM is an index that describes distance between the current operation state point and the maximum voltage stability limit point. The IEEE 14-bus standard test system is used for

***Corresponding author:** Amir Sharifian, Department of Electrical Engineering, Ahrar Institute of Technology and Higher Education, P.O.Box 41931-63584, Rasht, Iran, E-mail: amir.sharifiyan@gmail.com

simulation. The obtained results from preformed simulation show effects of PSS and governor on voltage stability.

In summary, the innovations of this paper are listed as follows:

- The modeling effects of PSS and governor on voltage stability is evaluated by applied accurate dynamic model of power system consists of the detailed models for the synchronous machines, automatic voltage regulators (AVRs), Prime mover and speed governor, Power system stabilizer (PSS).
- Voltage stability assessment is done by VSM index.

The rest of paper is classified as follows:

Section 4 defines power system model and equations. Section 5 introduces VSM index that is used to assess voltage stability. Section 6 describes the simulation results and discussion and finally, conclusions are outlined in Section 7.

Formulation of Multi-Machine Power System with SVC

In this section, details modeling of the multi-machine power system with SVC are described. It is assumed that the power system has *m* busses and *n* generators. It is mentioned that the assumed power system is simulated by using the software tool PSAT (Power System Analysis Toolbox) [20]. The notations of parameters in power system equations are described in Appendix 1.

Synchronous machine model

In this paper, we apply the two-axis model for description the synchronous machines. The ith synchronous machine fourth-order differential equations are formulated as follows [20-25]:

$$\frac{d\delta_i}{dt} = \omega_i - \omega_s, \quad i = 1,2,\dots,n \tag{1}$$

$$\frac{d\omega_i}{dt} = \frac{1}{M_i}[P_{mi} - (E'_{qi} - X'_{di}I_{di})I_{qi} - (E'_{di} - X'_{qi}I_{qi})I_{di} - D_i(\omega_i - \omega_s)], \quad i = 1,2,\dots,n \tag{2}$$

$$\frac{dE'_{qi}}{dt} = \frac{1}{T'_{doi}}[E_{fdi} - E'_{qi} - (X_{di} - X'_{di})I_{di}], \quad i = 1,2,\dots,n \tag{3}$$

$$\frac{dE'_{di}}{dt} = \frac{1}{T'_{qoi}}[-E'_{di} + (X_{qi} - X'_{qi})I_{qi}], \quad i = 1,2,\dots,n \tag{4}$$

The equation between power network and machine voltages is described as follows:

$$E'_{qi} = V_i\cos(\delta_i - \theta_i) + R_{si}I_{qi} + X'_{di}I_{di}, \quad i = 1,2,\dots,n \tag{5}$$

$$E'_{di} = V_i\sin(\delta_i - \theta_i) + R_{si}I_{di} + X'_{qi}I_{qi}, \quad i = 1,2,\dots,n \tag{6}$$

The ith machine currents I_{di} and I_{qi} can be obtained as follows:

$$I_{di} = [R_{si}E'_{di} + E'_{qi}X'_{qi} - R_{si}V_i\sin(\delta_i - \theta_i) - X'_{qi}V_i\cos(\delta_i - \theta_i)]A_i^{-1}, \quad i = 1,2,\dots,n \tag{7}$$

$$I_{qi} = [R_{si}E'_{qi} - E'_{di}X'_{di} - R_{si}V_i\cos(\delta_i - \theta_i) + X'_{di}V_i\sin(\delta_i - \theta_i)]A_i^{-1}, \quad i = 1,2,\dots,n \tag{8}$$

$$A_i = R_{si}^2 + X'_{di}X'_{qi}, \quad i = 1,2,\dots,n \tag{9}$$

Automatic voltage regulator

The IEEE model 1 is applied for the excitation system (Figure 1). The mathematical formulation of the AVR can be written as follows [20-25]:

$$\frac{dV_{ri}}{dt} = \frac{1}{T_{ai}}[-V_{ri} + K_{ai}(V_{refi} - V_i + V_{si} - V_{fi})], \quad i = 1,2,\dots,n \tag{10}$$

$$\frac{dE_{fdi}}{dt} = \frac{1}{T_{ei}}[V_{ri} - (1 + S_{ei}(E_{fdi}))E_{fdi}], \quad i = 1,2,\dots,n \tag{11}$$

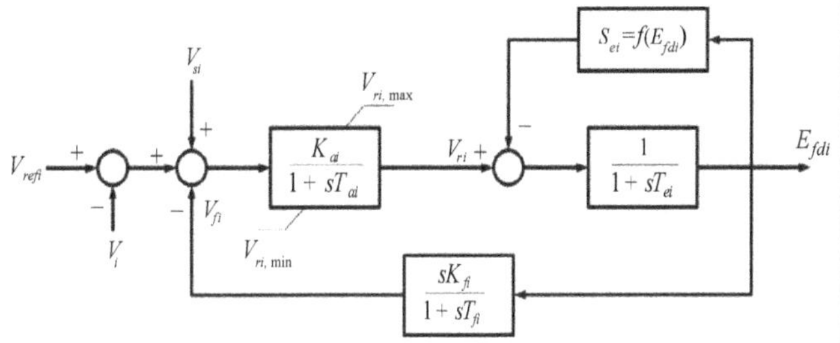

Figure 1: IEEE type 1 excitation system.

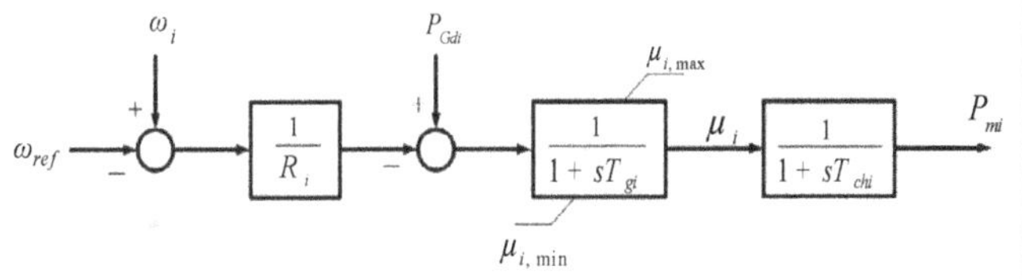

Figure 2: The simplified speed governor and prime mover.

$$\frac{dV_{fi}}{dt} = \frac{1}{T_{fi}}[-V_{fi} - (1 + S_{ei}(E_{fdi}))\frac{K_{fi}E_{fdi}}{T_{ei}} + \frac{K_{fi}V_{ri}}{T_{ei}}], \quad i = 1,2,\dots,n \tag{12}$$

$$S_{ei}(E_{fdi}) = A_{ei}(e^{B_{ei}|E_{fdi}|} - 1), \quad i = 1,2,\dots,n \tag{13}$$

Prime mover and speed governor

The block diagram of a simplified prime mover and speed governor [24,25]. The mathematical formulation can be expressed as follows (Figure 2):

$$\frac{dP_{mi}}{dt} = \frac{1}{T_{chi}}[\mu_i - P_{mi}], \quad i = 1,2,\dots,n \tag{14}$$

$$\frac{d\mu_i}{dt} = \frac{1}{T_{gi}}[P_{Gdi}(\lambda) - \frac{1}{R_i}(\omega_i - \omega_{ref}) - \mu_i], \quad i = 1,2,\dots,n \tag{15}$$

$$P_{Gdi}(\lambda) = P_{Gdi0}(1 + \lambda K_{Gi}), \quad i = 1,2,\dots,n \tag{16}$$

$$\mu_{i,\min} \le \mu_i \le \mu_{i,\max}, \quad i = 1,2,\dots,n \tag{17}$$

It should be noted that the value of K_{Gi} in Eq. (16), can be determined by EDC, AGC or other system operating practice as varies.

Power system stabilizer (PSS)

The PSS is an additional control block that is added to the AVR in order to enhance the system stability [20,21]. A PSS uses stabilizing feedback signals such as shaft speed, terminal frequency and/or power to correct the input signal of the AVR (Figure 3).

SVC

The SVC is applied to maintain or control bus voltage [23,26]. The popular structure of SVC is a parallel combination of thyristor controlled reactor (TCR) with fixed capacitor (Figure 4).

In addition to the main job of the SVC, it is maybe used to damp system oscillations, as denoted by "SVC-sig" (Figure 5).

Network power equations

The power network equations can be described as follows [24]:

$$\begin{cases} P_{Gi} - P_{Li}(\lambda) - P_{Ti} = 0 \\ Q_{Gi} - Q_{Li}(\lambda) - Q_{Ti} = 0 \end{cases}, \quad i = 1,2,\dots,m \tag{18}$$

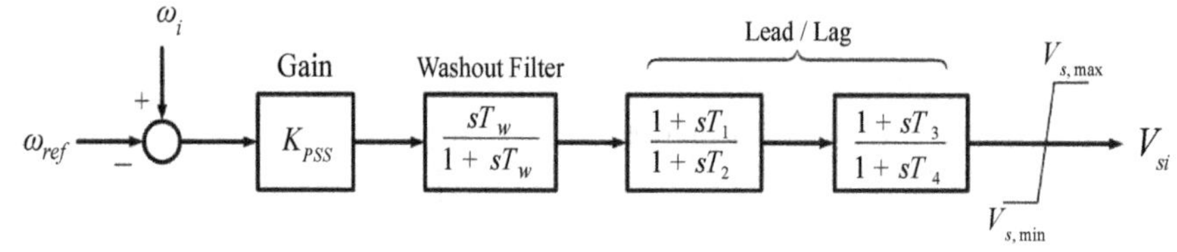

Figure 3: The Basic block PSS diagram.

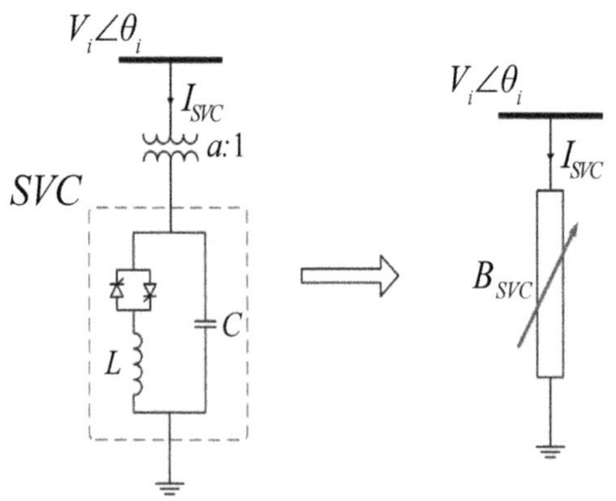

Figure 4: The structure of SVC.

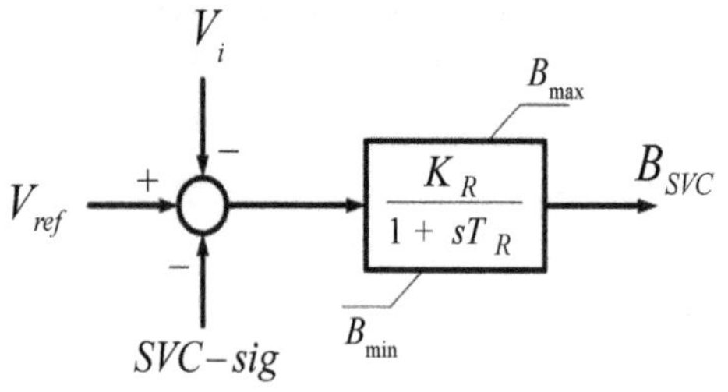

Figure 5: The block diagram of SVC.

$$\begin{cases} P_{Ti} = \sum_{k=1}^{m} V_i V_k y_{ik} \cos(\theta_i - \theta_k - \gamma_{ik}) \\ Q_{Ti} = \sum_{k=1}^{m} V_i V_k y_{ik} \sin(\theta_i - \theta_k - \gamma_{ik}) \end{cases}, \quad i = 1,2,\ldots,m \tag{19}$$

$$\begin{cases} P_{Gi} = I_{di} V_i \sin(\delta_i - \theta_i) + I_{qi} V_i \cos(\delta_i - \theta_i) \\ Q_{Gi} = I_{di} V_i \cos(\delta_i - \theta_i) - I_{qi} V_i \sin(\delta_i - \theta_i) \end{cases}, \quad i = 1,2,\ldots,n \tag{20}$$

P_{Gi} and Q_{Gi} are primarily obtained by the intrinsic characteristics of the AVR regulations and the speed governor. The values of $P_{Li}(\lambda)$ and $Q_{Li}(\lambda)$ for different load scenarios can be described as follows:

$$\begin{aligned} P_{Li}(\lambda) &= P_{Li0} + \lambda[K_{Li} S_{\Delta base} \cos(\phi_i)] \\ Q_{Li}(\lambda) &= Q_{Li0} + \lambda[K_{Li} S_{\Delta base} \sin(\phi_i)] \end{aligned} \tag{21}$$

Where

$$\begin{aligned} S_{\Delta base} \cos(\phi_i) &= P_{Li0} \\ S_{\Delta base} \sin(\phi_i) &= Q_{Li0} \end{aligned} \tag{22}$$

Substituting Eq. (22) in Eq. (21), we have;

$$P_{Li}(\lambda) = P_{Li0}(1 + \lambda K_{Li}) \tag{23}$$

$$Q_{Li}(\lambda) = Q_{Li0}(1 + \lambda K_{Li}) \tag{24}$$

Notice that the values of K_{Li}, K_{Gi}, and φ_i can be uniquely determined for every bus.

Test system

In this paper an IEEE standard 14-bus test system is applied for voltage stability assessment (Figure 6). The IEEE 14 Bus Test Case represents a portion of the American Electric Power System (in the Midwestern US). This electric network is constituted of 14 buses and 5 generators (at buses No. 1, 2, 3, 6 and 8) injecting their powers for a system nourishing 11 loads through 20 lines of transportation with 11 loads totaling 259 MW and 73.5 MVAR at base case. In addition, this system has 1 static VAR compensator (SVC) that is connected at bus 4.

Voltage Stability Assessment

Nowadays, the voltage stability problem is one of the fundamental challenges in design and exploitation of modern power systems. The important factors causing voltage instability are load characteristics, the inability of the power system to generate the demanded reactive power, characteristics of reactive compensation devices and the action of voltage control devices and the generator reactive power limits. Voltage instability leads to uncontrollable voltage decrease in some bus of power system. There are many methods for voltage stability assessment. In general, these methods can be divided into dynamic methods and static methods. In this paper we combine both of dynamic method and static method to voltage stability accurate assessment [27,28].

Voltage stability margin (VSM)

The Voltage stability margin (VSM) is a practical index for voltage stability assessment. The VSM index is defined as the distance from the current operating point to the voltage collapse point according to the system loading parameter λ. It should be noted that in first step, the saddle-node bifurcation (SNB) point or nose point should be located for computing the VSM index. In this paper the continuation methods [24,29,30] is applied to compute the SNB point (Figure 7).

The VSM index is computed using a continuation power flow method. The VSM index is defined as:

$$VSM = \frac{|S_{Max}| - |S_{base}|}{|S_{base}|} \tag{25}$$

For a power system with m buses, the VSM index can be computed as:

$$VSM = \frac{\sum_{i=1}^{m} \sqrt{P_{Li\max}^2 + Q_{Li\max}^2} - \sum_{i=1}^{m} \sqrt{P_{Li0}^2 + Q_{Li0}^2}}{\sum_{i=1}^{m} \sqrt{P_{Li0}^2 + Q_{Li0}^2}} \tag{26}$$

Where

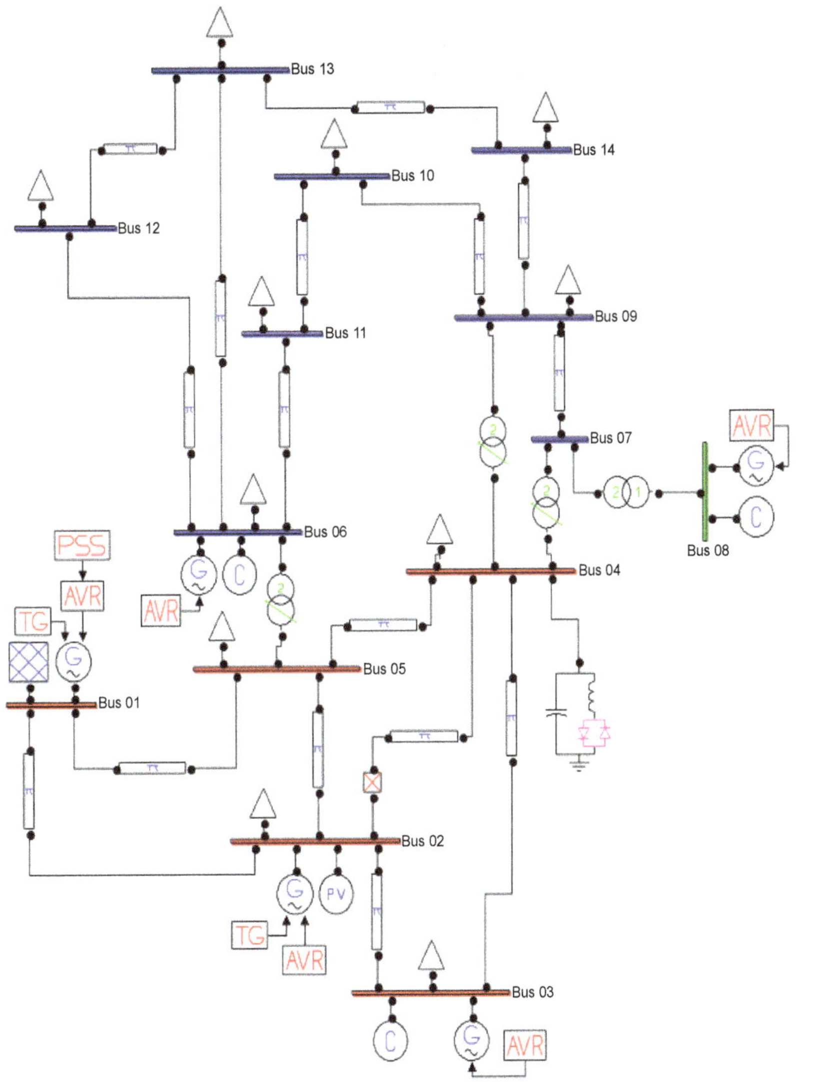

Figure 6: Diagram of the IEEE standard14-bus test system.

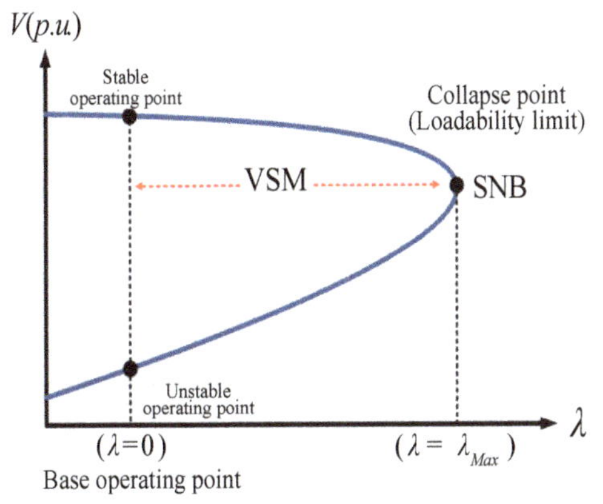

Figure 7: Illustration of the voltage stability margin (VSM).

$$P_{Li\max} = P_{Li0}(1+\lambda_{\max}K_{Li}) \quad (27)$$

$$Q_{Li\max} = Q_{Li0}(1+\lambda_{\max}K_{Li}) \quad (28)$$

The continuation methods that is applied to compute the VSM index, is precisionist and strong.

Results and Discussion

In this section, we evaluate effects of power system modeling on voltage stability margin (VSM). The P-V curve is used for achievement to this aim. It should be noted that, The P-V curve is drawn for PQ buses which load is connected to it.

None of the previous works did not evaluate the effects of PSS and governor modeling on voltage stability margin. In this paper, four different cases of power system modeling are assumed to evaluate effects of power system modeling on voltage stability margin. For 4 assumed cases, the P-V curve is drawn by simulation of power system equations, and then the value of voltage stability margin is computed (Table 1).

It should be noted that, in this section, in order to better evaluation of power system modeling effects on voltage stability margin; the connected SVC at bus 4 is disregard.

Case number	Used model for power system modeling
1	Static model for power system
2	Dynamic model of power system includes the 4th order dynamic model for the synchronous machines with automatic voltage regulators (AVRs).
3	Dynamic model of power system includes the 4th order dynamic model for the synchronous machines with automatic voltage regulators (AVRs) and Prime mover and speed governor.
4	Dynamic model of power system includes the 4th order dynamic model for the synchronous machines with automatic voltage regulators (AVRs), Prime mover and speed governor and Power system stabilizer (PSS).

Table 1: Used model for power system modelling.

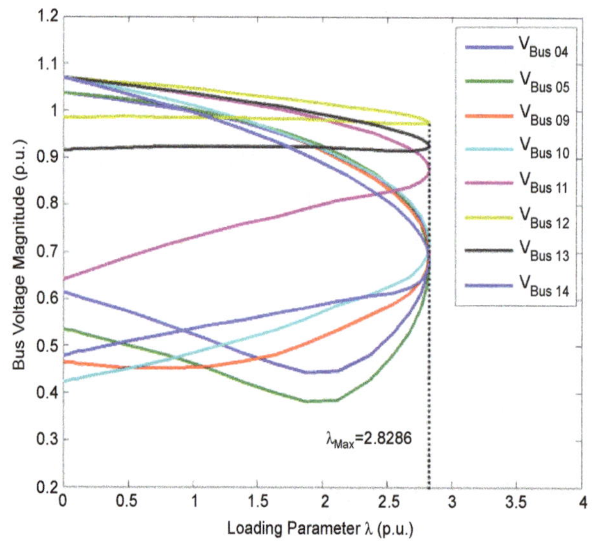

Figure 8: The P-V curve for case study 1.

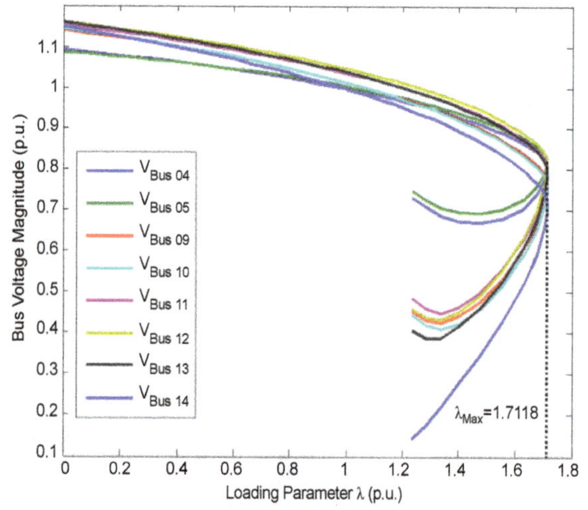

Figure 9: The P-V curve for case study 2.

Case study 1

In this case the static model of power system is used. Therefore, the effects of power system dynamics are ignored.

According to the Figure 8, the computed value of VSM for this case is equal to 2.8286.

Case study 2

In this case the dynamic model of power system includes the 4th order dynamic model for the synchronous machines with automatic voltage regulators (AVRs) is applied.

According to Figure 9, the computed value of VSM for this case is equal to 1.7118. It should be noted that in this case the value of VSM has decreased in compared to case study 1.

Case study 3

In this case the modeling effects of Prime mover and speed governor on voltage stability margin are assessed. For achievement to this purpose, the effects of Prime mover and speed governor are modeled for generators connected at buses 1 and 2. It should be noted that, the generators connected at the buses 3, 6 and 8 are synchronous condensers.

According to Figure 10, the computed value of VSM for this case is equal to 1.7086. It should be noted that in this case the value of VSM has decreased in compared to case study 2.

Case study 4

In this case the modeling effects of Power system stabilizer (PSS) on voltage stability margin are assessed. For achievement to this purpose, the effect of Power system stabilizer (PSS) is modeled for generator connected at buses 1.

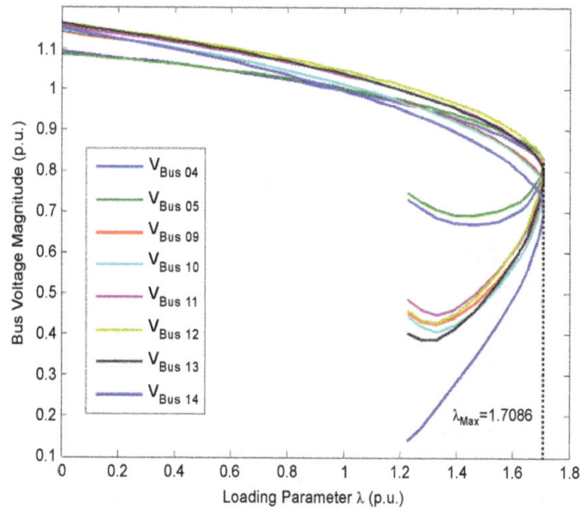

Figure 10: The P-V curve for case study 3.

According to Figure 11, the computed value of VSM for this case is equal to 1.7119. It should be noted that in this case the value of VSM has increased in compared to case study 3.

In order to better assessment of simulations results, the value of VSM and simulation time associated to four assumed case study are compared together (Table 2).

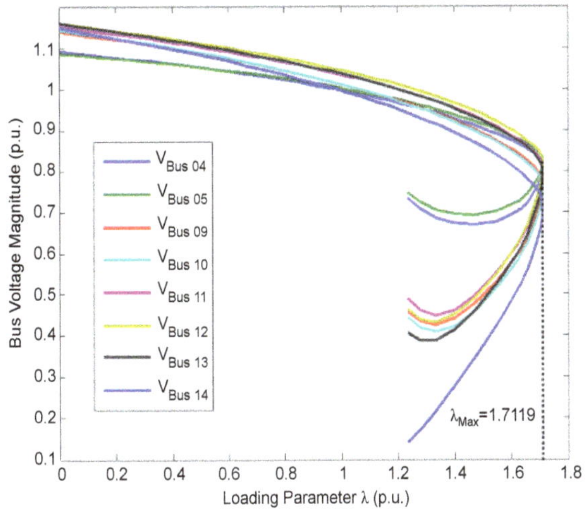

Figure 11: The P-V curve for case study 4.

Case number	Value of VSM	Simulation time
1	2.8286	2.1494 s
2	1.7118	3.5343 s
3	1.7086	4.0845 s
4	1.7119	4.5202 s

Table 2: The value of VSM and simulation time associated to four assumed case study.

According to Table 2, the obtained value of VSM and simulation time for four assumed case study are different. The simulation time has increased by modeling more details of a power system. On the other hand, modeling more details of power system leads to more accurate calculation of VSM value.

Nowadays, modeling effects of PSS on voltage stability assessment cannot be overlooked because all generators of a practical power system are equipped to PSS. Therefore, the presented model of power system in case study 4 is an operational model for voltage stability assessment. This model is highly accurate and reliable but its simulation is very time-consuming. In this paper proposed model of power system in case study 4 is used to assess voltage stability.

Conclusion

In this paper, we assess the modeling effects of PSS and governor on voltage stability of power system by applied accurate dynamic model of power system. This model consists of the detailed models for the synchronous machines, automatic voltage regulators (AVRs), Prime mover and speed governor, Power system stabilizer (PSS) is used for voltage stability assessment.

Voltage stability assessment is done by calculation of voltage stability margin (VSM). The obtained results from preformed simulation show effects of PSS and governor on voltage stability.

References

1. Jayasankar V, Kamaraj N, Vanaja N (2010) Estimation of voltage stability index for power system employing artificial neural network technique and TCSC placement. Neurocomputing 73: 3005-3011.
2. Wang L, Morison K (2006) Implementation of online security assessment. IEEE Power Energy Mag 4: 47-59.
3. Acharjee P (2012) Identification of maximum loadability limit and weak buses using security constraint genetic algorithm. Int J Electr Power Energy Syst 36: 40-50.
4. Padma Subramanian D, Kumudini Devi RP, Saravanaselvan R (2011) A new algorithm for analysis of SVC's impact on bifurcation, chaos and voltage collapse in power systems. Int J Electr Power Energy Syst 33: 1194-202.
5. Demarco CL, Overbye TJ (1990) An energy based security measure for assessing vulnerability to voltage collapse. IEEE Trans Power Syst 5: 419-427.
6. Rajagopalan C, Lesieutre B, Sauer PW, Pai MA (1992) Dynamic aspects of voltage/power characteristics. IEEE Trans Power Syst 7: 990-1000.
7. Bahmanyar AR, Karami F (2014) Power system voltage stability monitoring using artificial neural networks with a reduced set of inputs. Int J Electr Power Energy Syst 58: 246-256.
8. Sharifian A, Sharifian S (2015) A new power system transient stability assessment method based on Type-2 fuzzy neural network estimation. Int J Electr Power Energy Syst 64: 71-87.
9. Mohammadzadeh A, Ghaemi S (2015) Synchronization of chaotic systems and identification of nonlinear systems by using recurrent hierarchical type-2 fuzzy neural networks. ISA Transactions 58: 318-329.
10. Lou CW, Dong MC (2015) A novel random fuzzy neural networks for tackling uncertainties of electric load forecasting Original Research Article. Int J Electr Power Energy Syst 73: 34-44.
11. Chakrabarti S (2008) Voltage stability monitoring by artificial neural network using a regression-based feature selection method. Expert Systems with Applications 35: 1802-1808.
12. Zhou DQ, Annakkage UD, Rajapakse AD (2010) Online monitoring of voltage stability margin using an artificial neural network. IEEE Trans Power Syst 25: 1566-1574.
13. Vakil-Baghmisheh MT, Razmi H (2008) Dynamic voltage stability assessment of power transmission systems using neural networks. Energy Convers Manage 49: 1-7.
14. Kamalasadan S, Thukaram D, Srivastava AK (2009) A new intelligent algorithm for online voltage stability assessment and monitoring. Int J Electr Power Energy Syst 13:100-109.
15. Jainn T, Srivastava L, Singh SN (2003) Fast voltage contingency screening using radial basis function neural network. IEEE Trans Power Syst 18: 1359-1366.
16. Chakrabarti S, Jeyasurya B (2008) Multicontingency voltage stability monitoring of a power system using an adaptive radial basis function network. Int J Electr Power Energy Syst 30: 1-7.
17. Hashemi S, Aghamohammadi MR (2013) Wavelet based feature extraction of voltage profile for online voltage stability assessment using RBF neural network. Int J Electr Power Energy Syst 49: 86-94.
18. Devaraj D, Roselyn JP (2011) On-line voltage stability assessment using radial basis function network model with reduced input features. Int J Electr Power Energy Syst 33: 1550-1555.
19. Yang CF, Lai G, Lee CH, Su CT, Chang GW (2012) Optimal setting of reactive compensation devices with an improved voltage stability index for voltage stability enhancement. Int J Electr Power Energy Syst 37: 50-57.
20. Ajjarapu V (2007) Computational techniques for voltage stability assessment and control. Springer.
21. Kundur P (1994) Power system stability and control. McGraw-Hill.
22. Sauer PW, Pai MA (1998) Power system dynamics and stability. New Jersey: Prentice-Hall.
23. Milano F (2005) An open source power system analysis toolbox. IEEE Trans Power Syst 20: 1199-1206.
24. Razmi H, Shayanfar HA, Teshnehlab M (2012) Steady state voltage stability with AVR voltage constraints. Int J Electr Power Energy Syst 43: 650-659.
25. Miller THE (1982) Reactive power control in electric systems. New York: Wiley interscience.
26. Kennedy J, Eberhart R (1995) Particle swarm optimization. IEEE Int Conf Neural network 4: 1942-1948.
27. Kayacan E, Khanesar MA (2016) Chapter 8 - Hybrid Training Method for Type-

2 Fuzzy Neural Networks Using Particle Swarm Optimization. Fuzzy Neural Networks for Real Time Control Applications 133-160.

28. Sydel R (1994) Practical bifurcation and stability analysis: from equilibrium to chaos. New York: Springer.

29. Canizares C, Alvardo FL (1993) Point of collapse and continuation methods for large AC/DC systems. IEEE Trans Power Syst 8: 1-8.

30. Khodabakhshian A, Hemmati R (2013) Multi-machine power system stabilizer design by using cultural algorithms. Int J Electr Power Energy Syst 44: 571-580.

Design of Multichannel Sample Rate Convertor

Jain V* and Agrawal N

Department of Electronics and Communication, College of Technology and Engineering, MPUAT, Udaipur, India

Abstract

The multiobjective design of multichannel sample rate convertor using Genetic optimization technique is considered in this paper. This new optimization tool is based on mechanism of biological evolution. It is characterized by design of natural system retaining its robustness and adaption properties of natural systems. The objectives of multichannel sample rate convertor design include matching some desired response while having minimum linear phase; hence, reducing the time response, constant group delay, increasing bandwidth. Genetic optimization technique is also used for reducing the power consumption of multichannel sample rate convertor by optimization of coefficient of filter by scaling which are used in implementation of multichannel sample rate convertor design in FPGA implementation. After applying genetic algorithm 1 to 128 channel sample rate convertor bandwidth increased by 150%, power reduced by 62% to 85%, dynamic power reduced by 31% to 54% of conventional sample rate convertor, constant and less group delay, linear phase response, reducing time response. In an extended work the authors have tried and successfully executed the model and system upto 128 channels. The proposed model is first designed on simulink platform using Xilinx Blockset and then it is transferred on FPGA platform using system generator. The complete circuit is synthesized, implemented, simulated using Xilinx design suite.

Keywords: Multiobjective design; Genetic optimization technique; Magnitude response; Minimum linear phase; Group delay; Pipelines

Introduction

The full utilization and practical realization of Multichannel sample rate convertor is restricted by increased power consumption and resource utilization of the interpolator and decimator design in the context of contemporary wireless broad band standards using. So we are trying to make all the multimedia devices compatible with all the wireless communication multiple standards. For that purpose we are about to develop sample rate converter which will provide the flexibility to change its sampling rate according to requirement of the different wireless multiple standards. The second challenge of resampler design is that the numbers of channels are limited. In such a case, we have to develop multichannel interpolator and decimator of high performance low power consumption with near optimal performance.

Generally conventional interpolator and decimator implemented by using direct form FIR filter structure. The problem with Implementing Sample Rate Convertor using direct form architecture was that filter length linearly increases with the decimation and Interpolation rate. Therefore resource Utilization also increases; this in turn increases Power consumption, Area requirement and Delay of the sample rate convertor. To overcome these problem CIC FIR filter structure to implement sample rate convertor. CIC filters can efficiently perform either decimation or interpolation, with two complementary structures being employed to implement these functions. The problem with implementation of Sample Rate Convertor using CIC filter was that CIC filter was used only for narrowband signals. So it would not be used for implementation of sample rate convertor for large bandwidth. MAC architecture was an efficient solution developed as a sample rate convertor for large bandwidth signal.

MAC Architecture can also be used for increase bandwidth of the input signal. This application note deals with single MAC filters, as the low sample rates at which the FIR filters operate allow high clock-per-sample ratios, allowing many taps to be calculated in a single multiplier in each sample period. Limitation of MAC Architecture is not support for large change in sampling rate conversion but allow large bandwidth signal. Our purpose model CMFIR filter is an efficient solution of that problem. On CMFIR implement cascading of CIC and MAC FIR filter [1,2].

We apply genetic algorithm on coefficient of CMFIR filter to achieve desired frequency and magnitude response while having minimum linear phase; hence, reducing the time response, constant group delay, increasing bandwidth. Genetic optimization technique is also used for reducing the power consumption of multichannel sample rate convertor by optimization of coefficient of filter by scaling which are used in implementation of multichannel sample rate convertor design in FPGA implementation.

Problem Formulations

1. Number of Input Channels limited the total filter length grows linearly with the decimation rate. So that resource Utilization also increases.

2. To achieve change in high sampling rate is not possible with existing polyphase structure. It is not reliable.

3. Sample rate Convertor is implemented by using the Cascaded Integrator Comb (CIC) filter for narrowband.

4. Processing speed is very low.

5. Power consumption is very large.

6. Current commercial Sample Rate Convertor typically supports 32 channels.

***Corresponding author:** Jain V, Department of Electronics and Communication, College of Technology and Engineering MPUAT, Udaipur, India
E-mail: vivekjain297@gmail.com

7. We cannot change the sampling during the run time.

8. Non Linear Phase response of sample rate convertor.

9. Group Delay of sample rate convertor is very high.

Objective

1. To design and optimize CIC FIR filter to prevent aliasing.

2. To design and optimize MAC FIR filter to smoothen out the signal transitions.

3. Cascading of CIC and MAC FIR filter to design and optimize CM FIR filter.

4. To design and optimize suitable decimator and interpolator using CM FIR filter, delay and advanced units.

5. Assembling of single channel sample rate convertor using decimator and interpolator designed in objective (d).

6. To develop and optimize 8 channel sample rate convertor using multiplexer unit and single channel sample rate convertor.

7. To develop and optimize 16, 32, 64, 128 channel sample rate convertor using 8 channel sample rate convertor and multiplexer.

8. FPGA implementation of multichannel sample rate convertor and performance analysis.

Genetic Algorithms

Genetic Algorithm is very flexible, no problem specific, and robust. It can explore multiple regions of the parameter space for solutions simultaneously. Owing to the heuristic nature of GAs, arbitrary constraints can be imposed on the objective function without increasing the mathematical complexity of the problem. Multiple objective functions can be optimized simultaneously.

Gene: A single encoding of part of the solution space, i.e., either single bits or short blocks of adjacent bits that encode an element of the candidate solution.

Chromosome: A string of genes that represents a solution.

Population: The number of chromosomes available to test. Start with a population of candidate solutions

Variation: Introduce variation by applying two operators: crossover and mutation

Survival of the fittest: Use a fitness criterion to bias the evolution towards desired features.

Representation:

1. GAs on primarily two types of representations:

2. Binary Coded [0110, 0011, 1101,]

3. Real Coded [13.2, -18.11, 5.72,]

4. Binary-Coded (genotype) GAs must decode a chromosome into a real value (phenotype), for evaluating the fitness value.

5. Real-Coded GAs can be regarded as GAs that operates on the actual real value (phenotype).

6. For Real-Coded GAs, no genotype-to-phenotype mapping is needed.

Selection: A proportion of the existing population is selected to bread a new bread of generation.

Crossover: It is a genetic operator that combines (mates) two individuals (parents) to produce two new individuals (Childs). The idea behind crossover is that the new chromosome may be better than both of the parents if it takes the best characteristics from each of the parents.

Mutation: It is a genetic operator used to maintain genetic diversity from one generation of a population of chromosomes to the next.

Steps of genetic algorithm

The genetic algorithm loops over an iteration process to make the population evolves (Figure 1). Each consists of the Following steps:

1. Selection: The first step consists in selecting individuals for reproduction. This selection is done

Randomly with a probability depending on the relative fitness of the individuals so that best ones are often Chosen for reproduction than poor ones.

2. Reproduction: In the second step, offspring are bred by the selected individuals. For generating new chromosomes, the algorithm can use both recombination and mutation.

3. Evaluation: Then the fitness of the new chromosomes is evaluated.

4. Replacement: During the last step, individuals from the old population are killed and replaced by the new ones [3].

Methodology to Develop Multichannel Sample Rate Convertor Approach

A multichannel sample rate convertor based solution is proposed to make all the multimedia devices compatible with all the wireless communication multiple standards. The Multichannel sample rate convertor time response must be included in the requirements. On one hand, the time domain requirement where both a high speed and accurate system response are needed. On the other hand, the frequency domain requirements (DC, sub-synchronous and harmonic components elimination) which are the magnitude response within small bandwidth including sharp frequency edges as well as an approximately constant group delay in this band are required too. Usually the best optimum value of all the objective functions of this multichannel sample rate convertor design can be obtained for some values of design variables. A compromise or a trade-off between the objective functions must be made to achieve a satisfactory multichannel sample design [1,2].

The considered CMFIR multichannel sample rate convertor

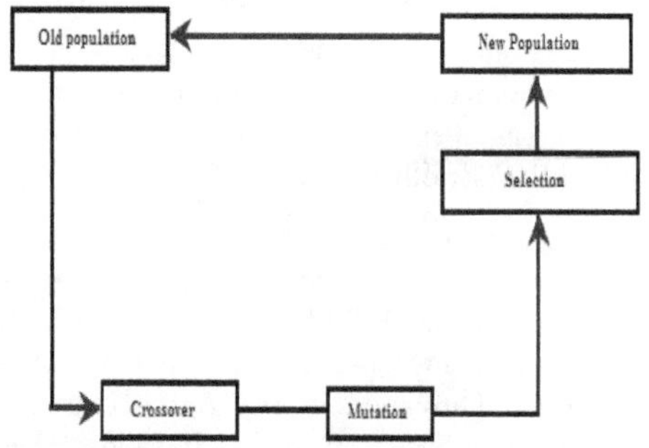

Figure 1: Genetic algorithm cycle.

satisfies four multi-objective functions. These functions are: 1) Meet a specified or a desired response specification; 2) An approximately constant group delay; and 3) A minimum time response or settling time which involves a minimum phase or a group delay. 4) For reducing the power consumption.

Transfer function of multichannel sample rate convertor

The two basic building blocks of a CIC filter are an integrator and a comb. CIC filters can efficiently perform either decimation or interpolation, with two complementary structures being employed to implement these functions. Decimation requires a cascade of a number of integrator units, followed by a down-sampling stage and finally a cascade of the comb filter units. Conversely, interpolation cascades several comb filters with an up-sampler and several integrators. A comb filter running at the high sampling rate, f_s, for a rate change of RM is an odd symmetric FIR filter described by [3].

$$y_1(n) = \sum_{k=0}^{RM-1} x(n-k) \tag{1}$$

$$y_1(n) = y_1(n-1) + x(n) - x(n-RM) \tag{2}$$

The second equality corresponds to a comb is given by:

$$c(n) = x(n) - x(n-RM) \tag{3}$$

Comb followed by an integrator

$$y_1(n) = y_1(n-1) + c(n) \tag{4}$$

$$y_1(n) = y_1(n-1) + x(n) - x(n-RM) \tag{5}$$

Taking Z transform of the equation

$$Y_1(Z) = Y_1(Z)Z^{-1} + X(Z) - X(Z)Z^{-RM} \tag{6}$$

$$Y_1(Z)(1-Z^{-1}) = X(Z)(1-Z^{RM}) \tag{7}$$

$$x = [a_{01}a_{11}a_{21}a_{01}b_{01}....b_{2M}] \tag{8}$$

$$H_1(Z) = \frac{(1-Z^{-RM})}{1-Z^{-1}} \tag{9}$$

To generate parameterizable, high-performance and area-efficient filter modules utilizing the Multiply-Accumulate (MAC) architecture. Multiple MACs can be used in achieving higher performance filter requirements, such as longer filter coefficients, higher throughput, or increased channel support. Output of the MAC FIR filter is given by:

$$y(n) = a_n \sum_{b=0}^{B-1} X_b(n) 2^b \tag{10}$$

Taking Z Transform:

$$Y_2(Z) = a_n \sum_{b=0}^{B-1} X_b(Z) 2^b \tag{11}$$

$$H_2(Z) = a_n \sum_{b=0}^{B-1} 2^b \tag{12}$$

Where

y(n) = Output of the MAC filters

$a_{n=}$ Coefficient of the filter

x_b = The bth bit of x[n].

B = The Total input width.

Cascaded multiple architecture finite impulse response (CMFIR) filter will be implemented by cascading of CIC filter and MAC architecture. The CMFIR filter (CMFIR) is an interpolating and decimator low-pass FIR filter. It provides a further increase in sample rate, reducing the resources requirements on the CIC and limiting the number of stages required, while also providing moderate pass band filtering of the QPSK modulated signal (although this requirement is less stringent where an effective QPSK modulator with good pulse-shaping properties has been used).

Final Response of CMFIR Filter:

$$H(Z) = H_1(Z).H_2(Z) \tag{13}$$

Magnitude response objective function

The amplitude and the phase responses of multichannel

Sample rate convertor are given by: [1]

$$H(Z) = \left(H\left(e^{jwT}\right)\right) \tag{14}$$

$$\Theta(x,w) = \arg(H(e^{jwt})) \tag{15}$$

Where w is the frequency and x is a column vector with 2M + 2N + 1 components, that is in Cartesian form [2].

$$x = [a_{01}a_{11}a_{21}a_{01}b_{01}....b_{2M}] \tag{16}$$

Group-delay and phase response objective function

The group delay is derived from the phase relation, as given in equation (11), and is defined as

GD(x,,w) = (dØ(x,w))/(d(w)) (17)

Where is Ø(x,w) the phase response of the filter [4].

Genetic Algorithm Implementation Methodology

The design flow is highlighted in below (Figure 2). Here are the steps that we follow:

Optimization of power dissipation

1. Given set of design specification, program them into MATLAB.

2. Create an ideal response, group delay of the desired multichannel sample rate convertor.

3. Use the order estimation function provided by MALTAB for each type of multichannel sample rate convertor.

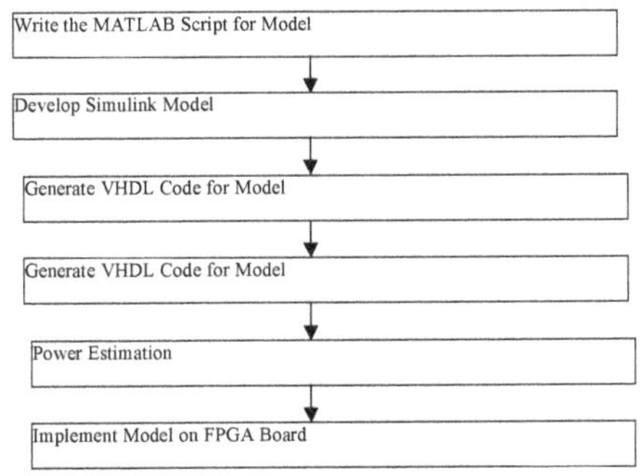

Figure 2: Flow chart of implementation.

4. Design the GA multichannel sample rate filter using MATLAB multichannel sample rate convertor function.

5. Compare the response, phase response, group delay of our design with the ideal multichannel sample rate convertor [2].

Optimization of power dissipation

The coefficient optimization is done in two phases:

In the first phase, all the coefficients are scaled uniformly. The advantage of such an approach is that it does not affect the sample rate convertor characteristics in terms of pass band ripples and stop band attenuation and phase response. The sealing results in the same gain/attenuation ratio.

In the second phase of optimization one coefficient is perturbed in the each iteration. In case of requirement to retain the linear phase characteristics, the coefficients are perturbed in pairs (Ai and An-1-i) so as to preserve coefficients symmetry. The selection of coefficient for perturbation and the amount of perturbation has the direct impact on overall optimization quality. Various strategies can be adopted for coefficient perturbation. The strategies adopted here include 'Genetic Algorithms'. The Genetic Algorithms are the evolutionary algorithm which generates the random numbers and selects the best fit value according to the fitness function and search the whole space to find the global value [5-7].

Implementation

a) Develop the MATLAB Subscript.

b) Developed the model on simulink using Xilinx Blockset.

c) Generate the HDL code for that simulink model using Xilinx System Generator.

d) Implement this model on FPGA with help hardware description language using ISE Design Suite.

e) Estimation of power consumption using Xilinx PLAN AHEAD.

f) Estimation of Resources utilization on FPGA using Xilinx ISE Design Suite.

Simulation Result

Tables 1-8 indicates simulation results.

Results and Discussion

The above line graph in Figures 3-5 lists the difference between the power (total, static, dynamic) consume in different fractional sample convertor which are implemented with different architecture of FIR filter. Here we have analyzed static power consume in all the available structure of fractional sample rate convertor, CMFIR filter with genetic algorithm provide minimum power consumption. The above line graph lists in Figures 6-10 the difference between the Resources

Number of Channel	MAC (mw)	CIC (mw)	CMFIR without genetic algorithm (mw)	CMFIR with genetic algorithm (mw)
1	429	63	62	61
2	536	70	69	68
4	782	91	89	87
8	802	129	127	118
16	881	204	196	171
32	1040	357	326	270
64	1222	489	465	403
128	1424	630	583	540

Table 1: Total Power consume in fractional sample rate convertor based on different architecture of FIR filter.

Number of Channel	MAC (mw)	CIC (mw)	CMFIR without genetic algorithm (mw)	CMFIR with genetic algorithm (mw)
1	33	17	16	15
2	42	24	23	22
4	56	45	43	41
8	95	82	80	71
16	173	156	148	123
32	329	306	276	221
64	508	436	413	353
128	706	574	529	489

Table 2: Dynamic Power consume in fractional sample rate convertor based on different architecture of FIR filter.

Number of Channel	MAC (mw)	CIC (mw)	CMFIR without genetic algorithm (mw)	CMFIR with genetic algorithm (mw)
1	396	46	46	46
2	494	46	46	46
4	706	46	46	46
8	707	47	47	47
16	708	48	48	48
32	711	51	50	50
64	714	53	52	50
128	718	56	54	51

Table 3: Static Power consume in fractional sample rate convertor based on different architecture of FIR filter.

Filter Structure	Bandwidth (MHz)
CMFIR with Genetic Algorithm	309.6
CMFIR without Genetic Algorithm	260.82
MAC	202.66
CIC	123.54

Table 4: Bandwidth for data transfer of fractional sample rate convertor based on different architecture of FIR filter.

Architecture (Number of Channel)	Number of Register
CIC (1)	2037
MAC(1)	1270
CMFIR Without Genetic Algorithm (1)	476
CMFIR With Genetic Algorithm (1)	472
CIC (2)	3991
MAC(2)	2476
CMFIR Without Genetic Algorithm (2)	934
CMFIR With Genetic Algorithm (2)	926
CIC (4)	7889
MAC (4)	4888
CMFIR Without Genetic Algorithm (4)	1856
CMFIR With Genetic Algorithm (4)	1850
CIC (8)	15715
MAC (8)	9712
CMFIR Without Genetic Algorithm (8)	3682
CMFIR With Genetic Algorithm (8)	3650

Table 5: Register vs. number of channels of fractional sample rate convertor based on different architecture of FIR filter.

Architecture (Number of Channel)	Number of LUT
CIC (1)	805
MAC (1)	441
CMFIR Without Genetic Algorithm (1)	104
CMFIR With Genetic Algorithm (1)	100

Table 6: LUT in implementation of fractional sample rate convertor based on different architecture of FIR filter.

Architecture (Number of Channel)	Number of LUT-flip flop pairs
CIC (1)	555
MAC (1)	312
CMFIR Without Genetic Algorithm (1)	102
CMFIR With Genetic Algorithm (1)	90
CIC (2)	1074
MAC (2)	599
CMFIR Without Genetic Algorithm (2)	196
CMFIR With Genetic Algorithm (2)	172
CIC (4)	2112
MAC (4)	1173
CMFIR Without Genetic Algorithm (4)	405
CMFIR With Genetic Algorithm (4)	384
CIC (8)	4188
MAC (8)	2321
CMFIR Without Genetic Algorithm (8)	760
CMFIR With Genetic Algorithm (8)	664
CIC (16)	8340
MAC (16)	4616
CMFIR Without Genetic Algorithm (16)	1512
CMFIR With Genetic Algorithm (16)	1320
CIC (32)	16644
MAC (32)	9208
CMFIR Without Genetic Algorithm (32)	3016
CMFIR With Genetic Algorithm (32)	2630
CIC (64)	33248
MAC (64)	18380
CMFIR Without Genetic Algorithm (64)	6432
CMFIR With Genetic Algorithm (64)	5454
CIC (128)	66436
MAC (128)	36715
CMFIR Without Genetic Algorithm (128)	13364
CMFIR With Genetic Algorithm (128)	11209

Table 7: LUT-flip flop pairs vs. number of channels of fractional sample rate convertor based on different architecture of FIR filter.

CMFIR With Genetic Algorithm (Samples)	CMFIR Without Genetic Algorithm (Samples)	MAC (Samples)	CIC (Samples)
10205.5	63	62.5	157.5

Table 8: Group delay vs. frequency of sample rate convertor based on different architecture of FIR filter.

(Register, LUT, LUT-Flip Flop pairs) are utilize in implement of different fractional sample convertor which are implemented with different architecture of FIR filter. Here we have analyzed the LUT-Flip Flop pairs are utilize in implement of all the available structure of fractional sample rate convertor, CMFIR filter with genetic algorithm provide minimum recourses are required in implementation. The above bar graph in Figure 6 lists the difference between the bandwidth

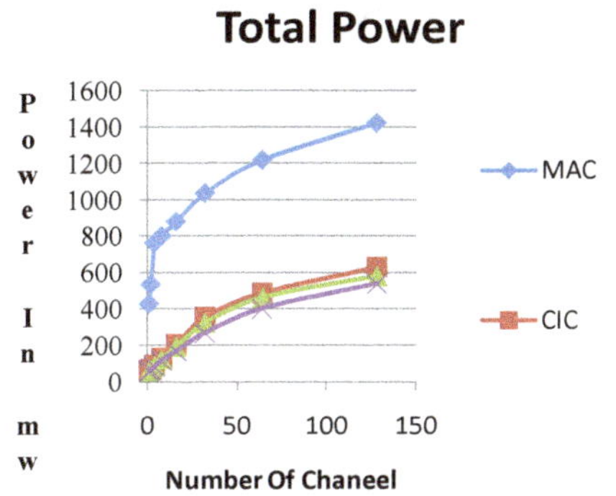

Figure 3: Total power consume in fractional sample rate convertor based on different architecture of FIR filter.

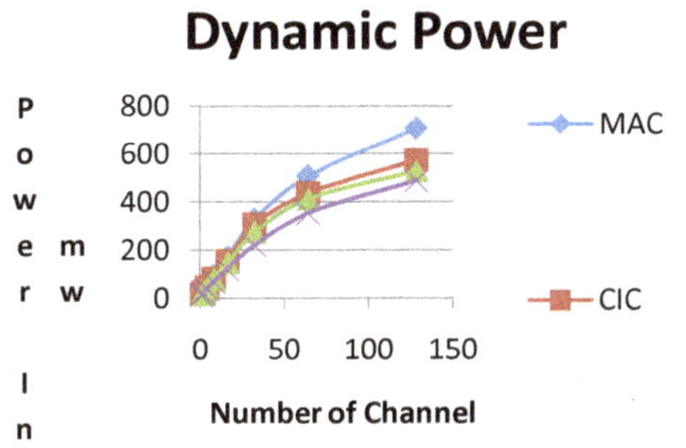

Figure 4: Dynamic power consume in fractional sample rate convertor based on different architecture of FIR filter.

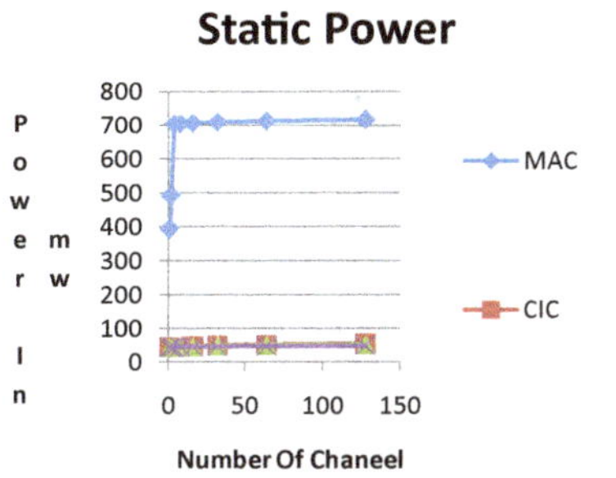

Figure 5: Dynamic power consume in fractional sample rate convertor based on different architecture of FIR filter.

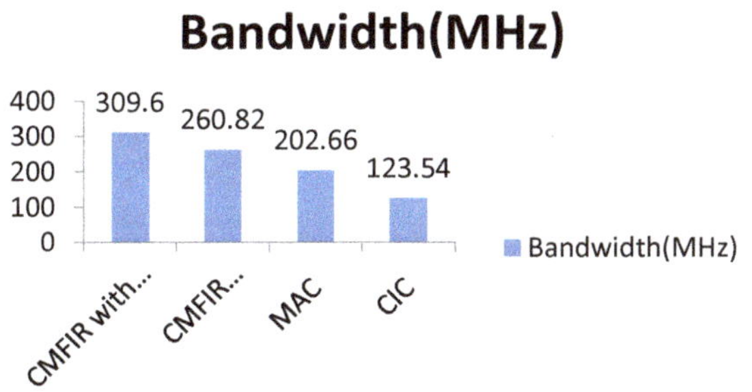

Figure 6: Bandwidth for data transfer of fractional sample rate convertor based on different architecture of FIR filter.

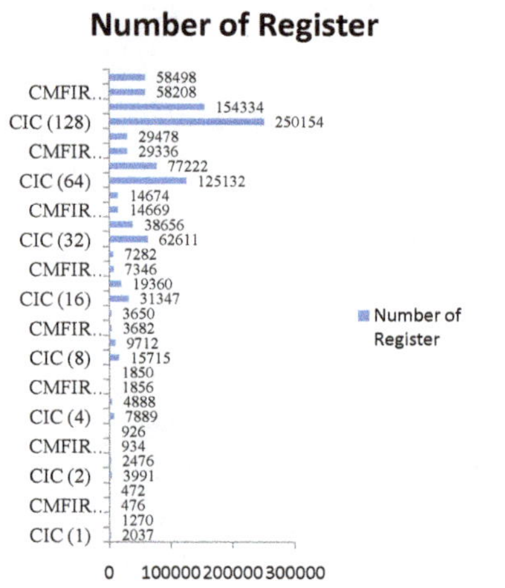

Figure 7: Register utilize in implement of factional sample rate convertor based on different architecture of FIR filter.

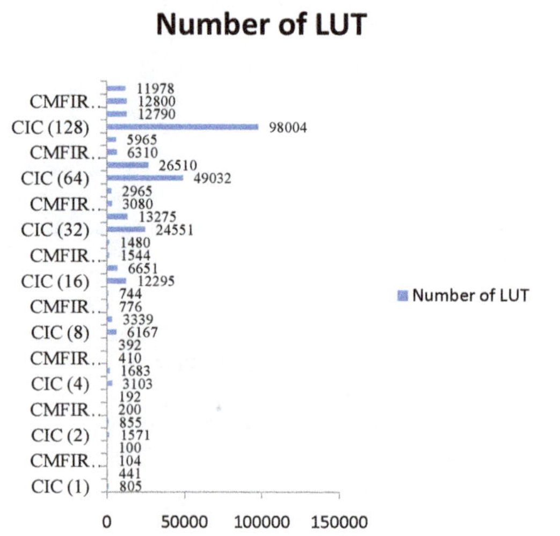

Figure 8: LUT utilize in implement of factional sample rate convertor based on different architecture of FIR filter.

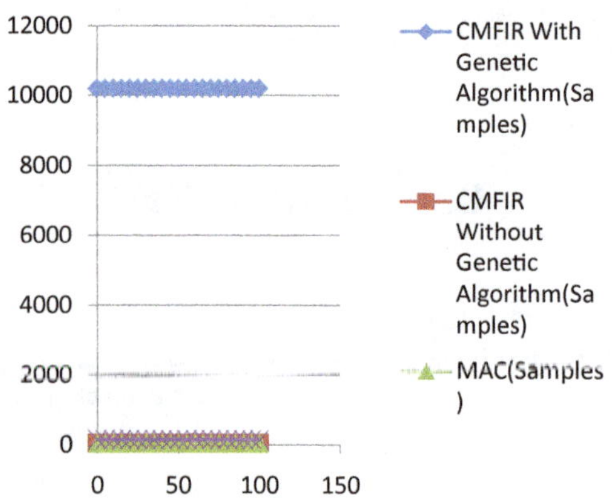

Figure 9: Group delay of sample rate convertor based on different architecture of FIR filter.

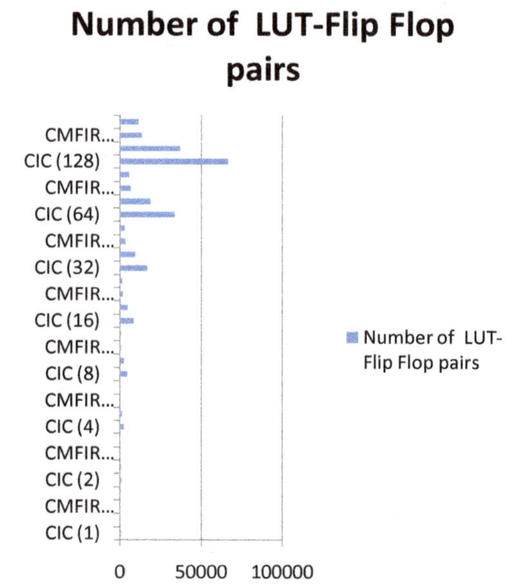

Figure 10: LUT-flip pairs utilize in implement of factional sample rate convertor based on different architecture of FIR filter.

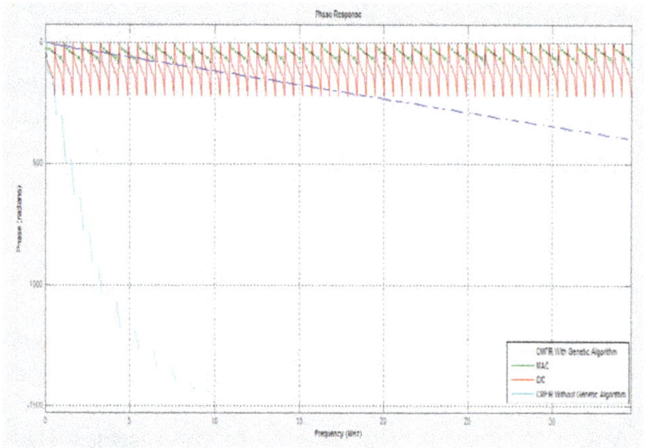

Figure 11: Phase response of sample rate convertor based on different architecture of FIR filter.

available for data transfer of different fractional sample convertor which is implemented with different architecture of FIR filter. Here we have analyzed bandwidth available for data transfer all the available structure of down sample rate convertor, CMFIR filter with genetic algorithm provide maximum bandwidth for data transfer. The above line graph in Figure 11 lists the difference between the Group delays of different fractional sample convertor which are implemented with different architecture of FIR filter. Here we have analyzed the group delay of all the available structure of sample rate convertor, CMFIR filter with genetic algorithm provides required maximum samples or minimum, constant group delay in second [8].

References

1. (2010) AN 623 Using the DSP Builder Advanced Blockset to Implement Resampling Filters, ALTERA.
2. Creaney S, Kostarnov I (2008) Designing Efficient Digital Up and Down Converters for Narrowband Systems. XAPP1113 (v1.0).

3. Ahmad S, Antoniou A (2006) Design of Digital Filters Using Genetic Algorithms. This research is part of the doctoral research of the first author.

4. Ouadi A, Bentarzi H, Recioui A (2014) Optimal multiobjective design of digitalfilters using taguchi optimization technique. Journal of Electrical Engineering 65.

5. Kaur P, Kaur S (2012) Optimization of FIR Filters Design using Genetic Algorithm IJETTC.

6. Goyal S, Raina JPS (2010) Design of Low Power FIR Filter Coefficients Using Genetic Algorithm (Optimization) International. Journal of Computer Science and Communication 1: 1-5.

7. Rezaee A (2010) Using genetic algorithms for designing of FIR digital filters. ICTACT Journal on soft computing.

8. Donadio M , Donadio P (2000) CIC Filter Introduction, Iowegian.

Optimal Placement of Distributed Generation Units for Constructing Virtual Power Plant Using Binary Particle Swarm Optimization Algorithm

Bahrami S* and Imari A

Department of Electrical Engineering, University of Isfahan, Isfahan, Iran

Abstract

The idea of applying distributed generation resources in distribution systems has become increasingly important due to changes in the distribution systems. Optimal sizing, location, type and installation time of DGs for constructing virtual power plant is one of the important subjects in applying distributed generation in the power system. In this paper, a new method is presented for optimal placement and type of distributed generation units and long term system planning aim to constructing virtual power plant. Minimizing the long term total cost of the system is considered as the objective function. The impacts of applying demand response on expansion planning is also investigated. The Binary particle Swarm Optimization (BPSO) method is used for solving this problem. In order to evaluate the efficiency of the proposed method, the method is tested on the IEEE 33 bus distribution system.

Keywords: Distributed Generation (DG); Binary particle Swarm Optimization (BPSO); Commercial Virtual Power Plant (CVPP)

Introduction

In recent years, the penetration of Distributed Generation (DG) units increased in power systems. Installation of DG units in distribution systems has several benefits such as reducing system losses, enhancing voltage profile, shaving peak demand, relieving overloaded distribution lines, reducing environmental impacts, increasing overall energy efficiency, and deferring investments to upgrade existing power systems [1]. In order to receive to mentioned advantages and better system planning, optimal placement and sizing, selecting appropriate type and the installation time of DGs are necessary.

Literature review

Many studied have been implemented with the aim of power loss reduction. Genetic Algorithm (GA) [2], Tabu search [3], particle swarm optimization [4,5] and combination of PSO and GA [6] are some of methods which their objective function is loss reduction. In [7] an improved reinitialized social structures PSO algorithm has been developed for optimal placement of multiple DGs in a micro grid to minimize the real power loss within voltage and power generation limits. The analytical approach has been demonstrated in [8,9] to find the optimal size and location of DG to minimize the real power losses and enhancement in voltage profile.

In [10], the optimal sizing of a small isolated power system that contains renewable and/or conventional energy technologies was determined to minimize the system's energy cost. In [11], DG units were placed at the most sensitive buses to improve the voltage stability. In [12], to determine the optimal locations of DG units in distribution system, a new multi-objective problem based on minimized power losses, enhanced reliability and improved voltage profile has been presented. Several papers have focused on the use of EAs (evolutionary algorithms), analytical methods or load flow methods to optimize the DG placement and its sizing [13,14].

The integration of DERs is another subject which affects the operation of a power system network. In this situation, distributed generation and controllable demand may have the opportunity to participate in the operation of transmission and distribution networks. However, the aggregation of many small-capacity generators into one large power generation project could improve the economics of DERs. If several DER units are linked together and are operated as one unit, the concept is often called a Virtual Power Plant (VPP). There are two types of a VPP. Commercial Virtual Power Plant (CVPP) is one type of VPP operation [15]. The CVPP is a competitive market actor that manages the DER portfolio(s) to make optimal decisions on participation in electricity markets. From the commercial point of view, VPP as a market agent seeks to obtain the maximum benefit from the generation and the demand portfolio without considering the network constraints. Technical Virtual Power Plant (TVPP) is another type of VPP operation [16]. The TVPP takes into consideration also the operation of the grid. The TVPP aggregates and models the response characteristics of a system containing DERs, controllable loads and networks within a single electric-geographical (grid) area. On the other hand, a TVPP consists of some DERs from the same geographic location. In this case, the impact of operation on the distribution network is also considered [17].

Demand Response (DR) is one of the important parts of a VPP. Demand response is established to motivate changes in electricity consumption by end users. The main benefit of DR is the improvement of power system efficiency, since a closer alignment between customers' electricity prices and the value they place on electricity is established [18].

Necessity of the virtual power plant

Determining the optimal place, size, capacity, type and installing time of DGs is one of the most important subjects. If the distributed generation units are not properly installed, it may lead to disadvantages in power quality and increasing the costs of system. Virtual power plant concept is a solution for the mentioned problems.

Virtual power plant (VPP) is a flexible representation of a portfolio

***Corresponding author:** Bahrami S, Department of Electrical Engineering, University of Isfahan, Isfahan, Iran, E-mail: s_bahrami@eng.ui.ac.ir

of DER that can be used to make contracts in the wholesale market and to offer services to the system operator. The VPP integrates the capacity of distributed energy resources and represents an operation strategy with the combination of operation parameters of each DER. Also, VPP can consider the effect of network in the output powers of integrated DGs.

Distributed generation units are the main part of virtual power plant. So, the optimal placement, type of generator, capacity and installing time of each DG unit in the system is the first stage for constructing virtual power plant. Demand response is the second element of VPP. Demand response is such a capacity which depends on the consumption patterns of consumers and means reducing the loads of consumers in response to high power market prices or emergency situations of system. Demand response programs have many advantages such as improving the reliability of the system, reducing the cost, improving the environmental problems, reducing market strength and giving better services to the consumers.

The goals of the research are including a method for determining the optimal placement, capacity and the type of DGs and the long term and short term system planning. In order to aim to the mentioned goals, the long term cost of the system must be minimized.

In this paper, a new method for optimal placement of distributed generation units for constructing virtual power plant is presented. Also, the long term planning of distribution system is performed. In order to receive to the mentioned goals, the long term total cost of system is considered as objective function and the problem is converted to an optimization problem. The Binary Particle Swarm Optimization (BPSO) method is used to solve in the proposed method.

The rest of the paper is organized as follows: Section 2 presents the problem formulation. The proposed method is described in Section 3. Section 4 describes the test system used in this paper. A brief summary of the simulation used to obtain the results, numerical results along with some observations and discussions are also included in this section. Finally, the contributions and conclusions of the paper are summarized in Section 5.

Problem Formulation

Optimal placement of distributed generation units in the distribution system is a very important subject for constructing virtual power plant. The problem to solve is to determine the optimal location and size of a given number of DG units. Since, the costs of the system must be optimized, in this paper, the long term total cost of system including the production cost of DGs, cost due to purchasing active power from the grid, DRs cost, upgrade cost of the system and the cost of installing smart meters in the system are considered as the objective function. The long time total cost of system is considered as following:

$$\text{minimize}: Total\ Cost = TC = \left(\sum_{y=1}^{Y_{\max}} \sum_{t=1}^{24} \sum_{i=1}^{G_n} C_{DG,i,y}(P_{DG,i,y}(t)) \right)$$
$$+ \left(\sum_{y=1}^{Y_{\max}} \sum_{t=1}^{24} (P_{st} \times MP_y(t)) \right) + \left(\sum_{y=1}^{Y_{\max}} \sum_{t=1}^{24} \sum_{j=1}^{N_{DR}} (P_{DR,y,j}(t) \times C_j \times MP_y(t)) \right)$$
$$+ \left(\sum_{y=1}^{Y_{\max}} (B_{num}(y) \times C_B(y)) \right) (1) + (DR_T \times C_{inst})$$

The objective function has five statements. The first statement represents the production cost of G_N distributed generations in all of hours for a long term period (Y_N). Second statement describes the cost due to purchasing energy from grid in ling term. The amount paid to customers for participation in demand response program is presented in the third part of the objective function. The fourth statement of the objective function represented the cost of system expansion in the long term. The required cost for installing smart metering devices on the customers is presented in the last statement of the objective function.

Where $(C_{DG,I,year}(P_{DG,I,year}(t))$ is the cost of ith generator in t^{th} hour of yth year in respect of generated active power $(P_{DG,I,year}(t)$,iof mentioned generator. The active power received from main bus and the market price in t^{th} hour of yth year are $P_s(t)$ and $MP_{year}(t)$, respectively. The number of lines in the network which should bundled in the yth year and the required cost for bounding one line are B_{num}(Year) and $C_B(year)$, respectively. DR_T is the total number of customers which installed smart meters and the cost of installing smart meters on a customer shows by C_{inst}. The production cost of each DG have both fixed and variable components. The fixed costs consist of capital and installation costs (*C & I*). It is including the cost to purchase and install a DG technology at a specified location [19]. The variable cost includes the operation and maintenance (*O & M*) and the fuel cost of DG technologies (*F*). The capital and investment cost of generator is defined as following:

$$C\&I = \frac{TIC(per\ \text{KW}) \times \frac{d(1+\text{d})^n}{(1+\text{d})^n - 1}}{CF \times 8760} \quad (2)$$

Where *d, n, CF* and *TIC* are interest rate, planning period, capacity factor and total investment cost, respectively. Also, CF is defined as following [19].

$$CF = \frac{(\text{working hour per day}) \times 365 \text{ days per hour}}{8760} \quad (3)$$

So, the cost of energy (*COE*) produced by a DG can be considered as following:

$$COE = (C\&I) + (O\&M) + F \ (3) \quad (4)$$

Constraints

In the optimal placement and sizing of DG units for constructing VPP, the following constraints should be satisfied.

i. Unbalanced three-phase power flow equations:

$$P_i = \sum_{i=1}^{N_{bus}} V_i V_j Y_{ij} \cos(\theta_{ij} - \delta_i + \delta_j) \quad (5)$$

$$Q_i = \sum_{i=1}^{N_{bus}} V_i V_j Y_{ij} \sin(\theta_{ij} - \delta_i + \delta_j) \quad (6)$$

ii. Active and reactive power constraints of DG units:

The output active and reactive power of each DG units must be between its minimum and maximum values:

$$P_{DG,i}^{\min} \le P_{DG,i} \le P_{DG,i}^{\max} \quad (7)$$

$$Q_{G,\min} \le Q_G \le Q_{G,\max} \quad (8)$$

iii. Active power constraints of distribution system lines:

$$|P_{n,m}(\delta,V)| \le P_{n,m,\max} \quad (9)$$

$$|P_{m,n}(\delta,V)| \le P_{m,n,\max} \quad (10)$$

iv. Voltage constraints

If the voltages along the feeder are not satisfied, optimal sizing and placement of DG are changed to the nearest values to take the feeder voltages to the voltage limits. The voltage must be kept within standard limits at each bus [20]:

$$V_{\min} \leq V \leq V_{\max} \tag{11}$$

Proposed Method

In this section, the proposed algorithm for optimal placement of DG units for constructing virtual power plant in distribution system is presented. The goals of the proposed method are determining the optimal site, capacity, and type and installation year of each DG in distribution system and long term expansion planning of distribution system. In order to aim to the mentioned goals, the proposed algorithm which shown in Figure 1 is presented. The proposed algorithm has 5 stages which are explained in following:

Stage 1: determining initial data

At the first stage of the proposed algorithm the initial data including the configuration of understudy distribution system, power market price, the operation cost of DG units and the peneteration value of demand response is determined.

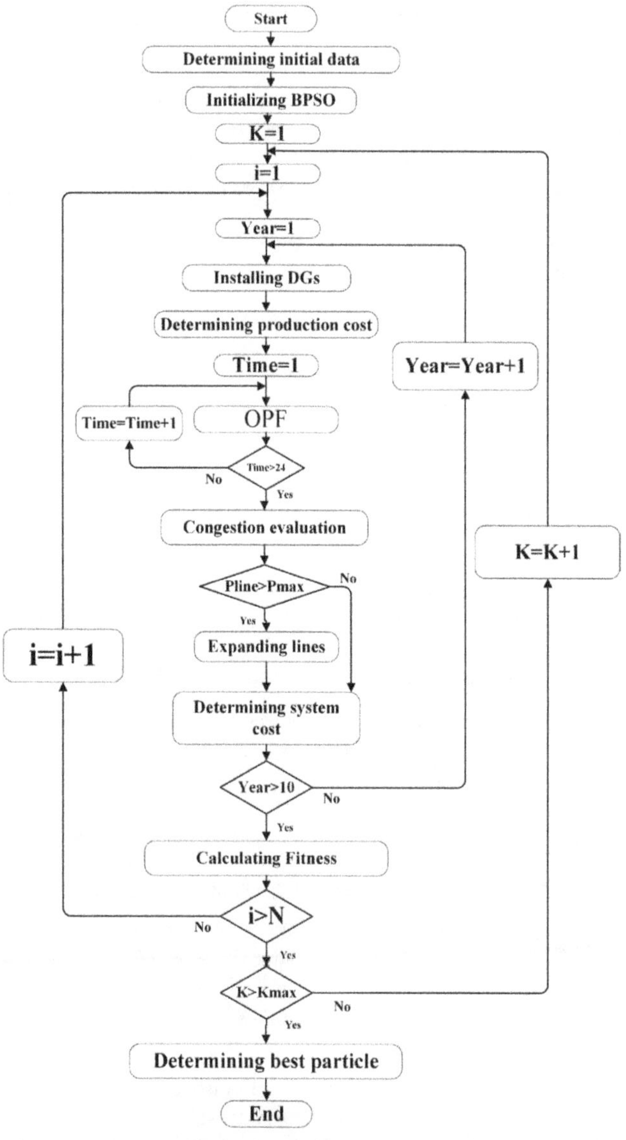

Figure 1: Algorithm flow chart.

Stage 2: initializing BPSO

In this stage, the number of particles (N), maximum iterations of algorithm (N_{max}) and the population of PSO method are determined.

Stage 3: each particle of population is including parameters required for installing DGs in system. In this stage, DG units are placed in distribution system based on the data of each particle. Also, the production cost of each DG is determined based on installing year and using (3).

Stage 4: Optimal Power Flow (OPF)

In this stage, optimal power flow is applied to determine the optimal output actvie power of DGs, power flow of lines, voltage of buses and total cost of system. The objective function of OPF is maximization of social benefit presented in [21].

Stage 5: evaluating congestion in distribution system

In this stage, the condition of distribution system lines is evaluated. if there is congestion in lines, it should upgrade in expansion planning of the system.

Stage 6: determining long term system costs

Since, the long term system cost is considered as objective function, the best planning of distribution system has minimum cost in long term. So, the long term system cost is considered as fitness in BPSO method.

Stage 7: determining the best particle

Stages 1 to 6 are performed for N particle in (N_{max}) iteration of BPSO method. The minimum fitness is obtained in each iteration. The best planning of the system including the site, size type and installing time of each generator has minimum cost in long term which is determined in the end of procedure shown in Figure 1.

Binary Particle Swarm Optimization (BPSO)

Conventional particle swarm optimization (PSO): The PSO algorithm models the behavior of a group of particles that randomly select the initial values. These particles search the problem space to find new solutions. The position and the velocity of every particle at the iteration k in the search space are described by X_k^i and V_k^i, respectively. Each particle records its best local position . Then, the velocity of particle I in the iteration k+1 P_{lbest}^i is obtained from the following equation:

$$V_{k+1}^i = \omega . V_k^i + C_1 . R_1 \left(P_{lbest}^i - X_k^i \right) + C_2 . R_2 \left(P_{global}^i - X_k^i \right) \tag{12}$$

Where R_1 and R_2 are the random functions that generate a random number between 0 and 1. Also, ω is the inertia weight factor and C_1,C_2 are the training coefficients. It should be noted that ω is decreased from 0.9 to 0.4 linearly. Also ω can be obtained as following:

$$\omega = \omega_{max} - ((\omega_{max} - \omega_{min}) / k_{max}) \times k \tag{13}$$

Where k_{max} is the number of the maximum iteration. At the end of each iteration, a new position for each particle is obtained by the summation of its old position and new velocity:

$$X_{k+1}^i = X_k^i + V_{k+1}^i \tag{14}$$

Binary PSO

Binary particle swarm optimization is presented in [22]. BPSO is used to solve discrete problems. The particle swarm formula (12) remained unchanged, except that now the position and speed of particles are integers in {0,1} and V_{k+1}^{i} ,since it is a probability, must be constrained to the interval [0,1]. A logistic transformation $S(V_k^i)$ is used to accomplish this modification.

$$S(V_{k+1}^{i}) = sig\,mode(V_{k+1}^{i}) = \frac{1}{1+\exp(V_{k+1}^{i})} \quad (15)$$

The resulting change in position then is defined by the following rule:

$$if\ \text{rand} \prec S(V_{k+1}^{i})\ \text{then} : X_{k+1}^{i} = 1\ ; \qquad else : X_{k+1}^{i} = 0\ ; \quad (16)$$

Where the function $S(V_k^i)$ is a sigmoid limiting transformation and rand is a quasi-random number selected from a uniform distribution in [0, 1].

Application of BPSO for solving optimization problem

In the proposed method four major parameters of each DG are determined using BPSO method. These major parameters are: site, size, type and installing year of each DG. In this paper, the goal is placement of G_N distributed generation in distribution system for constructing virtual power plant. Figure 2 represents the dimensions of each particle in PSO method. The number of dimension of each particle for placement G_N DG unit is 4* G_N. Equations (17) to (20) describe the limits of the dimensions of particles.

$$1 \prec B_i \prec B_{\max} \quad (17)$$

$$0 \prec P_i \prec P_{i,\max} \quad (18)$$

$$T_i = \{1, 2, ..., T_f\} \quad (19)$$

$$0 \prec Y_i \prec Y_{\max} \quad (20)$$

Where B_{max} is the maximum bus of system, $P_{i,max}$ represents the maximum allowable active power of ith bus of system. It should be noted that T_f is general index for the types of DGs and y_{max} is the last year of the planning of the system.

Simulation Results

In this paper, in order to evaluate the performance of the proposed method, the method is applied on IEEE 33 bus distribution system [23]. For this purpose, four DG is placed in distribution system by using the proposed method and an expansion planning for distribution system is performed. The results of proposed method are compared to case which no DG are placed in the system. Then, in order to evaluate the impact of demand response in distribution system, two penetration value of demand response (10% and 13%) is considered in the system and the system planning is performed.

Network information

The understudy system (IEEE 33 bus distribution system) is presented in Figure 3. The market price is in [24]. It should be noted

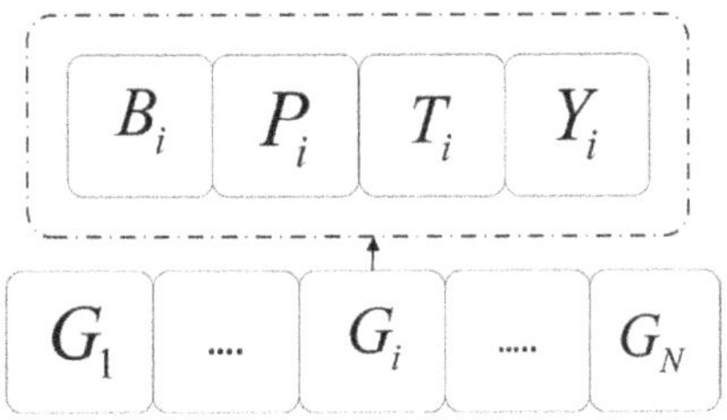

Figure 2: The dimensions of each particle in the PSO Method.

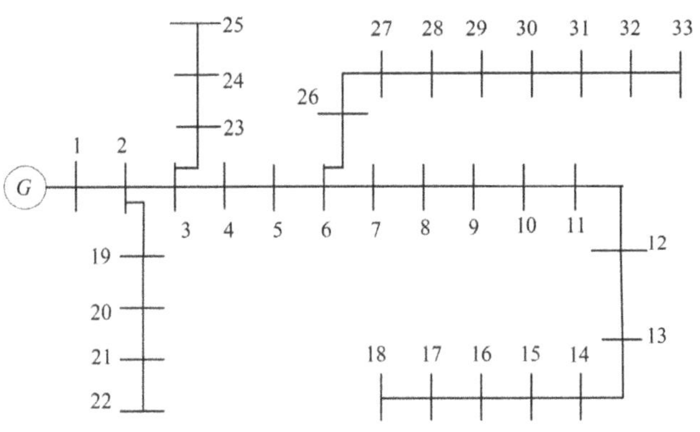

Figure 3: IEEE 33 bus distribution system.

that the rate of annual load growth is considered as a fixed rate and equal to 5%. Also, it is supposed that all of DRs have a contract with the owner of VPP. During the contract, the amount of fines paid by customers is considered 10 percent more than the market price of electricity. All of the lines of distribution systems are considered as candidates for upgrading.

Network planning without presence of DG in the system

If in the studied system, there is no DG, all the required energy will be received from the main grid. Accordingly, long term total cost of system in 10 year period is 17.66036 million dollars. Also, the lines between buses 1-2 and 2-3 are expanded at 3rd and 10th years, respectively.

Network planning by presence of DG and absence of DR in the system

In order to evaluate the performance of the proposed method, the algorithm which is presented in Figure 1 is used to optimal placing of several DG units in the system. This case is under the conditions that there is no demand response in the system. Four DG units are placed in the 33 bus distribution system. Micro turbine and combustion turbine are considered as the types of DGs. Results of the placing are presented in Table 1. In this table, the placing, capacity, type and installing year of each DG is presented. As it is demonstrated in the Table 1, installation of one combustion turbine unit at 3rd year prohibits expansion of distribution line between 1st and 2nd buses. Therefore, the long term total cost of the system for optimal placing of four DGs in the system without presence of DR is 16.68435 million dollars.

Placement of 4 DG and network planning with 10% penetration of DR in system

For estimating the effect of demand response on 10 year planning of the system, some parts of the system loads are equipped with smart meters. Hence, in order to using DR for constructing VPP, the customers

DG number	site (Bus number)	Capacity (KW)	Types of DG	Installation time (year)
1	30	500	Micro-Turbine	5
2	18	500	Micro-Turbine	5
3	14	500	CombustionTurbine	3
4	25	500	Micro-Turbine	5

Table 1: Results of 10 years planning of the system for 4 DG placement without presence of DR.

of 6th to 23rd buses (D_{RT}=18) are equipped with telecommunicating and measuring instruments. The cost of installation of instruments on every bus (C_{inst}) is considered $3000 [22]. Penetration rate of DR in the every mentioned bus is 10%. Therefore, while the congestion occurs in the lines of the system, the owner of the VPP can cancel utmost to 10% of the loads of every buses which equipped with DR. The result of 10 year planning of system for 4 DG placement with 10% penetration of DR in the system is presented in Table 2. In this case, no transferring line needs to be expanded. Also, the long term total cost of the system is 16.18549 million dollars.

Planning of the network by presence of DG and 13% penetration of DR in the system

The proposed method is applied for optimal placement of four DG 13% penetration rate of demand response in the system. The long term planning of the system and the results of the four DG placement is presented in Table 3. In this case, the total cost is 16.01971 million dollars and no distribution line needs to be expanded. It should be noted that the number of particles and iterations in BPSO method are considered 200 and 40, respectively. Figure 4 represents the convergence diagram of BPSO for the three investigated cases.

DG number	site (Bus number)	Capacity (KW)	Types of DG	Installation time (year)
1	30	500	Micro-Turbine	5
2	18	500	Micro-Turbine	5
3	14	500	CombustionTurbine	4
4	25	500	Micro-Turbine	5

Table 2: Results of 10 years planning of the system for 4 DG placement with 10% penetration of DR in the system.

DG number	site (Bus number)	Capacity (KW)	Types of DG	Installation time (year)
1	30	500	Micro-Turbine	5
2	18	500	Micro-Turbine	5
3	14	500	CombustionTurbine	5
4	25	500	Micro-Turbine	5

Table 3: Results of 10 years planning of the system for 4 DG placement with 13% penetration of DR in the system.

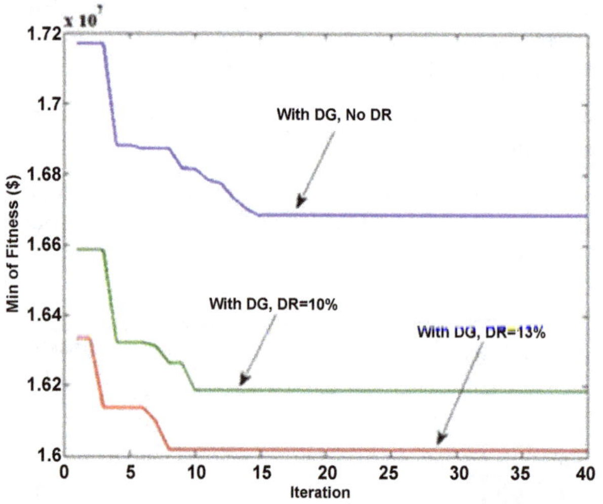

Figure 4: The convergence diagram of BPSO for the three investigated cases.

Discussion

The cases which are studied in this paper evaluated the impact of DG and DR presence in VPP. While there is no DG in the 33 bus distribution system, the lines between 1-2and 2-3buses needs to be bundled at 3rd and 10th years, respectively. But in the case that four DG units is placed in the system, the need for expanding the lines of VPP is completely canceled. This matter shows that installing DG units in the system can lead to delay in expansion of the system. In the case presented in section 4.5, the long term total cost of the VPP is decreased at 5.53% in respect to case which no DG installed in the system. If penetration rate of DR at a part of system loads is 10%, the expansion planning of lines is delayed to 5th year. The delay is resulted by decreasing the load DRs at 1-4 years. Also, it will be completely canceled by installing one DG unit at 4th year. Also, the total cost of system in case presented in 4.5 reduced in respect to the cases NO DG-NO DR and 4 DG-NO DR 8.35% and 2/99%, respectively. In condition that rate of penetration of DR in system increases to 13%, expansion of line between 1st and 2nd buses will be delayed by load peak shaving to 5th year of planning. Finally, it will be completely canceled by installing one micro-turbine unit in 5th year at 14th bus of VPP. The total cost of the system During 10 year planning in case DG-13%DR in respect to cases NO DG-NO-DR, DG-NO DR, DG-10% DR is decreased to 9.29%, 3.99% and 1.02%, respectively.

Conclusion

In this paper a new method is presented for optimal placement of distributed generation units for constructing virtual power plant. Also, the impacts of presence of demand response have been investigated in the system and on VPP planning. Minimizing the long-term cost of VPP is considered as objective function. Hence, the production costs of distributed generation units (consist of capital and investment cost, operation and maintenance and fuel cost), the costs of expanding the lines and purchasing energy from the grid is considered. The Binary Particle Swarm Optimization (BPSO) method is used to solve the optimization problem and minimizing the objective function. The proposed method is applied for placing 4 distributed generation units in IEEE-33 bus distribution network. In order to evaluate the impact of demand response on system planning, some parts of loads of the system have been equipped with DR. Then by means of the proposed method, 10 year planning of the system has been repeated with the penetration rates of 10% and 13% penetration of DR in the system. The results demonstrate that in case of correct placing of DGs in VPP, the need for developing VPP will be delayed or removed during a specific planning and the total cost of VPP will be reduced. The cost of VPP will significantly decrease in simultaneous use of DG and DR in the system. The more penetration rate of DR in the system, the more power of owner of VPP for using of DR and as a result cost of VPP will decrease more.

References

1. Rao RS, Ravindra K, Satish K, Narasimham SVL (2013) Power Loss Minimization in Distribution System Using Network Reconfiguration in the Presence of Distributed Generation . Power Systems, IEEE Transactions 28: 317-325.

2. Moradi MH, Abedini M (2010) Optimal load shedding approach in distribution systems for improved voltage stability. 4th International Power Engineering and Optimization Conference (PEOCO) 196-200.

3. Hamedi H, Gandomkar M (2011) Evaluation of reliability, losses and power quality considering time variations of load in presence of distributed generation. Int J Acad Res 3: 55-60.

4. Wong LY, Abdul Rahim SR, Sulaiman MH, Aliman O (2010) Distributed

generation installation using particle swarm optimization. In; Proc Inter Power Engineering and Optimization Conf PEOCO, Shah Alam, Selangor, Malaysia 159-163.

5. Lalitha MP, Reddy VCV, Usha V (2010) Optimal DG placement for minimum real power loss in radial distribution systems using PSO. J Theoretical and Applied Information Technology 107–116.

6. Moradi MH, Abedinie M (2010) A combination of Genetic Algorithm and Particle Swarm Optimization for optimal DG location and sizing in distribution systems. IPEC, conf 858-862.

7. Prommee W, Ongsakul W (2011) Optimal multiple distributed generation placement in microgrid system by improved reinitialized social structures particle swarm optimization. Euro Trans Electr Power 21: 489-504.

8. Acharya N, Mahat P, Mithulananthan N (2006) An analytical approach for DG allocation in primary distribution network. Elect Power Energy Syst 28: 669-678.

9. Gozel T, Hocaoglu MH (2009) An analytical method for the sizing and sitting of distributed generators in radial systems. Electr Power Syst Res 79: 912-918.

10. Katsigiannis YA, Georgilakis PS (2008) Optimal sizing of small isolated hybrid power systems using tabu search. J Optoelectron Adv M 10: 1241-1245.

11. Moravej Z, Akhlaghi A(2013) A novel approach based on cuckoo search for DG allocation in distribution network. Int J Elec Power 44: 672-679.

12. Khalesi N, Rezaei N, Haghifam MR (2011) DG allocation with application of dynamic programming for loss reduction and reliability improvement. Int J Elec Power 33: 288-295.

13. Gözel T, Hocaoglu MH (2009) An analytical method for the sizing and siting of distributed generators in radial systems. Electric Power Systems Research 79: 912-918.

14. Hung DQ, Mithulananthan N, Bansal RC (2010) Analytical Expressions for DG Allocation in Primary Distribution Networks. Energy Conversion, IEEE Transactions 25: 814-820.

15. Mashhour E, Tafreshi SMM (2011) Bidding Strategy of Virtual Power Plant for Participating in Energy and Spinning Reserve Markets,Part I: Problem Formulation, Power Systems, IEEE Transactions 26: 949-956.

16. Giuntoli M, Poli D (2013) Optimized Thermal and Electrical Scheduling of a Large Scale Virtual Power Plant in the Presence of Energy Storages. Smart Grid, IEEE Transactions 4: 942-955.

17. The FENIX vision: The virtual power plant and system integration of distributed energy resources. Contract No: SES6 - 518272.

18. Palensky P, Dietrich D (2011) Demand Side Management: Demand Response, Intelligent Energy Systems, and Smart Loads. Industrial Informatics, IEEE Transactions 7: 381-388.

19. California distributed energy resources guide.

20. Chakravorty M, Das D (2001) Voltage stability analysis of radial distribution networks. Int J Elec Power 23: 29-135.

21. Xie K, Song Y, Stonham J,Yu E, Liu G (2000) Decomposition model and interior point methods for optimal spot pricing of electricity in deregulation environments. Power Systems, IEEE Transactions 15: 39-50.

22. Kennedy J, Eberhart RC (1997) A discrete binary version of the particle swarm algorithm. Systems, Man, and Cybernetics, Computational Cybernetics and Simulation, IEEE Int Conf 5: 4104-4108.

23. Kashem MA, Ganapathy V, Jasmon GB, Buhari MI (2000) A novel method for loss minimization in distribution networks. Int Conf on Electric Utility Deregulation and Restructuring and Power Technologies (DRPT), 251-256.

24. Li T, Shahidehpour M, Li Z (2007) Risk-Constrained Bidding Strategy with Stochastic Unit Commitment. IEEE Trans Power Syst 22: 449-458.

29

QPP-MAC: A Greener Algorithm for Single Sink Wireless Sensor Networks

Xavier Fernando* and Sajjadul Latif

Department of Electrical and Computer Engineering, Ryerson University, Toronto, Ontario M5B 2K3, Canada

Abstract

Wireless sensor networks (WSN) with a single data collection node or 'sink' play major role in data acquisition (DAQ) and control systems such as smart buildings. These 'many-to-one' networks have unique challenges in addition to regular energy concern issues. For example, nodes closer to the sink experience high traffic and drain faster. In this paper, a priority based, distributed, quasi-planned Medium Access Control (MAC) scheduling algorithm is suggested for such network architectures. The proposed Quorum-Pattern-Priority (QPP) MAC approach reduces energy wastage due to idle-listening, overhearing and transmission of unnecessary overhead. The algorithm is tested with multiple sensor classes each with a different data trans-mission pattern. The proposed protocol shows significantly low energy consumption while providing almost ideal throughput for steady traffic. The algorithm also better handles varying traffic compared to many conventional fixed scheduling algorithms.

Keywords: Wireless sensor networks; Energy wastage; Collision; Synchronous; Asynchronous

Introduction

Major sources for energy wastage in wireless sensor networks (WSN) are Idle-listening, overhearing, collision, energy-hole and the transmission of unnecessarily large control-overheads. These issues cause fast energy drainage and low throughput that result in reduced network functionality. The idle-listening phenomenon occurs when the sensor nodes continue to listen although no data is expected to arrive [1]. Previous research shows that the idle:receive:send power ratios for a sensor can be typically 1:1.05:1.4 [2]. That means a receiving node may consume close to 95% of the energy needed to transmit. In addition, during the idle listening period, the sensors may pick up data packets not intended for them, resulting in the overhearing phenomenon consuming even more energy.

There have been many synchronous protocols developed by authors to reduce the idle-listening and overhearing problems. These techniques typically reduce the energy usage by scaling down the time the sensors are awake. Figure 1 illustrates a very basic Sensor MAC (S-MAC) protocol where the sensor node periodically sleeps to conserve energy [1]. It can be readily seen this is a very basic static arrangement.

The T-MAC algorithm was introduced [3] to improve the performance further. In T-MAC, if no incoming data is found within a predetermined threshold period during sensing, then the sensors will go to sleep. Although the T-MAC is an adaptive scheme that reduces the energy consumption, still the nodes have to periodically wake-up. More protocols have been suggested to address the shortcomings of the S-MAC and T-MAC [4-6]. Most of them have high latency and bulky synchronization overhead.

Many asynchronous protocols have also been suggested by researchers in order to reduce the synchronization overhead [7-9]. These asynchronous protocols use preambles to detect free channels. These preambles generally introduce longer latency.

Collision happens when multiple nodes try to use a single channel simultaneously. Collision requires re-transmission, resulting in wastage of energy and bandwidth. Many of the above mentioned protocols also try to solve the collision issue. Good scheduling is the key to avoid collision. Note that over-scheduling will waste bandwidth and energy while under-scheduling will limit throughput.

Figure 1: S-MAC listen/sleep scheduling technique.

The P-MAC algorithm was proposed [10] as an attempt to handle variable traffic loads effectively. The PMAC is an adaptive pattern based scheduling scheme where, the wireless sensors dynamically create sleep/wake-up schedules based on their expected traffic loads. In P-MAC, the central sink can also override the scheduling commanding particular nodes to be more or less active. The P-MAC system performs very well in variable traffic load; however, it may not outperform synchronous protocols due to added overhead when the expected traffic load is constant.

Given the above background, it is obvious that a protocol performing well in both the steady and variable traffic conditions is necessary. In addition, in a WSN with a central data collection sink, the energy depletion rate for the sensors closer to the sink is much higher than sensors far away from the sink. This is because the closer nodes have to relay additional traffic coming from outer nodes. This issue is known as the energy-hole problem. There have been efforts to resolve the energy-hole problem by introducing number of distribution techniques that increase the number of nodes closer to sink [11-13]. However, these methods may not be suitable for all network scenarios.

A Quorum based MAC (Q-MAC) algorithm is a superior distributed scheduling protocol, which uses a grid-quorum approach as shown in Figure 2. This grid based scheduling algorithm reduces the energy-hole problem along with some other issues discussed above

***Corresponding author:** Xavier Fernando, Department of Electrical and Computer Engineering, Ryerson University, Toronto, Ontario M5B 2K3, Canada
E-mail: fernando@ee.ryerson.ca

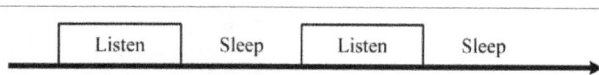

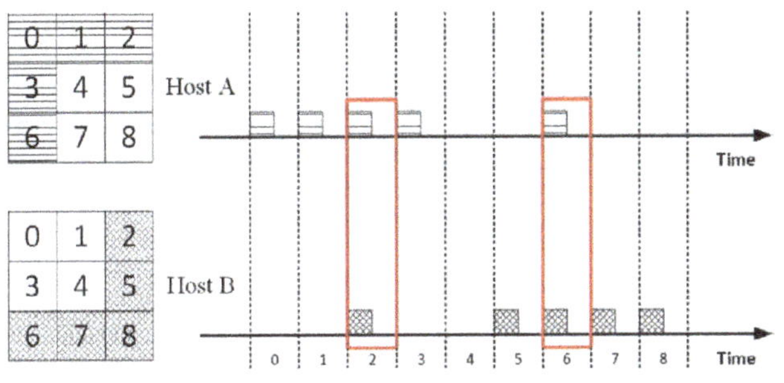

Figure 2: Grid based scheduling algorithm of Q-MAC protocol.

[14]. In the Q-MAC protocol, the total time frames are arranged in a grid form and each sensor node is assigned to a row and a column. Two nodes can communicate only when the row and the column is intersecting. For example, in the given figure, time-frames 0, 1, 2, 3 and 6 are designated for node A and time-frames 2, 5, 6, 7 and 8 are assigned for node B. Therefore, nodes A and B can communicate only during the common time frames 2 and 6. The sensors will sleep during the other time-frames to conserve energy. The detail of the grid quorum protocol is described [14], this protocol performs very effectively in steady traffic. However, the performance degrades with time varying traffic because the quorum assignments are static.

Another drawback in all the above mentioned protocols is that they only consider a single sensor type (same data rate, traffic pattern and QoS requirements etc.). For a more general study, different classes of sensors with varying service attributes need to be considered. Some sensors may transfer periodic data steadily while other nodes might emit bursty traffic on-demand basis.

Therefore, in this paper, a distributed, quasi-planned scheduling algorithm with priority control is developed. The proposed QPP-MAC algorithm is uses the grid approach from the Q-MAC, algorithm. However, the proposed algorithm dynamically optimizes the scheduling based on the expected traffic considering multiple classes of sensors. The protocol performs much better than Q-MAC and S-MAC and designed to handle both steady and time varying traffic loads effectively. It shows good energy efficiency, high throughput, low latency with minimal increase in complexity.

Network Model

Our network model is similar to this system [14], there is a central sink and uniformly distributed immobile sensor nodes. The nodes transmit data to the sink node in a unidirectional manner. The entire network is divided into multiple coronas (or concentric circles) based on the transmission range of the nodes. Sensors in one corona can typically transmit to the nodes in the immediate inward corona with one hop. The area of the ith corona C_i depends on the range of a sensor within the ith corona. Each hop distance creates a new corona in the system (i.e. if there are 5 hops in the shortest path to any sensor node from the sink, then the sensor is located in the 5th corona). Figure 3 illustrates the corona distribution concept. This corona architecture helps scheduling the network better.

The protocol design is based on the different traffic patterns of the different classes of sensors. Let each sensor node has a unique name and location identifier. Three classes of sensors are considered as follows to begin with. More classes can be added to the algorithm as required.

Class A: Small number of nodes that produce bursty traffic.

Class B: Data acquisition (DAQ) sensors [15] that wake-up less frequently than Class A and generate constant traffic. There are more class B sensors than class A sensors.

Class C: DAQ sensors for collecting slowly varying data. These wake-up even less frequently. The majority sensors in the network belong to class C. These have the lowest in transmission activity.

We consider different quality of service (QoS) requirements too [16]. For example, class A sensors need rapid data transfer without delay while, class B sensors can tolerate more delay and require less data transfer. Class C sensors are even more patient and slow than class B sensors.

It is also assumed that time is divided into equal length time-frames and all the nodes are time synchronized. We assume the simple synchronization technique described in S-MAC [1] with little overhead. All the sensor nodes have the same transmission range of 'R' meters. The area of a corona is C_i where 'i' stands for corona number [14].

Protocol Design

We take three steps for the protocol design; planning the sensor and time distributions, sensor classification, and planning the scheduling. The first step ensures proper geo-graphic distribution of sensors and the appropriate distribution of time-frames to them. Sensor classification is self-explanatory. Finally we develop the adaptive scheduling protocol that ensures optimal sleep/wake-up timings for a given data transmission pattern that maximizes the performance.

Planning the sensor distribution

In this stage, the network area is divided into multiple coronas as follows. During the network initialization, the sink node sends a counter packet called 'NET INIT' to find the lowest number of hops to reach each node. After this initialization step the sink will be able to estimate the corona (number of hops) of each node along with the total number of coronas present in the network. The sink will then broadcast this message to all the nodes. Therefore, all the nodes will know all other nodes' locations. Especially each node will identify its single hop neighbours toward the direction of the sink.

If a corona C_i has a total 'n' nodes, then the system will generate an 'n n' grid quorum for that corona and each sensor will have n^2 time-

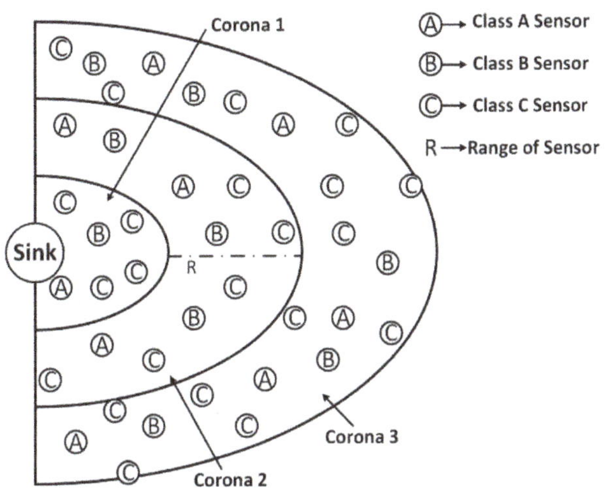

Figure 3: Corona distribution with sensor classification.

frames. Note that each corona will have to use at least 'n' time-frames for its own (self-generated) traffic. The rest of the time-frames can be used for forwarding on-going traffic or may be unused (i.e. sleeping). For example, in Figure 2 host A and B only communicate with each other during 2nd and 6th time-frames only and the remaining time-frames can be used for ongoing traffic. Therefore, the outermost corona will require only 'n' time frames.

At this stage, let us assume initially all the nodes are fully loaded. This means all the sensor nodes always use their allocated time frames irrespective of its class. Each node of an outer level corona is assigned to the time-frames in such a way that it can communicate to multiple next hop nodes (known as the next hop (n_h) group) of inner corona during their wake-up times. This ensures less delay. The time cycle C_y can be distributed among the nodes based on the grid quorum system. The traffic loads can be calculated for a node in C_i as

$$Tc_i = 1 + \frac{C_{i+1}}{C_i} Tc_{i+1} \quad (1)$$

The nodes in a lower level corona will always have higher traffic than the nodes in outer level corona because; they have to transmit both their own traffic and traffic from upper level coronas. Hence, if the time-frame duration is t_R, then receive, transmit and total active durations can be calculated as follows

Total active duration = Receive duration + Transmit duration

$$= t_R(Tc_{i+1} + Tc_i) \text{ (seconds)} \quad (2)$$

Therefore, active ratio for C_i becomes

$$\text{Active Ratio } (\text{AR}_i) = \frac{\text{Total Active Duration}}{\text{Total Length of Time Period}}$$

$$= \frac{(Tc_{i+1} + Tc_i)}{n^2} \quad (3)$$

From the calculation of active ratio (AR_i) and next hop group, the probability of finding a next hop neighbour in C_{i-1} awake as seen from C_i corona can be estimated as

$$P_i = 1 - \left(1 - \frac{Tc_i + Tc_{i-1}}{n^2}\right)^{n_h} \quad (4)$$

Here, n_h stands for the number of next hop neighbours in the designated group. This type of planned distribution ensures that all the nodes will have allotted time-frames for data transmission and the collision and retransmission will be minimal. This allocation is optimal for steady state traffic.

However, the system needs to be adjusted further for varying traffic situation.

Sensor classification

Let us define three priority constants, δ_A, δ_B, δ_C for nodes of class A, B, and C respectively. This classification is merely a way to differentiate between the nodes which will be used for scheduling the next section. The reason behind this assignment is to enhance planned distribution method described in section III-A to multi classes of sensors.

Planned schedulinge

A planned scheduling scheme is generated to ensure the nodes are following wake/sleep schedule according to the priority constant. The scheme follows basic algorithm of pattern generation of P-MAC as described in [10]. The nodes will generate the patterns based on their traffic load and priority constants. A node can increase its priority constants value to make it more energy efficient or decrease its priority constants value to transmit more data. The nodes will train themselves using the neighbour's traffic load and previous traffic history. Figure 4 shows the pattern generation method. A basic difference of the new protocol from P-MAC is that the priority constant δ_x; x ∈ A, B, C is determined directly from the traffic history and time-frame distribution information of section III-A. Diverse classes of sensors with dissimilar priority constants will generate the scheduling plans differently. Also, the new protocol generates the patterns based on the scheduling information of quorum. Thus, the pattern is generated only for the time-frames when there is a possible free next hop neighbour available in the network.

Hence, in this new scheme, the sensors will generate a tentative sleep/wake plan for the node and broadcast a pattern repeat time-frame (PRTF) packets [10]. The initial PRTF will be generated based on the next hop neighbour's availability and free channel in the network. The sensors will use the δ_x; x ∈ A; B; C value from the sensor classification to set its own wake-up sequence. It mainly determines how aggressively a sensor will be saving energy. When a sensor receives similar PRTFs from the next hop neighbours, it adjusts the sleep/wake schedule accordingly and transmits a pattern exchange time-frame (PETF). The neighbours will save the final wake-up schedule of this node as PETF. The illustration in Figure 5 shows the actual wake/sleep schedule based on planned distribution information. The grey overlay is the new sleep time-frame which will otherwise be idle-listen period in a quorum based system without priority control.

In Figure 5, each grey overlaid time-frame is a measure of saved energy by the sensors. The energy saving is really high for any class B or class C sensors. This protocol can effectively increase the network lifetime to a great extent. In this pattern generation scheme, only large time scales (in order of hundred milliseconds) is involved [10]. Hence, the system can perform very well with the simple synchronization method derived in S-MAC [1]. This ensures reduced complexity and control overhead of the network. The total active duration of sensor nodes in corona C_i and probability of a node awake can be found from

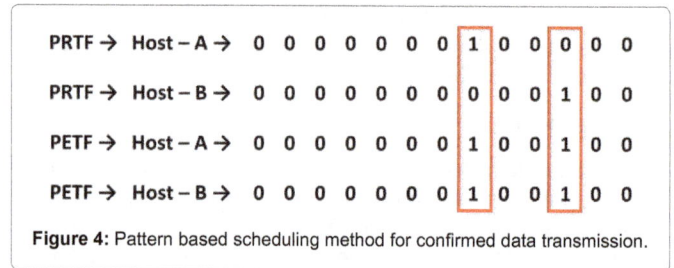

PRTF →	Host – A →	0	0	0	0	0	0	0	1	0	0	0	0	0
PRTF →	Host – B →	0	0	0	0	0	0	0	0	0	0	1	0	0
PETF →	Host – A →	0	0	0	0	0	0	0	1	0	0	1	0	0
PETF →	Host – B →	0	0	0	0	0	0	0	1	0	0	1	0	0

Figure 4: Pattern based scheduling method for confirmed data transmission.

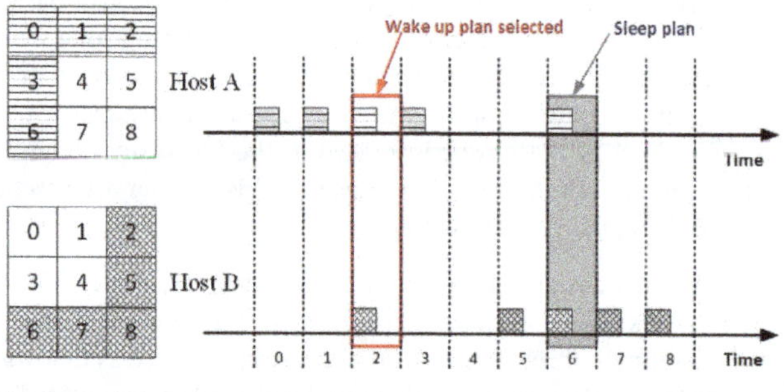

Figure 5: Planned scheduling combined with planned distribution in the proposed QPP MAC algorithm.

(3) and (4) of section III-A. The active ratio with pattern generation can be calculated using,

$$AR_i(n) = \frac{(Tc_{i+1}\delta_x) + (Tc_i\delta_x)}{n^2} \quad (5)$$

and the probability of a next hop node awake for a node in C_i will be,

$$P_i(n) = 1 - \left(1 - \frac{(Tc_i\delta_x) + (Tc_{i-1}\delta_x)}{n^2}\right)^{n_h} \quad (6)$$

This type of pattern generation scheme permits data aware network, where the networking nodes can choose their energy efficiency based on the traffic requirements. This planned scheduling scheme, combined with previously discussed planned distribution and sensor classification steps, can potentially provide high confidence data aware network with high energy performance. The nodes or the sink can change the x value anytime to manage network's energy efficiency.

Energy Savings

The conserved energy for this algorithm can be calculated from sleeping duration. The length of sleeping interval for any time-frame with t_R duration is (1- $wake_{time}$)t_R. In the proposed system, the nodes will be sleeping the entire time when there is no data to send or receive. The average number of sleep scheduled frames in a pattern is calculated in [10] as $E(0) = \sum_{i=0}^{M}(2^i P^i{}_2)$ Here, E(0) is the average sleep due to pattern. If C_y is the cycle duration and d is the duty cycle length, then the time interval becomes

$$t_R = \frac{E(0)t_R}{C_y}d \quad (7)$$

In effect, the additional energy saving by each node in the new system can be given by,

$$E_{save} = \frac{E(0)t_R dP_{idle}(1 - P_{i-1})^{n_h}}{C_y} \quad (8)$$

Here, P_{idle} is the idle power consumption for any node. For a total 'n' number of nodes in the network, the total energy saving is,

$$E_{save-total} = \frac{E(0)t_R dP_{idle}(1 - P_{i-1})^{n_h}}{C_y} \quad (9)$$

Simulation Results

A simulation model is developed to validate the performance of the new protocol. Table 1 shows the parameters used in the simulation. Most parameters used in the simulation are based on the actual data from smart homes. The scheduling scheme is generated according to the pattern generation algorithm outlined in section III-C. The simulation results in Figure 6 show the change in a node's wake-up ratio with the proposed algorithm. It can be observed that the new protocol has lower active ratio as compared to Q-MAC protocol. The low active ratio will result in high energy conservation.

Transmission power	0.69 Watts
Receiving power	0.36 Watts
Sleep power	0.03 Watts
Idle power	0.24 Watts
x	varying
Data packet length	32 byte
ACK packet length	3 byte
Channel rate	15 kbps
Length of a time-frame	100 ms
Number of corona	5
Number of iterations	10000

Table 1: Simulation parameters.

Figure 7 illustrates the average activity levels by various classes of sensors over multiple cycles. It is evident that class A sensors are more active than class B or C. However, the number of class A sensors are the lowest in the system. The simulation results show, the class A sensors will deplete energy faster if the priority scheduling is not deployed. In such a case, the network manager can change the priority constant for class B or class C sensors to facilitate the forwarding traffic. Hence, all the forwarding traffic only flows through 'B' or 'C' class sensors. This will reduce the potential energy-hole issue to some extent.

It can be observed from Figure 8 that class A sensors consumed the highest amount of energy as compared to other classes. This figure also indicates the importance of the sensor classification. If there were no classification, then class B and C sensors would wake-up at the same frequency as class A sensors. Hence, they will have the same energy consumption as class A. This shows the use of sensor classification and priority based scheduling manage the energy consumption better.

Figure 9 shows per node energy consumption for the proposed QPP-MAC protocol is much lower compared to QMAC and SMAC protocols. This proves the effectiveness of the new algorithm. The per node energy consumption shown in the figure is based on the highest energy consuming node (class A) in each corona. Similarly, nodes from other sensor classes will also eventually consume less amount of

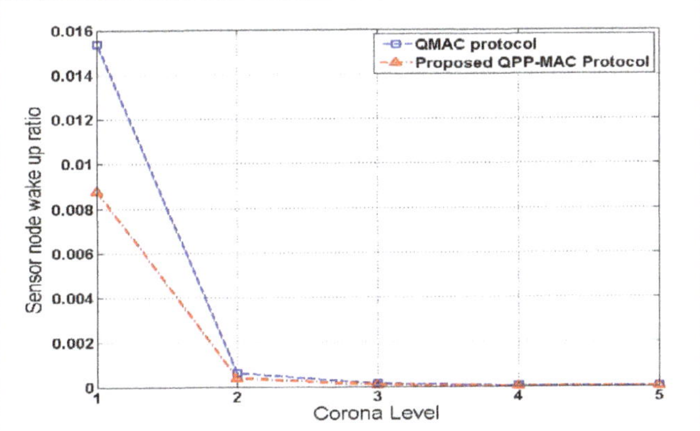

Figure 6: Wake-up ratio of a sensor irrespective of its class.

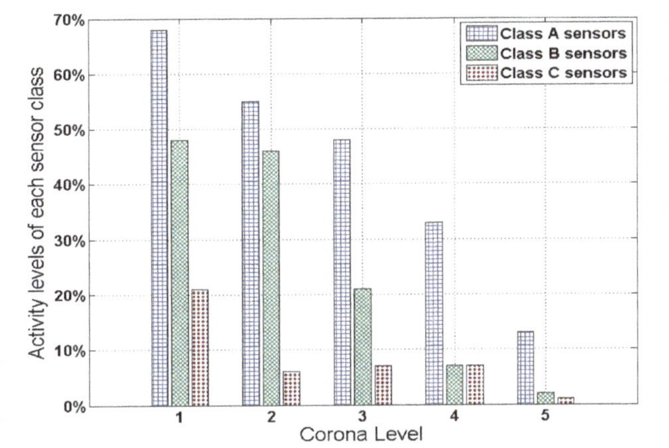

Figure 7: Activity levels of each sensor class.

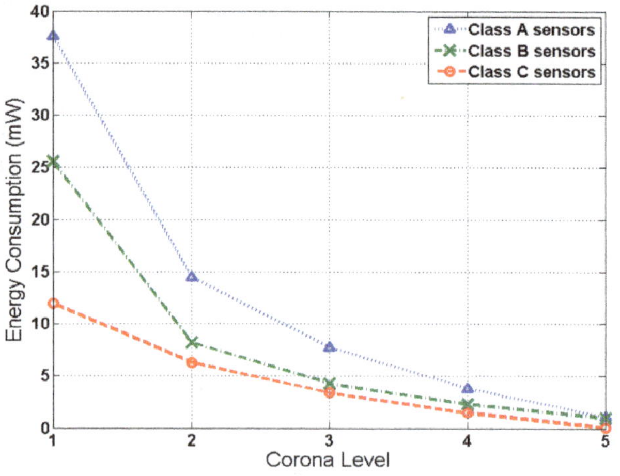

Figure 8: Energy usage by different classes of sensors.

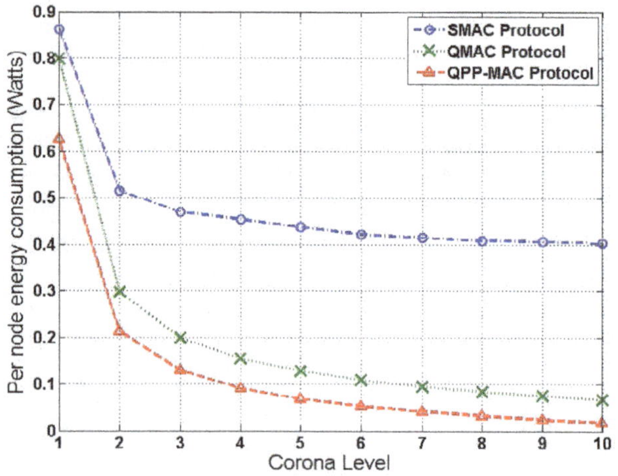

Figure 9: Energy performance of each corona according to the new protocol.

energy depending on the δ_x value. This will result in overall reduction of energy usage by the network.

The Figure 10 shows receive and transmit duration of a node during a duty cycle. It shows that the receive duration is lower than the transmit duration, due to the fact that the node need to transmit both its own traffic and forwarding traffic. Hence, total duration is increased. The new protocol shows a reduction in both transmit and receive durations as compared to QMAC. The overall simulation results show the proposed algorithm has performance improvement over S-MAC, QMAC and P-MAC.

Conclusion

An energy efficient MAC layer protocol named QPP-MAC is proposed in this paper for a central sink based wireless sensor networks. This new protocol reduces the total energy requirement by each node using adaptive scheduling. The quorum based distribution of nodes ensures confirmed communication for steady traffic. The pattern based scheduling gives adaptive control during time varying traffic. Combination of quorum and pattern approaches improves the overall performance of the network. The ability to control energy saving adaptively by changing δ_x gives extended flexibility without increasing

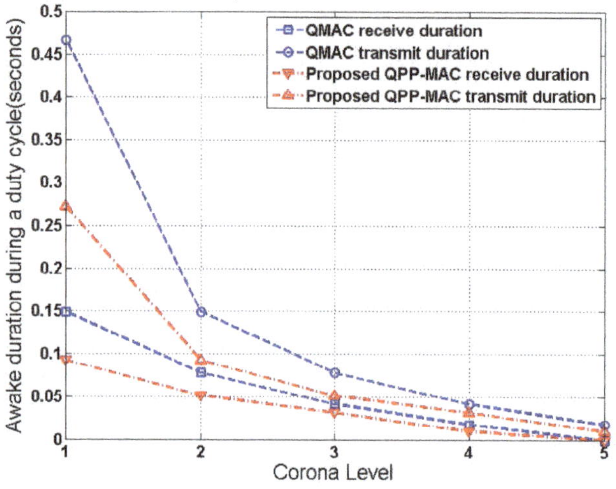

Figure 10: Awake duration of a node.

complexity. The sensors are trained to wake-up only when necessary. This traffic aware scheduling can be altered for any part of the network without affecting the whole system.

The novel classification method of the sensor nodes also gives options for new adaptive control and data acquisition methods. All these features make the algorithm robust and usable in most central sink based sensor network with fixed node distribution.

The proposed QPP-MAC algorithm has distributed nodes with planned scheduling. Therefore, theoretically the new system should perform well in both static and variable traffic load situation.

Acknowledgment

The authors would like to thank Centre for Urban Energy (CUE) at Ryerson University, Toronto and Region Conservation Authority (TRCA), and Toronto Hydro Electric Systems Ltd. for their support in this project.

References

1. Ye W, Heidemann J, Estrin D (2002) An energy-efficient MAC protocol for wireless sensor networks. Proceedings of 21st Annual Joint Conference of the IEEE Computer and Communications Societies (INFOCOM) 3: 1567-1576.
2. Stemm M, Katz RH, Katz YH (1997) Measuring and reducing energy consumption of network interfaces in hand-held devices. IEICE Transactions on Communications.
3. Dam TV, Langendoen K (2003) An adaptive energy-efficient MAC protocol for wireless sensor networks. Proceedings of 1st International Conference on Embedded Networked Sensor Systems.
4. Lu G, Krishnamachari B, Raghavendra CS (2007) An adaptive energy-efficient and low-latency MAC for tree-based data gathering in sensor networks. Energy Efficient MAC 7: 863-875.
5. El-Hoiydi A, Decotignie JD (2005) Low power downlink MAC protocols for infrastructure wireless sensor networks. Mobile Networks and Applications 10: 675-690.
6. Xu Q, Rong L, Fang S, Du Y (2009) Energy-efficient scheme for IEEE 802.15.4 compliant device. Progress in Electromagnetics Research Symposiam.
7. Buettner M, Yee GV, Anderson E, Han R (2006) X-MAC: A short preamble MAC protocol for duty-cycled wireless sensor networks. Proceedings of 4th International Conference on Embedded Networked Sensor Systems.
8. Polastre J, Hill J, Culler D (2004) Versatile low power media access for wireless sensor networks. Proceedings 2nd International Conference on Embedded Networked Sensor Systems. New York, NY, USA: ACM.
9. Javaid N, Hayat S, Shakir M, Khan MA, Bouk SH, et al. (2013) Energy efficient

MAC protocols in wireless body area sensor networks - a survey. Computing Research Repository (CoRR) 1303: 1-17

10. Zheng T, Radhakrishnan S, Sarangan V (2005) PMAC: an adaptive energy-efficient MAC protocol for wireless sensor networks. Proceedings of 19th IEEE International Parallel and Distributed Processing Symposium.

11. Wu X, Chen G, Das S (2008) Avoiding energy holes in wireless sensor networks with nonuniform node distribution. IEEE Transactions on Parallel and Distributed Systems 19: 710-720.

12. Li J, Mohapatra P (2005) An analytical model for the energy hole problem in many-to-one sensor networks. Proceedings of 62nd IEEE Vehicular Technology Conference 4: 2721-2725.

13. Stojmenovic I, Olariu S (2005) Data-centric protocols for wireless sensor networks. Handbook of Sensor Networks.

14. Chao CM, Lee YW (2010) A quorum-based energy-saving MAC protocol design for wireless sensor networks. IEEE Transactions on Vehicular Technology 59: 813-822.

15. Al-Ali AR, Zualkernan I, Aloul F (2010) A mobile GPRS-sensors array for air pollution monitoring. IEEE Sensors Journal 10: 1666-1671.

16. Shah G, Gungor V, Akan O (2013) A cross-layer qos-aware communication framework in cognitive radio sensor networks for smart grid applications. IEEE Transactions on Industrial Informatics 9:1477-1485.

Microcontroller Based Smart Control System with Computer Interface

Bamisaye, Ayodeji James and Ademiloye Ibrahim Bunmi

Department of Electrical and Electronic Engineering, The Federal Polytechnic, Ado-Ekiti, Nigeria

Abstract

In the fast growing world it is necessary to control the home appliances from remote locations through some set of instructions inputted into the computer. Electrical appliances need to be protected against over voltage in order to avoid appliance damage, this is essential as the reliability of the appliances will be improved. With the advancement of technology things are becoming simpler and easier for consumers. This paper presents an automation system that switches on/off electrical appliances at a specified time using PC, which can be placed in any location in the house (room). The on/off system can be programmed in advanced to perform a specific assignment at the required time. The aim of developing this system is to save time and manpower along with maintaining security and convenience. PIC18F4550 microcontroller acts as the 'intelligence' for this system in executing the tasks and operations according to the user's wish. The system's Graphical User Interface (GUI) was developed using Microsoft Visual Basic.Net to enable the user to easily control and monitor the appliances remotely.

Keywords: Microcontroller; Smart control; Computer; Electrical appliances

Introduction

The use of computer interfacing systems for controlling devices is spreading at an increasingly fast pace. The computing and information technology that is equipped by a residence will responds to the needs of user, working to promote their comfort, convenience, security and entertainment, this type of system is known as Smart Control System (SCS) or Intelligence Control System (ICS) As the technology progresses many control systems have been developed ranging from high end stuff to our common daily life.

The specialty of this convenient way of controlling electrical system is that the operator will be able to control different appliance at home/industry by using a Personal Computer (PC). An added advantage of this is that the operator will be able to control home appliances using timer option, by setting up a turn on time and turn off time. Four different light points are controlled as a test case and also monitor the voltage. Moreover; it is also possible to control appliances using Graphical User Interface. The USB port is used for data to be transferred from computer to the particular device to be controlled. A typical home and industry has many systems where an intelligent home or Industry system ties all of these systems together so that the user(s) may interface with these systems from a point of contact home and industry has many systems where an intelligent home or Industry system ties all of these systems together so that the user(s) may interface with these systems from a point of contact.

In references [1,2] propose Ubiquitous Access to Home Appliance Control System using Infrared Ray and Power Line Communication, and Bluetooth. This system cannot monitor the voltage level, time and using infrared ray and Bluetooth are not as reliable as computer interface.

In reference [3] presents an Internet based wireless Home Automation System for Multifunctional Devices. This approach is limited to internet based environment and cannot be employed outside the internet range.

In reference [4], two different approaches to control the home appliances; timer option and voice command was presented. The timer option provides control based on timer, and the voice command provide control by using voice commands to control the appliances. This is also limited in voltage monitoring.

Reference [5] proposes a Control of Home Appliances Remotely through Voice Command. He shows a model for home appliances control. This is also limited to voice command; there was no system security in case of high voltage and timer control.

Short Message Service (SMS) Based Wireless Home Appliance Control System (HACS) for Automating Appliances and Security was presented in reference [6]. This proposal was network provider-based and cannot have a high reliability, no visual display monitoring among others.

Reference [7] presents an electrical equipment control using PC, he explained the idea of using the printer port of a PC, for control application using software and some interface hardware. He further explained the reason he employed parallel printer port to control electrical appliance that it was inexpensive and available. The system is limited in monitoring the voltage levels and time control.

This paper designed and developed a microcontroller based system interfaced with a computer system using USB port to control domestic electric appliances such as light, fan, heater, washing machine, motor, TV, among others. The Paper adopt MPLAB IDE and MikroC Pro to program microcontroller PIC 18F2550, build hardware for the system, Interfacing the hardware to computer by using USB port communication and Authentication of program using Visual Basic dot net at computer to observe the performance of the system. These will reduce man power, Increase safety measures that prevent the electrical appliances from being easily damage and the user from sudden accidents, increase the security standards by denying any unidentified person to have accessed.

***Corresponding author:** James A, Department of Electrical and Electronic Engineering, The Federal Polytechnic, Ado-Ekiti, Nigeria
E-mail: ayobamisaye@gmail.com

The work is presented into four sections; the first section introduces the smart control system, the second section describes the system methodology, the third section presents the circuit construction and assembly, while the conclusion is drawn in section four.

Methodology

The basic block diagram of a Smart Control System was developed, which reflects each section of the desire system. The sections are; Computer System with Controller Software, Microcontroller, Voltage Monitoring Section, Current Monitoring Section and Relay Section.

Computer system with controller

Software: Visual Basic dotNet software for the controller which include Splash Screen Window, Log In Window and Controller Window was developed (Figure 1a and 1b).

i. Splash Screen Window: This is the window that shows the program function and name. This window will display for 10 seconds before log in window then disappear.

ii. Log In window: This is user authentication window. With this window user will be required for the username and the password to access the control window. If the username and password is incorrect the window will generate error which will not allow the user to access the controller window. The username use for this work is admin while admin123 for the password.

iii. Controller window: This is the window responsible for the controlling and monitor of the load. We have four controlling Section which was labelled according to the device to be controlled that make it easy for the user. The smart control system must be connected before you can use any of this section. Though there is a label to indicate if the device is detached or attached. One can know if the device is connected or not in this window.

The controller window has two sub windows which are Timer Option and Power Option: Timer Option allows the user to set turn on time and turn off time for the appliance while in Power Optionuser can be able to set the maximum power consumed level and maximum voltage level. When the software detect that the power or voltage is beyond the set level, the socket will be switch off and generate sound from the buzzer.

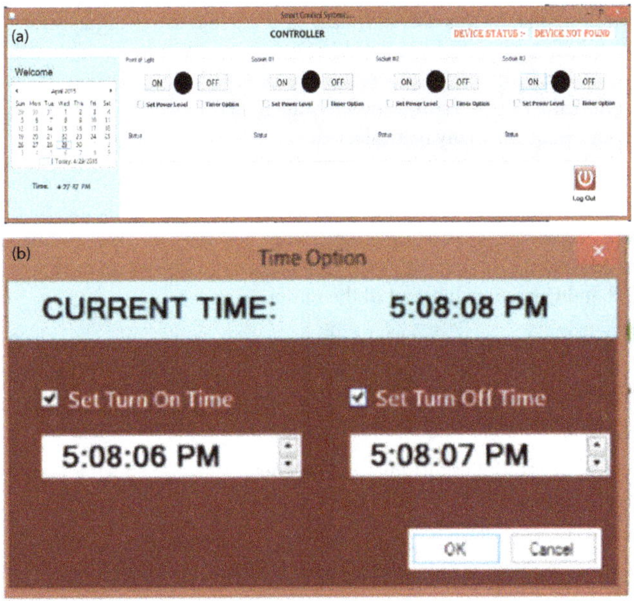

Figure 1: a) Window view of Controller b) Window view of timer option.

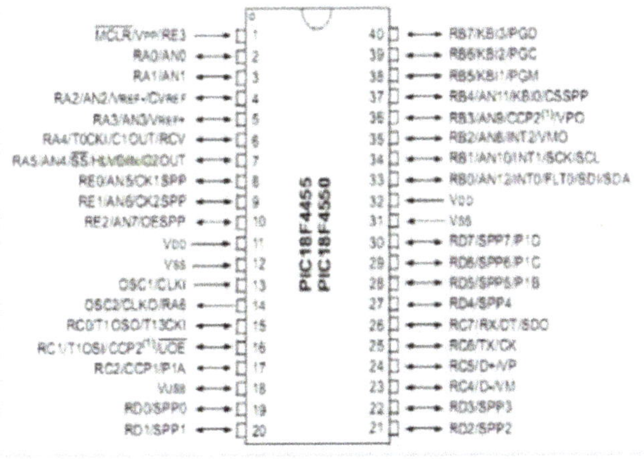

Figure 2: Microcontroller PIC18F4550.

Microcontroller section

This work makes use of PIC18F4550. Microchip PIC 18F4550 microcontroller family is also known as microcomputer, MCU, or µC. This is an integrated circuit that part of an embedded system which contains 200 nanosecond instruction execution and only 35 single word instruction make it more powerful yet easy-to-program. we consider this microcontroller because of its characteristic: Flash memory: 34KB (16kwords), Ram: 256byte, Max.CPU Frequency: 48MHz, internal Oscillator: 8MHz, 32KHz, A/D Converters: 1(13 Channels, 10Bit), USB: 1 (Full Speed), Voltage: 2-5.5V. Pins: 40 (Figure 2).

Relay to switch from microcontroller to mains output

A relay is an electromagnetic switch which is used to switch High Voltage/Current using Low power circuits. Relay isolates low power circuits from high power circuits. It is activated by energizing a coil wounded on a soft iron core. A relay should not be directly connected to a microcontroller, it needs a driving circuit. A relay should not be connected directly to a microcontroller because:

a) A microcontroller is not able to supply current required for the working of a relay. The maximum current that a PIC Microcontroller can source or sink is 25mA while a relay needs about 50 – 100mA current and

b) A relay is activated by energizing its coil. Microcontroller may stop working by the negative voltages produced in the relay due to its back emf.

Interfacing relay with pic microcontroller using ULN2003: work we use ULN2003A because it support more relay, up to Seven Relay can be used. These IC is monolithic ICs consists of High Voltage High Current Darlington transistor arrays. Five relays were connected using these ICs as shown in Figure 3. When using these driver ICs there is no need to connect freewheeling diode as they have built in clamp diodes.

Sampling: Of both analog inputs (AN4 for the voltage, AN0, AN1,AN2,AN3 for the current of light, socket1, socket2, socket3 respectively 100 samples are taken, one every 400 microseconds. This means a total measuring time of 40 milliseconds, which is 2 full

50Hz cycles. 2 full AC cycles is the minimum suitable for this type of measurement. In this work, voltage and current sampling cannot be taken simultaneously (as it should be), they are taken in sequence. This gives a little phase If more relays are required,using transistors will be difficult. In this error between voltage and current measurement, which can be normally be neglected.

Scaling: First the samples taken are translated (scaled) to actual volts and amperes, which are no "words" but "reals". This means the 2.5V4 offset has to be subtracted (value=nominal 511, the middle of the ADC range), and multiplied with a constant value to get the correct Volt and Current values out of the samples.

For the voltage the multiplication factor is 230.0, for the current the factor is 15.1515 according to the Current IC data sheet. Furthermore both have to be multiplied with the voltage per ADC step, in both cases the value 5/1024 (5V, 1024 steps).

So, the multiplier formulas here become:Vdd=5.0; which is the PIC supply voltage is 5V nominal

VMultiplier=Vdd/1024.0 * 230.0; AMultiplier=Vdd/1024.0 * 15.1515; The final formulas to become Volts and Amperes are:

VReal:=VMultiplier * real(integer(VRaw[I] VOffsct));

AReal:=AMultiplier * real(integer(ARaw[I] - AOffset));

Now we have the translated ADC values into real Volt and Ampère values.

Power supply/charger unit

The switching unit, microcontroller and current require a well-filtered and regulated DC power to drive their individual components. The power supply is made up of step down transformer, which steps the input 220Vac down to 15Vac.

The bridge rectifier converts the AC signal to DC of the same voltage level. The rectifier consists of diodes D1-D4. The circuit arrangement is such that at any point in time, two diodes are conducting while the other two are reverse biased.

The filter capacity removes the AC ripples from the DC voltage. The IC regulator regulates the DC signal to give a steady, well- regulated dc output voltage.

Circuit construction and assembly

The circuit board consists of the Vero board and all other components mounted on it. In its construction, the Vero board was cleaned with an iron brush to remove dirt from its surface which might affect soldering quality. Subsequently, following the circuit diagram, the components were mounted on the board one after the other and soldered. The IC was not directly soldered to the board but was mounted on an IC socket. This is to prevent heat damage and for ease of replacement. Units like the power switch, display etc. were connected to the board via flexible wires as shown in Figure 3. In the soldering process, care was taken to ensure that the soldered joints have good mechanical and electrical contact. Also great care was taken to ensure that the components were not damage from excess heat from the soldering iron.

In assembling the system, the circuit board was firmly fixed and screwed in the enclosure such that there was no conducting object like lead ball; nail etc. inside the enclosure and also the enclosure was not too small for the circuit board since this might cause compression which might result to breakage or the Vero board track as shown in Figure 4. Having constructed the circuit board and the enclosure, the functionality of the system was tested and confirmed okay. The interfacing of the Smart Control System with Computer is shown in Figure 5.

The Smart Control System consist of: Full Security with Authentication password, Automatic appliances control with load

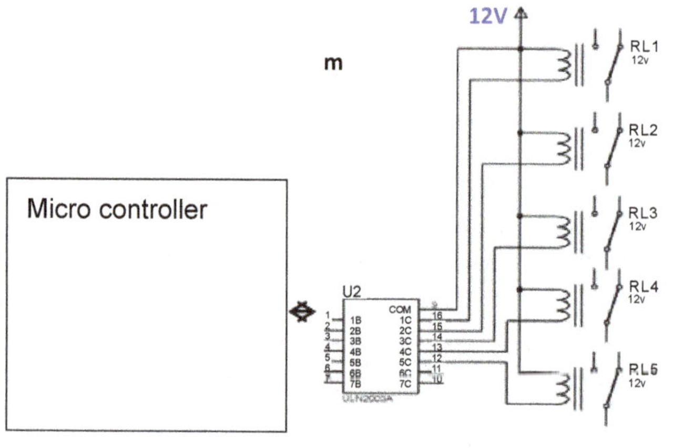

Figure 3: Circuit diagram for relay connection using ULN2003.

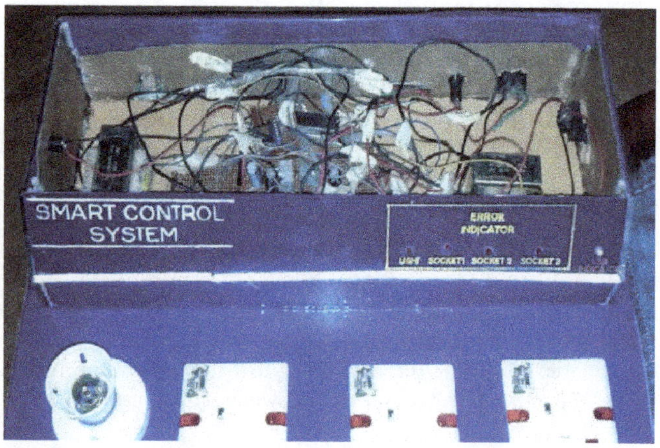

Figure 4: Circuit construction and assembly of smart control system.

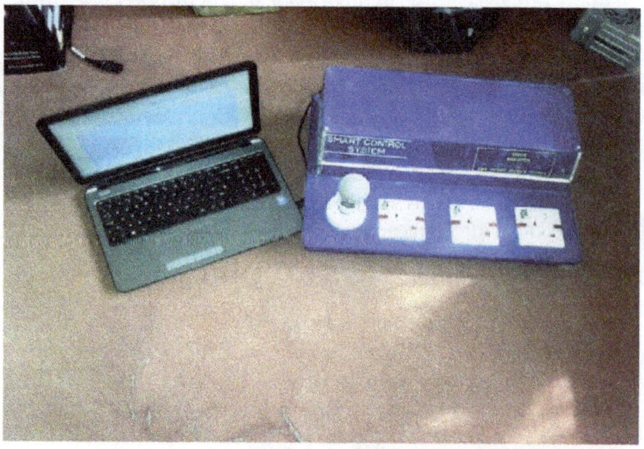

Figure 5: Computer interfaced with smart control system.

sensing system, Automatic Appliances control with real time clock timing, Switching on and off of all the appliances at once, Over voltage and low voltage automatic control with override method, Power Monitoring System with overload protection, Logs and Reports of operator and Input voltage and output in digital format.

Conclusion

Microcontroller Based Smart Control System with Computer Interface was designed to monitor and control electrical appliance, this was done by connecting the system to computer through USB cable. This smart controller has different controlling option via: timer controlling option, maximum voltage controlling option, maximum power controlling option. The current module determines excess current flow and switch off in case of excess current above the threshold. The design of this system can control several devices depending on the appliance to be controlled.

This system is designed to counter appliance attack, controlling of the appliance using computer program specially design for the system. It also provides efficient security management and resolves vulnerabilities or counter measure. SCS is an important control system, for ensuring continuity monitoring and switching of any electrical appliance. This system can be use in national grid system, offices, homes, industries etc.

References

1. Chien JRC, Chi CT (2004) "The information home appliance control system-a bluetooth universal type remote controller" Proceedings of the 2004 IEEE. International Networking, Sensing & Control. Taipei, Taiwan 1: 399-400.

2. Nguyen TV, Lee DG, Seol YH, Yu MH, Choi D (2007) "Ubiquitous access to home appliance control system using infrared ray and power line communication". ICI IEEE/IFIP International Conference in Central Asia, Tashkent, Uzbekistan 1: 1-4.

3. Krusienski D (2006) 'An Internet based wireless Home Automation System for Multifunctional Devices'. IEEE Consumer Electronics, 51: 1169-1174.

4. Haque ASM, Kamruzzaman SM, Ashrafullslam Md (2006) A System for Smart-Home Control of Appliances Based on Timer and Speech Interaction" Proceedings of the 4th International Conference on Electrical Engineering & 2nd Annual Paper Meet 26-28.

5. Jawarkar NP, Ahmed V, Thakare RD (2007) "Remote Control using Mobile through Spoken Commands". IEEE - International Consortium of Stem Cell Networks (ICSCN): 622-625.

6. Khiyal MSH, Khan A, Shehzadi E (2009) "SMS Based Wireless Home Appliance Control System (HACS) for Automating Appliances and Security". Issue in Information Science and Information Technology 6: 887-894.

7. Ahmed Z, Ali M, Majeed S (2011) "Implementing Computerized and Digitally Mobile. Home Automation System towards Electric Appliance Control and Security System". Int J Emerg Sci 1: 487-503.

31

Density based Traffic System and Collision Alert at Intersection of Roads

Vaishali B[1]* and Jeyapriya A[2]

[1]*Department of Electrical Engineering, Universite du Quebec a Trois-Rivieres, Québec, Canada*
[2]*CSE, Sri Ramakrishna Engineering College, Coimbatore, Tamil Nadu, India*

Abstract

Internet of things is the collection of gadgets and embedded objects to gather and exchange data by the usage of internet. The internet of things with intelligent transport system (ITS) has become the elucidation for the navigation of vehicles, traffic signal control systems, and automatic number plate identification. It has been evolved in applications for providing contemporaneous data and feedback from diverse sources such as weather information, parking guidance and also the gentrify traffic information from time to time. The proposed system focuses on dynamic traffic light sequence based on the traffic density in order to decrease the rate of accidents at intersection of roads. The obstacles that may or may not occur at intersection are recognized and warned by usage of an alert system and a notification is generated. The traffic management is used to supervise and retrieve current status of traffic and violators are discovered. This is mainly done to evacuate the traffic congestion and accidents at peak hours.

Keywords: Intelligent transport system; Red light; Runner; HMM; SVM; Hot spot

Introduction

In recent years, vehicle utilization has increased tremendously. Due to this, road traffic ambience has become more problematic and confusing. Accidents that occur at intersections of roads are due to some failure in traffic signal systems and drivers' knowledge. When the operations of traffic management systems are ameliorated with the help of security and proficiency of transportation systems, the general traffic system utilizes stable signalling times at intersections and does not provide any importance to emergency vehicles such as ambulances, fire-fighters and police cars thereby leading to a loss of lives, damage or destruction of property, and rise in fuel costs, pollution and congestion. The requirement of the sensor node is to manage traffic in a specific area by the usage of different devices that can measure and determine various physical traffic parameters like density and collision. A collision warning system operates as follows: a sensor installed at the front end of a vehicle constantly scans the road ahead for vehicles or obstacles. If such an obstacle is found, the system determines whether the vehicle is in imminent danger of crashing, and if so, a collision warning is given to the driver. The sensors used, fulfil the tasks of obstacle detection and tracking, which is the basis of collision warning sensing techniques (Figure 1).

Here we propose a system that generates the traffic light signals based on the density or the number of vehicles in the road which is a contradict to the old method. Where there are regular time intervals to all roads irrespective of their traffic voluminous. This type of traffic light signalling system is incorporated in almost of the metropolitans' cities. In this method to monitor and manage traffic, the density of traffic is measured using IR sensors; these sensors are placed on either sides of the road. The sensors output is given to a microcontroller then this data is processed in it. Thus, depending on density of traffic the timing of traffic lights are appropriately set. The ultrasonic sensors are also used to detect any collision or obstacles and if any, is detected warned by an alarm and warning message.

Figure 1: Collision at intersection of roads.

Literature Survey

The vehicle collision avoidance as well as the RLR [1] activity intersection is identified. When the RLR is detected the neighbouring vehicles are instructed to slow down. It uses on-scooter solution using smartphones. Using the GPS present in the smartphone behaviour of the scooter driver is known.

The red signal violators are predicted by infrastructure based and vehicle to roadside communication. The collision occurs basically because of last second turns and also because of the RLR violators. This [2] uses two schemes (1) for the enhancement of prediction rate (2) to find the status about the drivers.

The areas that high at risk are called as hot spot. The hot spots [3] are identified so the risk exposure would be minimized. The hot spot technique is not very effective for minor collisions.

The traffic exposure is considered as the incidents that occur on the roadway. The risk factor [4] for the occurrences of the incident is calculated. The calculation will help in the observation of traffic. This uses a hypotheses function. It's based four stages were each stage focus on simple, redundant and complex incidents

The collisions between the motor vehicles and pedestrians are more intense in regions where there is no traffic signal. It uses novel based

***Corresponding author:** Vaishali B, Department of Electrical Engineering, Universite du Quebec a Trois-Rivieres, Québec, Canada
E-mail: vaishalibabuskct@gmail.com

approach which uses the data about the characteristics of road and the surroundings of neighbourhood [5].

The double intersection roads have become cause to traffic congestion especially at rush hours. These kind of traffic that occur at double intersections include the roads are frequently used with not often used roads that have a wide median strip [6]. This type of intersections has a non-prior section in the middle to connect the prior road at the two sides, roads that are close to the banks of a river. The double-intersection should have a usual signalized intersection and for maximum capacity of vehicles there should be linked traffic signal control.

Proposed System

Traffic congestion is another problem that occurs now in most of the advanced cities around the world. Traffic congestion has induced many censorious problems and challenges in the most of the highly populated cities. The tour to different places within the city is becoming more problematic for the travellers because of traffic. Because of the clogging problems, people lose time, miss possibilities, and get frustrated. Due to traffic clogging there is a loss of yield from workers, trade possibilities are lost, delivery gets retarded, and thereby the costs are spiked. As result traffic is becoming one of important and major inconvenient problems in huge developed cities. Some of the most important traffic investigations are clogging and accidents which have causes an ample waste of time, property destruction and environs pollution. This research presents a novel intelligent traffic administration system, based on Internet of things, which is featured by low cost, high scalability, high compatibility, easy to upgrade, to replace standard traffic management system. The proposed system can improve road traffic immensely. The Internet of Things is based on the Internet, network wireless sensing and detection technologies to realize the intelligent recognition on the tagged traffic object, tracking, monitoring, managing and processed automatically. This paper includes the following list of components ARM7 microcontroller, IR sensor, ULTRASONIC sensor, LCD display, BUZZER.

IR sensor

An infrared sensor is an electronic appliance that discharges in order to sense some aspects of the environment. An IR sensor can estimate the heat of an object as well as detects the motion. This kind of sensors measures evaluate only the infrared radiation rather than emitting it, which is called as a passive IR sensor. The infrared spectrum generally radiates some kind of radiations with the objects it contains. Infrared sensor can detect these kinds of radiations which are not visible to our eyes. The emitter is generally an IR LED detector and the detector is generally an IR photodiode which is sensitive to IR light of the same wavelength as that emitted by the IR LED. The immensity of the IR light receives the changed proportion of photodiode, resistance and the output voltages when the IR light falls on them (Figure 2).

Ultrasonic sensor

The alternate transmission and reception of sound waves are allowed by the ultrasonic propinquity sensors that are usually used in the special sound transducers. The transducer discharges the sonic waves and are reflected by an object and received back in the transducer. The ultrasonic sensor switches to the receive mode after the emission of the sound waves, the measure of the distance of the object from the sensor is proportional to the time advanced between the release and receive of sound waves (Figure 3).

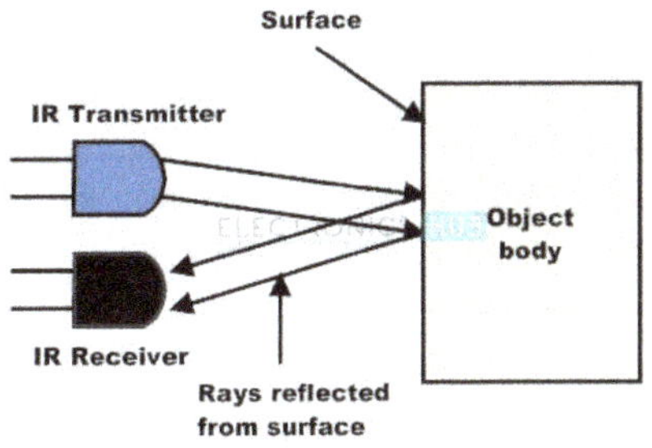

Figure 2: IR sensor.

Figure 3: Ultrasonic sensor.

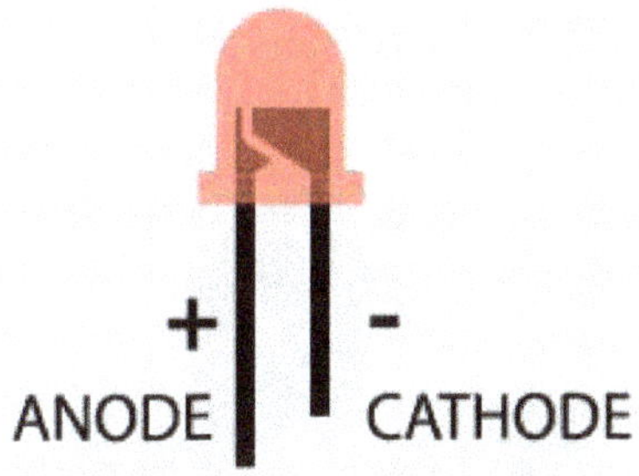

Figure 4: LEDs.

LED

LEDs are specifically designed so they make light of a certain wavelength and they're built into rounded plastic bulbs to make this light brighter and more focused. LEDs of red color generate light with a wavelength of about 630-660 nanometers-which happens to look red when we see it, while blue LEDs generate light with shorter wavelengths of about 430-500 nanometers, which we see as blue. The proficiency of simple tungsten lamp is almost 50 times lesser than an LED. When compared with 100 milliseconds for a tungsten lamp, the response time of the LED are 0.1 milliseconds faster (Figure 4).

LCD display

LCD is a technology used for small computers. It displays about 224 alphanumeric characters. The 8 data pins in the LCD are fetched to the ASCII 8-bit code of the character to LCD. However, the data can be converted to a character type array and it's sent one by one to the LCD. Data can be sent by the usage of LCD either in 8-bit or 4-bit mode. Two nibbles of data (First high four bits and then low four bits) are to be sent to complete a full eight-bit transfer in case of 4-bit mode. The 8-bit mode is preferred for use when the speed is required in an application 4-bit mode requires a minimum of seven bits (Figure 5).

Buzzer

Buzzers are often employed to grant the user or operator an audio indication of the phase of a mechanical device. For instance, buzzing someone in|| refers to opening an electromechanical door lock, which generates a buzz, stating that the lock has been moved to the open position. Buzzers are mainly being utilized on computers, generally to stipulate error conditions. These buzzers may go off if there is a fault present in the component of the device. Buzzers are also utilized for alarm systems (Figure 6).

ARM Microcontroller

It's an enhanced RISC machine in the RISC framework. There are small in size, less hardware cost, power consumption is minimum. The memory operation is performed by the general purpose register. Most commonly used in all mobile phones (Figure 7).

Figure 5: LCD display.

Figure 6: Buzzers.

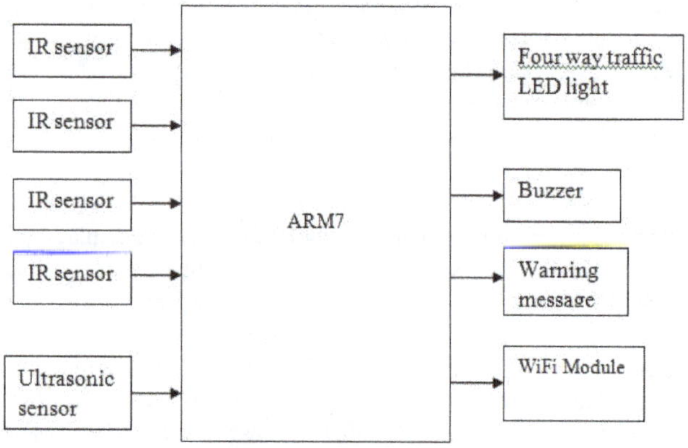

Figure 7: ARM Microcontroller.

Implementation Phase

The implementation phase of collision avoidance at intersection of roads consists of 3 modules: dynamic traffic light sequence, obstacle detection and monitoring traffic.

Figure 8 represents the architecture of the proposed system. The Infrared (IR) sensor are used for count the number of vehicles i.e., to determine the maximum capacity for dynamic control of signals. The ultrasonic sensors are used to detect the obstacles that may occur at intersections. The ultrasonic sensors are used since it detects the obstacles based on the sound so that the false alarm or false negative can be reduced. The warning message is displayed in a LCD display and alarm is given using a buzzer. The traffic management with internet of things is done to identify the violators, monitor the current state of traffic and to reduce the accidents at intersection.

The components that are used in the architecture are explained in Section III. The overall prototype of the system is depicted in Figure 9.

Dynamic traffic light sequence

In this module, the traffic signal at each side of the road is of 20 sec. The density or voluminous of each road is the total number of vehicles is calculated. The count of the total number of vehicles is measured using the IR sensor. Based on the highest count of the vehicles at each side of the road the traffic signal is turned green. This is done mainly minimize traffic clogging at rush time hours. All the traffic signal LEDs are connected to the GPIO pins of ARM controller. At normal traffic signal works on 2 modes are Time-basis mode and Density based mode. In time basis mode the traffic signals changed based on some time interval. This time interval is determined in the coding part of the

Figure 8: Architecture of the proposed system.

Figure 9: Prototype of the collision avoidance system.

arm controller. This mode will active only when the all ways of traffic will be normal condition and weather they have equal traffic in all the ways. In density basis mode the traffic signals changed based on density of traffic present in each way. This mode will only activate when the traffic will be exited the normal condition (Figure 10).

Obstacle detection

In this module, the ultrasonic sensor is used to detect whether there is traffic jam or collision between vehicles or any obstacle has occurred suddenly is identified. The obstacles are identified based upon their sound. If an obstacle is detected an alarm is generated from the alert system. The alert system includes alarm and LCD display. The detection of ultrasonic sensor is alerted using a alarm and also an warning message is sent (Figure 11).

Monitoring traffic

The traffic information includes about whether there is any collision occurred or any obstacle detected. It also includes the information about the traffic whether it's high or low. The traffic information can be viewed in the browser. This information is updated using wifi module with UART part (Figure 12).

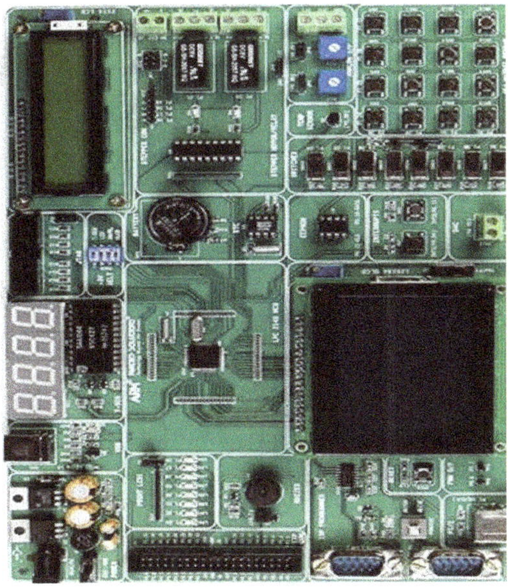

Figure 10: Density of traffic.

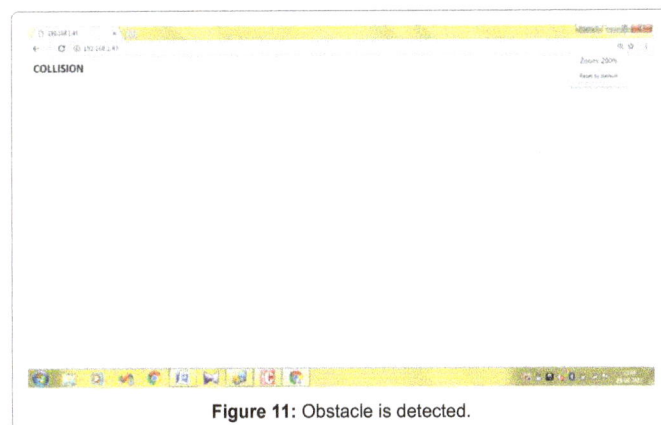

Figure 11: Obstacle is detected.

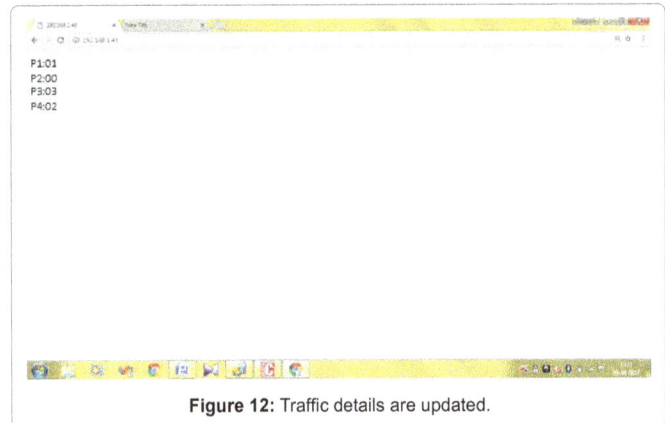

Figure 12: Traffic details are updated.

Existing system	Proposed system
Less accuracy	Accurate
Expensive	Economical
Manual control	Manual hurdles are avoided

Table 1: Comparison of existing and proposed system.

Results

In order to minimize the number of accidents those exist at the intersection of road for the welfare of vehicle drivers, pedestrians. This is not only caused by the collision of vehicles but also by the disobeyers of the traffic rules. The traffic clogging is depleted by the usage of IR sensor. This calculates the voluminous of vehicles at each side of the road and the traffic light sequence is changed accordingly. To ensure the safety of the vehicle drivers an ultrasonic sensor is used. If any obstacle is detected an alarm is generated thereby alerting the neighbouring vehicles. The identified obstacles and traffic information are updated to the webpage through Wi-Fi module. The human intervention is avoided to minimize accidents (Table 1).

Conclusion

The traffic system is used for decreasing the rate of accidents at intersection and the traffic is monitored. The traffic congestion and maximum capacity of vehicles are depleted by designing the signalized roads at intersections of roads. The identification of the critical regions is to be monitored in order to decrease the rate of the occurrence of accidents.

References

1. Huang KS, Chiu PJ, Tsai HM, Kuo CC, Lee HY, et al. (2016) RedEye: Preventing collisions caused by Red-light-Running scooters with smartphones. IEEE Transaction on Intelligent Transportation System 17: 1243-1257.
2. Wang L, Zhang L, Zhang WB, Zhou K (2009) Red Light Running Prediction for dynamic All-red Extension at signalized Intersection. Proceedings of the 12th International IEEE conference on Intelligent Transportation system.
3. Washington S, Haque M, Oh J, Lee D (2014) Applying Quntile regression for modelling equivalent property damage only crashes to identify accident blacspots. Accident Analysis and Prevention 66: 136-146.
4. Elvik R (2015) Some implications of an event-based definition of exposure to the risk of road accident. Accident Analysis and Prevention 76: 15-24.
5. Quistberg DA, Howard EJ, Ebel BE, Moudon AV, Saelens BE, et al. (2015) Multilevel models for evaluating the risk of pedestrian-motor vehiclecollisions at intersections and mid-blocks. Accident Analysis and Prevention 84: 99-111.
6. Vu MT, Nguyen VP, Nguyen VB, Nguyen TA (2016) Methods for deigning Signalized double-intersection with mixed Traffic in Vietnam. Procedia engineering, pp: 131-138.

An Efficient Bat Algorithm for Series-parallel Power System Optimization

Benaissa A, Zeblah A, Belafdal A and Chaker A*

Djillali Liabes University of Sidi Bel Abbes, Algeria

Abstract

Optimization techniques tend towards new methods, these methods are based on the nature of the impact thereof on the lifestyle of human beings, engineers working on new optimization approaches to achieve the objective of maximizing reliability of the electric current and with the minimal cost which means that our structure of network must answer this requirement, a new approach proves to solve major problems using a heuristic méta - which is called algorithm of beater (BA).

Keywords: Bat algorithm; Optimization; Design of the power system; Reliability; Universal moment generating function

Introduction

The reliability of the system, depends on each electric components which constitutes the system, and must answer the latter, it is the reason for which the good performance of our system must reach this level, therefore to always reduce the investment costs of the elements of our structure which constitutes our system. However the problem in the design of structure of network is to obtain a hand to maximize reliability of the system and with the minimal investment prone to the constraints of the tolerance and the reliability and the total costs of the fuel system.

In our problem, the function objectifies which is, the reduction of the total costs and that under maximum influence of reliability for the only reason to build a structure of the reliable network defines that under certain constraint for the systems design of food. The whole of manufacturers know this problem well and tend to optimize the redundancy (safety device in the event of an imbalance of the system) [1,2]. Once the application of the reliability of all the systems is applied, one regards it as a measurable size for our design and must correspond has the satisfaction of the request namely by provide a sufficient quantity of electrical energy [3]. These problems and unambiguous of failure because according to the specifications proposed by the customer the rated capacity generators will be never defective.

The load diagram of the consumer and that of the system are taken into account of made that it intervenes in the process of the evaluation of reliability. Our system is composed of a structure based on components technologies for the design of our electric system, we consider another time the load diagram of the consumer it is taken into account because it enters the constitution of the reliability of the system as all the components are taken in consideration because it is a problem of optimization of the feeding system of energy. The same problem is quoted in [4], which was formulated to optimize the reference in the basic approach [5], the component costs are taken in consideration and the capacities which compose the system, and the application was considered using a curved load. The disadvantage of the approach inside [2,5] is that the cost of components is defined as analytical explicit according to their capacities and of the values of the index of reliability are assignees with all the type indicating of the components independently of their capacity. In [6,7], the different meta-heuristic ones used in it modest work for a comparison, genetic algorithm, colony of ant, harmony seeks and the algorithm of the bats are employed to solve problems of technical order of optimization. we propose one algorithm news which is the algorithm of the bats (BA) to provide the design a structure of the feeding system with a choice inexpensive and optimal of the suitable electric components which compose (of the generators, of transformers and of the electric lines of technology) product range available on the electric market of the components for each type of electric components. The true one encloses practical, it is that there exists a great diversification of choice of products is available, and each components is characterized by a technology that its is in performance, cost, and reliability. Our algorithm (BA) selects by simple combinations of the components to arrive at an optimization of system with a reduced cost, however, the production, transfers it and the routing of this production is to feed a whole networks with a condition quoted in top under constraint cost and with a performance also to book with one tolerance of so powerful reliability. For the evaluation of the reliability of the structure of the feeding system in our case in parallel series, the development of one démarche one develops the fast and effective method which is based on Function of universal production (WMU) [4,8].

Notations:

c_i: Cost of electrical component *i*.

A_i: Available of electrical components technologies.

g_i: power components performance.

r_i: power components reliability.

r_i: power components reliability.

ROP: redundancy optimization problem.

B.A: Bat Algorithm

AC: Ant colony algorithm.

GA: Genetic algorithm.

HS: Harmony Search algorithm.

Description and Problematic Formulation of the Structure of the Network

On Figure 1 one presented the whole of the sub-system series-parallel

***Corresponding author:** Chaker A, Djillali Liabes University of Sidi Bel Abbes, BP 89 Sidi Bel Abbes, 22000, Algeria, E-mail: chaker.abdelkrim@live.fr

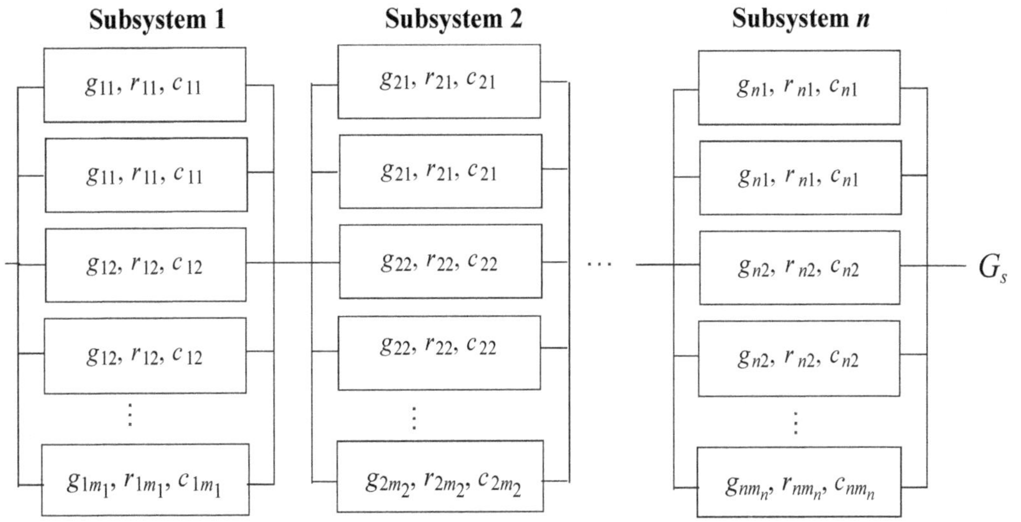

Figure 1: Series-parallel power system.

for "n" component which constitute it. Our system requirements are composed for not distorted the equation of "n, m" electric components assembled in parallel series knowing that all elements of kind, moreover $i=1,.., n$ components of powers are selected of kind i in the structure of which there are a multiplicity of technologies who different ones with the others by quality that its is performance, cost and reliability and capacity. These elements are defines in our system by names chosen to answer the equation these indices are C_{im}, r_{im}, g_{im}. The structure of our system parallel series is defined by a component vectorial $k_i = \{ k_{im_i} \}$ $(1 \le i \le n, 1 \le j \le m_i)$and the total costs of the system for the whole of the system are given $k_1, k_2, \ldots, k_n$ by the equation which is written so below:

$$C = \sum_{i=1}^{n} \sum_{j=1}^{m_i} k_{im} c_{im} \quad (1)$$

Electrical energy sudden continuously of the losses what leads us inevitably has a loss in the index of the probability of load (LOLP) and moreover the period *T* energy to deploy (EENS) in this function, are two great parameters which contribute has the estimate of reliability [3]. The calculation of the index of this probability allows us of carried out a load whose whole did not tighten not taken into account. The chart of the random curve discrete and often taken by the request of the load If the period of time of the load is the whole of *M* intervals, with the duration, d_j and T_j each level of request T_j $(J=1,\ldots M)$ has the duration, the LOLP is calculated as follows:

$$LOLP = \frac{1}{\sum_{j=1}^{M} T_j} \sum_{j=1}^{M} P(g_s \succ d_j)\, T_j \quad (2)$$

and

$$EENS = \frac{1}{\sum_{j=1}^{M} T_j} \sum_{j=1}^{M} P(d_j \ge g_s)\, T_j \quad (3)$$

The expression of the probability $P(g_s \le d_j)$represent the probability of all the system and d_j is equal to or higher than g_s compared to the level of the request. All the productions and requires of capacities are defined like a calculation proportional of their total face value. The graph of the cumulative load diagram is determined by vectors $d=\{d_i\}$ and $T=\{T_j\}$ who is parameters determined for each element of the system. The calculation of reliability is carried out by the indexing of *R* in the reference [9], given by the expression *R=1-LOLP*. This indexing is compared compared to R_0 and must be not less values inferior of the specific initial value.

The difficulty of the optimization of reliability of the system and the design of a network lies primarily in the choice of the elements which constitute the system which is perhaps formulated as follow K_1, K_2... K_n: what pushes us with find a design of system who certifies a cost in totality is minimal under constraint of reliability. This difficulty can be known is carried out like below:

$$\text{Minimize } C = \sum_{i=1}^{n} \sum_{j=1}^{m_i} k_{im} c_{im} \quad (4)$$

Subject To R(d, t, k1, k2,…., kn) ≥ R0

Description Assessment of Reliability

Our question quoted with by before is a question of optimization, it is essential of city several states possible of the system. It is the reason which he is obligatory employee a fast and effective procedure to consider reliability structural As mentioned previously, the major question is to estimate reliability R for the arbitrary continuations and the parallel circuit, and it is essential that the probability of all the structure of the electric system of refill is equal to a level of detail of the request for load, it will be calculated in the following way:

$$R(d) = P\{g_s \succ d\} = 1 - P\{g_s \le d\} \quad (5)$$

This evaluation of reliability is based on modern methods and mathematical techniques as observed above: technique of UGF (or U-transform) in [10-12]. The latter was applied for the estimate and the optimization of the solidity of the power circuit in [13,14], and interpret an occurring prolongation of function one ordinary moment [15]. The UGF, in our case, of a discrete variable *E* is thus given polynomial:

$$u(z) = \sum_{j=1}^{J} P_j z^{g_j} \quad (6)$$

where the discrete random variable g has *J* possible values and P_j is the probability that g is equal to g_j. Under consideration if only the components with total failures are considered. For instance for each element of type i and technology m has reliability R_{im_i} and nominal capacity g_{im_i}, then we denote by:

$P(g=g_s)=R_{im}$ and $P(g=0)=1-R_{im}$. The UGF can be defined of such an element has only two terms as:

$$u_{im} = (1-R_{im})z^0 + R_{im}z^{g_{im}} \quad (7)$$

A brief overview on UGF method with respect to its applications for multi-states system (MSS) which has a finite number of states, there can be m different levels of output performance at each time t:

$g(t) \in g=\{g_h, 1 \le j \le m\}$ and the system output performance can be defined by two finite vectors g and $p=\{p_h(t)\}=\Pr\{g(t)=g_h\}$ $1 \le h \le m$, here the UGF, represented by the polynomial $u(z)$ can define all the MSS output performance, i.e. it represent all the possible states of the system by relating the probability of each state to performance of MSS in that state in the form:

$$u_{MSS}(t,z) = \sum_{h=1}^{m} p_h(t) z^{g_h} \quad (8)$$

Having the MSS output performance, the system reliability for arbitrary time t and demand d can be obtained using the following operator Ω_A

$$\begin{aligned} R(t,d) &= \Omega_A(u_{MSS}(t,z), d) \\ &= \Omega_A\left(\sum_{h=1}^{m} p_h(t) z^{g_h}, d\right) \\ &= \sum_{h=1}^{m} p_h(t)\alpha(g_h - d) \end{aligned} \quad (9)$$

where $\alpha(x) = \begin{cases} 1, & x \ge 0 \\ 0, & x < 0 \end{cases}$

In clearer, the probability that all the faculty of the circuit of refill is not less than one precise request of level of loading; it is possible to write it thus,

$P\{g \ge d\} = \Omega(u(Z)Z^{-d})$

Where Ω is a distributive operator defined by the following expression:

$$\Omega(pz^{g-d}) = \begin{cases} p, & if \ g \ge d \\ 0, & if \ g < d \end{cases}$$

and

$$\Omega\left(\sum_{j=1}^{J} pz^{g_j-d}\right) = \sum_{j=1}^{J} \Omega(pz^{g_j-d})$$

For the installation of the feeding system containing of the components laid out of various manners, in parallel, all the capacity is equal to the sum of all the components of the capacity There with front, the U-function can be calculated using the Operator Γ:

$$u_s(z) = \Gamma(u_1(z),\dots,u_n(z)) = \prod_{i=1}^{n} u_i(z)$$

where

$$\Gamma(g_1,\dots,g_n) = \sum_{i=1}^{n} g_i \quad \text{so that}$$

$$\begin{aligned} \Gamma(u_1(z)\,u_2(z)) &= \Gamma\left(\sum_{i=1}^{n} p_i z^{a_i}, \sum_{j=1}^{m} q_j z^{b_j}\right) \\ &= \sum_{i=1}^{n}\sum_{j=1}^{m} p_i q_j z^{a_i+b_j} \end{aligned}$$

We notice that the operator Γ is with simplicity a polynomial production and representative u-function of the system in particular and containing m elements assembled in series. Another example if the circuit comprises attached components in a chain, mitigated operation is established by the worst condition observed for any of its components it is the reason why in employment another coefficient β and which leads us to:

$u_s(z)=\beta(u1(z), \dots, u_m(z))$

Whose value of is:

$\beta(g_1,\dots, g_m)=\min\{g_1,\dots, g_m\}$ in manner,

$$\begin{aligned} \beta(u_1(z)\,u_2(z)) &= \beta\left(\sum_{i=1}^{n} p_i z^{a_i}, \sum_{j=1}^{m} q_j z^{b_j}\right) \\ &= \sum_{i=1}^{n}\sum_{j=1}^{m} p_i q_j z^{\min(a_i,b_j)} \end{aligned}$$

The MSS reliability was presented and $p\{g \ge d\}$after time has passed for this probability becomes constant.

The Bat Optimization Approach

The bats are flying mammals and which proceed of the wings, and they have also the ability to perceive sounds which emit one calls it "echolocation". Whose most microchiroptères use this means which is echolocation for nourished because they are insectivorous, for protected from predatory and to locate to perch cracks in the dark darkness to them. It is estimated that there are approximately 1,000 various species of bat, which represents up to 20% of all the mammalian species. Their size varies the tiny ones bats bumblebee (from approximately 1.5 to 2 G) with the giant bats with a scale of approximately 2 m and weight going until approximately 1 kg. The microchiroptères generally have a front armlever length from approximately 2.2 to 11 cm.

These bats emit a very strong noise and to listen to the echo which rebounds starting from the surrounding objects. Their impulses vary in properties and can be in correlation with their strategies of hunting, according to the species Most bats use, of short signals modulated in frequency to sweep approximately an octave, others generally use signals at constant frequency for echolocation. Their signal with band-width varies according to the species and increases by using often more harmonics.

Studies show that microchiroptères uses time delay of the emission and the detection of the echo, time difference between their two ears, and the variations of sound intensity of the echoes to build a scenario in three dimensions of surrounding or they are. They can detect the distance and the orientation from the target, the type of prey, and even the rate of travel of the prey, such as the small insects. Indeed, studies suggest that the bats seem to be able to distinguish the targets by the variations from the Doppler effect induced by the rates of the target insects wing undulation.

Algorithm (BA) beats is a method méta-heuristics based on an algorithm of optimization which was initially inspired by the life of the bats to find their food. The bald people mouse emit signals has place or they are and of écoûté the echo of recalls this process known as of echolocation to locate itself by report has the prey. BA is mainly built by the use of four principal ideas.

1) Echolocation allows the bat of distinguished food the difference between the preys (prey and object).

2) Each bat in position X_i steals at the speed of production of one

V_i particular impulse with the frequency and the intensity of the f_i and A_i respectively.

3) The volume of Have exchange in various A_i ways such as the reduction of a great value to a low value.

4) The frequency r_i, f_i and laughed of rate of each impulse is controlled automatically.

Initially, all the bats fly randomly within the space of research producing of the random impulses. After each flight, the position of each bat is put up to date in the following way:

$$V_i^{new} = V_i^{old} + f_i(X_G - X_i); i = 1,\dots,N_{Bat}$$
$$X_i^{new} = X_i^{old} + V_i^{new}; i = 1,\dots,N_{Bat} \quad (10)$$
$$f_i = f_i^{\min} + Q_i(f_i^{\max} - f_i^{\min}); i = 1,\dots,N_{Bat}$$

Where X_G has the best overall solution. The limit of the upper frequency and the lower frequency sounds of the nth bat are represented by $f_i^{\max}$ and $f_i^{\min}$. The population size is the total number of designated snowshoe N_b, ϕ_1 and is a number generated randomly between 0 and 1

The second position of the movement of the bat is simulated as follows:

$$X_i^{new} = X_i^{old} + \varepsilon A_{mean}^{old}; i = 1,\dots,N_{Bat} \quad (11)$$

Epsilon "ε" is a random number in the beach of the segment [-1.1] and for the improvement of the amplitude for all the bald people mouse. Once the position of the bald people mouse is improved by the adjustments above Xinew and the new random individual is produced if it the level of signal is larger than a random value β.

One insert the new solution found by the new member in the population has condition which one observes the constraint:

$$[\beta < A_i] \ \& \ [f(x_i) < f(\text{Gbest}) \quad (12)$$

As previously mentioned the value of the amplitudes of signal generated by the bats a progressive reduction formulated has by

$$A_{inew=}\alpha A_{iold} \quad (13)$$

$$\Gamma_{\text{iltrer}}+1=\Gamma[1-\exp(-\gamma.t]$$

Constants α and are the approachs important for the bald people mouse, and represent the iteration count in the algorithm.

Step 1: Initialize the bat population or their position X_i^{old} and their velocities V_i^{old}. Define pulse frequency f_i at X_i^{old}. Initialize pulse rates r_i and the loudness A.

Step 2: Generate new solutions by adjusting frequency, and updating velocities and locations/solutions (Equation (10)).

Step 3: If (rand $>r_i$) Select a solution among the best solutions, Generate a local solution around the selected best solution.

Step 4: Else generate a new solution by flying randomly.

Step 5: If ([$\beta < A_i$] & [$f(x_i) < f$(Gbest))Accept the new solutions, increase r and reduce A.

Step 6: Rank the bats and find the current best X_i^{new}.

Step 7: while (iteration < Max number of iterations) Post process results and visualization. The algorithm stops with the total-best solution.

Power Design Example

In Table 1 summarizes the different technology components that will consevoir then our system vien Table 2 contains the cumulative power of the data request of the Year To illustrate the proposals of the Mete-heutistique that different bat BA algorithms, PSO, GA and HS for the construction and design of our system as shown in Figure 2, a numerical example is solved by the use of the data provided in Table 3. Every electrical component of the subsystem is considered a unit with total failures.

With a true simulation delivered by an example real of (G. Levitin) for each algorithm, knowing that the structure of the network is described with a number maximum g_{max}, for components which are placed parallel one take account of the values and of the parameters of this one the results are compared for all these meta-heuristic we summarize in the Table 3 which represents different meta-heuristic thus the configuration from the networks for the design, with lower costs and better a reliability. Knowing that the data of the availability of different component technologies are listed in.

Discussion for Results

The implementation of the bi-inspired bat algorithm to the serie parallel heterogeniousystem has proposed to determine the best configuration with the minimum investment cost.

Sub-System #	Components #	# 1	# 2	# 3	# 4	# 5	# 6	# 7	# 8	# 9
Generators 1	Reliability (%)	0.89	0.977	0.982	0.978	0.983	0.92	0.984	/	/
	Cost (%)	0.59	0.535	0.47	0.42	0.4	0.18	0.22	/	/
	Performance (%)	120	100	85	85	48	31	26	/	/
MT/HT Transformers 2	Reliability (%)	0.995	0.996	0.997	0.997	0.998	/	/	/	/
	Cost (%)	0.205	0.189	0.091	0.056	0.042	/	/	/	/
	Performance (%)	100	92	53	28	21	/	/	/	/
Lignes HT 3	Reliability (%)	0.971	0.973	0.971	0.976	/	/	/	/	/
	Cost (%)	7.525	4.72	3.59	2.42	/	/	/	/	/
	Performance (%)	100	60	40	20	/	/	/	/	/
HT/MT Transformers 4	Reliability (%)	0.977	0.978	0.978	0.983	0.981	0.971	0.983	0.982	0.977
	Cost (%)	0.18	0.16	0.15	0.121	0.102	0.096	0.071	0.049	0.044
	Performance (%)	115	100	91	72	72	72	55	25	25
Lignes MT 5	Reliability (%)	0.984	0.983	0.987	0.981	/	/	/	/	/
	Cost (%)	0.986	0.825	0.49	0.475	/	/	/	/	/
	Performance (%)	128	100	60	51	/	/	/	/	/

Table 1: Data of available different power components technologies.

The proposed problem is subject to the given high level of reliability taken as decision variable constraint. The goal optimization process was based on the combination of various variables decisions as (version of components and algorithm parameters). Also the universal moment generating to evaluate the corresponding constraint by verifying for each iteration the feasibility solution. In Second part the above Table 3 shows the best optimal power design obtained by the suggested meta heuristic's (Bat, harmony search, ant colony and genetic algorithms) for one desired reliability levels R_0 (0.97-0.990). The illustrates the index of evaluation of the price and the availability to the corresponding development of energy. In algorithm beats a series characterized by data are tested. For several parameters of data of the algorithm corresponds to a design of system of power: Mass of the population=500; iteration number=500; loudness=0.02; and the pulse ratean=0.45. The diversification of the influential values forcing gave better results.

Power Demand level (%)	100	80	50	20
Duration (h)	4203	788	1228	2536
Probability	0.479	0.089	0.14	0.289

Table 2: Parameters of the power demand curve.

The comparison and all the more effective if one measures the coefficient of quality of the result by NN (Nakagawa and Nakachima) and that one takes only to the best result given by the algorithm and taken one consideration NN $\lambda = \dfrac{Optimal\_Cost}{reliability}$

The results shows that the NN coefficient of optimal design given by bat algorithm is (λ=11.82/0.97) is very low than HS, ACO and GA.

For the different meta-heuristic ones proposed for a comparison in this modest work which is beats it algorithm, the colonies of ants, the antigens and the harmonies seek (BA, ACO, GA and HS), allows us to conclude that BA after having to start the programme leads us to spectacular solution. If one compares with those found by different the méta heuristic used and an optimal design of the network of power gives us. The difference between these comparative méta-heuristics in

Reliability Constraint R0	Sub-System	Optimal Power Design	Corresponding Availability R	Corresponding Cost C	Optimization Method
0.97÷0.990	Sub-System: 1	6-6 6-6-6-6-6	0.97	11.828	Bat Algorithm
	Sub-System: 2	4-3-1-5-4			
	Sub-System: 3	5-5-1-5			
	Sub-System: 4	6-8-9-7-9-8-2-4-9			
	Sub-System: 5	3-4-4-4			
0.97÷0.990	Sub-System: 1	4-4-6-7	0.992	13.175	Harmony Search
	Sub-System: 2	4-4-4-4-4-4-4			
	Sub-System: 3	01-Apr			
	Sub-System: 4	7-7-7-9			
	Sub-System: 5	04-04-2004			
0.97÷0.990	Sub-System: 1	3-4-4-6-7	0.9906	14.302	Ant Colony
	Sub-System: 2	5-5-5-5-5-5-4			
	Sub-System: 3	01-Apr			
	Sub-System: 4	7-7-7-8-8-9			
	Sub-System: 5	3-4-4-4			
0.97÷0.990	Sub-System: 1	04-04-2006	0.992	15.87	Genetic Algorithm
	Sub-System: 2	03-Mar			
	Sub-System: 3	02-02-2003			
	Sub-System: 4	07-07-2007			
	Sub-System: 5	04-04-2004			

Table 3: Optimal solution obtained by different algorithms, bat, ant colony, harmony search and genetic algorithm.

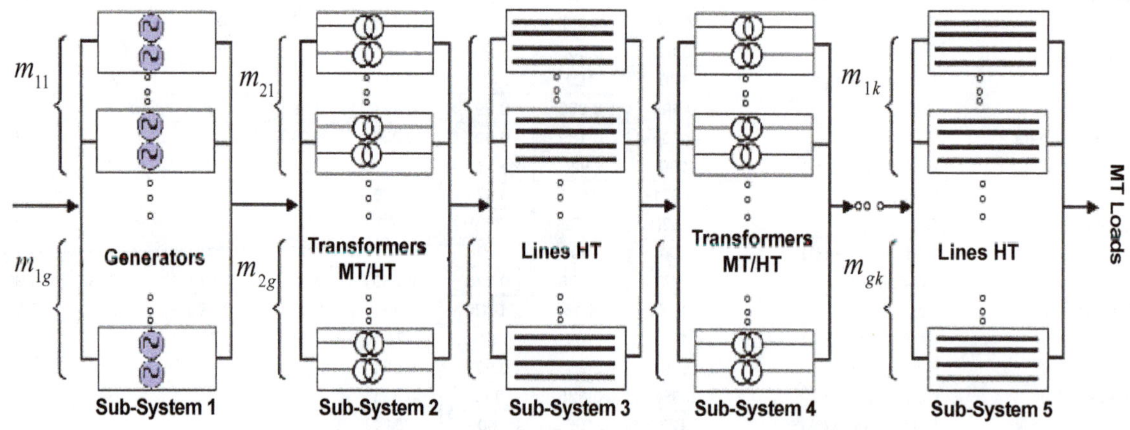

Figure 2: Series- parallel power design.

the Table 1 shows us well that for a beach of reliability gives and its constraint and the cost and its constraint gives us good performance and of this made optimal solutions for the design of the network.

Conclusion

This modest work deals thanks to a new meta-heuristic solving the optimization of the design of an electrical system which is a very interesting question often posed in the energy industry. It is formulated as a problem of redundancy. This new meta-heuristic for solving this problem, however, is based on an algorithm bats our algorithm creates a design such that it is optimal cost and reliable for the system of electricity supply, our system is composed a model structure. this under certain stress algorithm allows us to have a minimum investment cost under constraints with maximum reliability under stress and the cost and reliability. So our algorithm we selected material from a list of products available in the market with technology components, and the list also includes the cost of these products so their performance, this also results in the definition of these components that we will introduce in each sub system with which it also sets the components of electrical power series-parallel to each subsystem when demand changed the consumers. All projects methods to solve part of the optimization problem namely the optimal design of networks (Cost and reliability) with a wide range of product components that make up these subsystems without limiting the various technologies electrical components in series – parallel to it was put a new generation algorithm that could have some time given the best results compared to other algorithms it is the bat algorithms who is compared to algorithms (HS), (ACO) and (GA) for the same problem.

References

1. Liang YC, Smith AE (2001) An ant colony approach to redundancy allocation. IEEE Transaction on reliability.
2. Coit DW, Smith AE (1995) Optimization approaches to the redundancy allocation problem for series-parallel systems. Proceeding of the fourth industrial engineering research conference (IERC).
3. Billiton R, Allan R (1984) Reliability of power systems, Pitman, London, Ushakov IA, Harrison A (Eds), Handbook of reliability engineering, Wiley and Sons, NY/Chichester/Toronto, 1994.
4. Nourelfath M, Nahas N, Zeblah A (2003) An ant colony approach to redundancy optimization for multi-state system. In: International Conference on industrial engineering and production management, Porto-Portugal.
5. Ushakov IA (1987) Optimal standby problems and a universal generating function. Sov. J Compt Syst Sci 25: 7-82.
6. Massim Y, Zeblah A, Ghoraf A, Meziane R (2005) Reliability Evaluation of Multi-State Series-Parallel Power System Under Multi-States Constraints, Electrical Engineering. Journal Springer Verlags 87: 327-336.
7. Levitin G, Lisniaski A, Ben-haim H, Elmakis D (1996) Power system structure optimization subject to reliability constraints. Electric Power System research 39: 145-152.
8. Su CT, Lin CT (2000) New approach with a Hopfield modeling framework to economic dispatch. IEEE Transactions on Power System 15: 541-545.
9. Chowdhury BH, Rahman S (1990) A review of recent advances in economic dispatch. IEEE Trans Power Syst 5: 1248-1259.9.
10. Niknam T, Mojarrad HD, Nayeripour M(2009) A new fuzzy adaptive particle swarm optimization for non-smooth economic dispatch, Energy. J energy 12: 1764-1778.
11. Walters DC, Sheble GB (1993) Genetic algorithm solution of economic dispatch with valve point loading. IEEE Trans Power Syst 8: 1325-1332.
12. Chiang CL (2005) Improved genetic algorithm for power economic dispatch of units with valve-point effects and multiple fuels. IEEE Trans Power Syst 20: 1690-1699.
13. Ongsakul W, Dechanupaprittha S, Ngamroo I (2004) Parallel tabu search algorithm for constrained economic dispatch. IEE Proc Gener Transm Distrib 151: 157-66.
14. Wang KP, Fung CC (2006) Simulate annealing base economic dispatch algorithm. IEE Proc C 140: 507-513.
15. Dodu JC, Martin P, Merlin A, Pouget J (1972) An optimal formulation and solution of short-range operating problems for a power system with flow constraints. IEEE Proc 60: 54-63.

Permissions

All chapters in this book were first published in JAME, by OMICS International; hereby published with permission under the Creative Commons Attribution License or equivalent. Every chapter published in this book has been scrutinized by our experts. Their significance has been extensively debated. The topics covered herein carry significant findings which will fuel the growth of the discipline. They may even be implemented as practical applications or may be referred to as a beginning point for another development.

The contributors of this book come from diverse backgrounds, making this book a truly international effort. This book will bring forth new frontiers with its revolutionizing research information and detailed analysis of the nascent developments around the world.

We would like to thank all the contributing authors for lending their expertise to make the book truly unique. They have played a crucial role in the development of this book. Without their invaluable contributions this book wouldn't have been possible. They have made vital efforts to compile up to date information on the varied aspects of this subject to make this book a valuable addition to the collection of many professionals and students.

This book was conceptualized with the vision of imparting up-to-date information and advanced data in this field. To ensure the same, a matchless editorial board was set up. Every individual on the board went through rigorous rounds of assessment to prove their worth. After which they invested a large part of their time researching and compiling the most relevant data for our readers.

The editorial board has been involved in producing this book since its inception. They have spent rigorous hours researching and exploring the diverse topics which have resulted in the successful publishing of this book. They have passed on their knowledge of decades through this book. To expedite this challenging task, the publisher supported the team at every step. A small team of assistant editors was also appointed to further simplify the editing procedure and attain best results for the readers.

Apart from the editorial board, the designing team has also invested a significant amount of their time in understanding the subject and creating the most relevant covers. They scrutinized every image to scout for the most suitable representation of the subject and create an appropriate cover for the book.

The publishing team has been an ardent support to the editorial, designing and production team. Their endless efforts to recruit the best for this project, has resulted in the accomplishment of this book. They are a veteran in the field of academics and their pool of knowledge is as vast as their experience in printing. Their expertise and guidance has proved useful at every step. Their uncompromising quality standards have made this book an exceptional effort. Their encouragement from time to time has been an inspiration for everyone.

The publisher and the editorial board hope that this book will prove to be a valuable piece of knowledge for researchers, students, practitioners and scholars across the globe.

List of Contributors

Kansal S
Department of Electrical Engineering, Baba Hira Singh Bhattal Institute of Engineering and Technology, Lehragaga-148031, Punjab, India

Kumar V
Department of Electrical Engineering, Indian Institute of Technology, Roorkee, India

Tyagi B
EED, Indian Institute technology, Roorkee, India

Marouan Elazzaoui
Department of Electrical Engineering, Electronic Power and Control Laboratory, Mohammedia School of Engineering Université Mohammed V-Agdal, Morocco

Yen-Hung Hu
Department of Computer Science, Hampton University, Hampton, Virginia 23668, USA

Raghav Puri and Navpreet Singh
Department of Electronics and Communication Engineering, Global Institute of Management and Emerging Technologies, Amritsar, Punjab, India

Sathyavathin S
Final Year M.E (VLSI design), Department of ECE, Adhiparasakthi Engineering College, Melmaruvathur, TN, India

Mr Ilanthendral J
Assistant Professor, ECE Department, Adhiparasakthi Engineering College, Melmaruvathur, TN, India

Harleen Kaur and Neetu Gupta
Department of ECE, GIMET, Amritsar, Punjab, India

Ezennaya SO, Ezechukwu OA and Anierobi CC
Department of Electrical Engineering, Nnamdi Azikiwe University, Awka, Nigeria

Akpe VA
Transmission Company of Nigeria (TCN)

Hajizade Kanafgorabi M and Dr. Karami A
Department of Electrical Engineering, Faculty of Engineering, University of Guilan, Rasht, Iran

Shaheen H and Nawaz I
Department of Mechanical Engineering, Jamia Millia Islamia, Delhi, India

Kargapol'tsev ES
Institute of Laser Physics SB RAS, Lavrentyeva ave. 13/3, Novosibirsk, 630090, Russia

Razhev AM and Churkin DS
Institute of Laser Physics SB RAS, Lavrentyeva ave. 13/3, Novosibirsk, 630090, Russia
Novosibirsk State University, Pirogova st. 2, Novosibirsk, 630090, Russia

Pawan Whig
Department of Electronics and Communication, Bhagwan Parshuram College of Engineering and Technology Rohini, Delhi, India

Syed Naseem Ahmad
Head of Department, Department of Electronics and Communication Engineering, Jamia Millia Islamia, New Delhi 110025, India

Nakkela H
Andhra University College of Engineering, Visakhapatnam, Andhra Pradesh, India

Ripunjoy Phukan
Indian Institute of Technology, Guwahati, India

Nezamoddin N. Kachouie
Department of Mathematical Sciences, Florida Institute of Technology, USA

Armin Schwartzman
Department of Statistics North Carolina State University Raleigh, NC

Omorogiuwa Eseosa
Electrical/Electronic Engineering, Faculty of Engineering, University of Port Harcourt, Nigeria

Samuel Ike
Electrical/Electronic Engineering, Faculty of Engineering, University of Benin, Nigeria

Vladimir V. Rumyantsev
A.A. Galkin Donetsk Institute for Physics and Engineering, National Academy of Sciences, Ukraine

Bamisaye AJ and Adeoye OS
Department of Electrical and Electronic Engineering, The Federal Polytechnic, Ado-Ekiti, Nigeria

Faten Grouz and Lassaad Sbita
Research Unit of Photovoltaic, Wind and Geothermal Systems (SPEG), the National Engineering School of Gabes (ENIG), University of Gabes, Av. Omar Ibn El Khattab, Zrig Eddakhlania (6072), Tunisia

Najib Essounbouli
Center for Research in Information and Communication Sciences and Technology (CReSTIC), IUT, 9 Av Quebec, 10000 Troyes, France

Nadia Mars
Research Unit of Photovoltaic, Wind and Geothermal Systems (SPEG), the National Engineering School of Gabes (ENIG), University of Gabes, Av. Omar Ibn El Khattab, Zrig Eddakhlania (6072), Tunisia
Center for Research in Information and Communication Sciences and Technology (CReSTIC), IUT, 9 Av Quebec, 10000 Troyes, France

Joseph Jintu K
VLSI & Embedded systems PESIT Bangalore, India

Purushotham U
Department of Electronics & Communications PESIT Bangalore, India

Maryam Minhas, Tanveer Abbas, Reeja Iqbal and Fatima Munir
Department of Electrical Engineering, Pakistan Institute of Engineering and Applied Sciences, PIEAS, Islamabad, Pakistan

Nezar M
Laboratoire d'Innovation en éco-conception, construction et génie sismique (LICEC_GS), Algeria

Aggoune N
Laboratoirede Mécanique des structures et Matériaux (LaMsM), Algeria

Nezar DJ
Laboratoire de Physique Energétique appliquée de Batna (LPEA), Algeria

Abdessemed R
Laboratoire d'Electrotechnique de Batna (LEB), Algeria

Nezar KS and Nezar A
Centre hospitalier universitaire de Batna CHU, Algeria

Ibraheem Nasiruddin and Saab B Altamimi
Department of Electrical Engineering, Engineering College, Qassim University, Kingdom of Saudi Arabia

Edris Mohsen, Lyndon J Brown and Jie Chen
Department of Electrical and Computer Engineering Western University, London, Ontario, Canada

Allam Maalla
School of Information Technology and Engineering, Guangzhou College of Commerce, Guangzhou, China

Omorogiuwa E
Department of Electrical-Electronic Engineering, Faculty of Engineering, University of Port Harcourt, Nigeria

Onohaebi SO
Department of Electrical-Electronic, Faculty of Engineering, University of Benin, Benin City, Nigeria

Amir Sharifian
Department of Electrical Engineering, Ahrar Institute of Technology and Higher Education, P.O.Box 41931-63584, Rasht, Iran

Jain V and Agrawal N
Department of Electronics and Communication, College of Technology and Engineering, MPUAT, Udaipur, India

Bahrami S and Imari A
Department of Electrical Engineering, University of Isfahan, Isfahan, Iran

Xavier Fernando and Sajjadul Latif
Department of Electrical and Computer Engineering, Ryerson University, Toronto, Ontario M5B 2K3, Canada

Bamisaye, Ayodeji James and Ademiloye Ibrahim Bunmi
Department of Electrical and Electronic Engineering, The Federal Polytechnic, Ado-Ekiti, Nigeria

Vaishali B
Department of Electrical Engineering, Universite du Quebec a Trois-Rivieres, Québec, Canada

Jeyapriya A
CSE, Sri Ramakrishna Engineering College, Coimbatore, Tamil Nadu, India

Benaissa A, Zeblah A, Belafdal A and Chaker A
Djillali Liabes University of Sidi Bel Abbes, Algeria

Index

www.ingramcontent.com/pod-product-compliance
Lightning Source LLC
Chambersburg PA
CBHW080647200326
41458CB00013B/4764